Fire Department Pumping Apparatus

Seventh Edition

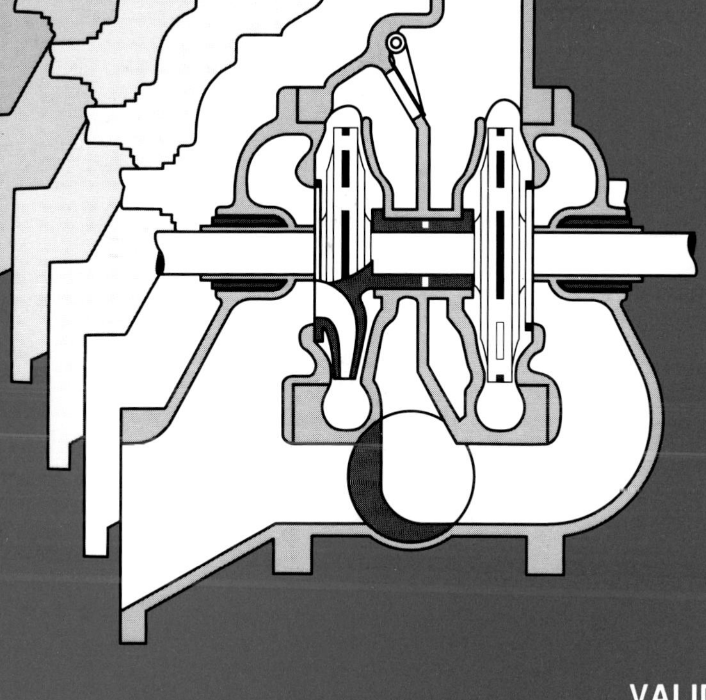

VALIDATED BY
THE INTERNATIONAL FIRE SERVICE TRAINING ASSOCIATION

PUBLISHED BY
FIRE PROTECTION PUBLICATIONS
OKLAHOMA STATE UNIVERSITY

Cover Photo Courtesy of: George Braun
Gainesville, Florida Fire-Rescue

Dedication

This manual is dedicated to the members of that unselfish organization of men and women who hold devotion to duty above personal risk, who count on sincerity of service above personal comfort and convenience, who strive unceasingly to find better ways of protecting the lives, homes and property of their fellow citizens from the ravages of fire and other disasters . . . The Firefighters of All Nations.

Dear Firefighter:

The International Fire Service Training Association (IFSTA) is an organization that exists for the purpose of serving firefighters' training needs. Fire Protection Publications is the publisher of IFSTA materials. Fire Protection Publications staff members participate in the National Fire Protection Association and the International Association of Fire Chiefs.

If you need additional information concerning our organization or assistance with manual orders, contact:

Customer Services
Fire Protection Publications
Oklahoma State University
930 N. Willis
Stillwater, OK 74078-8045
1 (800) 654-4055

For assistance with training materials, recommended material for inclusion in a manual, or questions on manual content, contact:

Technical Services
Fire Protection Publications
Oklahoma State University
930 N. Willis
Stillwater, OK 74078-8045
(405) 744-5723

First Printing, January 1989
Second Printing, August 1989
Third Printing, October 1991
Fourth Printing, May 1994
Fifth Printing, September 1996
Sixth Printing, February 1998

ISBN 0-87939-078-6
Library of Congress 88-83645
Seventh Edition
Printed in the United States of America

Table of Contents

10-91

5-89

List of Tables

Preface

This is the seventh edition of **Fire Department Pumping Apparatus.** The information on the operation of fire department pumping apparatus in this manual is greatly expanded from the previous edition. In fact, it was expanded so much that it became impossible to cover the operation of both pumping apparatus and aerial apparatus in the same book. For this reason, IFSTA's **Fire Department Aerial Apparatus** manual has been created for those who are concerned with the operation of aerial devices. The previous edition covered both topics in the same text.

Acknowledgement and special thanks are extended to the members of the validating committee who contributed their time, wisdom, and knowledge to this manual:

Chairman
Bill Cooper
Huntington Beach, California

Vice-Chairman
Lester Lukert
Oshkosh, Nebraska

Secretary
Wayne Sandford
Meriden, Connecticut

Dan Cotten
Shreveport, Louisiana

John Hoglund
College Park, Maryland

T.J. Sanborn
Pierre, South Dakota

Jerry Wiencek
Garden Grove, California

Oren Dowdy
Oklahoma City, Oklahoma

Richard Kleiman
Merced, California

Charles Scott
Lubbock, Texas

Larry Williams
Smyrna, Georgia

It would not be possible to publish a book of this scope without the assistance of many people and organizations. To the following and others listed in captions, we owe a great debt as they gave freely of time, advise, personnel, and/or equipment.

Hale Fire Pump Company
Waterous Fire Pump Company
W.S. Darley and Company
Gainesville (Florida) Fire Department
Palmer (Alaska) Fire Department
Hampshire (Illinois) Fire Protection District
Miami (Florida) Fire Department
Saulsbury Fire Equipment Corporation
Anderson Engineering, LTD.
Span Instruments, Inc.
Joel Woods, Maryland Fire and Rescue Institute
Bob Esposito, Pennsburg (Pennsylvania) Fire Co. No.1
Ed Prendergast, Chicago (Illinois) Fire Department
Ron Jeffers, Union City, New Jersey

3M Safety and Security
Oklahoma State University Fire Protection Society
Fire Research Corporation
American Petroleum Institute
National Fire Protection Association
John B. Lockwood, Bowie, Maryland
Thomas B. Reinish, Elkins Park, Pennsylvania
Robert Foote, Hyattsville, Maryland
Merl Gillette, Eau Claire, Wisconsin
United States Coast Guard
Warren Isman
Tuolumne County (California) Fire Department
California Division of Forestry Training Academy
American Godiva

One person and one organization deserve special mention for providing substantial portions of this manual. The seven appendices covering the major pump manufacturers were prepared by William Eckman of LaPlata, Maryland. They reflect the considerable knowledge and the many years of training that Bill has accumulated on the topic of fire pumps. We also wish to thank the City of Merced (California) Fire Department for their assistance in obtaining many of the procedural pictures included in this manual. Chief Kenneth Mitten gave freely of his personnel and equipment for our use.

The cover photo is courtesy of the Gainesville (Florida) Fire Department.

Gratitude is also extended to the following members of the Fire Protection Publications staff, whose contributions made the final publication of this manual possible.

Michael A. Wieder, Senior Publications Editor
Lynne C. Murnane, Senior Publications Editor
Bill Westhoff, Senior Publications Editor
Carol Smith, Associate Editor
Cynthia Brakhage, Publications Specialist
Robert Fleischner, Publications Specialist
Don Davis, Publications Production Coordinator
Ann Moffat, Graphic Designer
Karen Murphy, Phototypesetter Operator II
Desa Porter, Phototypesetter Operator II and Layout
Kevin M. Roche, Research Technician
Terri Jo Gaines, Senior Clerk/Typist

Gene P. Carlson
Editor

Glossary

A

ACCEPTANCE TESTS — Preservice tests performed at the factory or after delivery to assure the purchaser that the apparatus meets bid specifications.

AIRCRAFT FIRE APPARATUS — Fire apparatus specifically designed for aircraft crash fire fighting/rescue operations.

AMMETER — Gauge that indicates either the amount of electrical current being drawn from or provided to the vehicle's battery.

APPLIANCE — Generic term applied to any nozzle, wye, siamese, deluge monitor, or other piece of hardware used with fire hose for the purpose of delivering water.

ATTACK LINES — Hoselines or fire streams used to attack, contain, or prevent the spread of a fire.

ATTACK PUMPER — That pumper which is positioned at the fire scene and is directly supplying attack lines.

B

BLEEDER VALVE — Valve on a gate intake that allows air from an incoming supply line to be bled off before allowing the water into the pump.

BOOSTER APPARATUS — *See* Brush Apparatus.

BOOSTER TANK — *See* Water Tank.

***BOURDON TUBE** — A pressure gauge that has a curved, flat tube that changes its curvature as pressure changes; this movement is then transferred mechanically to a pointer on the dial.

BRUSH APPARATUS (BOOSTER APPARATUS) — Fire department apparatus designed specifically for fighting ground cover fires.

C

CATCH BASIN — *See* Portable Tank.

***CAVITATION** — A condition in which vacuum pockets form in a pump and cause vibrations, loss of efficiency, and possible damage.

***CENTRIFUGAL PUMP** — A pump with one or more impellers that utilizes centrifugal force to move the water.

CERTIFICATION TESTS — Preservice pump tests conducted by an independent testing laboratory before delivery of an apparatus. These tests ensure that the apparatus will perform as expected after being placed into service.

CHASSIS — The frame upon which the body of the fire apparatus rests.

***CIRCULATOR VALVE** — A device in a pump that routes water from the pump to the supply to keep the pump cool when hose lines are shut down.

***CLAPPER VALVES** — A hinged valve that permits the flow of water in one direction.

COMPOUND GAUGE — A gauge connected to the intake side of the pump that is capable of measuring positive or negative intake pressures.

D

DEAD-END MAIN — A water main that is not looped and in which water can flow in only one direction.

**Fire Terms,* Copyright © 1980, National Fire Protection Association, Quincy, MA 02269. Reprinted with permission.

DISCHARGE VELOCITY — The rate at which water travels from an orifice.

DRAFTING OPERATION — The process of drawing water from a static source into a pump that is above the level of the water supply.

DRAIN VALVE — A valve on pump discharges that facilitates the removal of pressure from a hoseline after the discharge has been closed.

DUAL PUMPING (also incorrectly called tandem pumping) — An operation where a strong hydrant is used to supply two pumpers by connecting the pumpers intake-to-intake. The second pumper receives the excess water not being pumped by the first pumper, which is directly connected to the water supply source.

E

ELEVATING WATER DEVICE — An articulating or telescoping aerial device added to some fire department pumpers to enable the unit to deploy elevated master stream devices. These devices range from 30 to 75 feet (9 m to 23 m) in height.

ELEVATION LOSS — *See* Elevation Pressure.

ELEVATION PRESSURE — The gain or loss of pressure in a hoseline due to a change in elevation.

EQUAL — In terms of specifying apparatus, it means the same level of quality, standard, performance, or design, but not necessarily identical.

EQUIPMENT — In this manual, used to denote portable tools or appliances carried on fire apparatus.

EXTERNAL WATER SUPPLY — Any water supply to a fire pump from a source other than the vehicle's own water tank.

F

FIRE APPARATUS — Any fire department emergency vehicle that participates in fire suppression or other emergency situations.

FIRE APPARATUS DRIVER/OPERATOR — A firefighter (Level II) charged with the responsibility of operating fire apparatus to, during, and from a fire or other emergency scene. The driver/operator is also responsible for routine maintenance of the apparatus and any equipment carried on it.

FIRE BOAT — A boat that carries large fire pumps and is capable of supplying master streams or supply line hoselines to land-based fire fighting apparatus.

FIRE DEPARTMENT CONNECTION (FDC) — The point at which the fire department can connect into a sprinkler or standpipe system to boost the water flow in the system. This connection consists of a clappered siamese with two or more 2½-inch (65 mm) intakes or one large diameter (4-inch [100 mm] or larger) intake.

FIRE DEPARTMENT PUMPER — Fire apparatus having a permanently mounted fire pump with a rated discharge capacity of 500 gpm (2 000 L/min) or greater. This apparatus may also carry water, hose, and other portable equipment.

FIRE DEPARTMENT SPRINKLER CONNECTION — *See* Fire Department Connection.

FIRE DEPARTMENT STANDPIPE CONNECTION — *See* Fire Department Connection.

***FIRE HYDRANT** — An upright metal casting that is connected to a water supply system and is equipped with one or more valved outlets to which a hose line or pumper may be connected to supply water for firefighting operations.

FIRE HYDRAULICS — The science that deals with water in motion as it applies to fire fighting operations.

FIRE STREAM — An unbroken stream of water applied to a fire for the purpose of controlling and extinguishing it.

FLOATING DOCK STRAINER — A strainer used for drafting operations that is designed to float on top of the water. This eliminates the problem of drawing debris into the pump and reduces the required depth of water needed for drafting.

FLOWMETER — Mechanical device installed in a discharge line that senses the amount of water flowing and provides a readout in units of gallons per minute (Liters per minute).

FLUID — Substance with no established shape but that is capable of flowing from place to place. Fluid will conform to the shape of the vessel in which it is contained. Fluids have very little adhesive quality and cannot be pulled or lifted, but must be pushed by having pressure applied and confined in some manner to direct the flow. Fluids are divided into liquids or gases.

FOLD-A-TANK — *See* Portable Tank.

FRICTION LOSS — Loss of pressure created by the turbulence of water moving against the interior walls of the hose or pipe.

G

***GATE** — A control valve for hose, a pump outlet, or a large caliber nozzle.

GRADABILITY — The ability of a piece of apparatus to traverse various types of terrain.

H

HARD SUCTION HOSE — A flexible, rubber length of hose, reinforced with a steel core to prevent it from collapsing. This type of hose is connected between a fire pump and a water supply source (usually a static supply) and must be used when drafting.

HEAD — Water pressure due to elevation. For every 1 foot increase in elevation, 0.434 psi is gained. (For every 1 meter increase in elevation, 10 kPa is gained.)

HEAD PRESSURE — *See* Head.

HOSE LAYOUT, COMPLICATED — Hose layout that includes the use of multiple lengths of unequal hoselines, unequal wyed or manifold lines, siamesed, or master stream devices. Such a layout requires the pump operator to perform complicated calculations in order to supply the lines properly.

HOSE LAYOUT, SIMPLE — Hose layout that includes the use of single hoselines or multiple, wyed, siamesed, or manifold lines of equal length.

HYDRAULIC CALCULATIONS — The process of using mathematics to solve a problem involving fire hydraulics.

I

***IMPELLER** — The vaned, circulating member of the centrifugal pump that transmits motion to the water.

***IMPELLER EYE** — The intake orifice at the center of a centrifugal pump impeller.

INCREASER — An adapter used to attach a larger hoseline to a smaller one. The increaser has female threads on the smaller side and male threads on the larger side.

INTAKE HOSE (HARD SUCTION HOSE) — A flexible, rubberized length of hose with a steel core that connects a pump to a source of water. Most commonly used for drafting operations.

**Fire Terms*, Copyright © 1980, National Fire Protection Association, Quincy, MA 02269. Reprinted with permission.

K

KILOPASCAL (kPa) — Metric unit of measure for pressure 1 psi = 6.895 kPa; 1 kPa = .1450 psi.

KINK — Severe bend in a hoseline that increases friction loss and reduces the flow of water through that hose.

L

LIFT, DEPENDABLE — The height a column of water may be lifted in sufficient quantity to provide a reliable fire flow. Lift may be raised through a hard suction hose to a pump, taking into consideration the atmospheric pressure and friction loss within the hard suction hose. Dependable lift is usually considered to be 14.7 feet (4.48 m).

LIFT, MAXIMUM — The maximum height to which any amount of water may be raised through a hard suction hose to a pump.

LIFT, THEORETICAL — The theoretical, scientific height that a column of water may be lifted by atmospheric pressure in a true vacuum. At sea level, this height is 33.8 feet (10 m). The height will decrease as elevation increases.

LIGHT ATTACK VEHICLE — *See* Minipumper.

LOADING SITE — The point in a tanker shuttle operation where apparatus tanks are filled from a water supply.

LUGGING — A condition that exists when the engine is operating at full throttle but below rated speed. Lugging can be eliminated by using a lower gear and proper shifting techniques.

M

MAINTENANCE — Keeping equipment or apparatus in a state of usefulness or readiness.

MANIFOLD — A) Hose appliance that divides one larger hoseline into three or more smaller hoselines. B) Hose appliance that combines three or more smaller hoselines into one large hoseline. C) Top portion of pump casing.

MANUFACTURER'S TESTS — Fire pump tests performed by the manufacturer prior to delivery of the apparatus.

MIDIPUMPER — Apparatus sized between a minipumper and a full-sized fire department pumper, usually with a Gross Vehicle Weight of 12,000 pounds (5 443 kg) or greater. The midipumper has a fire pump with a rated capacity generally not greater than 1,000 gpm (4 000 L/min).

MIDSHIP PUMP — Fire pump mounted on the apparatus frame between the front and rear axle.

MINIPUMPER — Small fire apparatus mounted on a pickup truck-sized chassis, usually with a pump having a rated capacity less than 500 gpm (2 000 L/min). Their primary advantage is speed and mobility, which enables them to respond to fires more rapidly than larger apparatus.

MOBILE WATER SUPPLY APPARATUS (TANKER, TENDER) — Fire apparatus with a water tank of 1,000 gallons (4 000 L) or larger whose primary purpose is transporting water. The truck may also carry a pump, some hose, and other equipment.

MULTISTAGE PUMP — Any centrifugal fire pump having more than one impeller.

N

NET PUMP DISCHARGE PRESSURE (NPDP) — The actual amount of pressure being produced by the pump. When taking water from a hydrant, it is the difference between the intake pressure and the discharge pressure. When drafting, it is the sum of the intake pressure and the discharge pressure. (**NOTE:** Intake pressure is credited for lift and intake hose friction loss and is added to the discharge pressure.)

NONCONFORMING APPARATUS — Apparatus that does not conform to the standards set forth by NFPA 1901, *Standard for Automotive Fire Apparatus.*

NOZZLE PRESSURE — The velocity pressure at which water is being discharged from the nozzle.

NOZZLE REACTION — The counterforce directed against the people or device holding a nozzle by the velocity of water being discharged.

O

ORIFICE — An opening through which a substance may pass.

P

PARALLEL OPERATION — *See* volume operation.

PERFORMANCE REQUIREMENTS — A written list of expected capabilities for new apparatus that is produced by the purchaser and presented to the manufacturer as a guide.

PORTABLE TANK (also port-a-tank, fold-a-tank, portable basin) — A collapsible storage tank used during a relay or shuttle operation to hold water from water tanks or hydrants. This water can then be used for supplying attack apparatus.

***POSITIVE DISPLACEMENT PUMP** — A pump that moves a given amount of water through the pump chamber with each stroke or rotation. The pumps are self-priming.

POWER PLANT — Another term for the apparatus engine.

***POWER TAKE-OFF (PTO)** — Device used to transmit power through a clutch arrangement from the power plant to the auxiliary equipment.

PRESERVICE TESTS — Tests performed on fire pumps before they are put into service. These tests are broken down into manufacturer's tests, certification tests, and acceptance tests.

PRESSURE GAUGE — Gauge that registers the pump discharge pressure.

PRESSURE GOVERNOR — Pressure control device designed to eliminate hazardous conditions resulting from excessive pressures by controlling engine speed.

PRESSURE OPERATION (SERIES) — The operation of a two or more stage centrifugal pump in which water passes consecutively through each impeller to provide high pressures at a reduced volume.

PRIME — To remove all air from a pump in preparation for receiving water under pressure.

PRIMING PUMP (primer) — A small positive displacement pump used to evacuate air from a centrifugal pump housing and hard suction hose. Evacuating air allows the centrifugal pump to receive water from a static water supply source.

PTO — *See* Power Take-Off.

PUMP AND ROLL — The ability of an apparatus to pump water while the vehicle is in motion.

PUMP CAPACITY RATING — The maximum amount of water a pump will deliver at the indicated pressure.

PUMP CHARTS — Charts carried on a fire apparatus to aid the pump operator in determining the proper pump discharge pressure to use when supplying hoselines.

PUMP DISCHARGE PRESSURE — The actual velocity pressure (measured in pounds per square inch) of the water as it leaves the pump and enters the hoseline.

PUMP DRAIN — A drain located at the lowest part of the pump to help remove all water from the

**Fire Terms,* Copyright © 1980, National Fire Protection Association, Quincy, MA 02269. Reprinted with permission.

pump. This is done to eliminate the danger of damage due to freezing.

PUMP OPERATOR — Firefighter charged with operating the pump and determining the pressures required to operate it efficiently.

PUMPING APPARATUS — Fire department apparatus whose primary responsibility is to pump water.

R

REDUCER — An adapter used to attach a smaller hose to a larger hose. The female end has the larger threads, while the male end has the smaller threads.

RELAY PUMPING — The process of using two or more pumpers to move water through hoselines over a long distance by operating the pumpers in series.

RELIEF VALVE — Pressure control device designed to eliminate hazardous conditions resulting from excessive pressures by allowing this pressure to bypass to the intake side of the pump.

REPAIR — To restore or put together that which has become inoperable or out of place.

***RESIDUAL PRESSURE** — The pressure remaining in a system while water is flowing.

RESPONSE DISTRICT — The geographical area that a particular apparatus has a first-due assignment for a fire or other emergency incident.

S

SAE — These initials stand for The Society of Automotive Engineers. Coupled with a number they indicate the viscosity of motor oil.)

SAFETY BAR — Hinged or sliding bar designed to protect firefighters from falling out of the jumpseat area of a fire apparatus.

SEDIMENT — Dirt and other foreign debris carried along in water; may collect in water-moving equipment.

SERIES OPERATION — *See* Pressure Operation.

SERVICE TESTS — Series of tests performed on the fire pump to ensure operational readiness of the unit. These tests should be performed at least yearly or whenever apparatus has undergone extensive repair.

***SIAMESE** — A hose appliance used to combine two or more hose lines into one. It generally has female inlets and a male outlet.

SINGLE-STAGE CENTRIFUGAL PUMP — A centrifugal pump with only one impeller.

***SIZE-UP** — The mental evaluation made by a fire fighter or officer that enables him or her to determine a course of action.

SOFT SLEEVE HOSE (also called soft suction) — A large diameter, collapsible piece of hose used to connect a fire pump to a pressurized water supply source.

SPECIFICATIONS — 1. Detailed information provided by a manufacturer on the function, care, and maintenance of equipment or apparatus. 2. A detailed list of requirements prepared by a purchaser and presented to a manufacturer or distributor when purchasing equipment or apparatus.

SPEEDOMETER — Dashboard gauge that measures the speed at which the vehicle is traveling. It may also indicate proper operation of the fire pump.

***SPRINKLER HEAD** — The water flow device in a sprinkler system. The sprinkler head consists of a threaded nipple that connects to the water pipe, a discharge orifice, a heat-actuated plug that drops out when a certain temperature is reached, and a deflector that creates a stream pattern that is suitable for fire control.

**Fire Terms,* Copyright © 1980, National Fire Protection Association, Quincy, MA 02269. Reprinted with permission.

SPRINKLER SYSTEM — System of water pipes and sprinklers installed in a structure to automatically control and/or extinguish fires.

STAGING — The process by which noncommitted units responding to a fire or other emergency incident report to a location away from the fire scene to receive their assignment.

STANDARD APPARATUS — Apparatus that conforms to the standards set forth by NFPA 1901, *Standard for Automotive Fire Apparatus.*

STANDARD OPERATING PROCEDURES (SOP's) — Standard methods by which a fire department carries out routine function. Usually these are in written form and all firefighters should be well versed in the content.

STANDPIPE SYSTEM — A wet or dry system of pipes in a large single-story or multistory building with fire hose outlets connected to it. The system is used to provide for quick deployment of hoseline during fire fighting operations.

STATIC PRESSURE — The pressure exerted in all directions at a point in a fluid at rest.

STATIC WATER SOURCE — A nonpressurized source of water that pumpers may draft from for fire fighting or other operations.

STEAMER CONNECTION — The 4½-inch (115 mm) or larger connection on a fire hydrant.

T

TACHOMETER — A dashboard and pump panel gauge that measures the engine speed in revolutions per minute.

TANDEM PUMPING — A short relay operation in which the pumper taking water from the supply source pumps into the intake of the second pumper. The second pumper boosts the pressure of the water even higher. This method is used when pressures higher than the capability of a single pump are required.

TANKER — Also, in the ICS System, tanker refers to a water transporting fixed wing aircraft. *See* Mobile Water Supply Apparatus.

TENDER — *See* Mobile Water Supply Apparatus.

TOTAL PRESSURE — The total amount of pressure loss in a hose assembly due to friction loss in the hose and appliances, elevation losses, or any other factors.

TRAFFIC CONTROL DEVICE — Mechanical device that automatically changes traffic signal lights to favor the path of responding emergency apparatus.

TRIPLE-COMBINATION PUMPER — Fire department pumper that carries a fire pump, hose, and a water tank.

TWO-STAGE CENTRIFUGAL PUMP — Centrifugal pump with two impellers.

U

UNLOADING SITE — The point in the tanker shuttle operation where portable tanks are located and the tankers unload their water.

V

VACUUM — A pressure that is less than normal atmospheric pressure.

VOLUME OPERATION — (Parallel) The operation of a two or more stage centrifugal pump in which each impeller discharges into a common outlet, thereby providing the maximum flow at the rated pressure.

***VOLUTE** — The spiral, divergent chamber of a centrifugal pump in which the velocity energy

**Fire Terms,* Copyright © 1980, National Fire Protection Association, Quincy, MA 02269. Reprinted with permission.

given to water by the impeller blades is converted to pressure.

W

WARNING DEVICES — Any audible or visual devices added to an emergency vehicle to gain the attention of drivers of other vehicles. These may include flashing lights, sirens, horns, or bells.

WATER DISTRIBUTION SYSTEM — A system designed to supply water for residential, commercial, industrial, and/or fire protection purposes. This water supply is delivered through a network of piping and pressure-developing equipment.

WATER SUPPLY PUMPER — A pumper that takes water from a source and sends it to attack pumpers operating at the fire scene.

WATER SUPPLY — Any source of water available for use in fire fighting operations.

WATER TANK (BOOSTER TANK) — A water storage receptacle carried directly on the apparatus.
10-91 NFPA 1901 specifies that Class A pumpers must carry at least 500 gallons (2 000 L) of water and minipumpers must carry at least 150 gallons (600 L) of water.

WYE — A hose appliance that divides one hoseline into two hoselines of equal or smaller size.

THE INTERNATIONAL FIRE SERVICE TRAINING ASSOCIATION

The International Fire Service Training Association (IFSTA) was established as a "nonprofit educational association of fire fighting personnel who are dedicated to upgrading fire fighting techniques and safety through training."

This training association was formed in November 1934, when the Western Actuarial Bureau sponsored a conference in Kansas City, Missouri. The meeting was held to determine how all the agencies interested in publishing fire service training material could coordinate their efforts. Four states were represented at this initial conference. It was decided that, since the representatives from Oklahoma had done some pioneering in fire training manual development, other interested states should join forces with them. This merger made it possible to develop training materials broader in scope than those published by individual agencies. This merger further made possible a reduction in publication costs, since it enabled each state or agency to benefit from the economy of relatively large printing orders. These savings would not be possible if each individual state or department developed and published its own training material.

To carry out the mission of IFSTA, Fire Protection Publications was established as an entity of Oklahoma State University. Fire Protection Publications' primary function is to publish and disseminate training texts as proposed and validated by IFSTA. As a secondary function, Fire Protection Publications researches, acquires, produces, and markets high-quality learning and teaching aids as consistent with IFSTA's mission. The IFSTA Executive Director is officed at Fire Protection Publications.

IFSTA's purpose is to validate training materials for publication, develop training materials for
10-91 publication, check proposed rough drafts for errors, add new techniques and developments, and delete
obsolete and outmoded methods. This work is carried out at the annual Validation Conference.

The IFSTA Validation Conference is held the second full week in July, at Oklahoma State University or in the vicinity. Fire Protection Publications, the IFSTA publisher, establishes the revision schedule for manuals and introduces new manuscripts. Manual committee members are selected for technical input by Fire Protection Publications and the IFSTA Executive Secretary. Committees meet and work at the conference addressing the current standards of the National Fire Protection Association and other standard-making groups as applicable.

Most of the committee members are affiliated with other international fire protection organizations. The Validation Conference brings together individuals from several related and allied fields, such as

— key fire department executives and training officers,
— educators from colleges and universities,
— representatives from governmental agencies,
— delegates of firefighter associations and industrial organizations, and
— engineers from the fire insurance industry.

Committee members are not paid, nor are they reimbursed for their expenses by IFSTA or Fire Protection Publications. They come because of commitment to the fire service and its future through training. Being on a committee is prestigious in the fire service community, and committee members are acknowledged leaders in their fields.

This unique feature provides a close relationship between the International Fire Service Training Association and other fire protection agencies, which helps to correlate the efforts of all concerned.

IFSTA manuals are now the official teaching texts of most of the states and provinces of North America. Additionally, numerous U.S. and Canadian government agencies as well as other English-speaking countries have officially accepted the IFSTA manuals.

COPYRIGHT LAW
AN EDUCATOR'S RESPONSIBILITIES

The law limits what you may copy, under what conditions you may copy, and for what purpose you may copy. Authors and publishers have specific rights under the law. However, the law permits educators access to information and to copy materials under clearly defined guidelines.

These guidelines include:

- The purpose and character of use: whether the purpose is commercial or nonprofit educational
- The nature of the copyrighted work (example: Is the work a textbook meant for classroom use?)
- The amount and substantiality copied in relation to the copyrighted work as a whole
- Whether the copied material will affect the potential sales or value of the copyrighted work

A single copy may be made by a teacher, by request, for research or teaching of a chapter from a book; an article from a periodical; short stories or essays; or a chart, diagram, or drawing. Multiple copies for the classroom must not exceed one copy per student and must meet the test of "brevity, spontaneity, and cumulative effect," which are defined as follows:

- Brevity is either a complete article, story, or essay of less than 2,500 words or an excerpt from any work of not more than 1,000 words or 10 percent of the work, whichever is less for example, (one chart, graph, diagram, drawing, cartoon, or picture per book or periodical issue).
- Spontaneity is (if permission is not sought prior to use) the decision to use the article or excerpt so close
10-91 in time that it would be unreasonable to expect a timely reply to a request for permission.
- Cumulative Effect is no more than one piece copied from the same author nor no more than three articles from the same periodical volume during one classroom term. There can be no more than nine instances of multiple copying per course, per class term.

Under no circumstances can there be copying of or from works intended to be consumable in the course. These include workbooks, answer sheets, and the like. Copying shall not substitute for the purchase of books, publisher's reprints, or periodicals.

The copyright law specifies a monetary penalty for legal damages for each violation. Even a defendant (individual and/or the organization) not found in violation must bear court costs and attorney's fees.

Permission to copy is obtained by writing to the publisher with the following information:

- Title, author(s), and editor(s)
- Edition and/or issue
- Exact amount of material to be copied
- Nature of use, including if it is for resale
- How material will be reproduced
- Number of copies to be made

Copies of copyrighted materials must include a credit line to the original work, author or editor, and the publisher with copyright notice or reprint permission.

Introduction

EARLY FIRE APPARATUS

Fire apparatus has been identified or referred to by several different names since its inception. One of the first recognized names was *engine* which still reflects traditions of our U.S. heritage. Engine was frequently pronounced "enjine" by its early users. Other terms used were *hook and ladder* and *steamer*, and those prefixed by the word *fire* included wagon, truck, and cart.

The old engines or "masheens" as they were affectionately called were, like so many things, not fully appreciated until they were gone. The early machines consisted of wooden boxes that had no wheels. Firefighters used handles to carry them to fires. A small hand-operated piston pump was mounted to the tub, and a gooseneck-shaped metallic pipe extended above the pump. The levers that operated the pump were designed so that several firefighters, usually volunteers, could move them up and down. These long handles were called brakes. A complete cycle of up-and-down action was called a stroke.

Wheels were later added to the sides of the tubs, as was flexible fire hose, to permit better control of the water supply. Two-wheeled carts with large reels for the hose were called *tenders* or *jumpers*. The introduction of four-wheeled ladder wagons called *trucks* eliminated the necessity of carrying ladders while running to a fire.

These early devices, all of which were pulled and operated by hand, were forerunners of steam-powered pumps and were known as *steamers*. Steam pumps were of the one-cylinder piston type, equipped with hose discharge outlets. As with other early equipment, steamers were generally pulled by horses.

With the growing use of chemical fire extinguishers, apparatus was designed to carry large soda-acid tanks. Over the years, these units were hand-drawn, horse-drawn, then motorized. Essentially, they are the origin of the booster line for the initial operations on small fires. These apparatus bore the name *chemical wagon* or *chemical engine* since the extinguishing agent was discharged by a chemical reaction, rather than a pump. A variety of styles and combinations were produced.

The turn of the century brought with it the internal combustion engine, which soon ended the horse-drawn era. The conversion from horsepower to automotive power was easily made, although increased efficiency ended what many have considered the most picturesque era in fire fighting.

PRESENT-DAY REQUIREMENTS FOR FIRE APPARATUS

Certain performance functions are required of present-day automotive fire apparatus, such as the ability to carry and pump large quantities of water, the ability to carry large quantities of hose and other equipment, and the ability to safely transport personnel to and from the emergency scene. Currently, suggested specifications for fire apparatus are published in NFPA 1901, *Standard for Automotive Fire Apparatus*. The NFPA committee membership is chosen so that representation is given to the general public through consultants, insurance organizations, manufacturers and trade associations. The fire service is also represented and provides the chairman. This standard is subject to annual review by the committee. All proposals for revisions must be approved by the committee and the membership of NFPA at their semiannual convention after general public comments.

Municipal officials are urged to work toward a long-range plan for systematic replacement of fire apparatus. Such a plan should be based upon the ability of fire apparatus to meet the requirements imposed by the jurisdiction it will be serving. Maintenance costs, parts replacement difficulties, and performance reliability should all be carefully considered before purchasing apparatus.

Underwriters Laboratories provides testing and certification of fire apparatus. These tests, although not mandatory, are universally accepted, and apparatus specifications should require such testing. Specifications should include the required performance of the pump, road acceleration and braking, piping flow requirements, and maximum speed. Several detailed performance requirements are outlined in NFPA 1901. Upon delivery, tests should be conducted to determine if all specification requirements have been met. Specific tests for a given locale may need to be added. The vehicle should also be weighed to determine if it meets gross vehicle weight and weight distribution limitations as set forth in the specifications.

PURPOSE AND SCOPE

This seventh edition of **Fire Department Pumping Apparatus** is designed to educate driver/operators who are responsible for operating apparatus equipped with fire pumps. This includes pumpers, tankers (tenders), brush apparatus, and aerial apparatus equipped with fire pumps. Operation of apparatus equipped with aerial devices is covered in IFSTA's **Fire Department Aerial Apparatus.**

Probably no other subject in the fire service covers a broader field of specialized training. Some of these areas include pump theory and operations, vehicle operation, maintenance and testing, and specifications of new apparatus. The purpose of this manual is to present general principles of pump operation, with application of those principles wherever feasible. It is also meant to guide driver/operators in the proper operation and care of apparatus.

This manual includes an overview of the qualities and skills needed by a driver/operator, safe driving techniques, types of pumping apparatus, positioning apparatus to maximize efficiency and water supply, fire pump theory and operation, apparatus maintenance and testing, and determining specifications for apparatus purchase. The seven appendices included at the end of the text provide specific information on each of the major U.S. manufacturers of fire pumps.

1

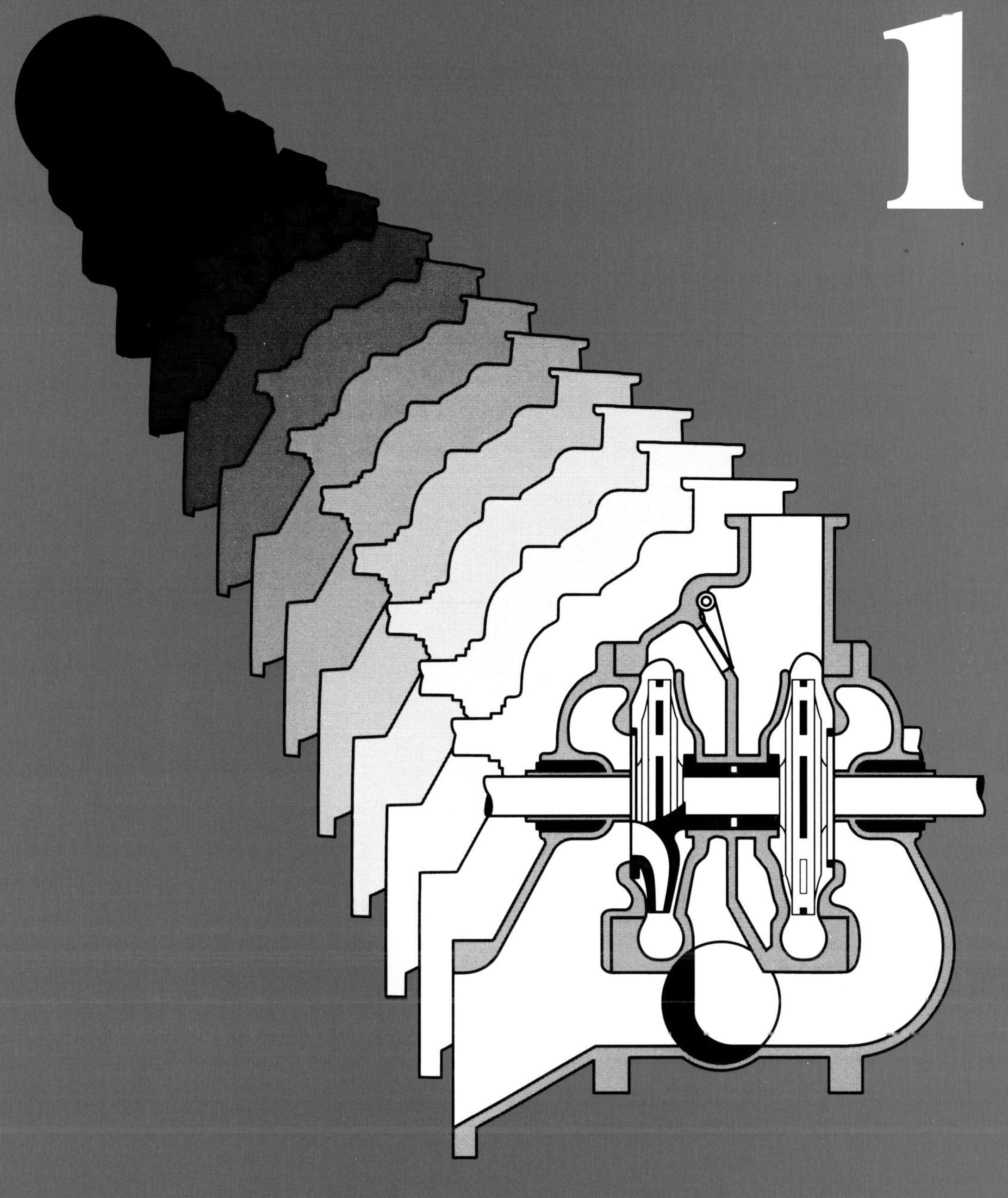

The Driver/Operator

LEARNING OBJECTIVES

After reading this chapter, the student should be able to accomplish the following learning objectives:

- Name the NFPA standards that apply to fire apparatus driver/operators.
- Identify the various physical skills necessary for driver/operators and explain their importance.
- Describe the process through which driver/operator candidates are selected.

Chapter 1

The Driver/Operator

The fire apparatus driver/operator is responsible for the safe transportation of firefighters, apparatus, and equipment to and from the scene of an emergency or other call for service. Once on the scene, the driver/operator must be capable of operating the apparatus properly and *swiftly*. The driver/operator must also ensure that the apparatus and the equipment carried on the apparatus are maintained in a ready condition at all times.

In general, the driver/operator must be a mature, responsible adult with a good attitude about safety. Psychological profiles, drug and sobriety testing, and background investigations may be necessary to ensure that the driver/operator is ready to accept the high level of responsibility that comes with the job.

In order to perform duties properly, every driver/operator must possess certain mental and physical skills. The levels of these skills needed by a job candidate are usually determined by each jurisdiction. In addition, minimum qualifications for driver/operators can be found in National Fire Protection Association (NFPA) 1002, *Standard for Fire Apparatus Driver/Operator Professional Qualifications*. Included in these qualifications is a requirement that the driver/operator meet the requirements of NFPA 1001, *Standard for Fire Fighter Professional Qualifications* for Fire Fighter II. Some basic mental and physical skills that may be required are discussed in the following sections.

SKILLS NEEDED BY THE DRIVER/OPERATOR

Reading

It is essential that all driver/operators be able to read and understand the written word because they are constantly required to utilize reading skills. Some examples of duties that call for good reading comprehension are as follows:

- Reading maps (Figure 1.1)
- Reviewing manufacturer's operating instructions
- Studying prefire operations plans

Figure 1.1 Good reading skills are necessary in order to quickly locate points on a map.

Writing Skills

The driver/operator must also possess a certain degree of writing skills so that written information can be conveyed completely and accurately. Some examples of job functions that require writing skills are completing maintenance reports, equipment repair requests, and fire reports. Each driver/operator should be evaluated for the ability to write clearly and concisely.

Mathematical Skills

Basic mathematical skills are important to the driver/operator. Math is used in hydraulic calculations, ladder load determinations, and in numerous other situations encountered by the driver/operator every day. The driver/operator should be able to add, subtract, multiply, divide fractions and whole numbers, and determine square roots. The driver/operator should also be able to solve simple equations, such as those used in friction loss problems. It is not the purpose of this manual to review basic mathematical skills. If a prospective candidate is deficient in these skills, attempts should be made to provide the assistance necessary to correct this situation. Often, a local high school or institution of learning will offer programs designed to increase math skills. These programs can be beneficial not only to the prospective candidate, but also to current driver/operators who need to brush up their skills (Figure 1.2).

Figure 1.2 Driver/operator candidates need a refresher course in basic mathematical skills before learning complex hydraulic calculations.

PHYSICAL REQUIREMENTS

Physical Fitness

The driver/operator is often required to set up the apparatus unassisted by other firefighters. Connecting to a hydrant with a soft sleeve hose or stretching a supply line to a hydrant by hand are examples of tasks that require a high level of physical fitness. The driver/operator must be prepared to perform the lifts, bends, and strenuous activity necessary to complete these tasks (Figure 1.3).

Figure 1.3 Many fire departments require firefighters to participate in physical fitness activities.

Vision Requirements

The safe operation of an apparatus depends greatly upon the driver/operator's ability to see. NFPA 1001, *Standard for Fire Fighter Professional Qualifications,* requires that the firefighter have an uncorrected visual acuity of 20/40 or better in one eye and 20/100 or better in the other eye. Corrected vision must be 20/20 or better in one eye and 20/40 or better in the other eye. The firefighter must also be able to identify the colors red and green. Eye and motor sensory tests should also be performed. These tests evaluate the driver/operator's ability in the following areas:

- Depth perception
- Peripheral vision
- Identification of traffic signs and signals
- Glare vision and glare recovery time

- Night vision
- Reaction time
- Hand-eye coordination

Hearing Requirements

High levels of noise are generated by emergency vehicles en route to and on the emergency scene. Despite the noise caused by engine operation, sirens, air horns, and radio traffic, the driver/operator must be able to recognize the different types of noises and their importance. Keen hearing is critical requiring the increased use of hearing protection devices.

The driver/operator must also be able to focus on particular sounds such as radio instructions for placing apparatus. Failure to hear such orders may result in the apparatus being placed in a less effective or even unsafe position. NFPA 1001 recommends rejection of the firefighter candidate who has a hearing loss of 20 decibels or more for the speech frequencies (500-1000-200 cycles) in either ear, or a loss of speech reception of phonetically balanced words below 90 percent normal reception for either ear.

Other Abilities

Several other abilities, although not required, will help the driver/operator perform well. Mechanical ability is helpful in order to understand the operation and maintenance of the apparatus. Since the driver/operator is often in command of the apparatus while the officer is absent, basic supervisory skills will help in coordinating activities on the fireground.

SELECTION OF DRIVER/OPERATORS

Driver/operators are most often promoted from the rank of firefighter. Promotion is frequently based upon written or performance tests, or a combination of these (Figure 1.4). Some departments are using assessment centers to determine promotional choices. Whatever the method used, promotions should be based upon skill and ability rather than seniority or position.

Firefighters being considered for a driver/operator's position should be allowed to practice their apparatus driving skills before the department driving test. Before driving fire apparatus on a public highway, it may be necessary to obtain an operator's license or learner's permit for the

Figure 1.4 Promotional exams for driver/operators often include both written and practical examinations.

particular type of apparatus. Some states and localities have exempted the driver/operator from the above requirements; however, every driver/operator candidate should investigate all state and local laws pertaining to the operation of motor vehicles. The fire department should provide a practice driving course so that driver/operator candidates can practice their driving skills in a controlled environment rather than on a busy city street (Figure 1.5). Apparatus that closely resembles front-line vehicles should be used for practice so that the driver/operator learns how to operate the type of vehicle currently used by the department.

Figure 1.5a Driver/operators should learn to operate the vehicle in a controlled environment before driving on public roads.

Figure 1.5b Large paved areas covered with water, also known as skid pads, are used to give driver/operators realistic training in controlling skids.

Chapter 1 Review

Answers on page 358

LISTING

1. List at least three different reading skills that are required of a driver/operator.

2. Name at least four tests that prospective drivers should be given to check their visual abilities.

SHORT ANSWER: Answer each item briefly.

3. What are the primary responsibilities of the driver/operator?

4. Why is it important for the driver/operator to have at least a basic understanding of supervisory skills?

5. How are driver/operators usually selected?

2

Operating Emergency Vehicles

NFPA STANDARD 1002
Fire Apparatus Driver/Operator Professional Qualifications

2-2 Driving/Operating.

2-2.1 The fire apparatus driver/operator shall identify all applicable state and local laws of the authority having jurisdiction, including rules and regulations governing the safe driving and operating of fire department vehicles.

2-2.2 The fire apparatus driver/operator, given a fire department vehicle, shall identify all automotive gauges and demonstrate their usage.

2-2.3* The fire apparatus driver/operator, given a fire department vehicle, shall demonstrate the following driving skills:

(a) serpentine
(b) alley dock
(c) opposite alley pull in
(d) diminishing clearance
(e) straight line
(f) turn around
(g) lane change
(h) stopping procedure
(i) parking procedures.

2-2.4* The fire apparatus driver/operator shall identify and demonstrate the theory and principles of defensive driving techniques, both emergency and nonemergency.

2-2.6 The fire apparatus driver/operator, under emergency response conditions, shall demonstrate the legal and safe driving, positioning, and operating of assigned fire department vehicles of the authority having jurisdiction.

2-2.7* The fire apparatus driver/operator shall describe the safety precautions necessary when driving during adverse environmental conditions.

2-2.8* The fire apparatus driver/operator shall describe the effects on vehicle control of (a) braking reaction time (b) load control factors (c) general steering reactions.

3-1 General.

3-1.7* The fire apparatus driver/operator, given a fire department pumper, shall demonstrate the following driving tests:

(a) serpentine
(b) alley dock
(c) opposite alley pull in
(d) diminishing clearance
(e) straight line
(f) turn around
(g) lane change
(h) stopping procedure
(i) parking procedures.

Chapter 2
Operating Emergency Vehicles

The ability to safely control and maneuver fire apparatus is a critical part of a driver/operator's responsibilities. The driver/operator is also responsible for the safety of any personnel riding on the apparatus. The driver/operator is responsible for the proper placement of apparatus and operation of equipment. Fire apparatus driver/operators must be thoroughly trained in the techniques of laying hose, pump hookup, pump operation, developing fire streams, and detecting pump malfunctions. To qualify as an apparatus driver/operator, a firefighter must be knowledgeable about traffic regulations, safe driving techniques, spotting apparatus, and general apparatus care and maintenance (Figure 2.1).

Figure 2.1 The driver/operator must be able to perform basic preventive maintenance.

This chapter discusses the many elements of safe operation of fire apparatus. Among these elements are knowledge of driving regulations, actual street operation of the vehicle, driving the vehicle under emergency conditions, and practical testing and evaluational exercises.

DRIVING REGULATIONS

Driver/operators of fire apparatus are regulated by state laws, city ordinances, and departmental policies. Copies of these laws and rules should be studied by all members of the fire department. Because regulations vary in different areas, those discussed here are general in nature.

Unless specifically exempt, fire apparatus driver/operators are subject to any statute, rule, regulation, or ordinance that governs any vehicle operator. Statutes usually describe those vehicles that are in the emergency category; this classification usually covers all fire department vehicles when they are responding to an emergency. Speed limits that are set for the general public may be exceeded within the limits of a local policy as long as the driver/operator does not endanger life or property. Excessive speed has also caused many accidents. A quick response is necessary, but using excessive speed with little regard for safety is never justified.

A driver/operator may also disregard regulations that apply to the general public concerning the direction of travel, direction of turns, and parking. Under these circumstances, the driver/operator must exercise care for the safety of others and must maintain complete control of the vehicle.

Most driving regulations pertain to dry, clear roads. Driver/operators should adjust their speed to compensate for conditions such as wet roads, darkness, fog, or any other condition that makes normal emergency vehicle operation more hazardous.

Emergency vehicles are not exempt from laws that require vehicles to stop for school buses that are flashing signal lights to indicate that children are boarding or disembarking. Fire apparatus should proceed only after a proper signal is given by the bus driver or police officer. All traffic signals and rules must be obeyed when returning to quarters from an alarm.

Recent legal decisions have held that a driver/operator who does not obey state, local, or departmental driving regulations can be subject to criminal and civil prosecution if the apparatus is involved in an accident. If the driver/operator is negligent in the operation of an emergency vehicle and becomes involved in an accident, both the driver/operator and the fire department may be held responsible.

SAFE DRIVING TECHNIQUES

The driver/operator's job is to keep the fire apparatus under control at all times. Maintaining control requires that the driver/operator take into account traffic regulations, road and weather conditions, and other vehicles. The driver/operator must be proficient in defensive driving techniques, including taking evasive action and recovering from skids.

Attitude

The first element in learning to drive safely is to develop the proper attitude. *It is critically important that the driver/operator remain calm and drive in a safe manner. Reckless driving, even in response to an emergency, is never acceptable.* The driver/operator who drives aggressively, failing to observe safety precautions, is a menace to other vehicles, pedestrians, and other firefighters on the apparatus.

Driver operators must realize that they cannot demand the right of way although they may legally have it. Such driver actions as ignoring approaching apparatus, refusing to yield, and driving erratically due to panic may be expected from the public. At all times, the driver/operator must be prepared to yield the right of way in the interests of safety. Further, it must be remembered that an accident that occurs en route to the emergency scene has many consequences:

- Emergency efforts are delayed at the original emergency scene as additional apparatus are dispatched.
- A second emergency scene is created, possibly involving injured firefighters or members of the public. Emergency apparatus must be dispatched to this scene as well.
- Damage to apparatus and other vehicles or property may be extensive.
- The public image of the fire department suffers.
- The fire department and the driver/operator may be subject to civil or criminal lawsuit.

Safety on the Apparatus

To drive safely, a driver/operator must have safe equipment. All functional parts — particularly the brakes, steering, and tires — must be in good condition. Faulty equipment has been responsible for many accidents.

Riders and driver/operators of emergency vehicles should dress fully before getting in the apparatus and should wear seat belts while the apparatus is in motion. When apparatus is purchased, seat belts large enough to accommodate a firefighter in full protective clothing should be specified for all seats. No firefighters should be in a standing position anywhere on the apparatus. *Riding on the tailboard is not safe and must be discontinued.*

Safety bars are available that may prevent a firefighter from falling out of a jump seat (Figure 2.2). These bars are not a substitute for safety procedures that require that firefighters ride in safe, enclosed positions wearing their seat belts. Safety bars do have value as an additional barrier between the firefighter and the road. When the bars are installed, they should be secured with bolts strong enough to withstand the full weight of a firefighter falling upon them. Safety bars that are held in an upright position by straps or ropes

Figure 2.2 Many apparatus have safety bars that are designed to minimize the chance of personnel accidentally falling out of the jump seat area. *Courtesy of Hampshire (Illinois) Fire Protection District.*

provide no extra security for the firefighter riding in the jump seat.

Whenever possible, the driver/operator should avoid backing up the fire apparatus. It is normally safer and sometimes quicker to drive around the block and start all over again. There are situations, however, when it is necessary to back up fire apparatus. This operation should be performed very carefully. When backing, there should be a firefighter assigned to clear the way and to warn the driver/operator of any obstacles obscured by blind spots (Figure 2.3). All fire apparatus should be equipped with an alarm system that warns others while apparatus is backing up.

Defensive Driving Techniques

Sound defensive driving skills are one of the most important aspects of safe driving. Every driver/operator should be familiar with these basic concepts of defensive driving. They include anticipating other drivers' actions, visual lead time, braking and reaction time, combating skids, evasive tactics, and knowledge of weight transfer.

Anticipating Other Drivers' Actions

The driver/operator should know the rules that govern the general public when emergency vehicles are on the road. Most laws or ordinances provide that other vehicles must pull toward the

Figure 2.3 Personnel should stand on both sides of the rear of a vehicle to assist the driver/operator in backing the vehicle safely.

right and remain at a standstill until the emergency vehicle has passed. These laws, however, do not mean that the apparatus driver/operator can ignore stopped vehicles. People may panic at the sound of an approaching siren and accidentally move into the lane of traffic or stop suddenly. Others may not hear warning signals because of car radios, closed windows, or air conditioning. It must also be remembered that some individuals simply ignore emergency warning signals.

Intersections are the most likely place for an accident involving an emergency vehicle to occur. When approaching an intersection, the driver/operator should slow the apparatus to a speed that allows a stop at the intersection if necessary. The apparatus should be brought to a complete stop if there are any obstructions, such as buildings or trucks, that block the driver/operator's view of the intersection. The apparatus should only proceed through the intersection if the light is green or if the driver/operator can account for all lanes and it is safe to proceed. Fire apparatus may travel with caution against a red traffic signal or stop sign if state regulations and department policy permit. However, it is strongly recommended that the apparatus be brought to a complete stop before proceeding through the intersection, even if local policies permit travel against a red light or stop sign. Blind and heavily traveled intersections should be approached and crossed at a reduced speed of 15 to 20 mph (24 to 32 km/h) (Figure 2.4).

Traffic waiting to make a left-hand turn may pull to the right or left, depending upon the driver. In situations where all lanes of traffic in the same direction as the responding apparatus are blocked, the apparatus driver/operator should move the apparatus into the opposing lane of traffic and proceed through the intersection at an extremely reduced speed. Oncoming traffic must be able to see the approaching apparatus. Full use of warning devices is essential. Driving in the oncoming lane is not recommended in situations where oncoming traffic is unable to see the apparatus, such as a freeway underpass. Be alert for traffic that may enter from access roads and driveways. Approaching traffic on the crest of a hill, slow-moving traffic, and other emergency apparatus must be closely monitored.

Even though use of warning sirens, lights, and signals is essential, fire apparatus driver/operators should realize that these signals may be blanketed by other warning devices and by street noises. Serious accidents and fatalities have been caused by overreliance on warning signals. Never assume what another driver's actions will be; expect the unexpected.

Anticipation is the key to safe driving. Always remember these control factors:

- Aim high in steering: find a safe path well ahead.
- Get the big picture: stay back and see it all.

Figure 2.4 Because approaching apparatus may not be plainly visible to other drivers, the driver/operator should always use extreme caution at busy or blind intersections.

- Keep your eyes moving: scan — do not stare.
- Leave yourself an "out." Be prepared by expecting the unexpected.
- Make sure others can see and hear you. Use lights, horn, and signals in combination.

Visual Lead Time

Visual lead time interacts directly with reaction time and stopping distances. By "aiming high in steering" and "getting the big picture" it is possible to become more keenly aware of conditions that may require slowing or stopping. The driver/operator is responsible for 360 degree driving.

Braking and Reaction Time

Speed directly affects the distance required to stop a vehicle. A driver/operator should know the total stopping distances for the particular fire apparatus. The total stopping distance is the sum of the driver/operator reaction distance and the vehicle braking distance (Figure 2.5). The driver reaction distance is the distance a vehicle travels while a driver is transferring the foot from the accelerator to the brake pedal after perceiving the need for stopping. The braking distance is the distance the vehicle travels from the time the brakes are applied until it comes to a complete stop. Table 2.1 shows driver reaction distances, vehicle braking distances, and total stopping distances for dif-

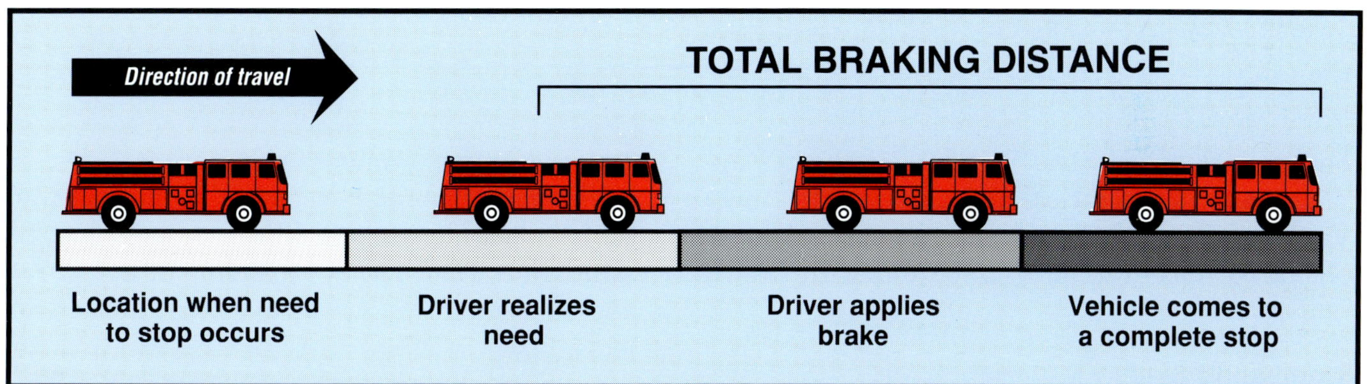

Figure 2.5 The distance a vehicle travels from the time the driver realizes the need to apply the brake until it comes to a complete stop is known as the total stopping distance.

TABLE 2.1 (U.S.)
BRAKING AND STOPPING DISTANCES
(dry, level pavement)

Typical Brake Performance
A — Average automobile
B — Light two-axle trucks
C — Heavy two-axle trucks
D — Three-axle trucks and trailers

Speed (mph)	Average Driver Reaction Distance (feet)	Braking Distance (feet) Vehicle A	Vehicle B	Vehicle C	Vehicle D	Total Stopping Distance (feet) Vehicle A	Vehicle B	Vehicle C	Vehicle D
10	11		7	10	13		18	21	24
15	17		17	22	29		34	39	46
20	22	22	30	40	50	44	52	62	72
25	28	31	46	64	80	59	74	92	108
30	33	45	67	92	115	78	100	125	148
35	39	58	92	125	160	97	131	164	199
40	44	80	125	165	205	124	169	209	249
45	50	103	165	210	260	153	215	260	310
50	55	131	225	255	320	186	280	310	375
55	61	165	275	310	390	226	336	371	451
60	66	202	350	370	465	268	426	436	531

TABLE 2.1 (metric)
BRAKING AND STOPPING DISTANCES
(dry, level pavement)

Typical Brake Performance
A – Average automobile
B – Light two-axle trucks
C – Heavy two-axle trucks
D – Three-axle trucks and trailers

		Braking Distance (meters)				Total Stopping Distance (meters)			
Speed (km/h)	Average Driver Reaction Distance (meters)	Vehicle A	Vehicle B	Vehicle C	Vehicle D	Vehicle A	Vehicle B	Vehicle C	Vehicle D
16	3.4		2.1	3	4		5.5	6.4	7.3
24	5.2		5.2	6.7	8.8		10.4	11.9	14
32	6.7	6.7	9.1	12.2	15.2	13.4	15.8	18.9	21.9
40	8.5	9.4	14	19.5	24.4	18	22.6	28	32.9
48	10.1	13.7	20.4	28	35.1	23.8	30.5	38.1	45.1
56	11.9	17.7	28	38.1	48.8	29.6	39.9	50	60.7
64	13.4	24.4	38.1	50.3	62.5	37.8	51.5	63.7	75.9
72	15.2	31.4	50.3	64	79.2	46.6	65.5	79.2	94.5
80	16.8	39.9	68.6	77.7	97.5	56.7	85.3	94.5	114.3
88	18.6	50.3	83.8	94.5	118.9	68.9	102.4	113.1	137.5
96	20.1	61.6	106.7	112.8	141.7	81.7	129.8	133	161.8

ferent sizes of vehicles. This table indicates approximates for vehicles, and the statistics may vary for different fire apparatus. Each department should conduct braking distance tests with its own apparatus.

Combating Skids

Avoiding conditions that lead to skidding is as important as knowing how to correct skids once they occur. The most common causes of skids are

- Driving too fast for road conditions
- Failing to properly appreciate weight shifts of heavy apparatus
- Failing to anticipate obstacles. These range from other vehicles to stray dogs.

If the apparatus goes into a skid, release the brakes, allowing the wheels to rotate freely. Turn the apparatus steering wheel so the front wheels face in the direction of the skid and let up on the accelerator gradually. If using a standard transmission, do not release the clutch until the vehicle is under control and just before stopping the vehicle. Once the skid is controllable, gradually apply power to the wheels to further control the vehicle by giving traction.

Proper maintenance of tire air pressure and adequate tread standards for tires are crucial for skid prevention. Some chassis manufacturers install computer brake systems that help prevent skids by not allowing all wheels to lock up at the same time. Their use is expanding at a slow rate due to cost considerations and operational problems.

Proficiency in skid control may be gained through practice at facilities having skid pads. In some instances, large open areas can be used where danger to the public, apparatus, and apparatus driver/operators can be controlled.

EVASIVE TACTICS

Remember that anticipation is the key to defensive driving. This sounds familiar, but this phrase cannot be overemphasized.

Now, a scenario: While responding on a three-lane highway during moderately heavy traffic conditions, the vehicle ahead, which you are rapidly overtaking, stops! Because there is insufficient space to stop without striking the vehicle, evasive tactics are necessary.

The best choice is to pass the stopped vehicle on the left side because the driver's likely next move is to pull to the right, as is generally required by law. Passing on the right of any vehicle invites serious exposure and possible liability.

During the evasive maneuver, hands should not leave the steering wheel. Do not lean or sway back and forth in the seat. Use your arms to steer the apparatus. Look ahead of the stopped vehicle, concentrating on where you want to be, not where you do not want to be.

Braking is generally not indicated, as a momentum skid could occur. Maintaining speed, accelerating, or in some cases, decelerating may be necessary to avoid skids and maintain control.

WEIGHT TRANSFER

The effects of weight transfer must be considered in the safe operation of fire apparatus. Weight transfer occurs as the result of physical laws that state that objects in motion tend to stay in motion; objects at rest tend to remain at rest. Whenever a vehicle undergoes a change in velocity or direction, weight transfer takes place relative to the severity of change. Apparatus driver/operators must be aware that the high weight carried on most fire apparatus can contribute to skidding or possible rollover due to excessive weight transfer. These hazardous conditions can result from too much speed in turns, harsh or abrupt steering action, or driving on slopes too steep for a particular apparatus. Use only as much steering as needed to keep weight transfer to a minimum. Steering should be smooth and continuous.

ADVERSE WEATHER

Weather is another factor to consider in terms of safe driving. Rain or ice will make roads slippery (Figure 2.6). A driver/operator must recognize these dangers and adjust apparatus speed according to the crown of the road, the sharpness of curves, and the condition of road surfaces. The driver/operator should decrease speed gradually, slow down while approaching curves, keep off low or soft shoulders, and avoid sudden turns. The driver/operator should recognize areas that first become slippery, such as bridge surfaces, northern slopes of hills, shaded spots, and areas where snow is blowing across the roadway.

Figure 2.6 Adverse weather conditions can create additional difficulties for units responding to an incident. *Courtesy of Dan Contini.*

Since the stopping distance is greatly increased on slippery road surfaces, it is sometimes a good policy to try out the brakes while in an area free of traffic to find out how slippery the road is. Speed must be adjusted to road and weather conditions so the apparatus can be stopped or maneuvered safely. Good windshield wipers and defrosters should keep the windshield clean and clear (Figure 2.7).

Figure 2.7 To keep large windshields clear, some apparatus are equipped with extra windshield wipers.

Snow tires or chains will reduce the stopping distance and considerably increase starting and hill climbing traction on snow or ice. Despite this added security, it is still necessary to maintain lower speeds on slippery roads. Brakes should be pumped, not jammed, for a gradual slowdown or stop. Sudden application of the brakes can lock the wheels and throw the apparatus into a dangerous skid. A driver/operator cannot feel the point at which the brakes begin to hold and so may be unable to release them at the proper moment. One good technique to use on glare ice is a series of rapid, but brief, brake applications, making sure that the brake is completely released on each application.

Apparatus equipped with air brakes may have a "dry road/slippery road" brake limiting valve. When the brakes are applied with the limiting valve in the dry road position, equal pressure is exerted on all brakes. When the brakes are applied with the limiting valve in the slippery road position, the pressure applied to the front brakes is less than the pressure applied to the rear brakes.

During slippery road conditions, the safe following distance between vehicles increases dramatically. Remember that it takes 3 to 15 times more distance to come to a complete stop on snow and ice as it does to come to a complete stop on dry concrete.

WARNING DEVICES AND CLEARING TRAFFIC

All fire apparatus should be equipped with audible and visible warning devices. Audible warning devices may include air horns, bells, mechanical sirens, or electronic sirens. Studies have shown that civilian drivers respond better to sounds that change pitch often. Short bursts with the air horns and the constant up-and-down oscillation of a mechanical or electronic siren are the surest ways to catch a driver's attention. At speeds above 50 mph (80 km/h), an emergency vehicle may "outrun" the effective range of its audible warning devices. A study conducted by the staff of Driver's Reaction Course concluded that a siren operating on an emergency vehicle moving at 40 mph (64 km/h) can project 300 feet (90 m) in front of the vehicle. At a speed of 60 mph (97 km/h), however, the siren is only audible 12 feet (3.7 m) or less in front of the vehicle. Driver/operators must operate within the effective range of their audible warning devices.

Expensive warning devices are of no value to the driver/operator or the public if they are not used. Warning devices should be operated from the time the apparatus begins its response until it arrives on the scene. The driver/operator does need to use some discretion in the use of sirens, though. When responding to sensitive situations, such as heart attacks, it is often better to turn off the siren as the apparatus nears the destination. The sudden use of an audible warning device immediately behind another vehicle may unnerve the driver. If this happens, the driver may suddenly stop or swerve, causing an accident. The use of warning devices is essential, but does not give the driver/operator the right or permission to disregard other drivers.

There have been numerous collisions involving fire apparatus and other emergency vehicles (Figure 2.8). It is not always possible to hear the warning devices of other vehicles when the apparatus features a similar signal. When more than one emergency vehicle is responding along the same route, units should travel at least 300 to 500 feet (90 m to 150 m) apart. Some fire departments rely upon designated response routes. This practice can be hazardous if a company is delayed or detoured for some reason. Standard operating procedures should call for radio reports of location and status. Regardless the system or pattern used, always take precautions to ensure a safe, accident-free response.

Figure 2.8 Most accidents involving emergency vehicles can be prevented by slowing down and using better judgment. *Courtesy of Ron Jeffers.*

White lights can be readily distinguished during daylight hours. For this reason, headlights should be turned on while responding. Some departments use white warning lights in conjunction with red or other colored lights, but some state laws prohibit the use of flashing white lights. A spotlight moving across the back window of a vehicle will rapidly gain the driver's attention. The spotlight should not be left shining on the vehicle, however, as this will blind the driver. Headlights should be dimmed and spotlights turned off in situations where they may blind oncoming drivers. Headlight flashers are an inexpensive and effective warning device; however, some states do not allow or consider them as warning lights.

Warning devices are an essential part of a safe emergency response. Fire departments and even individual driver/operators have been held responsible for damage caused in accidents involving apparatus responding without the use of a siren.

Traffic Control Devices

Some cities have installed traffic control devices to assist responding emergency vehicles. Some cities have apparatus that utilize predetermined response routes and then control all intersections along those routes in favor of the responding apparatus. Other cities place strobe lights on each emergency vehicle and sensors on each traffic light. The sensors detect the approaching fire apparatus from the strobe and then change the lights to favor the responding apparatus (Figure 2.9).

These devices have increased efficiency for all emergency vehicles in these cities. Traffic control devices should not be considered fail-proof and driver/operators should use normal caution at all times.

STARTING AND DRIVING THE VEHICLE

The fire apparatus driver/operator should start the engine with the drive transmission in neutral and with the clutch disengaged. If the vehicle has an automatic transmission, it should be started in the "park" or "neutral" position. The

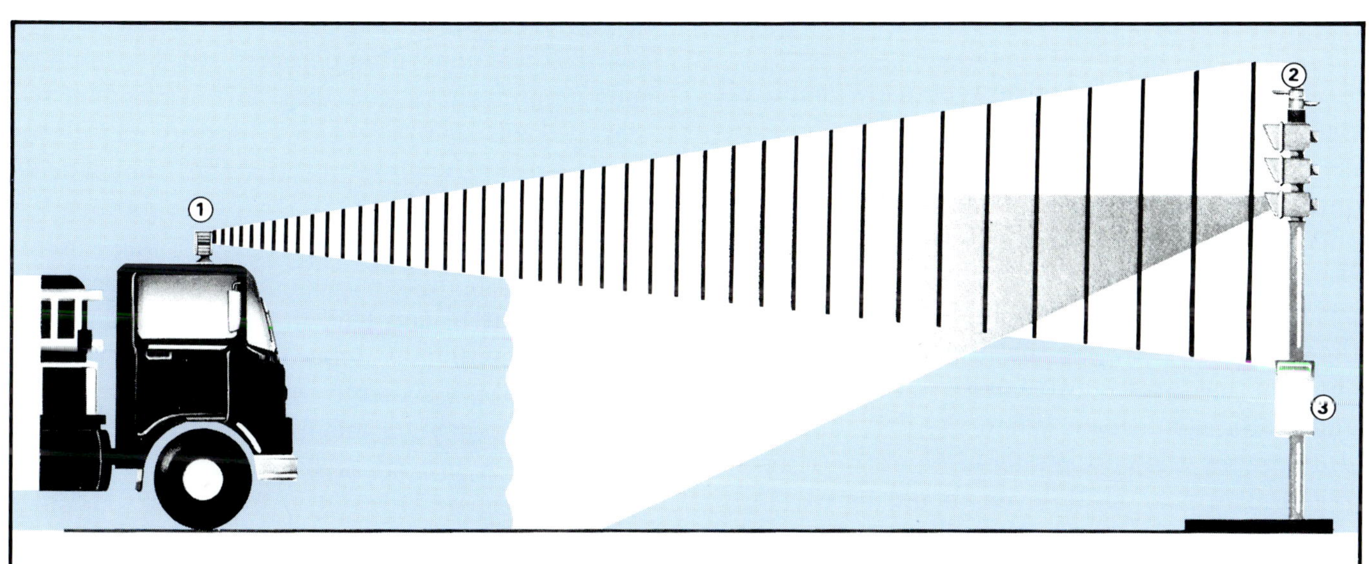

Figure 2.9a How the Opticom system works: (1) Emitter generates an optical signal to the Detector (2), which converts the optical signal to an electronic impulse. The Phase Selector (3) in the control cabinet processes and then manipulates the controller to provide a green signal for the emergency vehicle. *Courtesy of 3M Safety and Security.*

Figure 2.9b The Opticom Emitter is mounted on top of the emergency vehicle. *Courtesy of 3M Safety and Security.*

driver/operator should take a moment to review the incident location, road closings, traffic conditions, and other factors that may affect the response route. The officer should signal the driver/operator to begin the response only after the entire crew have secured themselves to the vehicle with seat belts. The transmission should be put into the proper gear and the emergency brake released before the driver/operator attempts to move the vehicle. Take care to avoid vehicle rollback before engaging the clutch.

A fire apparatus driver/operator should never attempt to start the apparatus moving while it is in a high or driving gear. This action causes the clutch to slip, which may damage the clutch facing. The clutch should be released slowly when starting from a standstill because a sudden, rapid release of the clutch throws a heavy load upon the engine, clutch, transmission, and drive components. This also includes sudden acceleration applied to automatic transmissions.

The apparatus should be kept in low gear until the proper speed or revolutions per minute (rpm) is reached for shifting to a higher gear. An engine develops its maximum power up to a certain speed, and excessive engine speed may result in decreased power and excessive use of fuel. It is good practice to keep the transmission in low gear until the apparatus is clear of the station and the driver/operator has an unobstructed view of the street and traffic conditions. When the driver/operator shifts gears, the clutch should be entirely disengaged. The gear shift lever should be carefully moved into proper position, not jammed.

Driver/operators of apparatus equipped with automatic transmissions should be aware that the pressure placed upon the accelerator will influence automatic shifting. When the pedal is fully depressed, the transmission will downshift or remain in a lower gear until there is sufficient engine speed to allow an upshift. The upshift will occur near the governed speed of the engine. This may not give the driver maximum acceleration, but it will deliver the maximum acceleration allowed by the manufacturers design. In normal driving conditions a slow continual pressure on the accelerator will cause the transmission to upshift at lower engine speeds decreasing fuel consumption and reducing overall wear on the vehicle.

USE OF A LOWER GEAR ON STEEP DOWNHILL GRADES WILL ASSIST IN CONTROLLING VEHICLE SPEED AND REDUCE EXCESSIVE WEAR ON BRAKES.

When climbing a hill or going down a steep grade, shift the transmission to a lower gear. This practice provides adequate driving power and enables the driver/operator to keep the apparatus under control. On sharp curves or when turning corners, shift standard transmissions into a lower gear before entering the curve or intersection. This action maintains peak engine power and apparatus control. When fire apparatus must be driven over rough or rugged terrain, use the lower gears. If the apparatus becomes stuck in mud or snow, maintain a constant low rpm (approximately 1200-1500) and slowly engage and disengage the clutch while keeping the front wheels parallel to the chassis, "straight." Spinning the tires may cause the apparatus to become further stuck. Sudden engagements of the clutch or "popping the clutch" can lead to mechanical damage. The combination of apparatus weight and soft ground causes excessive torque in the driveline components. If your apparatus is stuck call for assistance. A tow truck bill is cheaper than an apparatus repair bill.

The process of braking fire apparatus to a standstill should be performed smoothly so the apparatus will come to an even stop. Before braking, the driver/operator should consider the weight of the apparatus and the condition of the brakes, tires, and road surface. An abrupt halt can cause a skid, injury to firefighters, and mechanical failure. Some apparatus employ engine brakes or retarding devices, which assist in braking. The engine brake and retarder are activated when pressure is released from the accelerator. Because they provide most of the necessary slowing action, these devices allow the driver/operator to limit the use of service brakes to emergency stops and final stops. Both devices save wear on the service brakes and make the apparatus easier to manage on hills and slippery roads. Driver/operators of units having retarders should become thoroughly familiar with all the manufacturer's recommendations regarding their operation prior to use.

Apparatus with automatic transmission may be slowed by downshifting to the next lowest gear range. Refer to the manufacturer's recommendations for the particular make and model.

Some air brake systems have limiting systems for varying road conditions. The clutch should not be disengaged while braking until the last few feet (meters) of travel. This practice is particularly important on slippery surfaces because an engaged engine allows more control of the apparatus.

DRIVING PRACTICES FOR DIESEL ENGINES

Starting

The procedure for starting an apparatus with a diesel engine is as follows:

Step 1: Be sure that the gear selector is in "park" or "neutral" for an engine with automatic transmission or in "neutral" if the engine has a manual transmission. The parking brake should be set for all apparatus.

Step 2: Turn the battery switch to "both" if it is so equipped and put ignition switches in the "on" position.

Step 3: Turn the ignition key or press the start buttons to crank the engine. Crank the engine no more than 30 seconds at a time. Allow sufficient cooling time between attempts. Long and frequent starting attempts result in rapid discharge of batteries, heat buildup in the starter and solenoid, and early starter failures.

The need to respond as soon as possible will prevent proper warm-up periods. When not responding to an alarm, allow three to five minutes of warm-up time. If responding, attempt to keep engine rpm's moderate and apply lighter-than-normal throttle until the engine has had a chance to warm up. Never move the apparatus until the oil pressure gauge indicates sufficient pressure to lubricate the engine.

Oil Pressure

Driver/operators should be familiar with the normal readings of the gauges on their equipment. In this way, abnormal readings will be noticed sooner and problems can be corrected.

If the oil pressure gauge flutters, the oil level may be low. Stop the engine and check the oil level. Start-up pressures may be higher than during normal driving. Higher-than-normal pressures while the oil is cold should not be cause for alarm.

Oil Temperature

Consult the operator's handbook for the permissible temperature range. If the engine temperature goes above this range during normal driving, reduce speed, pull off the road, check the oil level, or check for damage.

Coolant Temperature

Consult the operator's handbook for minimum operating temperature. If the coolant temperature does not reach this level, the gauge or thermostat may be defective or another device may be needed to assist in maintaining minimum temperature.

The apparatus may be driven from the station and in city traffic at lower-than-minimum highway temperature, but operating cold causes high piston ring wear, increased oil consumption, and shorter engine life. Do not operate the engine continuously below the minimum highway driving temperature. Low coolant temperature may result in poor lubrication because of cold oil; inefficient combustion, evident by lower power and smoke, caused by low cylinder temperatures; increased valve deposits; and increased fuel consumption.

Sudden temperature increase is a sign of trouble. Heavy operating loads may increase operating temperatures. Shut down as soon as possible and investigate. Typical causes include a broken water pump belt; coolant flow clogging; coolant loss; or a faulty thermostat, shutter thermostat, or gauge.

Accelerating

Start the vehicle from a standstill at the slowest engine speed that will move the load. Select a gear ratio that will enable the vehicle to move without slipping the clutch. A sudden application of power may result in a torque condition overloading the clutch, transmission, U-joints, ring and pinion gears on the axle, or the axle shaft. This overtorquing can cause severe damage to any or all of these components.

Use progressive shifting. When progressive shifting is used, the engine need not be accelerated to the governed speed before shifting. This lowers engine wear and noise exposure.

City Operation

Accelerate the vehicle gradually. Do not try to reach rated speed in the low gears. Going to rated speed in the low gears can be very noisy and will increase engine wear. Stay in the highest gear that allows the truck to keep up with traffic and still have some power in reserve for acceleration.

Operators of fire service vehicles can significantly reduce drive train damage and extend apparatus service life by adopting proper operating habits. When driving a fire service vehicle, attempt to maintain engine rpm control through correct throttling. Keeping the engine operating within its power curve will ensure adequate power and optimum fuel economy for a given set of conditions.

Whenever possible, avoid lugging, which results in overthrottling. Lugging results from the engine operating at full throttle (accelerator fully depressed) below rated speed. When the engine is lugging, there is no response to the accelerator. When climbing a hill with the accelerator to the floor, the engine is lugging if engine speed is dropping or cannot be increased because all available power is being used to maintain speed.

Overthrottling occurs with a diesel engine when more fuel is being injected than can be burned. This results in an excessive amount of carbon particles issuing from the exhaust (black smoke), oil dilution, and additional fuel consumption.

Another point that requires consideration is maximum engine rpm. The fire service maintains apparatus for a much longer period of time than commercial vehicles are maintained. Over this period of time, the valve springs can become weakened. For this reason, allowing an engine to overspeed as the result of improper downshifting or hill descent should be avoided in an effort to prolong engine life.

Cruising

Choose a gear that will cruise the engine at 200 or 300 rpm lower than recommended rpm. This reduces engine wear; power losses caused by the fan, driveline, and accessories; noise; and fuel consumption.

Uphill Operation

Do not allow the engine rpm to drop below recommended operational speed when climbing. As engine speed decreases torque (power) also decreases presenting the need to shift. Automatic transmissions will downshift accordingly. Standard transmissions must be downshifted to avoid stalls and prevent lugging. When ascending a steep grade, momentary unavoidable lugging takes place. Although this brief lugging is unavoidable, time spent in the lugging state should be minimized to avoid possible damage to the power plant. Engine rpms drop even with heavy throttle settings. Progressively lower gears are selected until the combination of available power and the torque multiplication of the transmission gears allow the hill to be taken in stride.

By familiarizing themselves with equipment and by adopting proper driving habits, apparatus operators can increase the life of expensive fire service vehicles considerably.

Downhill Operation

When driving downhill, select a lower gear before descending and remain in gear at all times. The engine provides braking power when the vehicle is left in gear.

To prevent engine damage, limit downhill speed to lower than maximum governed rpm. This can be achieved through a combination of gearing and braking. The engine governor cannot control engine speed downhill; the wheels turn the drive shaft and engine. Engine rotation faster than rated rpm can result in valves hitting pistons, increased oil consumption, damage requiring overhaul, and injector seizures.

Never attempt to shut down the engine while the apparatus is in motion because this action cuts off fuel flow to the injectors. Fuel flow through the injectors is required for lubrication anytime the injector plunger is moving.

Engine Idling

Shut the engine down rather than leave it idling for long periods of time. Long idling periods can result in use of 1/2 gallon (2 L) of fuel per hour; buildup of carbon in injectors, valves, pistons, and valve seats; misfiring because of injector carbonizing; and damage to the turbocharger shaft seals.

When the engine *must* be left to idle for a long time because of extremely cold weather or clean-up operations, set it to idle at 900 to 1100 rpm rather than at lower speeds.

Engine Shutdown

Never shut down immediately after full-load operation. Shutting down the engine without a cooling-off period results in immediate increase of engine temperature from lack of coolant circulation, oil film "burning on" hot surfaces, possible damage to heads and exhaust manifolds, and possible damage to the turbocharger that may result in turbo seizure. Allow the engine temperature to stabilize before shutdown. A hot engine should be idled until it has cooled.

General

Know the machine; be thoroughly familiar with the operator's handbook. Consult the manufacturer when in doubt. Apparatus driver/operators and other fire service personnel can find a considerable amount of information about an apparatus in the literature provided by the apparatus and component manufacturers. Local product service agencies and commercial fleets using similar drive components can be another valuable source of information. These contacts can also be useful in supplying updated product information such as service bulletins and other pertinent data. The practical application of this material can enhance the overall efficiency of apparatus operations.

DRIVING EXERCISES AND EVALUATION METHODS

After driver/operators have been selected and trained, their performance should be evaluated by some standard method. These evaluations should occur before the driver/operator is allowed to operate the apparatus under emergency conditions.

Written Test

The written test for driver/operators should include questions pertaining to the following areas:

- State and local driving regulations for emergency and nonemergency situations
- Departmental regulations
- Hydraulic calculations
- Specific operational questions regarding pumping

Pumping Apparatus Operational Tests

The fire apparatus driver/operator candidate should be required to perform simulated fireground tasks using front-line equipment. These tasks should include, but are not limited to, setup, initial attack, and pumping.

The driver/operator's first road test should be conducted only after the firefighter has had sufficient preliminary training. A road test should be part of promotional examinations when the promotion involves apparatus driving. A road test should also be given annually to all driver/operators. An annual road test may be used as a portion of the driver/operator's annual evaluation and included in the driver/operator's personnel file. If the road test is given as part of a promotional examination, the same route should be used for each driver/operator.

The route for the test should cover at least 5 miles (8 km) of variable terrain, road conditions, and traffic situations. Try to plan a route that includes hills, a variety of corners (slow curves, right and left turns, sharp angle turns), and rail-

DRIVER'S ROAD TEST RECORD

0-9 Satisfactory = POOR, not passing
10 Satisfactory = FAIR, passing
11 Satisfactory = AVERAGE, passing
12-13 Satisfactory = EXCELLENT, passing

Driver's Name ______________________ Driver's Station ______________________

Driver's License Number and Classification ______________________

Apparatus Number ______________

Evaluator's Name ______________________ Test Date ______________

Special Information (route, etc.) ______________________

____

Criterion	Rating
1. Driver's manner and ability to handle fire apparatus.	______
2. Driver's ability to shift up smoothly and without clashing gears.	______
3. Driver's ability to shift down smoothly.	______
4. Driver's use of hydraulic retarder or engine brake.	______
5. Driver's treatment of engine (unnecessary speed increases, lugging under load, revving before shutting down).	______
6. Driver's ability to stop apparatus smoothly, without sliding wheels.	______
7. Driver's ability to spot apparatus at curb and out of traffic lane.	______
8. Driver's ability when backing.	______
9. Caution in backing; positioning look-out at rear.	______
10. Driver's exhibition of responsibility and concern for safety of apparatus and personnel while driving apparatus.	______
11. Control of apparatus by manually shifting automatic transmission and keeping engine speed correct.	______
12. Automatic transmission shifted correctly.	______
Number Satisfactory	______

Drivers in the POOR class should be given additional training before being allowed to drive apparatus for other than training purposes.

> Drivers receiving below 70 percent should be considered poor drivers and should go through driver training again. Upon completion of driver training, the driver must be retested and pass before being allowed to drive emergency vehicles to emergency situations. Every year, *all* drivers should be tested and any weakness in their skills should be brought to their attention and corrected.

Figure 2.10 The sample road test record sheet shown above should be modified to reflect local needs.

road crossings that do not have automatic signals. In city traffic (try to avoid rush hour), include as many variables as possible, such as stop lights and signs, busy intersections, and alleys.

The road test should be as objective as possible. Do not try to assess such things as attitude, which calls for a subjective judgment. This test is meant only to measure the ability of the driver/operator to get the apparatus, its equipment, and its personnel to a destination in a safe manner. Whether the driver/operator is psychologically able to handle the stress and excitement of driving in a real emergency situation can be determined only in a real emergency situation. For this reason, and because such an attempt could endanger firefighters and others, the road test should not be a mock emergency run. For the same reasons, the road test should not include the factor of time.

A sample road test record sheet is shown in Figure 2.10. It is presented here as a guide for the development of the local test. It should not be used as a standard test. Note, for instance, that two of the criteria involve automatic transmissions; if the apparatus being driven has a manual transmission, the scoring would be incorrect.

A department may want to make its test more specific than the sample shown. For instance, points could be given for various turns, stopping and starting on hills, backing in different types of areas, and approaching intersections. However the test is structured, the easiest and fairest method of scoring is a simple "satisfactory" or "unsatisfactory." Each portion of the test should carry about equal weight.

Braking Exercise

In the braking exercise, the driver/operator must maneuver the vehicle at a constant speed through a set of markers. They must stop the vehicle with the front bumper as close as possible to a line of markers without running over them (Figure 2.11). This exercise simulates situations re-

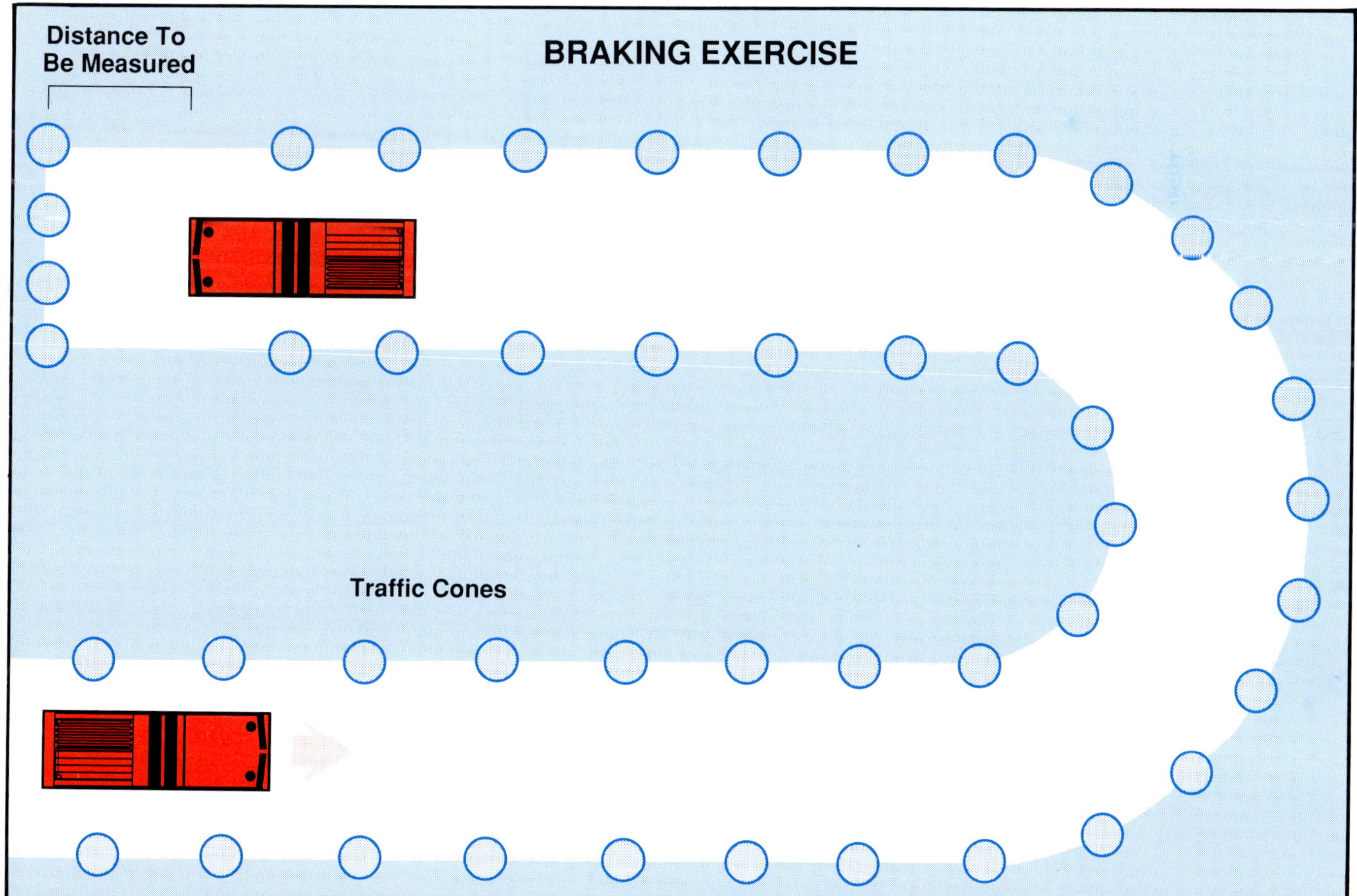

Figure 2.11 In the braking exercise, the distance between the front bumper and the line of traffic cones at the end of the course is measured.

quiring rapid braking. Upon coming to a complete stop, the distance from the front of the front bumper to the stop line is measured. Scoring is as follows: (50 points are possible)

0 to 6 inches (0 to 150 mm) 50 points
6 to 9 inches (150 to 225 mm) 45 points
9 to 12 inches (225 to 300 mm) 40 points
12 to 15 inches (300 to 375 mm) 35 points
15 to 18 inches (375 to 450 mm) 30 points
18 inches or more (450 mm +) No points

Penalties may also be assessed for driving errors. These are subtracted from the original score. For example, the driver/operator must maintain a normal driving position within the cab of the vehicle. Failure to do so results in a loss of 25 points. Each time the vehicle stops within the markers, 5 points are subtracted. Striking any course markers also results in a 5-point penalty.

Serpentine Course

In the serpentine exercise, markers are placed an equal distance apart in a line (approximately 34 feet [10 m] for pumpers, tankers, and smaller emergency-response vehicles). (Figure 2.12) The driver/operator is required to maneuver the vehicle forward and backward. This exercise simulates maneuvering around parked and stopped vehicles and tight corners. The course must be traveled in one continuous motion without touching any of the course markers. Adequate space must be provided on each side of the markers for the apparatus to move freely.

First, the driver/operator is required to drive the apparatus along the left side of the markers in a straight line and stop just beyond the last marker (Figure 2.13). Then the driver/operator must back the apparatus between the markers by passing to the left of No. 1, to the right of No. 2, and to the left of No. 3. At this point, the driver/operator must stop the vehicle and then drive it forward between the markers by passing to the right of No. 3, to the left of No. 2, and to the right

Figure 2.13 A serpentine course laid out on a training site.

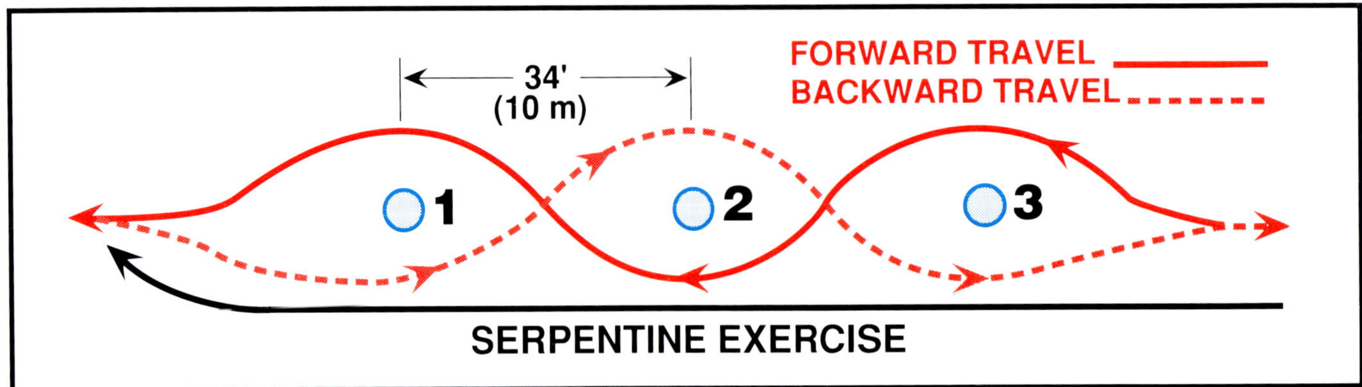

Figure 2.12 The serpentine exercise is designed to simulate driving around tight corners and parked vehicles.

of No. 1. The grading scale for this exercise is as follows: (50 points are possible)

- Passing course marker on the wrong side: 5-point penalty
- Each time the vehicle stops during the exercise: 5-point penalty
- Hitting or running over any course markers: 5-point penalty

0 to 6 inches (0 mm to 150 mm) 50 points
6 to 9 inches (150 mm to 225 mm) 45 points
9 to 12 inches (225 mm to 300 mm) 40 points
12 to 15 inches (300 mm to 380 mm) 35 points
15 to 18 inches (380 mm to 450 mm) 30 points
10-91 18 inches or more (450 mm +) No score
Hitting or running over any course markers: 5-point penalty

Alley Dock

The alley dock exercise tests the driver/operator's ability to move the vehicle backward within a restricted area and into an alley, dock, or fire station without striking the walls, and to bring the vehicle to a smooth stop close to the rear wall (Figure 2.14). The boundary lines for the restricted area should be 40 feet (12 m) wide, similar to curb-to-curb distance. Along one side and perpendicular is another simulated area 10 feet wide and 20 feet deep (3 meters wide by 6 meters deep). The test procedure has the driver/operator moving past the alley, which is on the left, backing the apparatus, making a left turn in reverse into the defined area, and stopping. The grading scale for this exercise is as follows: (50 points are possible)

Offset Alley or Opposite Alley

The purpose of this exercise is to test the driver/operator's ability to maneuver the vehicle at a constant speed within a confined space. To do this, the driver/operator must be able to accurately judge the distance between the vehicle and the barriers. This exercise simulates responding to an emergency and making a rapid lane change without striking any nearby vehicles.

In the offset-alley exercise, two sets of markers are set off center and one vehicle length apart. These alleys can be arranged from pylons or barricades and are 10 feet (3 m) wide. The 10-foot (3 m) alley from which the driver/operator must exit is arranged 10 feet (3 m) out of line to the opposite 10-foot (3 m) alley into which the apparatus must be maneuvered. No set speed should be established for this exercise, but the driver/operator should

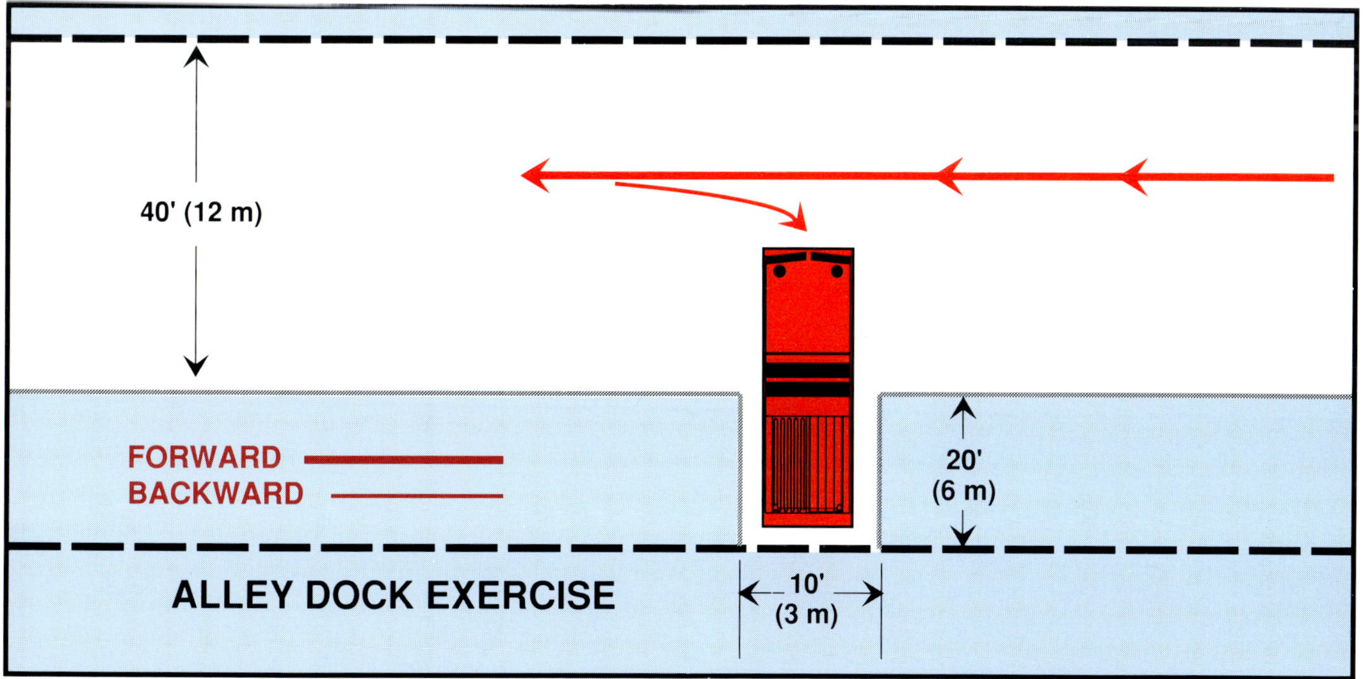

Figure 2.14 The alley dock exercise is designed to test the driver/operator's ability to back up a vehicle without striking any objects.

not stop or back the apparatus during the maneuver (Figure 2.15). It will be necessary to adjust the dimensions of the course for apparatus with greater wheel base lengths. Scoring for this exercise is as follows: (50 points are possible)

- Each time the vehicle stops or fails to maintain constant motion: 5-point penalty.
- If the vehicle strikes any course markers, 10 points are subtracted. After each such incident, the driver/operator must realign the vehicle before continuing with the exercise. Failure to do so will result in the subtraction of an additional 25 points.

Lane Change Exercise

The lane change exercise tests the driver/operator's ability to change lanes while moving at a constant speed. Sets of markers are set up as shown, with each set being 50 feet long and 10 feet wide (15 meters long and 3 meters wide)(Figure 2.16). The distance from one set of markers to the next is 30 feet (10 m). This area is called the change space.

If the first lane assigned to the driver/operator is Lane A, the second lane assigned must be either Lane 1 or 2. If the first lane assigned to the driver/operator is Lane B, the second lane assigned must

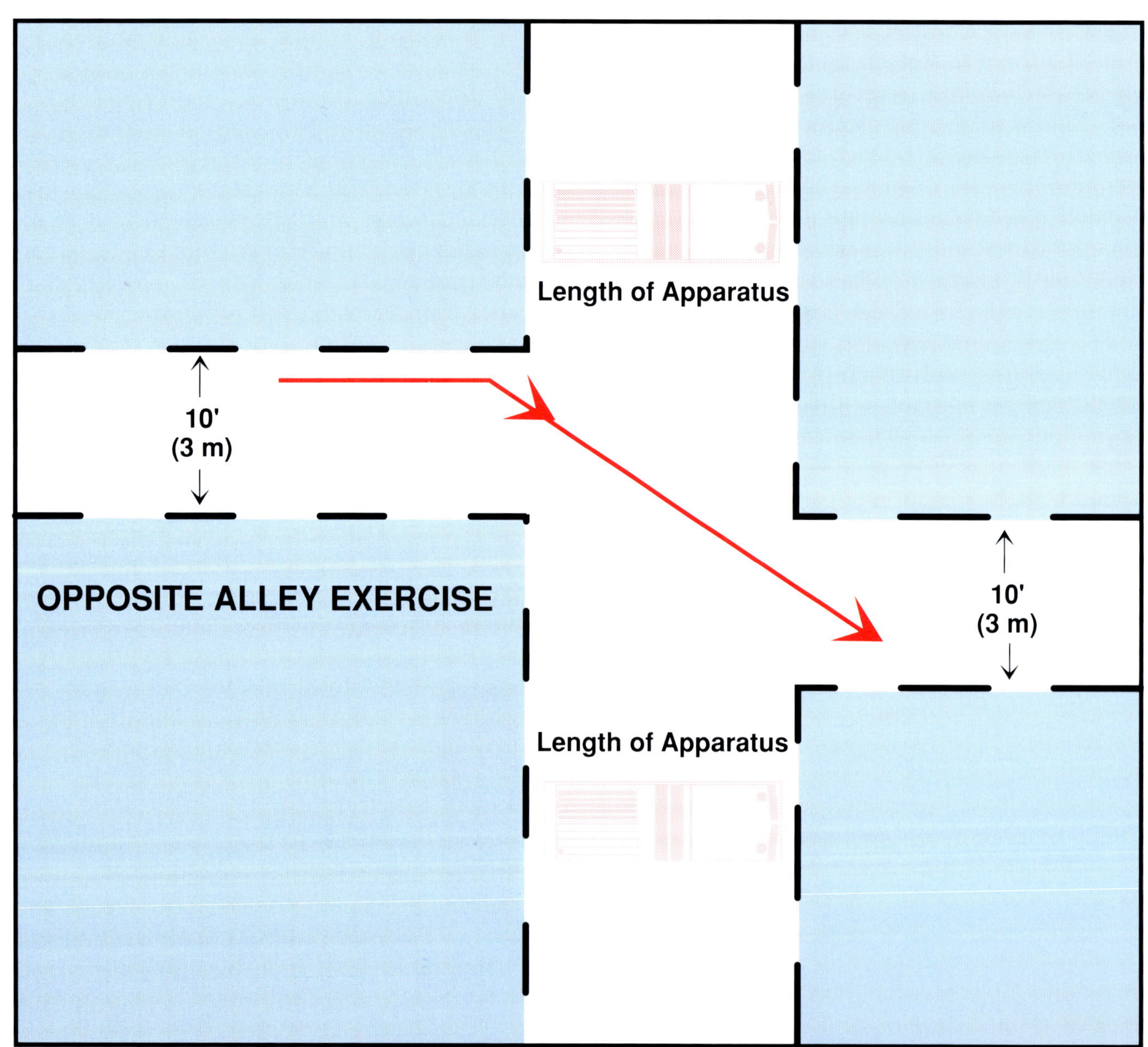

Figure 2.15 The purpose of the offset alley exercise is to test the driver's ability to maneuver the vehicle at a constant speed within a confined space.

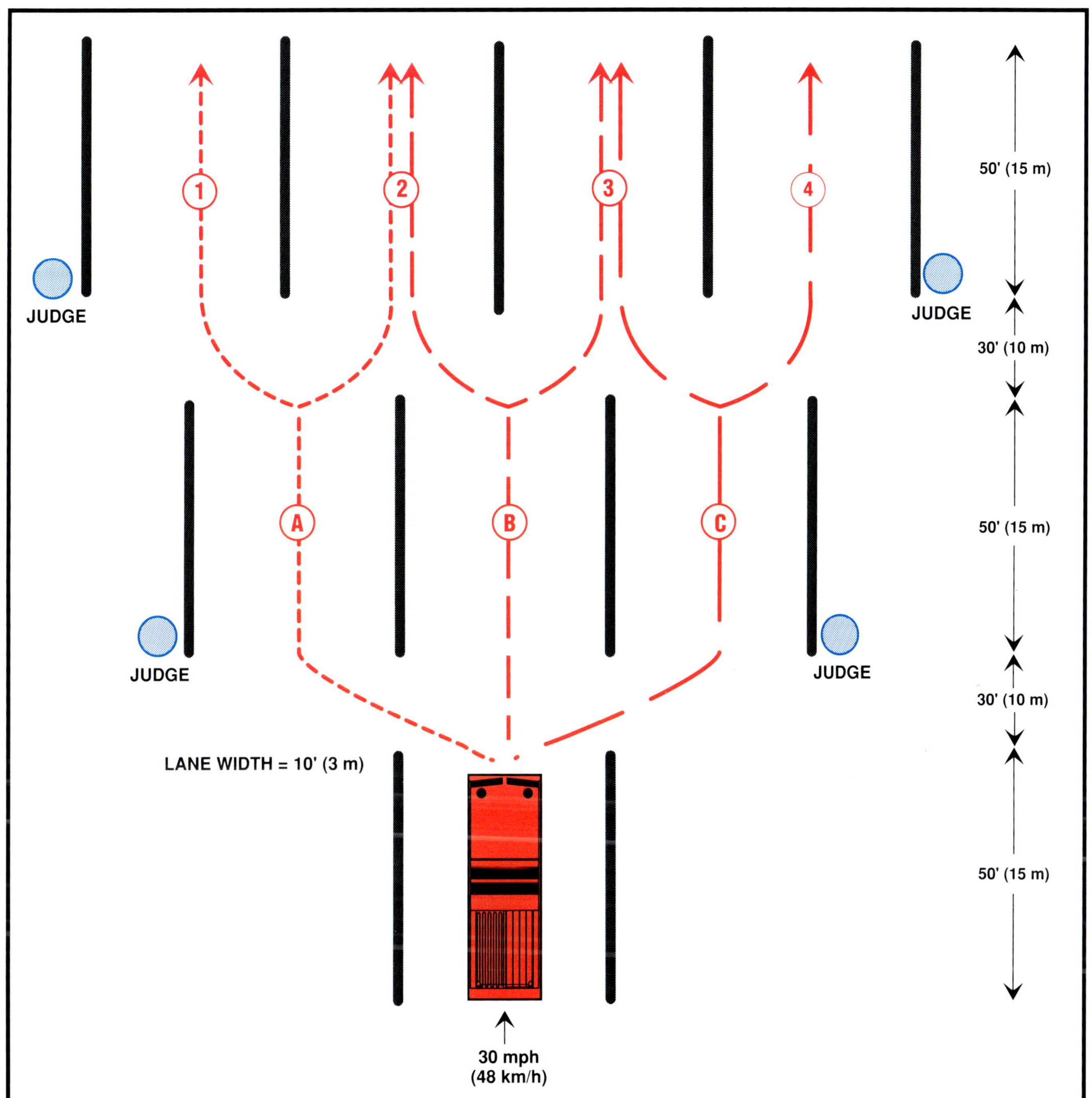

Figure 2.16 During the lane change exercise, the driver/operator must change lanes on signal without stopping or striking any markers.

be either Lane 2 or 3. If the first lane assigned to the driver/operator is Lane C, the second lane assigned must be either Lane 3 or 4.

During the exercise, judges stand on both sides of each change space and hold up flash cards telling the driver/operator which lane to take. The judges may change the lane required anytime the vehicle is still between the set of markers. The driver/operator must adjust accordingly and take the designated lane. Grading for this exercise is as follows: (50 points are possible)

- The driver/operator approaches the first lane at a speed of 30 mph (48 km/h) and maintains this speed throughout the entire exercise. Failure to maintain semiconstant speed: 10-point penalty.
- Hitting or running over a course marker: 5-point penalty.

- Anytime the driver/operator stops the vehicle during the exercise, 25 points will be subtracted.
- Failure to take the lane signaled by the judges: 25-point penalty.
- Failure to maintain control of the vehicle: no score.

Diminishing-Clearance Exercise

The diminishing-clearance exercise measures a driver/operator's ability to steer the apparatus in a straight line, to judge distances from wheel to object, and to stop at a finish line. The speed at which the apparatus is driven is optional, but it should be fast enough to require the driver/operator to exercise quick judgment. The course for this exercise is arranged by two rows of stanchions that form a lane 75 feet (23 m) long (Figure 2.17). The lane narrows from a width of 9 inches (225 mm) wider than the vehicle to a diminishing clearance of 1 inch (25 mm). The driver/operator must maneuver the apparatus through this lane without touching stanchions. At a point 50 feet (15 m) beyond the last stanchion, the driver/operator must stop with the front bumper within 6 inches (150 mm) of the finish line. Scoring for this exercise is as follows: (50 points are possible)

- The driver/operator must maintain a normal driving position behind the wheel and cannot lean out the window. Failure to do so will result in a 25-point penalty.
- Each time a course marker is touched: 5-point penalty.
- Each stop within the stanchions will result in a 5-point penalty.
- A driver/operator who stops while approaching the finish line and then permits the apparatus to move forward or backward before a measurement is obtained shall receive no score for this exercise.
- A driver/operator who stops between the time the front of the apparatus leaves the last stanchion and the stop line must remain stopped and will be scored on the distance the unit is from the finish line.
- If the driver/operator stops the apparatus with the bumper past the finish line, no points are awarded for the exercise.
- After the apparatus has come to a complete stop, the distance from the finish line to the front bumper is measured and scored as follows:

0 to 6 inches (0 to 150 mm) 50 points
6 to 9 inches (150 to 225 mm) 45 points
9 to 12 inches (225 to 300 mm) 40 points
12 to 15 inches (300 to 375 mm) 35 points
15 to 18 inches (375 to 450 mm) 30 points
18 inches or more (450 mm +) No score
All other penalties are subtracted from this score.

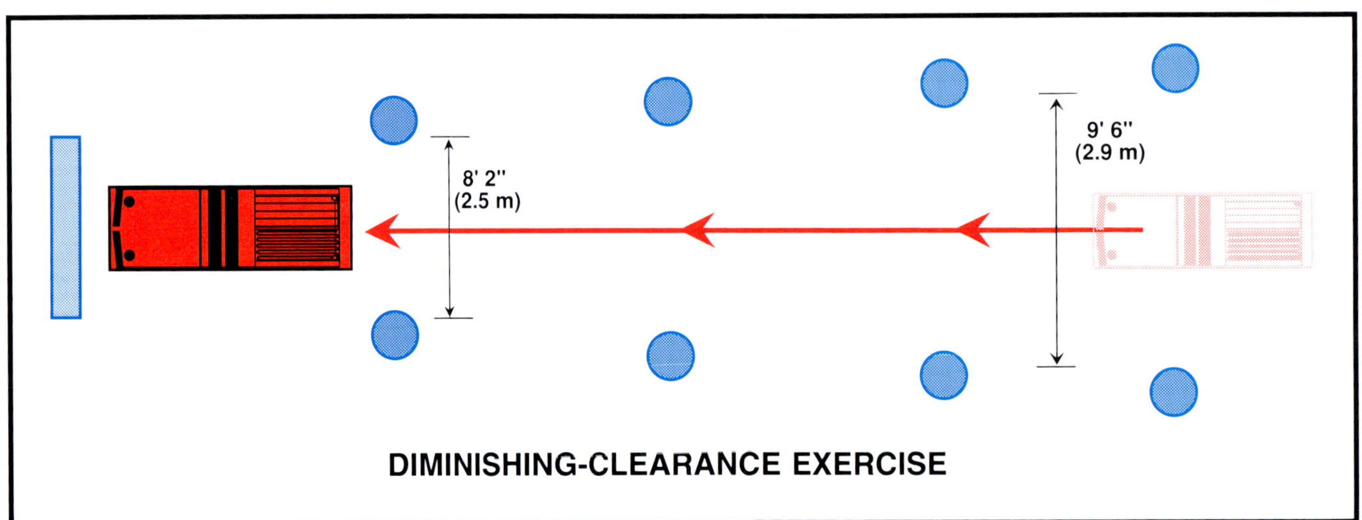

Figure 2.17 The diminishing-clearance exercise measures the driver/operator's ability to steer the apparatus in a straight line and to judge the distance from the wheel to the various objects.

Straight-Line Driving

The straight-line exercise has two variants. In the first method, a set of markers forming two rows 2 1/2 feet (0.76 m) apart and at least 25 feet (8 m) long is used (Figure 2.18). The driver/operator must keep the right wheel of the apparatus between the set of markers while moving at a constant speed. The vehicle cannot touch any part of the markers. The markers should be offset slightly so the driver/operator must judge the distances and direction by using them as a guide.

The second variant is outlined in NFPA 1002, *Professional Qualifications for Fire Apparatus Driver/Operator*. For this test, the driver/operator must steer the entire apparatus between a set of markers forming two rows at least 100 feet (30 m) long (Figure 2.19). The space between the rows is 4 inches (100 mm) wider than the apparatus body, and the markers are placed 20 feet (6 m) apart in their rows. A minimum distance of 100 feet (30 m) should be used for this exercise. The driver/operator may be required to maintain a constant speed or accelerate through the gears during the exercise. Scoring for this exercise is as follows: (50 points are possible)

- Each time the vehicle stops or fails to maintain a constant forward motion, 25 points are subtracted.
- If the driver/operator stops and aligns the apparatus before starting the exercise, 25 points are subtracted.
- Each time a course marker is touched: 5-point penalty.
- Anytime a set of markers is bypassed or straddled, no points are given for the exercise.
- If the driver/operator does not maintain a normal driving position within the cab or the truck, 25 points are subtracted.

Turning-Around Exercise

Turning-around exercises develop a driver/operator's ability to properly spot apparatus for operations. Fire apparatus, particularly fire de-

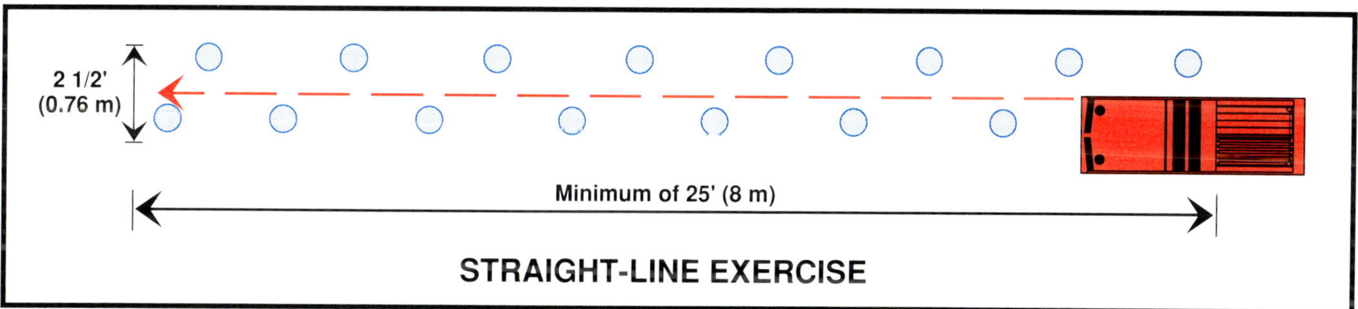

Figure 2.18 The driver/operator must steer a straight course, keeping the right wheel of the apparatus between a set of markers.

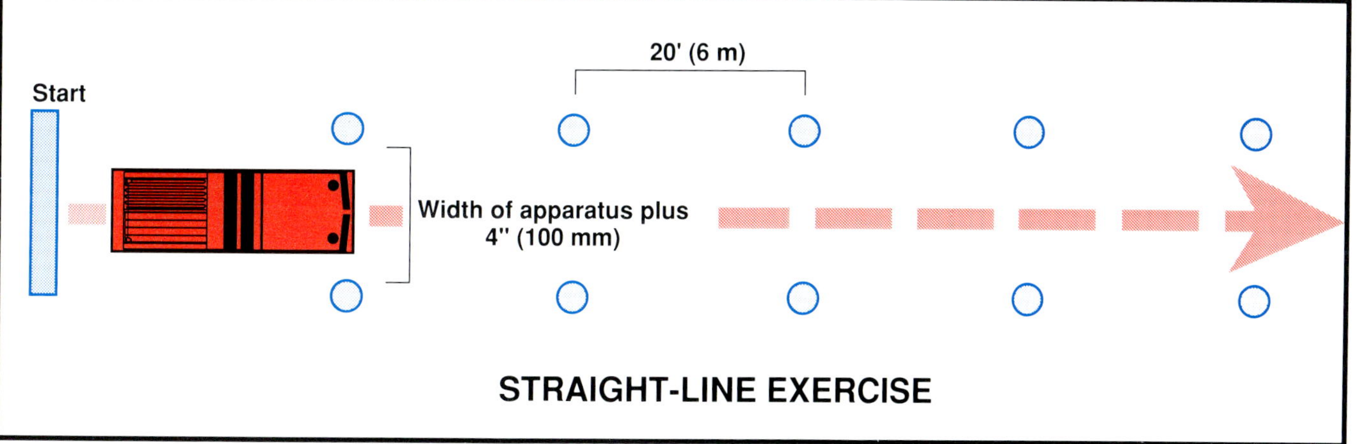

Figure 2.19 In another type of straight-line exercise, the driver/operator must steer the apparatus between a set of markers for at least 100 feet (30 m).

partment pumpers, often need to turn around to complete an operation. Turning around is frequently necessary when laying fire hose, and care must be taken to avoid backing over the hose.

If streets are wide enough and if traffic permits, the U-turn may be used. Although turning fire apparatus around may not be difficult in adequate space, it becomes more complicated in streets or intersections (Figure 2.20).

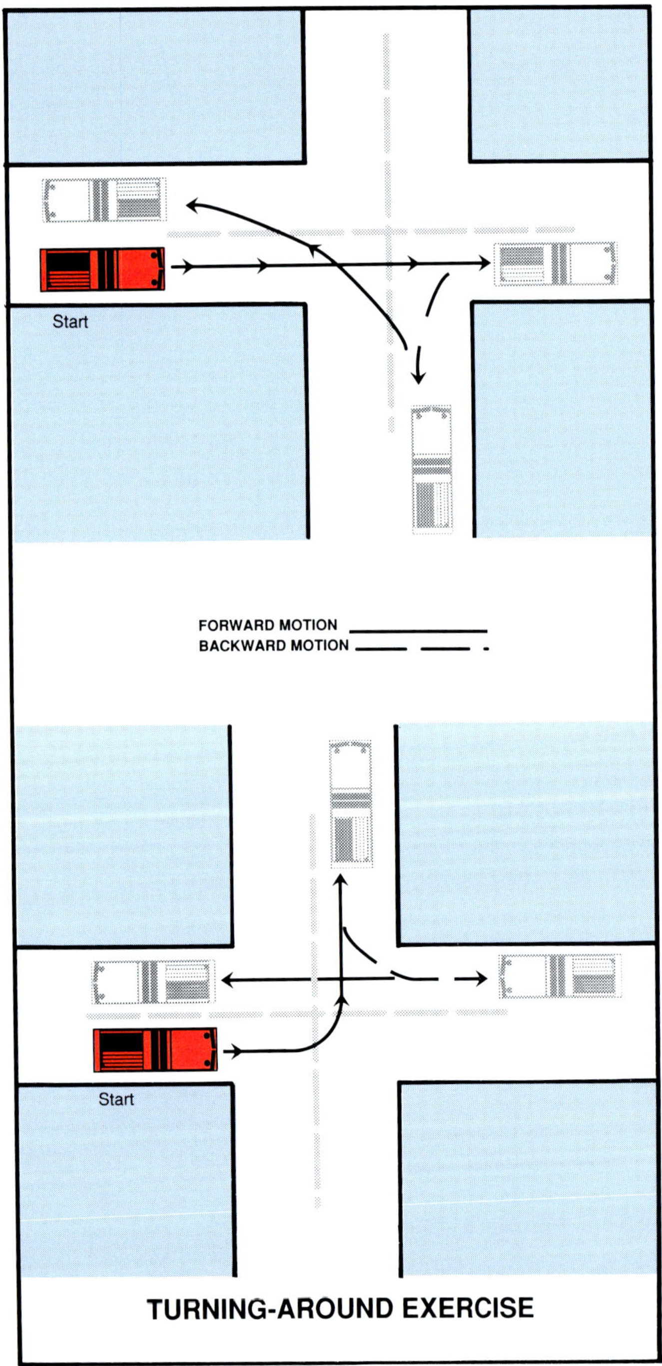

Figure 2.20 Turning-around exercises are valuable for learning to spot apparatus properly.

SUMMARY OF GOOD DRIVING PRACTICES

An accident or vehicular failure caused by irresponsible driving has many repercussions and is inexcusable. Lives that could have been saved may be lost; property that should have been protected will be destroyed; firefighters on the apparatus may be injured or killed; innocent bystanders and other drivers may be injured or killed, leaving the department and the driver/operator open to lawsuits and civil prosecution for such things as negligent homicide or manslaughter; and the apparatus will be useless for an indefinite time, leaving citizens with less protection.

The following are some critical points to remember for safe operation and driving of fire apparatus:

- Speed is less important than arriving safely at the destination.
- Slow down for intersections and be prepared to stop. Anticipate the worst possible situation.
- Drive defensively. Be aware of everything that is happening or likely to happen 360 around the apparatus.
- Expect that some motorists and pedestrians will neither hear nor see the apparatus warning devices.
- Be aware of the route's general road and traffic conditions. Adjust this expectation with the season, weather, day of the week, and time of day.
- Remember that icy, wet, or snow-packed roads increase braking distance.
- Do not clash the gears.
- Do not use the clutch pedal as a footrest.
- When leaving the station, do not exceed 10 mph (15 km/h).
- Racing the engine when the apparatus is standing still is unnecessary and abuses the engine.
- Always use low gear when starting from a standstill. Using second or third gear and slipping the clutch damages the clutch and causes unnecessary, rapid wear.

- Keep the apparatus under control at all times.
- Take *nothing* for granted.

Driver/operators are responsible for keeping their fire apparatus in good operating condition. Good operating condition refers to the care and minor maintenance necessary to keep the apparatus in a condition of readiness. Major repairs or replacement of parts should be done only by a qualified mechanic. The driver/operator is responsible for maintaining an accurate inventory of the equipment on the apparatus. Driver/operators must also be familiar with their own district and response routes, as well as the location of all water sources, including hydrants, lakes, streams, and other static water sources.

In addition to all these responsibilities, a fire apparatus driver/operator should keep accurate records concerning the operation of the vehicle. These records may include minor maintenance as well as apparatus servicing. A driver/operator should know how to properly fill out reports, which should be neat and legible. Before filling out a report, a careful check should be made to ensure the accuracy of facts and information.

Chapter 2 Review

Answers on page 358

TRUE-FALSE: Mark each statement true or false. If false, explain why.

1. The driver/operator or the fire department may not be held liable for negligent acts incurred while responding to an emergency call.

   ☐ T ☐ F ____

   ____

2. The officer riding the apparatus is responsible for the safety of the personnel riding on it.

   ☐ T ☐ F ____

   ____

MULTIPLE CHOICE: Circle the correct answer.

3. When responding along the same route, emergency vehicles should maintain a distance of at least ______ feet (______ m) between vehicles.
   A. 100 to 300 feet (30 m to 90 m)
   B. 200 to 400 feet (60 m to 120 m)
   C. 300 to 500 feet (90 m to 150 m)
   D. 500 to 600 feet (150 m to 180 m)

4. Blind or well-traveled intersections should be approached and crossed at a maximum speed of ______?
   A. 5 to 10 mph (8 km/h to 16 km/h)
   B. 15 to 20 mph (24 km/h to 32 km/h)
   C. 20 to 25 mph (32 km/h to 40 km/h)
   D. 25 to 30 mph (40 km/h to 48 km/h)

5. Where are apparatus most likely to be involved in an accident?
   A. Freeways
   B. Alleys
   C. Off-ramps
   D. Intersections

6. A driver/operator's road test should cover various types of terrain and road conditions. What is the minimum distance it should cover?
   A. 2 miles (3.2 km)
   B. 5 miles (8 km)
   C. 7 miles (11.2 km)
   D. 10 miles (16 km)

7. In what position should the truck's transmission be when starting the unit?
   A. Park or neutral
   B. Low gear
   C. Reverse
   D. Pump gear

LISTING

8. Identify the five factors that are important to defensive driving.

   A. ______________________________

   B. ______________________________

   C. ______________________________

   D. ______________________________

   E. ______________________________

9. List at least three items that should be included on a written driver's examination.

   ____

   ____

   ____

FILL IN THE BLANK: Fill in the blanks with the correct response.

10. __________ directly affects the distance required to stop a vehicle.

11. When employing evasive tactics, the driver/operator should attempt to pass the overtaken vehicle on the __________ side.

12. When cruising, the engine speed should be __________ to __________ rpm lower than the recommended rpm.

SHORT ANSWER: Answer each item briefly.

13. Under what circumstances may the driver/operator disregard traffic regulations that apply to the general public?

    ____

    ____

    ____

14. Name one particular circumstance in which the driver/operator is not exempt from obeying traffic regulations that pertain to the general public. Explain the recommended procedure for handling this situation.

    ____

    ____

    ____

15. Under what conditions may emergency apparatus travel against a red light or stop sign?

    ____

    ____

    ____

    ____

16. As you approach an intersection, all lanes of traffic in your direction of travel are blocked. How should you cross the intersection?

17. Explain the best way to combat a skid.

18. In what manner should brakes be applied in adverse weather conditions?

19. What is the rule of thumb for remembering the stopping distance of a vehicle on snow and ice as compared to dry concrete?

20. A fluttering oil pressure gauge is a possible indication of what problem?

21. What should your reaction be to a sudden increase in coolant temperature not caused by heavy pulling?

3

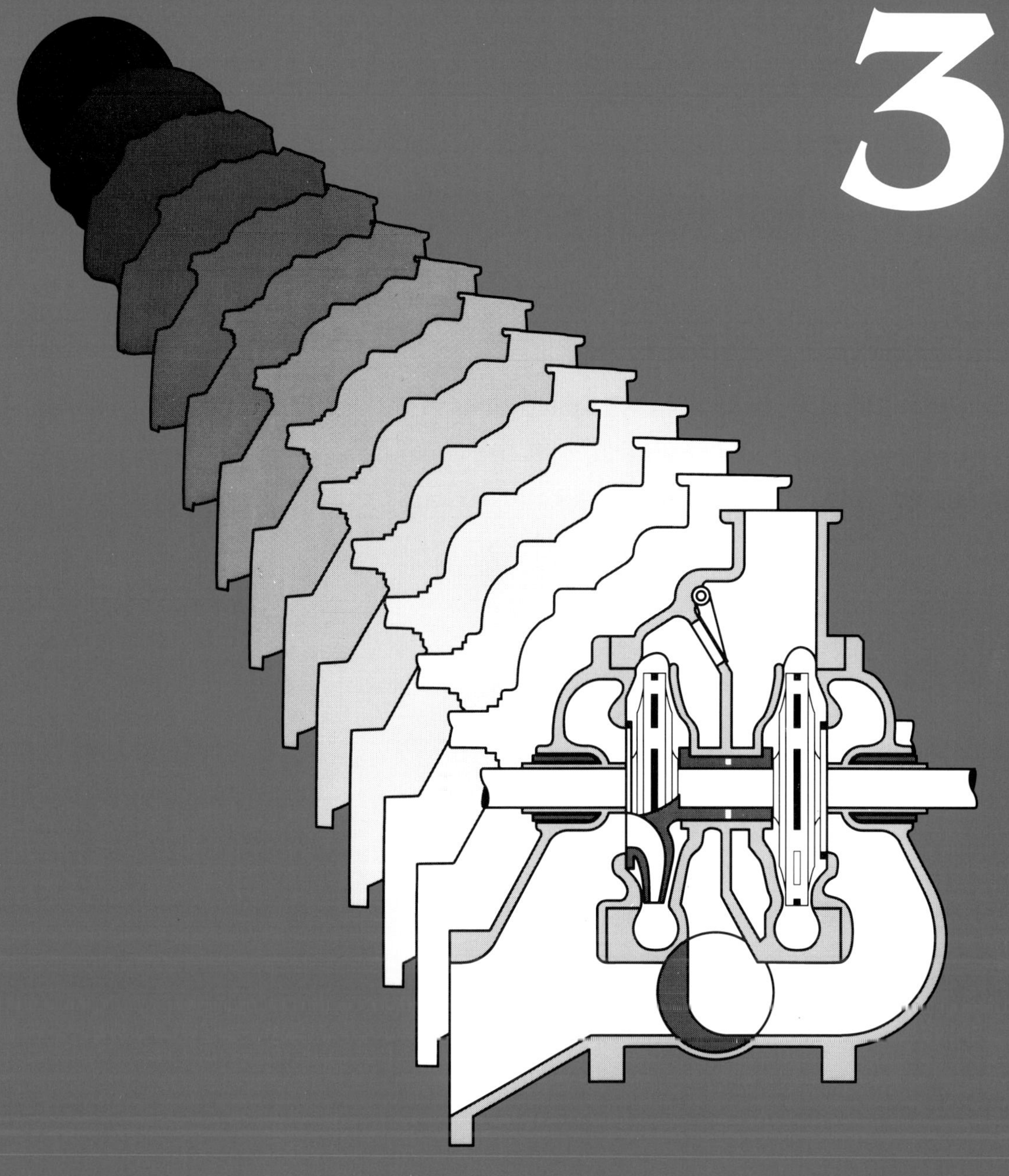

Types of Fire Apparatus

LEARNING OBJECTIVES

After reading this chapter, the student should be able to accomplish the following learning objectives:

- Explain the difference between standard and nonstandard apparatus.
- Describe the features of a triple-combination pumper.
- Describe the differences between minipumpers and midipumpers.
- State the variables that determine the size of the water tank on a mobile water supply apparatus.
- List five construction features of a safe and efficient water supply apparatus.
- Describe the characteristics of brush apparatus and the manner in which it is used.
- Describe the purpose and design of aircraft fire apparatus.
- State the three main functions of fire boat apparatus.

Chapter 3
Types of Pumping Apparatus

Fire apparatus are classified according to the functions for which they are designed. This chapter discusses different types of pumping apparatus, including pumpers, mobile water supply apparatus, minipumpers and midipumpers, grass fire apparatus, brush or booster apparatus, aircraft fire apparatus, fire boat apparatus, and other specialized pumping apparatus.

When apparatus conform to the specifications set forth in NFPA 1901 *Standard on Automotive Fire Apparatus*, they are classified as standard apparatus. Apparatus that does not comply with NFPA 1901 is considered to be nonstandard. For the purposes of this manual, the factors that constitute standard apparatus are clearly defined in NFPA 1901, and reference to the standard denotes agreement with it.

FIRE DEPARTMENT PUMPERS

The main purpose of the fire department pumper is to provide water at adequate pressure for fire streams. The water supplied by the pumper may come from the apparatus water tank, a fire hydrant, or a static supply such as a lake or a pond. The functional parts of a pumper are described in Chapters 5 and 6 of this manual.

Present recognized capacities of pumps for fire department pumpers are 500, 750, 1,000, 1,250, 1,500, 1,750, and 2,000 gpm (2 000 L/min, 3 000 L/min, 4 000 L/min, 5 000 L/min, 6 000 L/min, 7 000 L/min, and 8 000 L/min), although some larger capacity pumps have been built. A fire department pumper must also feature adequate intake and discharge pump connections, pump and engine controls, gauges, and other instruments (Figure 3.1).

Figure 3.1 The pump panel can be a confusing array of knobs, levers, and gauges if the driver/operator is not properly trained in pump operation.

Some fire department pumpers are referred to as triple-combination pumpers, which means that they have a water tank, hose bed, and fire pump

(Figure 3.2). A triple-combination pumper usually carries an extension ladder, a roof ladder, forcible entry tools, hose appliances, and other accessories. The capacity of the three components of a triple-combination pumper will vary according to the manufacturer's design or specifications written by the purchaser.

Figure 3.2 Fire department pumpers come in a variety of styles and may be mounted on either a commercial or custom chassis. *Courtesy of Joel Woods.*

Courtesy of Joel Woods.

Courtesy of Joel Woods.

Courtesy of Joel Woods.

Courtesy of Anderson Engineering, Ltd.

Pumpers with Foam Capability

Fires involving flammable liquids are a challenge faced by every fire department. To meet this challenge, some fire departments have purchased special pumpers equipped with aerating foam nozzles mounted on the roof of the cab and/or foam proportioning equipment built directly into the main fire pump. These foam nozzles are similar to those found on airport crash-rescue vehicles and are capable of discharging large amounts of foam in a short period of time. The nozzle and the pump can be operated from inside the cab, so there is little delay in setting up the nozzle for use. Another advantage of these nozzles is that firefighters can achieve quick knockdown without being unnecessarily exposed. (Figure 3.3).

Figure 3.3 Some fire department pumpers are also equipped to supply moderate to large quantities of foam for fighting large flammable liquids fires. These units can also be used on other fires. *Courtesy of Joel Woods.*

Pumpers with foam capability typically carry the full complement of tools and equipment regularly found on a pumper, as well as special foam nozzles and foam equipment.

Pumpers with Elevating Water Devices

Pumpers equipped with elevating water devices are becoming common (Figure 3.4). These units provide a means for discharging water from elevated nozzles. Through the use of articulating or telescoping booms, hydraulically operated towers may be mounted on pumpers to form a combination unit. Elevating water devices can also be used to apply water to the lower floors of a building. The operation of elevating water devices is covered in IFSTA's **Fire Department Aerial Apparatus** manual.

Figure 3.4 The elevated water devices found on fire department pumpers may be (a) telescoping or (b) articulating. *Courtesy of Joel Woods.*

Minipumpers

Smaller, quick-attack pumpers, known as minipumpers, are designed to handle small fires that do not require the capacity and personnel needed of a larger pumper. Minipumpers also enable a fire department to save on fuel and apparatus replacement costs while still having larger pumpers ready for action.

Minipumpers are most often mounted on pickup chassis with custom-made bodies. They carry pumps usually no larger than 500 gpm (2 000 L/min) (Figure 3.5). Most of the equipment carried on a larger pumper is also carried on a minipumper, although in small numbers. Some minipumpers also carry medical equipment, enabling them to serve as a rescue unit as well as a fire fighting unit. Many minipumpers are also equipped with a turret gun that can be supplied directly from another pumper. The small size and maneuverability of these pumpers allows them to get into small spaces and set up a master stream where a larger pumper would not be able to fit.

While minipumpers are very cost effective, fire departments should not place total reliance on them as attack pumpers. While they are very useful for handling small fires, they are not capable of mounting a sufficient attack on large fires. Since it is difficult to estimate a fire's size and potential upon receipt of an emergency call, it is good practice to dispatch a full-sized pumper along with a minipumper. These units, like all fire apparatus, have their limitations and should be used within those limitations.

Figure 3.5 Minipumpers offer the advantage of being highly maneuverable and quick in response. *Courtesy of Joel Woods.*

Midipumpers

Midipumpers are used in the same type of situations in which minipumpers are used. A midipumper is well suited for such small fires as grass and dumpster fires, and for service calls that do not require the capacity and personnel of a larger pumper. Midipumpers also have the ability to start an initial attack on larger fires.

Midipumpers are built on chassis usually over 12,000 pounds (5 443 kg) Gross Vehicle Weight (GVW). The main differences between a midipumper and a minipumper are size, pump capacity, and the amount of equipment carried. Midipumpers have been equipped with pumps as large as 1,000 gpm (4 000 L/min). These units typically carry the same type of equipment as a full-size pumper such as hose, ground ladders, and other fire fighting equipment. Some midipumpers also carry emergency medical equipment so they can double as rescue and fire fighting vehicles (Figure 3.6).

Figure 3.6 Midipumpers are somewhat larger than minipumpers and usually carry the same type of equipment as a full-size pumper. *Photos Courtesy of Joel Woods.*

MOBILE WATER SUPPLY APPARATUS

Mobile water supply apparatus, known as tankers or tenders, are widely used to transport water to areas beyond a water system or to areas where water supply is inadequate. Most attack pumpers carry water, but not in large enough quantities to sustain an extended attack. Mobile water supply apparatus have water tanks that are larger than those generally found on standard pumpers.

The size of a tanker's water tank depends upon a number of variables:

- Terrain — The tanker may be required to climb steep hills or to operate on winding roads.
- Bridge weight limits — Bridges in the protected area may be too old or may not be designed to bear the weight of heavy tankers. This presents a danger to firefighters as they drive over these bridges if alternate routes are not available.
- Monetary constraints — The fire department may not have enough money to purchase a large tanker.
- Size of other tankers in the area — Tanker shuttles flow more easily when tankers of the same or similar size are used.

If an approved mobile water supply apparatus is desired, the requirements of NFPA 1901 should be met. The road tests and weight distribution requirements generally limit tank capacity to 1,500 gallons (6 000 L) or less for single rear-axle vehicles. When tanks of capacities greater than 1,500 gallons (6 000 L) are desired, either tandem rear axles, or a tractor-trailer design should be considered. Figures 3.7-3.10 show examples of various types of mobile water supply apparatus.

A mobile water supply apparatus should conform to the following construction requirements in order to be able to move water safely and efficiently:

5-89

- Adequate but reasonable water tank capacity
- Adequate filling and dumping rates
- Proper front to rear weight distribution
- Adequate suspension and steering
- Properly sized chassis
- Properly sized engine for tank size and terrain
- Sufficient braking ability
- Proper tank mounting
- Proper and safe tank baffling

Courtesy of Saulsbury Fire Apparatus.

Courtesy of Bob Esposito.

Courtesy of Bob Esposito.

Figure 3.7 Single rear axle tankers usually have a tank capacity of 1,500 gallons (6 000 L) or less.

Figure 3.8 Larger tankers require dual, or tandem, rear axles. *Courtesy of Joel Woods.*

Courtesy of Joel Woods.

Courtesy of Joel Woods.

Courtesy of Bob Esposito.

Figure 3.9 Examples of pumper-tankers.

Courtesy of Joel Woods.

Courtesy of Bob Esposito.

Figure 3.10 Fire departments in areas with a severe shortage of water supply sources may require large, tractor-trailer tankers.

Tankers are used as support vehicles for attack pumpers. The tanker can act as a reservoir or "nurse tanker" for some fires. It can also take part in a shuttle operation at extended fires and those fires requiring more water than the tanker carries. Tanker shuttles call for organization and equipment, such as portable tanks, that permit tanks to unload quickly (Figure 3.11). Large capacity pumpers should be placed at designated filling sites to fill tankers as quickly as possible. If pumpers are not available to proceed to the fill site, mutual aid or special-called apparatus should be requested and sent to the site. For more information on proper utilization of mobile water supply apparatus, see IFSTA's **Water Supplies for Fire Protection** manual.

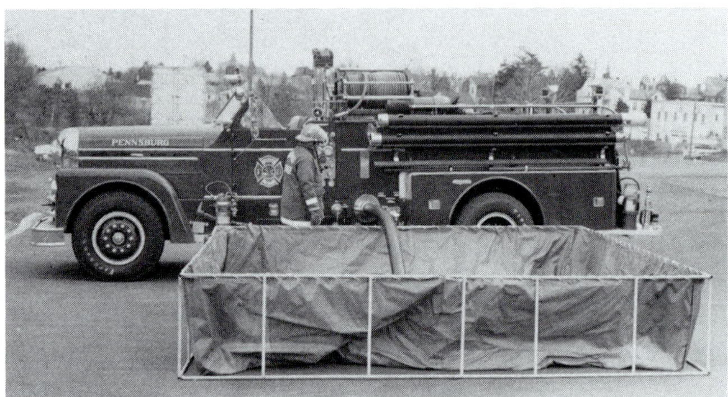

Figure 3.11 Portable water tanks greatly increase the efficiency of water shuttle operations. *Courtesy of Bob Esposito.*

Figure 3.12 Brush vehicles commonly have low-capacity pumps and all-wheel drive capability. *Courtesy of Joel Woods.*

BRUSH OR BOOSTER APPARATUS

The control of ground cover fires often requires a lightweight, highly maneuverable vehicle that can go places inaccessible to larger apparatus. Fire apparatus specifically adapted for fighting ground cover fires are designed to fulfill these requirements. These units are usually built on utility-type vehicle chassis and most have all-wheel drive. Most of these specially designed vehicles carry a comparatively small water tank with a small capacity pump and very little hose (Figure 3.12).

The ability to "pump and roll" is a tremendous advantage when combating ground cover fires. Vehicles with the ability to pump and roll use a separate motor or a power take-off (PTO) for the pump. This arrangement enables the apparatus to move and put water on the fire at the same time. Many ground cover fires can be initially controlled by firefighters riding on the apparatus. All firefighters should be secured to the apparatus by a seat belt while sitting. Appropriate speed limits and some form of communication should be established between the driver and riders of the apparatus.

Most brush fire vehicles use booster hose as attack lines. However, using short sections of 1½-inch (38 mm) hose and adjustable flow nozzles reduces friction loss and gives the nozzleman a choice of flows to combat the volume and intensity of fire. Ground sweep nozzles may be useful if the number of personnel is limited and the fuel is short and slow burning.

Ground cover fire apparatus is usually equipped with either a portable pump, auxiliary-engine-driven pump, or a PTO-powered pump. PTO-powered pumps on ground cover apparatus normally have a larger gpm (L/min) capacity than portable pumps, but a smaller capacity than auxiliary

Figure 3.13 Ground cover apparatus with an auxiliary-powered pump do not have the problem of maintaining pump speed when operating in rough terrain.

pumps. PTO-powered pumps are easily put into operation and require little maintenance of the power source other than normal apparatus engine service. Auxiliary-powered pumps deliver a constant flow regardless of the apparatus engine speed (Figure 3.13). Portable pumps may serve as the main pump for ground cover apparatus or may be carried as a backup unit for PTO- or auxiliary-driven pumps.

Booster tanks for ground cover apparatus vary from approximately 50 gallons (200 L) on jeeps to around 1,000 gallons (4 000 L) on larger apparatus. Care should be taken to select a vehicle suitable for the terrain and ground conditions of the area that the vehicle will protect. Water tanks should be baffled into compartments to maintain vehicle stability when cornering or driving on rough terrain. Tanks and equipment should be mounted so that the vehicle is not top heavy.

AIRCRAFT FIRE APPARATUS

Aircraft fire apparatus is usually built to perform a specific function or a series of functions for aircraft crash fire fighting and rescue. This type of apparatus may differ greatly in shape, size, purpose, and design from other fire fighting apparatus. The size of the airport, the types of aircraft that constitute normal traffic, the amount of air traffic, and the purpose of the airport are all factors that influence the design of this type of apparatus.

Aircraft crash fire fighting vehicles are designed to be totally self-sufficient. They carry large quantities of water, foam concentrate, and may also be equipped with Halon or dry chemical extinguishing systems. They also have large-capacity fire pumps and carry a variety of rescue tools and equipment. Aircraft fire apparatus on a military base may be very different from the same type of apparatus used in a municipal airport. These differences are due to heavy fuel loads, different cargos, and possible munitions.

Aircraft fire fighting often requires that several different units operate in combination without conflicting procedures (Figure 3.14). This kind of complementary operation is similar to the way in which fire department pumper and aerial apparatus function at a fire. For more information

Courtesy of Joel Woods.

Courtesy of Joel Woods.

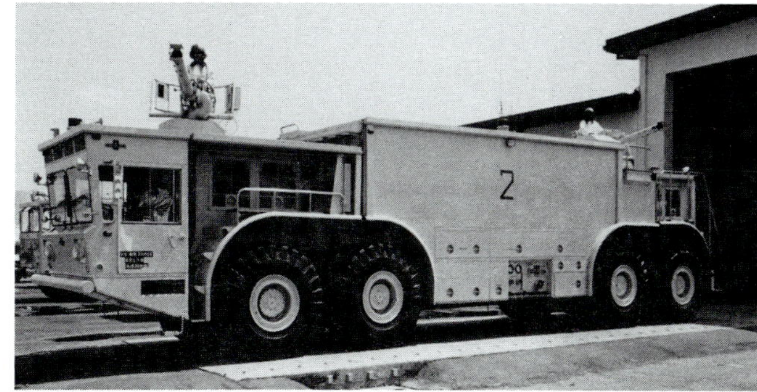

Figure 3.14 Aircraft crash fire fighting vehicles feature large-capacity fire pumps and large quantities of water and foam concentrate. They also carry a variety of rescue tools and equipment.

on the design and operation of airport fire apparatus, see IFSTA's **Aircraft Fire Protection and Rescue Procedures** manual.

FIRE BOAT APPARATUS

Fire boats are used in waterfront cities to protect docks, wharves, piers, and boats. A fire boat may be a small, high-speed, shallow-draft vessel, or it may be the size of a river, harbor, or ocean-going tug, depending upon its duties and the area to be covered (Figure 3.15). The number of personnel will vary with the size of the apparatus. The functions performed by a fire boat may include water rescue, fire fighting, and relaying water to land-based apparatus. The two phases of fire fighting that fire boats are most suited for are pumping through large master stream devices and providing additional water for on-shore fire fighting operations. Fire boats have been built to deliver as much as 26,000 gpm (104 000 L/min). Individual master stream turrets that discharge 2,000 to 3,000 gpm (8 000 L/min to 11 500 L/min) are common.

Figure 3.15 Large fire boats are ideally suited to provide additional water for onshore fire fighting operations. *Courtesy of Joel Woods.*

Some smaller fire boats are propelled by water jets or are amphibious for operation on land and in the water (Figure 3.16). Most heavy-duty fire boats are powered by marine-type diesel engines. Dual purpose engines for propulsion and pumping have also been built. At least one auxiliary power generator is needed on all fire boats, and receptacles must be provided for receiving shore power, communications, and fresh water.

Courtesy of Miami (Florida) Fire Department.

Courtesy of Joel Woods.

Figure 3.16 Examples of amphibious fire vehicles.

OTHER TYPES OF APPARATUS WITH FIRE PUMPS

This book covers only those apparatus whose primary function is to pump water. Other types of apparatus, such as aerial apparatus, foam or chemical units, squad or rescue apparatus, and utility vehicles, may also be outfitted with fire pumps. The operation of the pumps, and in most cases operation of the vehicle itself, is similar to the procedures described in this manual. To operate other features on those types of apparatus, see the manufacturer's instructions. For more information concerning operation of fire department aerial apparatus, see IFSTA's **Aerial Apparatus Procedures** manual.

Chapter 3 Review

Answers on page 359

TRUE-FALSE: Mark each statement true or false. If false, explain why.

1. The smallest recognized pumping capability for a fire department pumper is 250 gpm (1 000 L/min).

☐ T ☐ F ______________________________

____

MULTIPLE CHOICE: Circle the correct answer.

2. What is generally considered to be the maximum amount of water that should be carried on a single rear axle mobile water supply apparatus?

A. 1,000 gallons (4 000 L)
B. 1,500 gallons (6 000 L)
C. 1,800 gallons (7 200 L)
D. 2,000 gallons (8 000 L)

LISTING

3. List the four main variables that dictate the size of the water tank in a mobile water supply apparatus.

A. ______________________________
B. ______________________________
C. ______________________________
D. ______________________________

4. List at least five construction features of a safe and efficient water mover.

____
____
____
____
____

5. Name three functions of fire boats.

A. ______________________________
B. ______________________________
C. ______________________________

6. Name the three elements required of a pumper for it to be considered a triple-combination pumper.

A. ______________________________
B. ______________________________
C. ______________________________

FILL IN THE BLANK: Fill in the blank with the correct response.

7. Apparatus that does not conform to NFPA 1901 is classified as ____________________.

SHORT ANSWER: Answer each item briefly.

8. When are apparatus said to be classified as standard apparatus?

____

____

____

9. What is the primary difference between minipumpers and midipumpers?

____

____

____

____

10. What is the primary responsibility of mobile water supply apparatus?

____

____

____

____

4

Positioning Apparatus

NFPA STANDARD 1002
Fire Apparatus Driver/Operator Professional Qualifications

3-6 Operations.

3-6.6 The fire apparatus driver/operator, given a fire department pumper and a simulated fire scene, shall demonstrate proper maneuvering and positioning of the apparatus to function from the given source of water.

Chapter 4

Positioning Apparatus

Fire incidents require that apparatus work in harmony in order to perform effectively. For fire control to be achieved efficiently and safely, apparatus must be positioned so that its use is maximized. The driver/operator's ability to properly position apparatus requires training, practice, and ingenuity. The driver/operator must also be able to execute certain maneuvers when called for by incident commanders or by prefire plans. Each type of apparatus will be positioned according to its purpose and overall strategic objectives. It must also function in coordination with other apparatus working at the same incident.

This chapter discusses proper positioning of pumping apparatus based on various functions for which it may be used such as fire attack or water supply. Hydrant operations, dual pumping, tandem pumping, supplemental pumping, relay operations, tanker fill and dump operations, brush fire apparatus, and staging are discussed. Positioning for maximum effectiveness and safety on limited-access highways is discussed at the end of the chapter.

Positioning a fire department pumper to supply hoselines from an apparatus water tank differs greatly from positioning a pumper to obtain water from a fire hydrant or drafting from a static source such as a lake. Some of the general factors that determine where and how driver/operators should position fire department pumpers are

- Availability and amount of water needed at a given fire
- The immediate need for effective fire streams
- Available hose and pumping capacity

ATTACK PUMPERS

There are no set rules for positioning attack pumpers on the fireground. This is because multiple factors must be reviewed when determining position for both first-due and late-arriving pumpers. The following section contains some guidelines that will assist driver/operators in determining effective placement. As in all fire situations, standard operating procedures and the judgment of the responsible officer should be the deciding factors when committing apparatus. The following factors must be evaluated for determining initial positions:

- Initial fire size-up and potential spread
- Rescue situations
- Water supply
- Apparatus exposure (radiant heat, building collapse, occupancy hazards, overhead power lines, backdraft)
- Method of attack
- Fire exposures
- Wind direction
- Terrain
- Relocation potential

Engine company officers should use their initial approach to a small structure to observe the fire conditions. If possible, they should quickly observe the side they approach from, the front of the building, and the far side.

At incidents where such size-up is not possible, attack apparatus should position for easiest access to the fire area. It is often better for pumpers to pull *past* the front door of the building to allow incoming aerial apparatus better position (Figure 4.1). Many departments position incoming attack apparatus alternately from the front to the rear, starting on the side with the worst fire condition. Once commitments to the front and rear are made, apparatus should begin to work the sides of the building. Fireground pumpers should also remember to lay hoselines in a manner that allows access for future companies (Figure 4.2).

Figure 4.1 The engine company is positioned in the outer lane to allow sufficient room for aerial apparatus to position close to the fire building.

Figure 4.2 Keep supply lines to the side of the street so later-arriving apparatus will not have to drive over the hose to reach the scene. This is especially critical when using large diameter hose.

WATER SUPPLY PUMPERS

Drafting Operations

Drafting pumpers may supply fireground apparatus directly or may serve as source pumpers for hose relays or tanker shuttles. These operations are common in rural areas but can just as easily be performed at urban incidents.

In both rural and urban areas, the drafting pumper should stop before reaching the water source and connect the hard suction hose and strainer to the pumper before moving to the final draft position (Figure 4.3). This prevents firefighters from having to stand in unsure footing or currents. The final draft position of the apparatus should be made after considering the following:

- Available hard suction hose and lift
- Wet ground approaches (entrapment)
- Stability and support for pumper
- Access for additional apparatus
- Tidal movement

Figure 4.3 It may be necessary to use a rubber mallet on hard suction hose connections to ensure that they are airtight for drafting.

Minimizing lift distances will provide better discharge abilities and should be of primary concern. In no case should the maximum lifts in Table 4.1 be exceeded if full pump capacity is to be used.

The hard suction hose strainer should not rest on the bottom during drafting. A rope may be tied off to the apparatus or a nearby object in order to hold the strainer off the bottom. Floats, such as a spare tire or plastic bucket, can be used with rope to hold the strainer at an appropriate depth.

TABLE 4.1 (U.S.)
MAXIMUM DISCHARGE AT VARIOUS LIFTS

Rated Capacity of Pump (gpm)	Intake Hose Diameter (inches)	Volume Discharged (gpm)										
		2 sections hose (20 feet)							3 sections hose (30 feet)			
		Feet of Lift										
		4	6	8	10	12	14	16	18	20	22	24
500	4	590	560	530	500	465	430	390	325	270	195	65
500	4½	660	630	595	560	520	480	430	370	310	225	70
750	4½	870	830	790	750	700	650	585	495	425	340	205
750	5	945	905	860	820	770	720	655	560	480	375	235
1,000	5	1,160	1,110	1,055	1,000	935	870	790	670	590	485	340
1,000	6	1,345	1,290	1,230	1,170	1,105	1,045	960	835	725	590	400
1,250	6	1,435	1,375	1,310	1,250	1,175	1,100	1,020	900	790	660	495
1,500	6	1,735	1,660	1,575	1,500	1,410	1,325	1,225	1,085	995	800	590
1,500	Dual 5	1,990	1,990	1,810	1,720	1,615	1,520	1,405	1,240	1,110	950	730
1,500	Dual 6	2,250	2,150	2,040	1,935	1,820	1,710	1,585	1,420	1,270	1,085	835

NOTE: Test conducted with net pump pressure of 150 psi, 1,000 feet altitude, 60°F water temperature, 28.94 inches Hg barometric pressure (poor weather).

Operation at lower than 150 psi net pump pressure will result in an increased discharge; at higher pressure, a decreased discharge.

This data based on a pumper with ability to discharge rated capacity when drafting at not more than a 10-foot lift. Many pumpers will exceed this performance and therefore will discharge greater quantities than shown at all lifts.

Source: American Insurance Association.

TABLE 4.1 (metric)
MAXIMUM DISCHARGE AT VARIOUS LIFTS

Rated Capacity of Pump (L/min)	Intake Hose Diameter (mm)	Volume Discharged (L/min)										
		2 sections hose (6 m)							3 sections hose (9 m)			
		Meters of Lift										
		1.2	1.8	2.4	3	3.6	4.3	4.9	5.5	6.1	6.7	7.3
2 000	100	2 233	2 120	2 006	1 893	1 760	1 628	1 476	1 230	1 022	738	246
2 000	115	2 498	2 385	2 252	2 120	1 968	1 817	1 628	1 401	1 173	852	265
3 000	115	3 293	3 142	2 990	2 830	2 650	2 461	2 214	1 874	1 609	1 287	776
3 000	125	3 577	3 426	3 255	3 104	2 915	2 725	2 479	2 120	1 817	1 420	890
4 000	125	4 391	4 202	3 993	3 785	3 539	3 293	2 990	2 536	2 233	1 835	1 287
4 000	150	5 091	4 883	4 656	4 429	4 183	3 956	3 634	3 161	2 744	2 233	1 514
5 000	150	5 432	5 204	4 959	4 732	4 448	4 163	3 861	3 407	2 990	2 498	1 873
6 000	150	6 568	6 284	5 962	5 678	5 337	5 016	4 637	4 107	3 615	3 028	2 233
6 000	Dual 125	7 533	7 533	6 852	6 511	6 113	5 754	5 318	4 694	4 202	3 596	2 763
6 000	Dual 150	8 517	8 139	7 722	7 324	6 889	6 473	6 000	5 375	4 807	4 107	3 161

NOTE: Test conducted with net pump pressure of 1 050 kPa, 300 m altitude, 15.6°C water temperature, 735 mm Hg barometric pressure (poor weather).

Operation at lower than 1 050 kPa net pump pressure will result in an increased discharge; at higher pressure, a decreased discharge.

This data based on a pumper with ability to discharge rated capacity when drafting at not more than a 3 m lift. Many pumpers will exceed this performance and therefore will discharge greater quantities than shown at all lifts.

Source: American Insurance Association.

Hydrant Operations

HARD SUCTION HOSE CONNECTIONS

Hard suction hose is seldom longer than 10 feet (3 m) unless special arrangements are made to carry one end preconnected. Fire department pumpers that are involved in drafting should carry at least two 10-foot (3 m) sections.

Intake Hose Connected to the Pumper First: The driver/operator must stop the pumper with the pump intake a few feet short of being in line with the hydrant outlet. The hard suction hose is connected to the pumper first. Bending the hose and moving the pumper forward enables the hard suction hose to be connected to the hydrant (Figure 4.4).

Figure 4.4 Many departments prefer to connect the hard suction hose to the pump first and then to the hydrant. This reduces the chance of damaging the pump panel.

Intake Hose Connected to the Hydrant First: The driver/operator must stop the front of the pumper just short of the hydrant outlet. The hard suction intake hose is removed and connected to the hydrant first (Figure 4.5). The pumper must then be moved to the best position for intake connection to the pump.

Figure 4.5 Hard suction hose may also be connected to the hydrant first and then to the pump intake.

SOFT SLEEVE HOSE CONNECTIONS

A fire apparatus driver/operator must know the length of the soft sleeve hose on the pumper. Soft sleeve hose sections are commonly 10 to 25 feet (3 m to 7.5 m) long. The driver/operator must be able, through practice, to judge the proper distance to position the vehicle from the hydrant. This distance must be judged from the hydrant, rather than the curb line, since most hydrants are located different distances from the curb. Whether the hydrant outlet faces the street or is parallel to the curb line is another regulating factor. If the front wheels of the apparatus are turned at a 45-degree angle, the driver/operator can easily adjust the distance to or from the hydrant by moving the unit forward or backward (Figure 4.6).

Figure 4.6 Stopping the pumper short of the hydrant with the wheels at a 45-degree angle allows the driver/operator to correct the distance between the hydrant outlet and the pumper intake.

SIDE INTAKE CONNECTIONS

The driver/operator must stop the pumper with the pump intake a few feet short of being in line with the hydrant outlet. Stopping short of the hydrant outlet permits the intake hose to be slightly curved, preventing kinks that drastically restrict flow (Figure 4.7).

Figure 4.7 Positioning the pumper intake short of the hydrant outlet allows for a smoother hookup.

A good way to minimize kinks in soft sleeve hose is to put two full twists in the hose when making the connection between the hydrant and the pumper. These twists prevent the formation of kinks, yet do not affect the hose's ability to flow water. Figures 4.8-4.11 illustrate the benefits of using this method. Practice with the hose is necessary to become proficient in making twists and in determining how many are required to avoid kinks.

Figure 4.8 An intake-to-hydrant connection made with no twists in the hose.

Figure 4.9 When the hose connected in Figure 4.8 was charged, several friction-increasing kinks were formed.

Figure 4.10 Soft sleeve hose should have at least two full twists in it before being charged with water.

Figure 4.11 When the hose connected in Figure 4.10 was charged, there were no kinks in the hose. This is the major advantage of putting two full twists in the hose before charging it.

FRONT AND REAR INTAKE CONNECTIONS

Similar precautions and judgment must be used when positioning pumpers with front and rear pump intakes. The driver/operator must stop the pumper either a few feet short or a few feet beyond the hydrant to permit the hose to be curved. Only practice with the individual pumper will develop proper positioning for intake hose connections. When using front or rear intake connections, the vehicle should be aimed or angled in the direction of the hydrant. This angle should be 45 degrees or less (Figure 4.12).

Figure 4.12 Front intake connections make it easier to judge the distance between the pumper intake and the hydrant discharge outlet. Note the rolled section of hose placed on the pavement to prevent chafing of the soft sleeve.

CONNECTION TO THE 2½-INCH (65 mm) OUTLETS

When the maximum flow is not needed, connection to the hydrant may be made with one or two of the 2½-inch (65 mm) outlets. This is done by connecting short sections of 2½- or 3-inch (65 mm or 77 mm) hose to the pump intake (Figure 4.13). This type of operation is by far the easiest to set up. The short sections of hose allow for maximum flexibility concerning the location of the pumper. They are also light enough to be easily handled by one person. With the ease of handling, maneuvering time is cut down. This allows the pumper to be connected and to begin supplying water with a minimum of delay.

Figure 4.13 Gated 2½-inch (65 mm) intakes are generally found to the side and slightly below the large diameter intake of a pump.

The main disadvantage of connecting to the 2½-inch (65 mm) outlets is that it limits the amount of water that can be supplied. This is mainly due to the high amount of friction loss in the smaller hoselines. This can be minimized by using 3-inch (77 mm) hoselines instead of 2½-inch (65 mm) lines. A kink in the hoseline can also reduce the maximum flow appreciably. Removing kinks is one of the easiest ways to ensure the maximum possible flow.

The pump intakes to which the hoselines are connected are also important in terms of friction loss. The gated 2½-inch (65 mm) intake connections to the fire pump drastically limit total flow. The maximum flow through one of these 2½-inch

(65 mm) intakes depends upon the way the piping is arranged; in general, 250 gpm (1 000 L/min) is an average capacity. A better arrangement is to bring the 2½- or 3-inch (65 mm or 77 mm) lines into the pump through the large intake connection. This can be done with a bell reducer or a suction siamese fitting that allows connection of more than one 2½- or 3-inch (65 mm or 77 mm) hose.

MULTIPLE INTAKE CONNECTIONS

Occasionally, a pumper will use both the soft sleeve and additional intake lines from one exceptionally strong hydrant. In these cases, the pumper position should be determined by the soft sleeve requirements since it is the shorter (and greater capacity) hose (Figure 4.14).

Figure 4.14 When a strong hydrant is used to make multiple connections, angle the pumper so the soft sleeve hose can be connected properly.

SPECIAL OPERATIONS

Relay Operations

The positioning of pumpers within a relay operation is dependent upon the type of relay to be used. The number of pumpers needed and the distances between them are determined by several factors:

- Amount of water needed at the fire
- Distance between water source and fire
- Hose size
- Pumper capacity
- Terrain

The first determination to be made in spacing the pumpers for a relay is the number of pumpers needed for the relay. This depends on the maximum allowable working pressure, the distance that hose must be laid, the size(s) of the hose available, and the capacities of the pumpers.

If the maximum allowable working pressure is 200 psi (1 400 kPa) based on a hose test of 250 psi (1 724 kPa) less 50 psi (350 kPa) for a safety margin, the available pump discharge pressure (PDP) is 180 psi (1 260 kPa). Note that 20 psi (140 kPa) is subtracted for necessary residual pressure at the intake side of the next pumper. This 180 psi (1 260 kPa) is the pressure available to overcome friction loss in the hose between pumpers. Table 4.2 lists the distances that 180 psi (1 260

TABLE 4.2 (U.S.)
MAXIMUM DISTANCE OF WATER FLOW AT 180 psi (In Feet)

Flow GPM	Hose Diameter (inches)				
	2½	3	4	5	6
100	9,000	22,500	90,000	225,000	360,000
200	2,250	5,625	22,500	56,250	90,000
250	1,440	3,600	14,400	36,000	57,600
300	1,000	2,500	10,000	25,000	40,000
400	563	1,406	5,625	14,060	22,500
500	360	900	3,600	9,000	14,400
750	160	400	1,000	4,000	6,400
1,000	90	225	900	2,250	3,600

TABLE 4.2 (metric)
MAXIMUM DISTANCE OF WATER FLOW AT 1 260 kPa (In meters)

Flow L/min	Hose Diameter (mm)				
	65	77	100	125	150
400	2 445	6 102	25 410	56 159	93 373
800	611	1 526	6 352	14 040	23 343
1 000	391	976	4 066	8 986	14 940
1 200	272	678	2 823	6 240	10 375
1 500	174	434	1 807	3 994	6 640
2 000	98	244	1 016	2246	3 735
3 000	43	108	452	998	1 660
4 000	24	61	254	562	940

kPa) will deliver at various flows through 2½-, 3-, 4-, 5-, and 6-inch (65 mm, 77 mm, 100 mm, 125 mm, and 150 mm) hose. The number of pumpers needed for a relay can be found by using this table. For more information about setting up relay operations, consult IFSTA's **Water Supplies** manual.

Dual Pumping Operations

With dual pumping (often incorrectly referred to as tandem pumping), one strong hydrant may be used to supply two pumpers. This type of operation has several advantages, including better use of available water and shorter hose lays (particularly if the hydrant is close to the fire). Additional hoselines can be placed in operation quicker and apparatus may be grouped close together, allowing easier coordination. The method for a dual pumping hookup is as follows:

Step 1: Pumper 1 connects to the hydrant steamer connection using a large intake hose. This pumper then pumps water through its lines to the fire (Figure 4.15).

Step 2: Pumper 2 is positioned intake-to-intake with Pumper 1. The hydrant is closed down until the compound gauge of Pumper 1 reads near 0 (about 5 psi [35 kPa]) (Figure 4.16). The throttle of Pumper 1 will need adjusting. This makes the volumes of discharges and intake equal, so the cap of the unused intake can be removed (Figure 4.17). (**NOTE:** If Pumper 1 is equipped with a keystone or gate valve on its unused intake, the hydrant need not be turned off.)

Figure 4.15 At the beginning of a dual pumping operation, one pumper is connected to a hydrant and begins to flow water.

Figure 4.16 The hydrant is closed down until the water being supplied to the first pumper is equal to the water being discharged.

Figure 4.17 Little water comes out of the intake when the large diameter intake cap is removed, even though a large amount continues to flow from the master stream device.

Step 3: Pumper 2 is connected by large intake hose to the unused steamer intake of Pumper 1 (Figure 4.18).

Step 4: The hydrant is opened completely.

Step 5: Pumper 2 pumps water through its lines to the fire. Its supply is the water not being used by Pumper 1 that is passing through it (Figure 4.19).

Figure 4.18 The second pumper is connected intake-to-intake with the first pumper.

Figure 4.19 After the hydrant has been fully opened again, both engines can now flow water.

Tandem Pumping Operations

Tandem operations may be used when higher pressures than a single engine is capable of supplying are required. The two engines are positioned in basically the same manner as for dual pumping, although the pumpers may be as much as 300 feet (90 m) apart. In dual pumping, the pumpers are connected intake-to-intake. In tandem pumping, the pumper directly attached to the water supply *pumps* water through its discharge outlet(s) into the intake(s) of the second engine. This enables the second engine to discharge water at a much higher pressure than a single engine could have supplied. The higher pressures result from the fact that the pumps are actually acting in series.

Tandem pumping operations are commonly used when the attack pumper is only a short distance from a hydrant. The second pumper is placed directly on the hydrant to support supply lines to the attack unit. Tandem pumping may also be needed to overcome friction loss problems that occur in large sprinkler or standpipe systems and in long hose layouts. It is important to use caution when supplying hoselines with a tandem pumping operation because it is possible to supply greater pressure than the hose can withstand. Pressure supplied to the hose should not exceed that pressure at which the hose is annually tested by the department (Figure 4.20).

Figure 4.20 In this example of tandem pumping, the pumper at the hydrant is boosting the intake pressure of the second pumper, which is supplying a master stream device.

Supplemental Pumping

The term "supplemental pumping" can be misleading because it is used to describe several different types of operations. All these operations have the same purpose, which is to provide additional water to a pumper when the hydrant it is already attached to is not supplying enough. This may become necessary when the hydrant is unable to supply the amount of water the pump is rated for. Supplemental pumping may also be necessary when the intake hose already in use is not capable of flowing the rated capacity of the pump or the water available at the hydrant.

The most common type of supplemental pumping is used when a pumper is unable to obtain the maximum amount of water available through its original connection to the hydrant. In this case, it may be desirable to add one or two 2½- or 3-inch (65 mm or 77 mm) lines between the hydrant and the pump. This is a simple task if a hydrant gate valve has been placed on the hydrant's 2½-inch (65 mm) connections before the hydrant is turned on the first time. Just connect the line(s) between the two and open the gate valves.

If no gate valves were attached, it will be necessary to shut down the hydrant so the connections can be made. This means that the flow of water through discharge lines will be interrupted, so attack crews should be warned beforehand. Hooking the hose to the pumper before shutting down the hydrant minimizes the time the water supply is interrupted.

An alternative to shutting down the hydrant is to run supplemental lines to another hydrant. Then the pumper can operate directly off the second hydrant or have another pumper supply the line from the second hydrant.

In extreme cases, water can be pumped from a hydrant on a strong water main into a hydrant on a weaker main in order to boost the available water. This is not a common procedure and is generally discouraged by water officials due to its potential for contaminating the water supply system. Figure 4.21 shows several different layouts for supplemental water supply operations.

Sprinkler and Standpipe Support

Driver/operators positioning to support sprinkler and standpipe must be aware of factors influencing the support operation:

- Location of water supply to be used
- Distance from street to fire department connection
- Positioning requirements of other apparatus
- Pump pressure readings that indicate malfunctions and damaged systems

Pumpers will generally position as close as possible to the sprinkler or standpipe fire department connection to help speed operations (Figure 4.22). The location of the water supply may be such that the pumper will have to position at the source. This is obviously the case when support pumpers are using a draft source. In extreme cases where there is no water supply available near the sprinkler or fire department connection, it may be necessary to establish a relay to supply water (Figure 4.23).

There are situations when pumpers supporting sprinklers or standpipes must give priority to

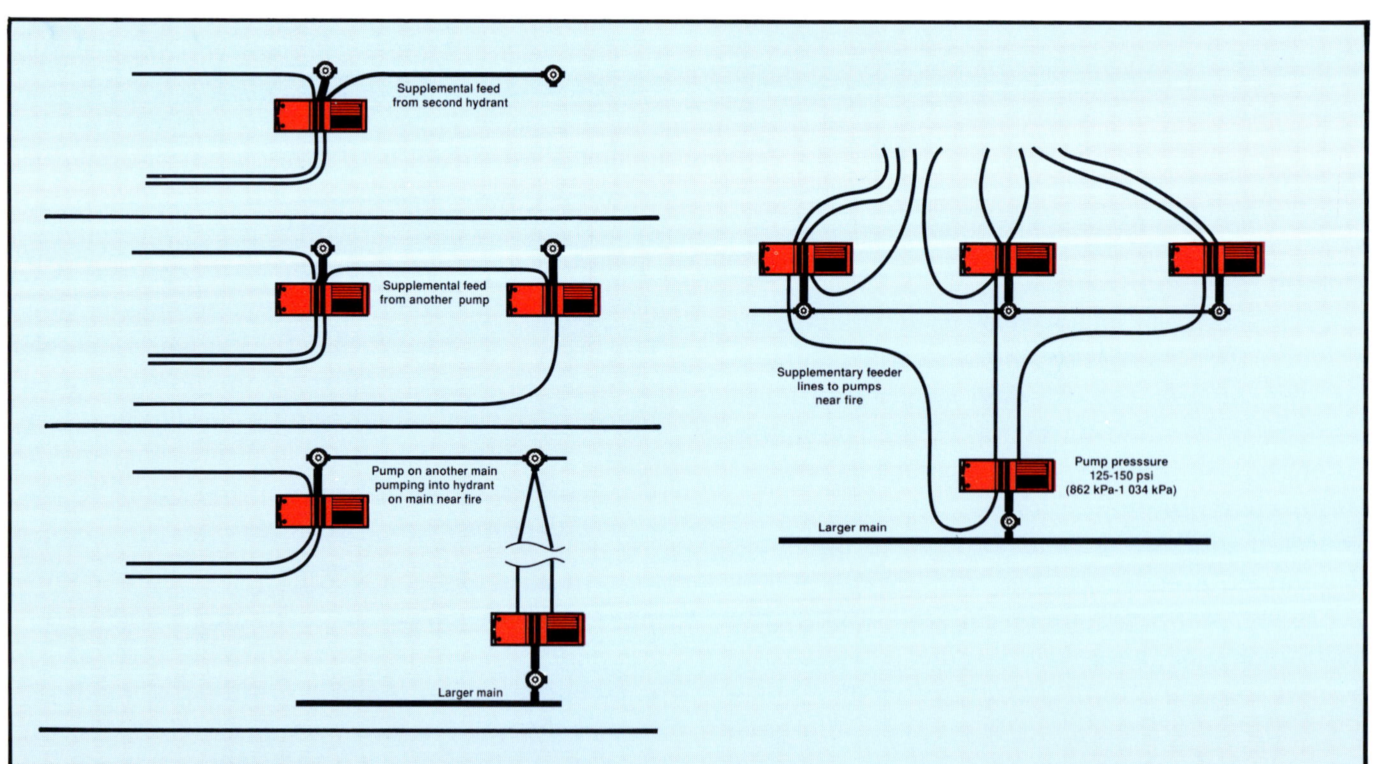

Figure 4.21 Examples of supplemental pumping hose layouts.

Figure 4.22 A water supply source located near the fire department sprinkler connection enables an engine to connect to both.

Figure 4.23 In some cases, it may be necessary to relay water to a fire department connection.

other apparatus. This is especially true when aerial rescue is required on the connection side of the building. Support pumpers can easily operate at varying distances, but aerial apparatus requires more precise locations.

Support for Aerial Apparatus

Pumpers providing water for elevated stream operations should position as close to the aerial apparatus as possible (Figure 4.24). Friction and elevation loss is a major consideration when supporting elevated streams and pumpers must be aware of this loss. Pumpers equipped with their

Figure 4.24 Aerial apparatus not equipped with a fire pump will require a pumper to supply elevated master stream devices.

own elevated stream device should position in the same manner as aerial apparatus providing fire suppression.

Tankers

FILL SITE LOCATION

Tanker positions at a shuttle fill site are dependent upon the number of fill pumpers, the fill pumper locations, and the available parking at the fill site. Tankers should park next to the fill pumper, allowing sufficient clearance for the pumper hoselines (Figure 4.25). Special fill opera-

Figure 4.25 A pumper has been set up at draft with discharge lines waiting to fill arriving tankers.

tions may require unique tanker positions (Figure 4.26). When fill sites have limited access, tankers should be backed into the fill site empty instead of attempting turnarounds when full.

Figure 4.26 Departments that frequently use tanker shuttles may have apparatus specially designed for fill operations. *Courtesy of Rich Mahaney.*

UNLOADING SITE LOCATION

Tanker positioning at the unloading site will depend upon the method of water transfer. Tankers dumping to portable tanks should position along one of the sides not occupied by the draft pumper. The tankers should be positioned in a manner that allows incoming tankers access to the other available sides (Figure 4.27).

Figure 4.27 It is most efficient for a tanker to dump its load through a large diameter dump valve.

Tankers being drafted from or pumping to an attack pumper or nurse tanker will position in a manner similar to that used for fill pumpers at the fill site. This unloading position may be affected by the location of the inlets on the pumper or tanker being supplied (Figure 4.28).

Figure 4.28 If no portable tank is available and a tanker is not equipped with a pump, it may be necessary for pumpers to draft from the tanker.

Brush Fire Apparatus

Brush fire apparatus are in a unique situation for fireground positioning requirements. Many brush companies are intended to be mobile, thus providing a moving fire attack. Secondary apparatus providing support will position according to function:

Command vehicles

Water supply units at draft or hydrant

Earth-moving equipment

General guidelines for positioning brush attack apparatus are as follows:

- Position should maximize protection from heat and fire. Take into consideration such hazards as overhead power lines, heavy fuel stands, and incoming air drops. Take advantage of natural breaks such as roads and orchards. Protect apparatus with 1- or 1½-inch (25 mm or 38 mm) line.

- Do not drive apparatus into unburned fuels higher than the bumper or running board without a spotter. Hidden stumps, logs, or other hazards can disable the vehicle. Spotters are also needed for nighttime driving when the terrain is not visible.

- All four- and six-wheel drive apparatus should be driven by experienced driver/operators who understand the vehicle's capability and limitations.
- Hilly or hillside terrain must be watched for loose and unstable ground that could cause apparatus to slide or overturn.
- Use areas of burned fuel whenever possible (Figure 4.29). Apparatus attacking from the unburned side must leave sufficient clearance distances from the fire line to allow for loss of water and mechanical failure.
- Be aware of fire conditions; travel and position accordingly at all times.
- Consider the location of operating crews when moving apparatus. Do not drive into smoke where crews may be operating. If you *must* drive through smoke, sound the horn or siren intermittently, use warning and headlights, and drive slowly.

Additional information about brush company operations can be found in IFSTA's **Ground Cover Fire Fighting Practices.**

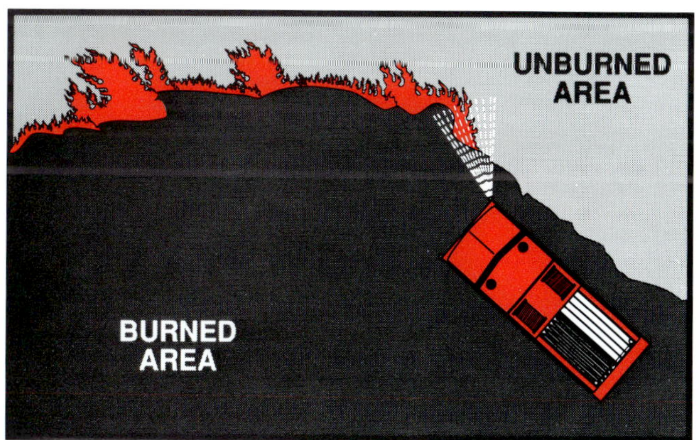

Figure 4.29 At a brush fire, it is usually best to operate from the side that has already been burned to prevent the vehicle from being caught in advancing fire.

STAGING

Often, apparatus placement at the scene of a fire or medical incident is limited by the order in which responding apparatus arrive. A late-arriving ladder truck may be blocked from a better position by earlier-arriving apparatus. Standard operating procedures governing apparatus placement are one way to prevent this type of situation from occurring. An apparatus staging procedure facilitates the orderly positioning of apparatus and allows the incident commander to fully utilize the potential of each unit and crew.

Through improvements in incident command strategies, an apparatus staging procedure in two levels has been developed that can be used for any multicompany response. Level I staging is applied to the initial response to a fire or medical incident involving more than one engine company. Level II staging is used in greater alarm situations where a large number of emergency vehicles are responding to an incident. Level II staging procedures must be initiated by the incident commander when requesting additional alarms or by a dispatcher when a large initial response is called for.

Level I staging is used on *every* emergency response when two companies performing like functions are dispatched (for example, two engine companies). The first-due engine company, ladder truck, rescue or squad, and command officer proceed directly to the scene. Later-arriving units park or stage away from the scene in their direction of travel (Figure 4.30). The incident commander may order the staged units to lay additional supply lines, send personnel to the scene, or proceed to the scene and set up. Engine companies should stage near hydrants or other sources of water. Staged apparatus should not allow their path to the scene to become blocked.

Level II staging is used when numerous emergency vehicles will be responding to an incident. Incidents that require mutual aid or that result in multiple alarms will need Level II staging. When the additional units are requested by the incident commander, an apparatus staging area is designated. Companies are informed of the staging area location when they are dispatched and respond directly to that location (Figure 4.31). A parking lot or open field can serve as a staging area. A fire department officer should take charge of the staging area and communicate with the incident commander. Apparatus officers should re-

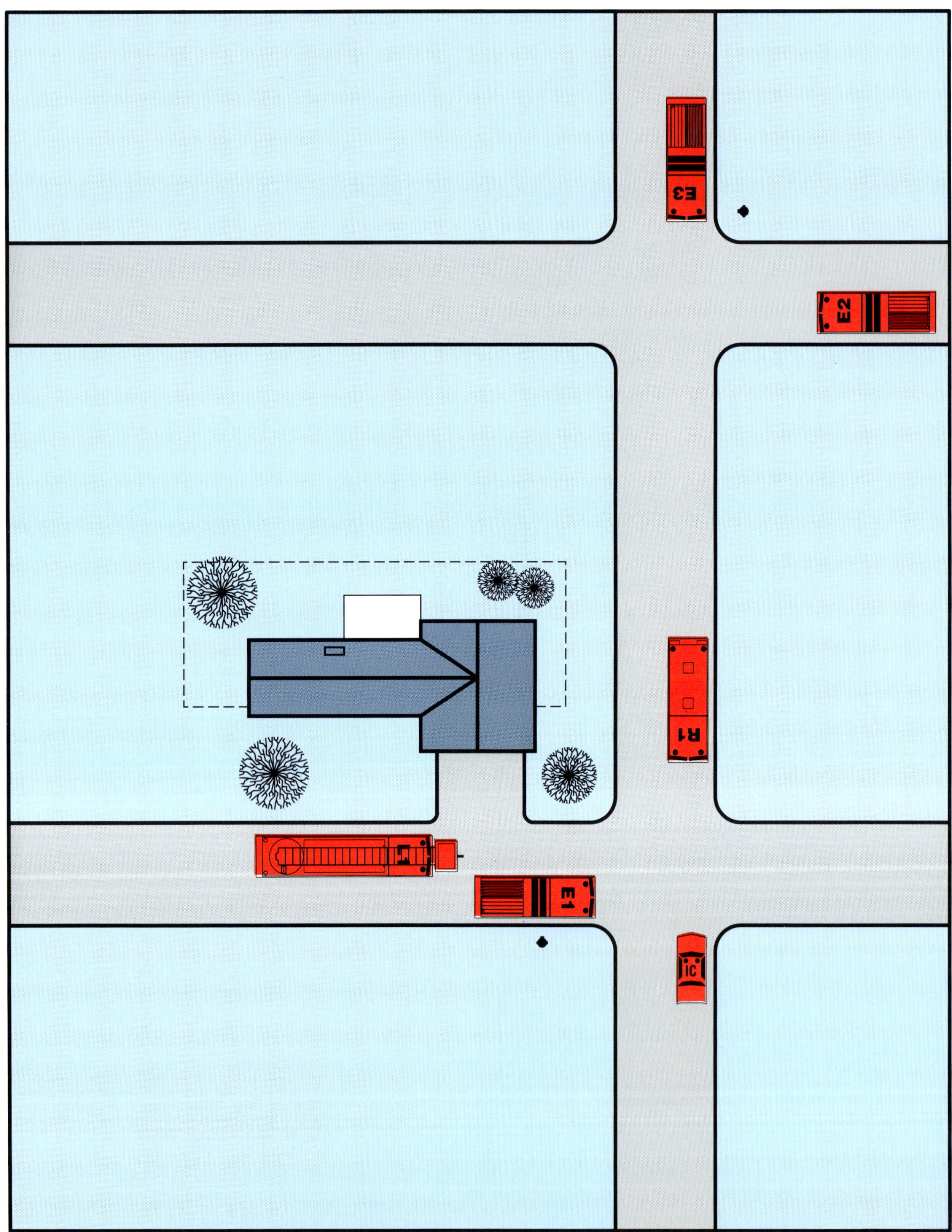

Figure 4.30 Level I staging should be used on every response.

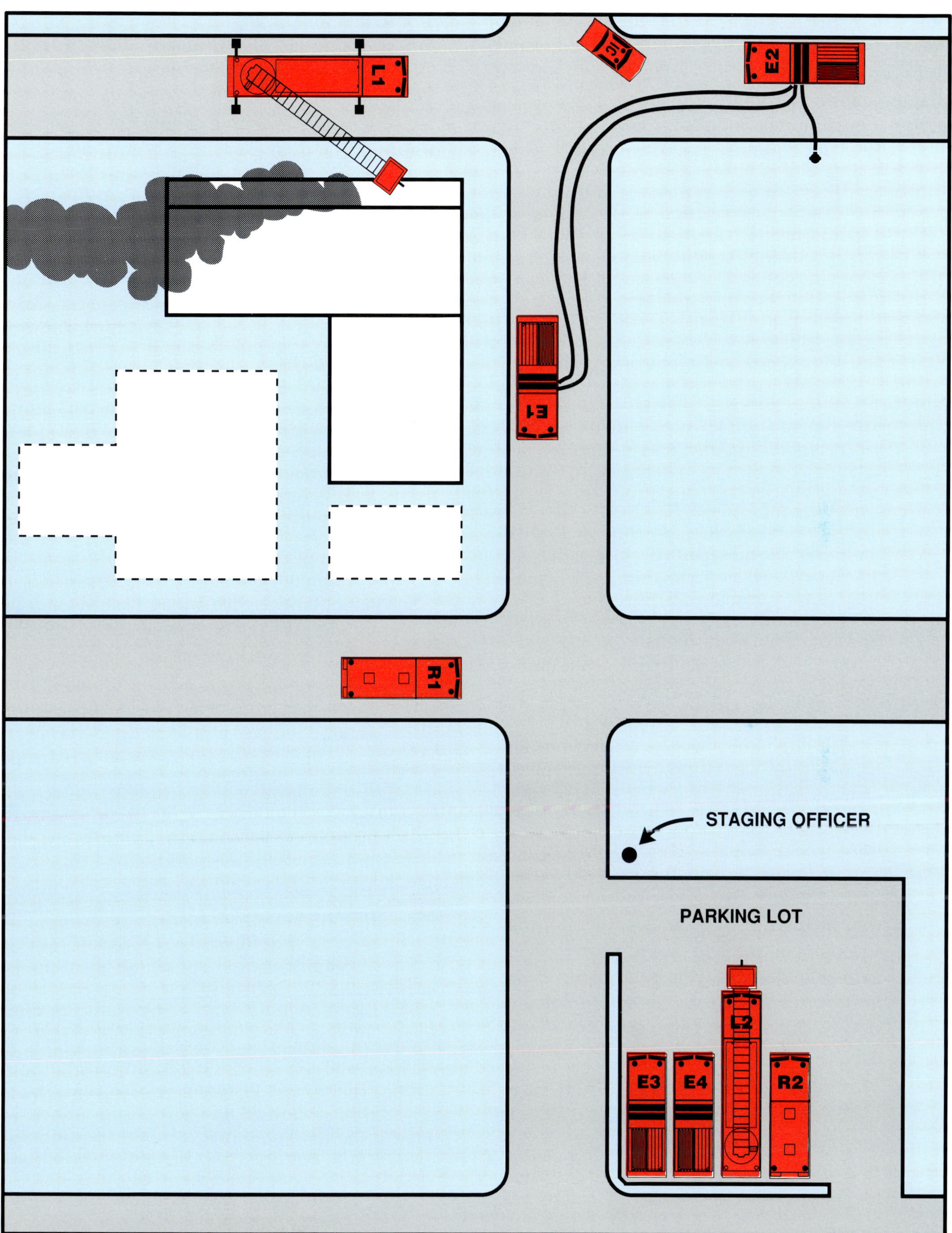

Figure 4.31 Level II staging is used on large scale responses.

port to the staging officer as they arrive and park. When the incident commander requires additional assistance, companies are summoned through the staging officer and sent to the scene. The staging officer should meet with police officials and ensure that parked apparatus are safe from traffic and in a secure area.

OPERATIONS ON LIMITED-ACCESS HIGHWAYS

Incidents on limited-access highways or freeways usually involve motor vehicle accidents and/or fires. The potential for multiple-injury accidents involving hazardous materials is also high. These operations are plagued by the following problems:

- *Access* — Apparatus may have to respond over long distances between exits to reach an incident. Apparatus should not be driven against the normal flow of traffic unless the road has been closed by police units. Incidents occurring on bridges may require the use of aerial apparatus or ground ladders in order to reach the scene from below.
- *Water Supply* — Long hose lays or tanker operations may be needed to supply water to the incident scene. Hydrant placement on the freeway may be infrequent or may not exist at all. Hydrants located on access roads may provide the needed water. Rather than carrying supply lines up an embankment, it is easier to lower them from the freeway than the access road if the freeway is higher. If the access road is higher, lower the lines from the access road to the freeway. When responding to a freeway incident, it is often desirable to have one pumper respond to the nearest overpass or underpass. This pumper can assist units on the freeway in establishing a water supply if the source is off the highway.
- *The Safety of Other Drivers* — Highway traffic on the same side or on both sides of the incident should be closed down if it is dangerous to the drivers or to firefighters working at the incident. Most warning lights on responding apparatus should be turned off when the apparatus enters the highway. The siren should not be used except to clear slow traffic. Fire apparatus usually travels slower than the normal flow of traffic and the use of warning lights and sirens may create traffic conditions that actually slow the fire unit's response. Warning lights should be used with discretion at the scene to prevent blinding other drivers or distracting them, possibly leading to another accident. In no case should all warning lights be turned off because that might compromise the safety of firefighters working at the scene.
- *Firefighter Safety* — Cooperation between police and fire department personnel at highway incidents is essential. Police should stop or slow highway traffic if fire department personnel are working in traffic lanes. Fire apparatus should be placed between the flow of traffic and the firefighters working on the incident to act as a shield (Figure 4.32). The apparatus should be parked on an angle so the operator will be protected from traffic by the tailboard. Front wheels should be turned away from the firefighters working highway incidents so the apparatus will not be driven into them if struck from behind. Also consider parking additional apparatus 150 to 200 feet (45 m to 60 m) behind the apparatus being used at highway incidents to act as an additional barrier between firefighters and flow of traffic.

All members must use extreme caution when getting off apparatus so they are not struck by passing traffic. Similarly, pump operators are extremely vulnerable to being hit by motorists if they step back beyond the protection offered by properly spotted apparatus.

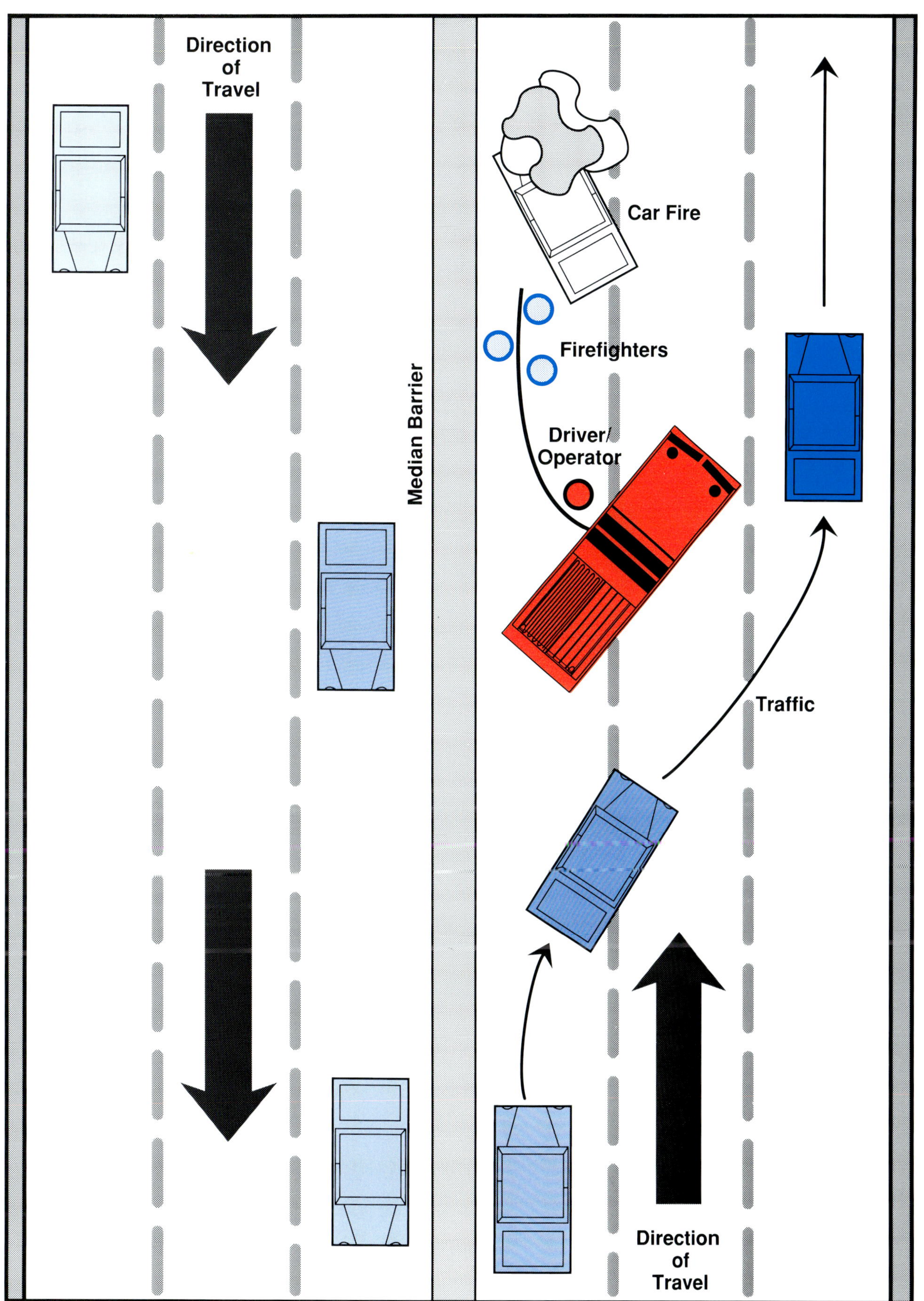

Figure 4.32 Apparatus placement on a limited-access highway should provide a barrier between the scene and oncoming traffic.

Chapter 4 Review

Answers on page 359

TRUE-FALSE: Mark each statement true or false. If false, explain why.

1. When positioning a vehicle for attachment to a fire hydrant, it is best to use the curb line to judge the distance to the hydrant.

   ☐ T ☐ F ______________________________

   ____

10-91

2. When connecting a soft sleeve between a fire hydrant and the front or rear intake of the apparatus, it is desirable to have the truck parked perpendicular to the discharge (90-degree angle).

   ☐ T ☐ F ______________________________

   ____

MULTIPLE CHOICE: Circle the correct answer.

3. On what incidents should Level I staging be used?
   A. All responses
   B. Two engine responses
   C. Greater alarm responses only
   D. Only as ordered by first-arriving company

LISTING

4. Name at least five factors that determine the placement of an attack pumper.

   ____

   ____

   ____

   ____

   ____

5. Name the five final considerations for determining the final draft position of a pumper.
   A. ______________________________
   B. ______________________________
   C. ______________________________
   D. ______________________________
   E. ______________________________

6. List at least five factors that determine the number of pumpers needed and the distances between them when establishing a relay operation.

   ____

   ____

   ____

   ____

   ____

7. List at least three factors driver/operators must be aware of when positioning to support sprinkler and standpipe systems.

SHORT ANSWER: Answer each item briefly.

8. What are some of the primary factors that determine the placement of a fire department pumper?

9. What tactical advantage can the fire officer gain by having the first engine pull past the front door of a building?

10. How is it possible to keep hard suction hose from resting on the bottom of a drafting location.

11. Explain the difference between dual pumping and tandem pumping.

12. Explain the difference between Level I and Level II staging.

13. What is the major disadvantage of using apparatus warning devices while responding on a freeway?

____

____

____

____

____

10-91

14. What should be the major consideration when placing apparatus at the scene of a freeway incident?

____

____

____

____

____

5

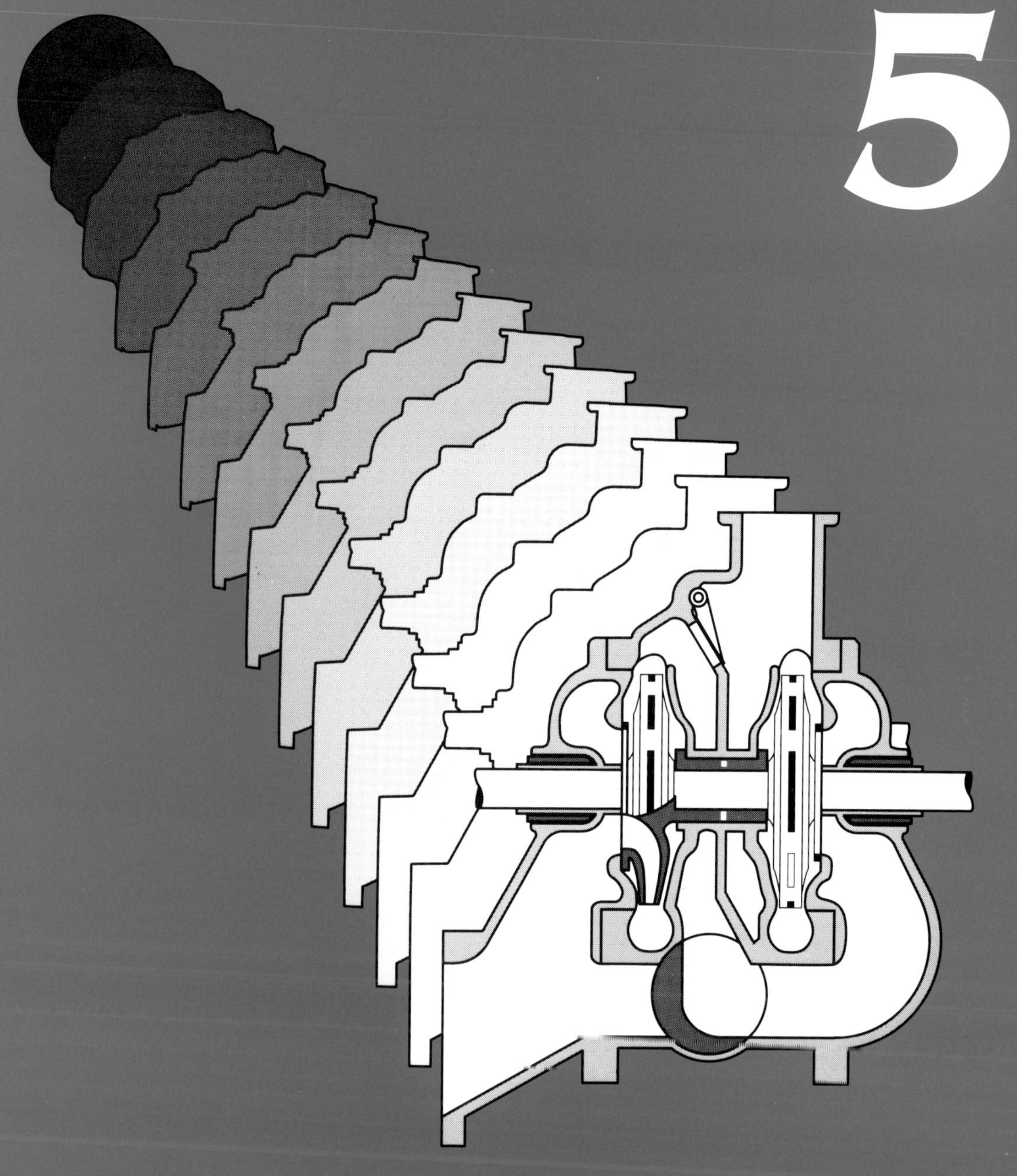

Fire Pump Theory

NFPA STANDARD 1002
Fire Apparatus Driver/Operator Professional Qualifications

3-1 General.

3-1.2 The fire apparatus driver/operator shall identify the operating principles of single-stage and multiple-stage centrifugal fire pumps.

3-1.3 The fire apparatus driver/operator, given pump models or diagrams, shall identify the major components and trace the flow of water through single-stage and multiple-stage centrifugal pumps.

3-1.4 The fire apparatus driver/operator shall identify the percentages of rated capacity, rated pressures, and the capacity in gallons per minute (gpm) at the rated pressures a fire department pumper is designed to deliver.

3-1.6 The fire apparatus driver/operator shall identify the following conditions that may result in possible pumper apparatus damage or unsafe operation, and identify corrective measures:

(a) cavitation
(b) leaking fuel, oil, or water
(c) overheating
(d) unusual noises
(e) vibrations
(f) water hammer.

3-5 Apparatus Systems.

3-5.1 The fire apparatus driver/operator shall identify three methods of power transfer from the vehicle engine to the pump.

3-5.2 The fire apparatus driver/operator shall identify the theory and principles of pumper priming systems.

3-5.3 The fire apparatus driver/operator shall identify the theory and principles of pumper pressure relief systems and pressure control governors.

3-5.4 The fire apparatus driver/operator, given a fire department pumper, shall identify all pump gauges and demonstrate their use.

3-5.5 The fire apparatus driver/operator shall identify the auxiliary cooling systems and explain their function.

Chapter 5
Fire Pump Theory

The basic physical characteristics of water dictate the way in which it is used for fire suppression purposes. In order to become a competent pumping apparatus driver/operator, it is important that the firefighter fully understand the principles and theory concerning the water that will be pumped, as well as the operation of the pump itself. A driver/operator who fully understands these characteristics will be able to operate the pump efficiently and effectively.

Just as important as knowledge of water theory is knowing the operating principles of the fire pump. If the driver/operator is to obtain maximum use from the fire pump, it is necessary to understand more than just which knob to turn or handle to pull. "Knob turners" are likely to be lost when faced with an unusual situation, while the driver/operator who understands the operation of the pump will be able to adapt to the situation.

After reading this chapter, the driver/operator should have a good knowledge about the way in which the fire pump operates. The first part of the chapter introduces the driver/operator to the principles of movement and pressure in regard to fluids (in this case, water). Included is information on the properties of water, types of water pressure, and measuring water pressure.

The latter part of the chapter introduces the driver/operator to the principles of fire pump operation. Included is a brief discussion of the types and principles of operation of positive displacement fire pumps. Following this section is a more detailed explanation of the types and principles of operation of centrifugal fire pumps. Power transfer and drive arrangements, pump mounting locations, piping, priming devices, and pressure control devices are among the topics discussed.

UNDERSTANDING FLUIDS

In order to more fully understand the operation of a fire pump, the driver/operator should be familiar with the properties of water that allow its use as an extinguishing agent. Fluids are substances that have no established shape, but are capable of flowing from place to place and will conform to the shape of their container. They have very little adhesive quality and cannot be pulled or lifted, but must be pushed by having pressure applied and confined in some manner to direct the flow. Fluids are divided into liquids or gases. While both liquids and gases are fluids and share certain characteristics, they are quite different and must be treated as such:

- Liquids have a certain specified volume and have free surfaces where the amount of liquid is insufficient to fill the container. They are practically incompressible and for this reason are unable to store pressure. Liquids have mass and are usually heavier than gases.
- Gases have no specific volume, but will expand to fill the total space available. As the gas expands, each cubic foot (cubic meter) weighs less. Gases can be compressed into a smaller space, with a corresponding increase in pressure. (**NOTE:** If gases are compressed enough, they will be converted into liquid form.) When the external pressure is removed, the gas expands, at-

tempting to regain its former volume, and the pressure gradually decreases as it expands. Gas can be used to store pressure.

Many substances can exist in more than one form. Carbon dioxide (CO_2) will serve as an example. It can be a solid (dry ice), a liquid (inside a CO_2 extinguisher), or a gas (once it is expelled from a CO_2 extinguisher). Two conditions help determine in what form a substance exists at a particular time: the temperature and the pressure to which the substance is subjected. Carbon dioxide is a liquid inside the extinguisher due to the pressure at which it is maintained. When it is allowed to escape through the hose and the pressure is removed, it converts instantaneously to a gas.

Properties of Water

Water is not only an effective coolant and very effective in fire extinguishment, but also has many other qualities that make it desirable for use in fire fighting operations. Water is composed of masses of molecules. Each of these molecules is composed of two atoms of hydrogen and one of oxygen. In their natural state, each atom is a gas, but once they have combined to form a molecule of water, they are not easily separated into their original elements. Although water changes from a solid to a liquid to a gas, each molecule of water retains its chemical structure. Other changes do, however, take place.

Between 32°F and 212°F (0°C and 100°C), water is usually in the liquid state, which is the form most useful to the fire service. As a liquid, water can be handled easily. It can be pumped through hoselines and applied very effectively to a fire. When it changes to a solid, problems are encountered (Figure 5.1).

When water freezes and changes from a liquid to a solid, it expands. This action exerts strong pressures and can cause extensive damage to equipment. For this reason, drains are provided on all fire apparatus, usually at the lowest points of the pump and piping. Their purpose is to ensure that all water can be removed when freezing temperatures are encountered.

The temperature required to change water to steam varies with the pressure applied. For example, a 14 psi (98 kPa) radiator cap on a truck engine increases the boiling point in the cooling system from 212°F (100°C) to 234°F (112°C). At the other extreme, high vacuum readings at the eye of the impeller of the fire pump can reduce the vaporization point below 100°F (30°C). This may result in serious damage to the pump and other equipment. Chapter 6 outlines ways to avoid this situation.

Figure 5.1 Ice buildup adds a substantial amount of weight to equipment. *Courtesy of Edward Prendergast.*

Water at Rest

Water at rest can be defined as water that is in a standing state and is undergoing no physical action. There are several characteristics of water at rest that are important. First of all, water is virtually incompressible. In its fluid state, it would take a pressure of 33,000 psi (231 000 kPa) to reduce its volume by 1 percent. For this reason, water will not store pressure. As soon as pressure is removed, there is not appreciable expansion of water, and the pressure is relieved. Water is pumped easily because it cannot compress when pressure is applied to it; therefore, it moves rapidly into an area where pressure is lower (hose, nozzle). This is also the reason water hammer results when hoselines are shut down suddenly.

Drops of water can be visualized as a container full of marbles, each coming in contact with each other, but with no physical adhesion. Pressure against one of these marbles would be communicated to each of the other with which it is in contact. In the same manner, water pressure can be communicated when water is confined within a container, pressure applied from without will be transmitted in all directions without decreasing. Where it comes in contact with a restraining surface, pressure will be applied perpendicular to the surface.

The downward pressure of a liquid in an open vessel is proportional to the depth of the liquid. Since a column of water 1 inch square (645 mm^2) and 1 foot high (0.3 m) exerts a pressure of .434 psi (3 kPa), each foot (0.3 m) of depth results in a pressure of .434 psi (3 kPa) and each 2.304 feet (0.7 m) of depth increases the pressure 1 psi (7 kPa) (Figure 5.2).

Because pressure is transmitted equally in all directions when the body of water is at rest, the shape of the container will not have any effect in the pressure being applied at the base of it (Figure 5.3).

Pressure exerted in all directions in any system with no water being moved is known as *static pressure.* Static pressure is stored potential energy that is available to force water through pipe, fittings, fire hose, and adapters.

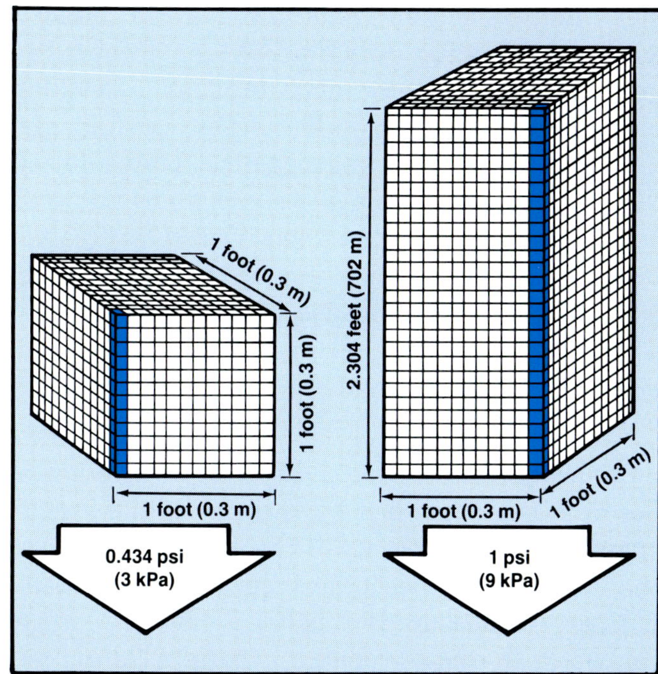

Figure 5.2 One cubic foot (0.3 m^3) of water contains 1728 cubic inches (11 150 000 cm^3) of water. The pressure exerted by a column of water one foot high is 0.434 pounds per square inch (3 kilopascals). To exert a pressure of 1 psi (7 kPa), a column of water must be 2.304 feet (0.70 m) high (about 27½ inches [70 mm]).

Water in Motion

There are several important characteristics of water in motion. These are discussed in the paragraphs that follow.

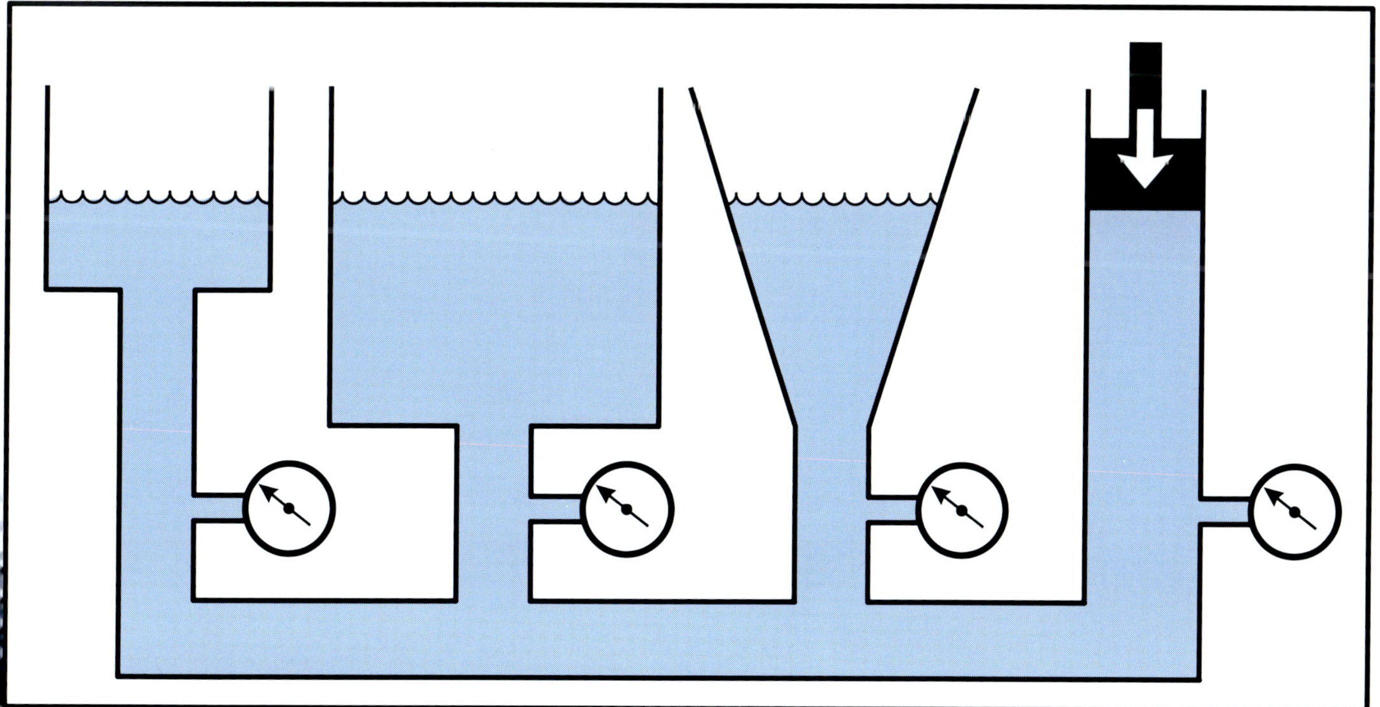

Figure 5.3 Since the level of the liquid in these three containers is equal, the pressure will be the same at all four gauges.

Water in motion tends to remain in motion, without changing direction. Water frequently moves through fire hose at a rate of 1,000 feet (300 m) per minute or more. If the flow is suddenly interrupted, the high velocity plus the weight of the water creates momentum that will cause a water hammer. Water hammer is the force created by the sudden acceleration or deceleration of water. This force can cause damage to hoselines and equipment and can also be dangerous for hose handlers. All valves should be closed slowly to prevent water hammer (Figure 5.4).

The amount of water that flows through a hoseline is determined by the size of the nozzle opening and the velocity of the water as it leaves the line. The velocity of the water leaving an opening can be measured by using a pressure gauge connected to a pitot tube inserted into the center of the stream. From this measurement, it is possible to calculate how much water is flowing from the opening.

When water moves through any type of waterway, the friction of the water against the surface of the waterway creates a resistance to its movement. This pressure loss due to this resistance, also known as friction loss, applies to fire hose as well as to pipe used in water systems. Friction loss increases with the amount of water flowing through a given size waterway and decreases with an increase in the size of the waterway with a constant flow.

Pressure differences due to gravity also occur as a result of change in elevation between different points in a water supply operation.

Water moving through a system loses pressure while in motion. The residual pressure of a water system is the static (no movement) pressure of the system minus the friction loss between the source and the point of measurement.

Example: A city water pump station can create 150 psi (1 050 kPa) in the water system

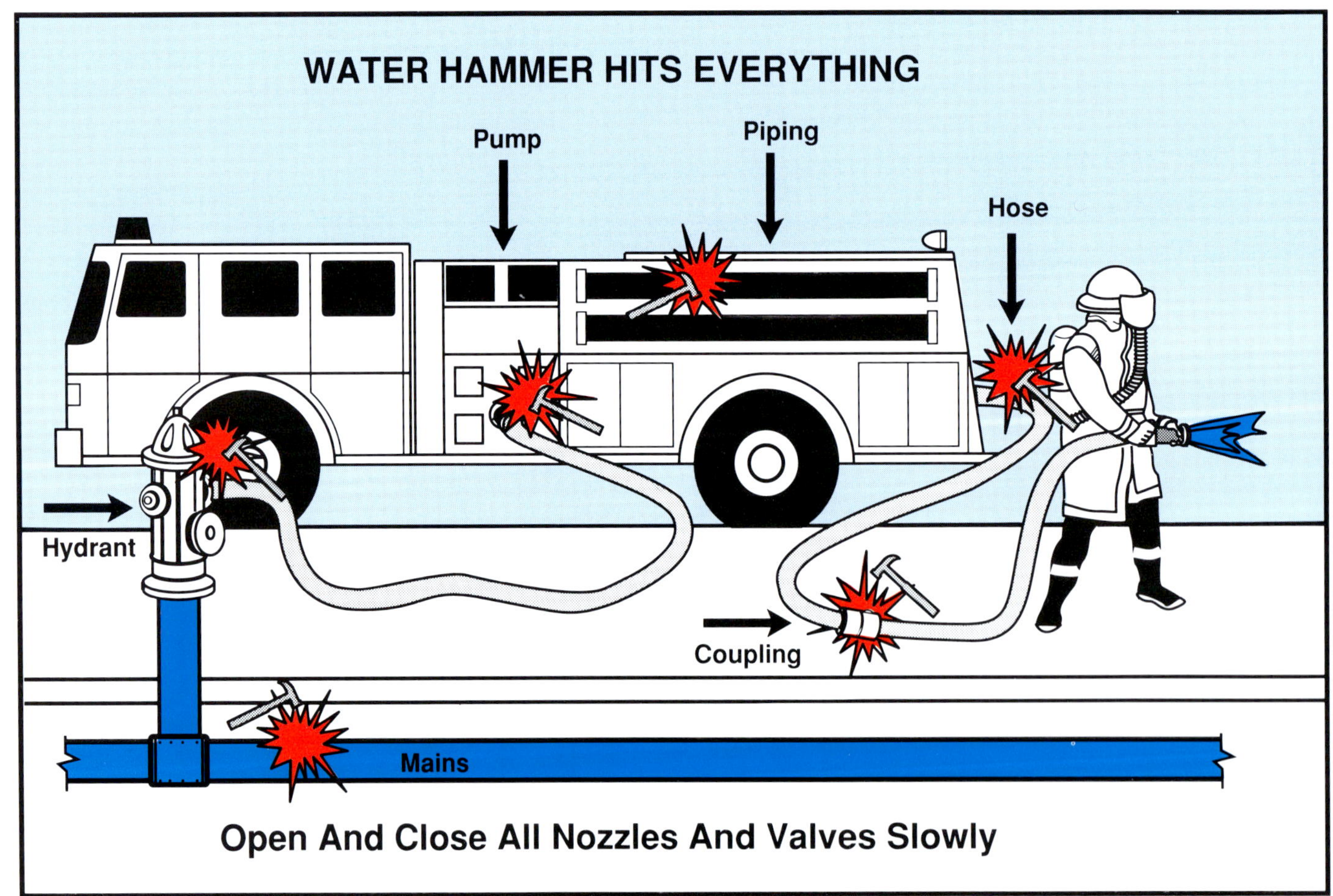

Figure 5.4 Water hammer can have damaging effects throughout the water supply system.

when no water is flowing (Figure 5.5). When water is flowed, the pressure at the hydrant is reduced to 100 psi (700 kPa) (Figure 5.6). The reduction in pressure is due to the friction loss between the water pump station and the hydrant.

Static pressure = 150 psi (1 050 kPa)
Residual pressure = 100 psi (700 kPa)
Friction loss = 150-100 = 50 psi (350 kPa)

Water Pressure

Pressure results when water is being forced into a confined area faster than it can be moved out. Anytime additional hoselines are placed in service or water flow is increased, pressure can be expected to drop. Increasing the supply without a corresponding increase in discharge will cause the pressure to rise.

Water pressure can be generated naturally, mechanically, or chemically. Naturally created pressure comes from the force of gravity and the mass of the water. When water is stored at a higher elevation than the point of discharge, 1 psi will be generated for each 2.3 feet (10 kPa/m) of height differential. Some commercial pumps are rated to pump against a specified head pressure. To determine how much head pressure the pump will handle, it is necessary to interpolate by dividing the feet (meters) of head rating by 2.3 (10) and convert it to psi (kPa).

Chemically created pressure is generated when two substances are combined. This principle was used in such early fire extinguishers as soda-acid chemical foam extinguishers. Chemical tanks were also used on early fire apparatus for rapid fire extinguishment before hoselines could be laid and the fire pumps put into service.

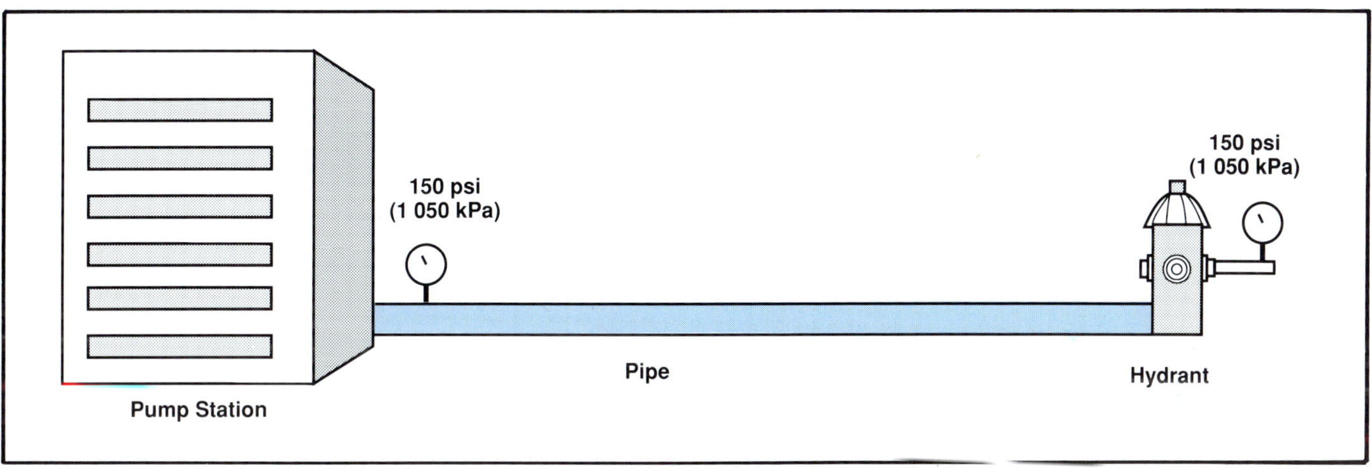

Figure 5.5 With no water flowing (static system) the pressure will be equal throughout the entire system.

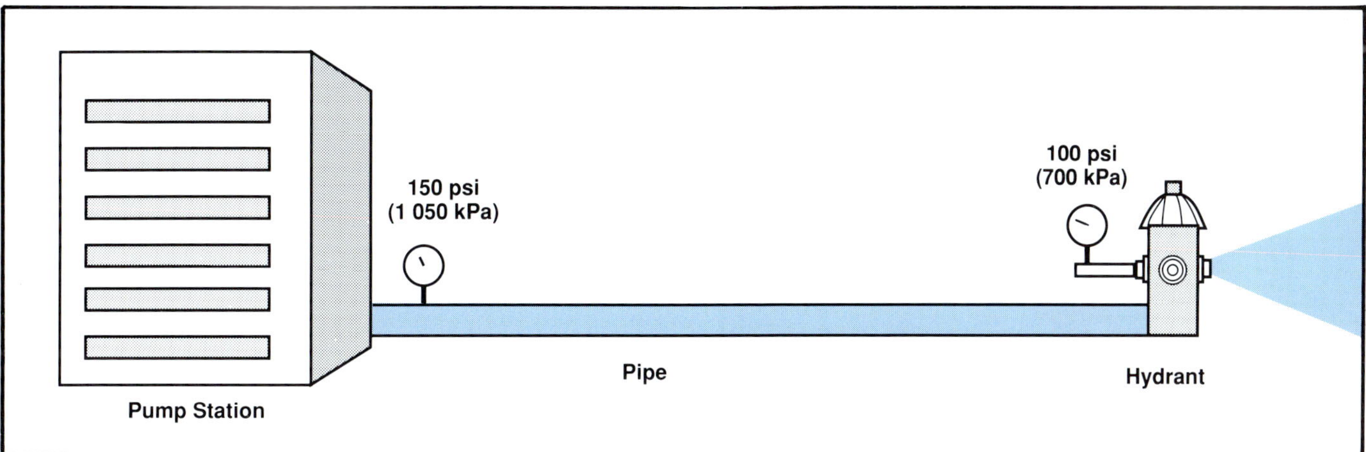

Figure 5.6 When water is flowing, the pressure at the hydrant drops below the discharge pressure of the pumping station. This is due to friction loss in the pipe between the pumping station and the hydrant.

Mechanically created pressure is generated by a fire pump, either positive displacement or centrifugal. Mechanical pressure is often used to pump water into an elevated tank for storage (Figure 5.7). This stored water can then be used to supply additional water using the naturally created pressure generated by the height differential. In a typical approved water supply system operating from wells or a surface supply, pumps are turned on automatically when the pressure drops to a predetermined level in the system. The pumps then provide a pressure higher than the stored pressure in the system, forcing the water into the elevated storage tanks. When the system reaches the desired pressure, the pumps are again turned off and the system operates from the natural pressure created by elevation differences.

Figure 5.7 Elevated water storage tanks utilize gravity to create pressure.

TYPES OF PRESSURE

Atmospheric Pressure

The atmosphere surrounding the earth has depth and density and exerts pressure upon everything on earth. Atmospheric pressure is greatest at low altitudes; consequently, its pressure at sea level is used as a standard. At sea level, the atmosphere exerts a pressure of 14.7 psi (100 kPa).

A common method of measuring atmospheric pressure is to compare the weight of the atmosphere with the weight of a column of mercury: the greater the atmospheric pressure, the taller the column of mercury. The pressure of 1 psi (7 kPa) will make the column of mercury about 2.04 inches (52 mm) tall. At sea level, then, the column of mercury will be 2.04 x 14.7, or 29.9 inches tall (52 mm x 14.7 = 759 mm) (Figure 5.8).

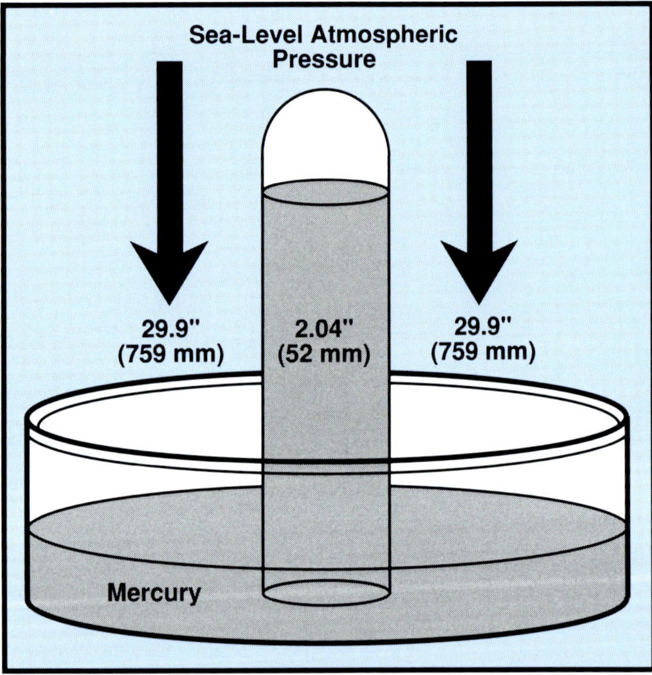

Figure 5.8 A simple barometer. Atmospheric pressure forces the mercury into the tube. A pressure of 1 psi (7 kPa) will cause the mercury column to rise 2.04 inches (52 mm).

The readings of most pressure gauges are psi *in addition* to the existing atmospheric pressure of 14.7 psi. For example, a gauge reading 10 psi (70 kPa) at sea level is actually indicating 24.7 psi (14.7 + 10 [170 kPa]).

NOTE: Engineers make the distinction between such a gauge reading and actual atmospheric pressure by writing *psig,* which means "pounds per square inch gauge." The notation for actual atmospheric pressure is *psia,* meaning "pounds per square inch absolute." Absolute zero pressure is called a perfect vacuum. Any pressure less than atmospheric pressure is called a vacuum. When a gauge reads -5 psig (-35 kPa) it is actually reading 5 psi (35 kPa) less than the existing atmospheric pressure (at sea level, 14.7 - 5, or 9.7 psia [101 - 35 or 65 kPa]). Throughout this manual, psi indicates psig.

Head Pressure

Head, in the fire service, refers to the height of a water supply above the discharge orifice. In Figure 5.9, the head is 100 feet (30 m) above the hydrant discharge opening. To convert head in feet to head in pressure, divide the number of feet (or meters) by 2.304 (0.1 if in meters), which is the number of feet (meters) that 1 psi (7 kPa) will raise a column of water. Note that the water source in Figure 5.9 has a head pressure of 43.4 psi (300 kPa). Table 5.1 gives head in feet and head pressure.

TABLE 5.1

(U.S.) HEAD IN FEET AND HEAD PRESSURE				(metric) HEAD IN METERS AND HEAD PRESSURE			
Feet of Head	psi	psi	Feet of Head	Meters of Head	kPa	kPa	Meters of Head
5	2.17	5	11.50	1	10	5	.5
10	4.33	10	23.00	2	20	10	1
15	6.50	15	34.60	3	30	15	1.5
20	8.66	20	46.20	4	40	20	2
25	10.83	25	57.70	5	50	25	2.5
30	12.99	30	69.30	6	60	30	3
35	15.16	35	80.80	7	70	35	3.5
40	17.32	40	92.30	8	80	40	4
50	21.65	50	115.40	9	90	50	5
60	26.09	60	138.50	10	100	60	6
70	30.30	70	161.60	15	150	70	7
80	34.60	80	184.70	20	200	80	8
90	39.00	90	207.80	25	250	90	9
100	43.30	100	230.90	30	300	100	10
120	52.00	120	277.00	40	400	200	20
140	60.60	140	323.20	50	500	300	30
160	69.20	160	369.40	60	600	400	40
200	86.60	180	415.60	70	700	500	50
300	129.90	200	461.70	80	800	600	60
400	173.20	250	577.20	90	900	700	70
500	216.50	275	643.00	100	1 000	800	80
600	259.80	300	692.70	200	2 000	900	90
800	346.40	350	808.10	300	3 000	1 000	100
1,000	433.00	500	1,154.50				

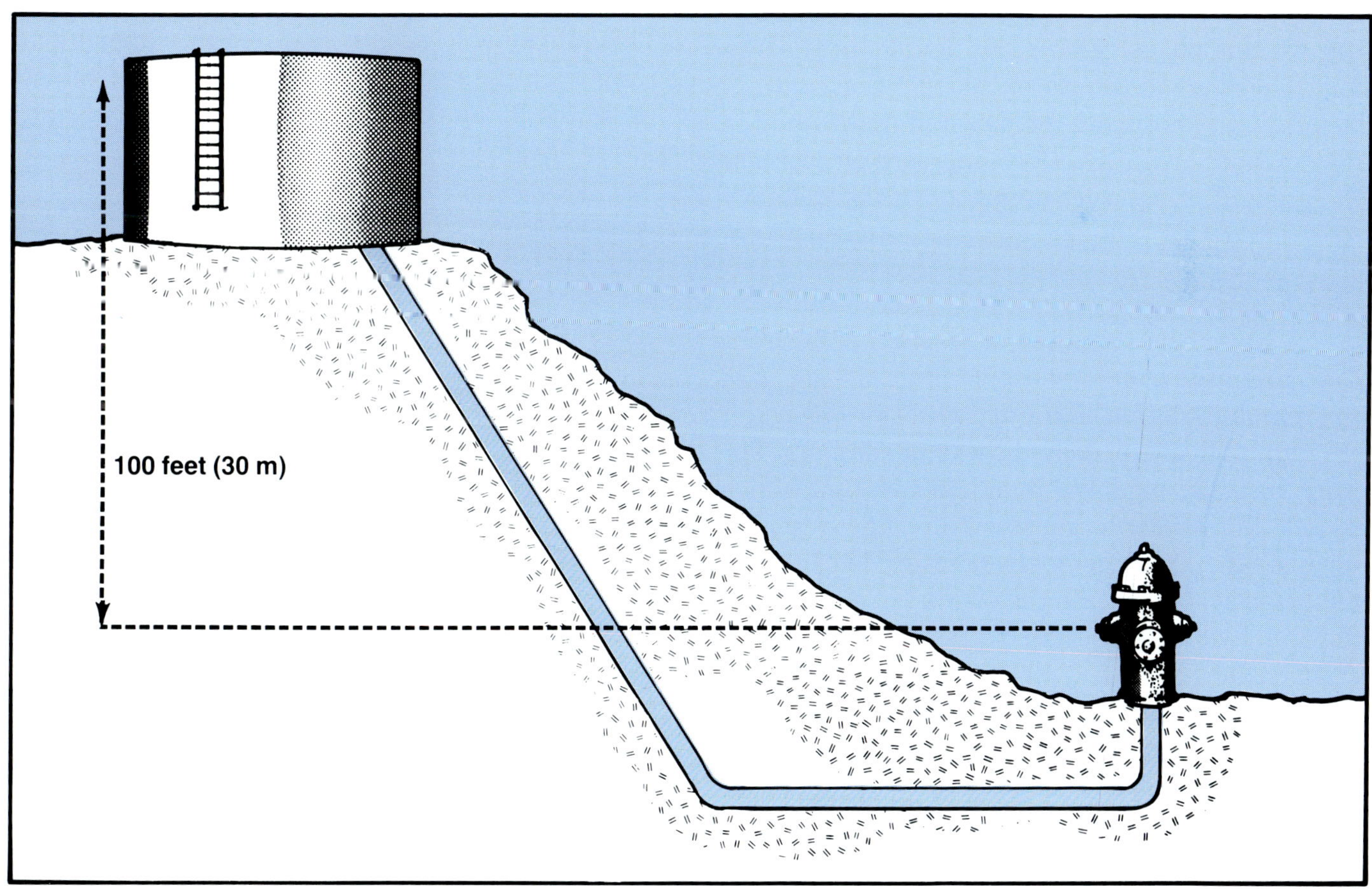

Figure 5.9 The head in this illustration is 100 feet (30 m). The head pressure is 43.4 psi (300 kPa).

Static Pressure

A waterflow definition of static pressure is stored potential energy available to force water through pipe, fittings, fire hose, and adapters. The word *static* means at rest or without motion. If water is not moving, the pressure exerted is static. A truly static pressure is seldom found in municipal water systems. There is always some flow in the pipes because of normal domestic or industrial needs. The pressure normally found in a water system before water flows from a hydrant is considered to be static pressure. Pressure on water may be produced by an elevated water supply, by atmospheric pressure, or by a force pump.

Normal Operating Pressure

A waterflow definition of normal operating pressure is pressure found in a water distribution system during normal consumption demands. Normal operating pressure is the pressure that occurs when some water is flowing (as soon as water starts to flow, static pressure no longer exists). The difference between static and normal operating pressure is the friction loss caused by water flowing through the various pipes, valves, and fittings in the system. The demands for water consumption continuously fluctuate during the day and night and the flow of water will, therefore, increase or decrease the amount of pressure available in the system.

Residual Pressure

A waterflow definition of residual pressure is that part of the total available pressure not used to overcome friction or gravity while forcing water through pipe, fittings, fire hose, and adapters. The word *residual* means a remainder or that which is left. For example, during a fire flow test, the term residual represents the pressure left in a distribution system within the vicinity of one or more flowing hydrants. Residual pressure in a water distribution system will vary according to a number of factors. These include water that may be flowing from one or more hydrants, water consumption demands, and the size of the pipe. An important point to remember about residual pressure is that it must be identified at the location where the reading is taken.

Flow Pressure (Velocity Pressure)

A waterflow definition of flow pressure is forward velocity pressure at a discharge opening when water is flowing. The rate of flow (velocity) of the water coming from a discharge opening produces a force called flow pressure or velocity pressure. Since a stream of water emerging from a discharge opening is not encased within a tube, it exerts pressure in a forward direction but not in a sideways direction.

The forward velocity can be measured by using a pitot tube and gauge as flow pressure. If the size of the opening is known, the flow pressure can be used to calculate the quantity of water flowing in gallons per minute (gpm) (liters per minute [L/min]). An example of flow pressure is one in which the forward velocity of a water stream exerts a pressure that can be read on a gauge.

Water Pressure Measurements

Water pressure is generally measured by a gauge containing a closed, curved tube known as a Bourdon tube. As pressure is applied to the open end of the tube, it will tend to straighten itself. This movement is communicated by a gearing arrangement to a pointer (Figure 5.10). The pointer indicates a relative pressure on a scale, usually

Figure 5.10 A Bourdon tube as it would be found inside a typical pressure gauge. *Courtesy of Span Instruments, Inc.*

calibrated in pounds per square inch or kilopascals. This pressure reading is referenced internally to atmospheric pressure (since the gauge is at rest or zero, the tube is actually at atmospheric pressure).

There are two general types of water pressure gauges: pressure gauge and compound gauge (Figure 5.11). A pressure gauge is capable of measuring positive pressure only. The needle will only move one way from zero. The compound gauge adds a second Bourdon tube and linkage, which enables it to measure either pressure or vacuum readings. In other words, it measures either positive or negative pressure. The compound gauge is normally used on the suction side of a fire pump.

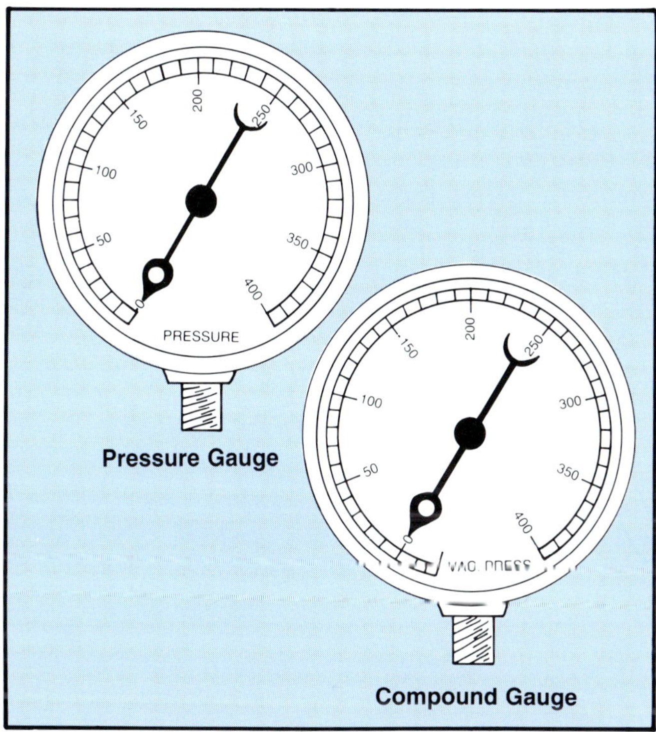

Figure 5.11 A pressure gauge measures only positive pressure. A compound gauge can measure either positive pressure or vacuum readings.

Both types of water pressure gauges are delicate instruments and are very susceptible to damage. Either freezing temperatures or an accumulation of dirt can make the gauge inoperative. More commonly, many gauges have gotten out of calibration, either from neglect and normal vibration of the motor pump apparatus, or from water hammer and strains that have damaged the mechanism itself. To make a quick check of the calibration, observe the gauge with a stopped, dry pump and be certain that it reads zero. This is especially important on a compound gauge since the scale from 0-30 inches (0 mm to 760 mm) of vacuum is very small and slight errors in the zero setting can create large errors in the vacuum reading.

Older gauges are equipped with gauge stabilizer valves, also known as dampening or weeper valves. If excessive fluctuations are noticed in reading the pressure gauges, the dampening valves can be partially closed to smooth out the reading. It is important to check frequently to make sure these valves have not been inadvertently closed and the gauge made inoperative. Many fireground pumping operations have been hampered when these valves have been left in a closed position. Newer gauges are filled with liquid to prevent these fluctuations. This reduces the need for any further dampening devices.

FIRE PUMPS — CONSTRUCTION AND PRINCIPLES OF OPERATION

High pressure water systems in industrial or specialized applications are capable of providing a sufficient volume of water at suitable pressure for fire fighting operations. A few well-developed municipal systems may also have this capability, but the majority of water systems are unable to maintain adequate pressure in the hydrant system for effective fire fighting. In order to produce effective fire streams, it is necessary to increase existing pressures with fire service pumps (Figure 5.12). Fire pumps are also necessary to provide

Figure 5.12 An example of a midship fire pump. *Courtesy of Hale Fire Pump Company.*

the pressures needed to supply attack lines from the water tank mounted on the apparatus. Modern pumpers must also be able to pump water from static sources such as lakes, streams, ponds, and rivers (Figure 5.13).

Figure 5.13 Static water supply sources, such as lakes and irrigation ditches, can supply large amounts of water for fire fighting efforts.

The earliest pumps used in the fire service were hand operated. To supply water under pressure, the firefighter pumped a handle, which operated a piston in a cylinder. This action forced the water out of the pump with enough velocity to carry some distance. These pumps were soon followed by the rotary pump, in which a hand crank caused a gear to rotate, again forcing the water out of the pump at a workable pressure. Both of these pumps are known as positive displacement pumps because a positive action takes place, forcing all the water and air out of the pump body with each operating cycle.

In most cases, modern fire apparatus is equipped with a centrifugal pump as its major pump. The centrifugal pump does not use positive action to force water from the pump; rather, it depends on the velocity of the water produced by centrifugal force to provide the necessary pump discharge pressure for effective operation. All fire pumps in use today fall into one of these two categories: centrifugal or positive displacement. Each type is discussed more fully in the sections that follow.

Positive Displacement Pumps

The positive displacement pump has largely been replaced by the centrifugal pump. Positive displacement pumps are still necessary, however, because they can pump air and centrifugal pumps cannot. For this reason, positive displacement pumps are used as priming pumps to get the water into the centrifugal pump during drafting operations. The basic principle of the positive displacement pump is a hydraulic law that stems from the near-incompressibility of water: *When pressure is applied to a confined liquid, the same outward pressure is equally transmitted within the liquid in all directions.* There are two basic types of positive displacement pumps: piston and rotary.

PISTON PUMPS

All piston pumps contain a piston that moves back and forth inside a cylinder. The pressure developed by this action causes intake and discharge valves to operate automatically and provide for the movement of the water through the pump. As the piston is driven forward, the air within the cylinder is compressed, creating a higher pressure inside the pump than the atmospheric pressure in the discharge manifold. This pressure causes the discharge valve to open and the air to escape through the discharge lines (Figure 5.14). This

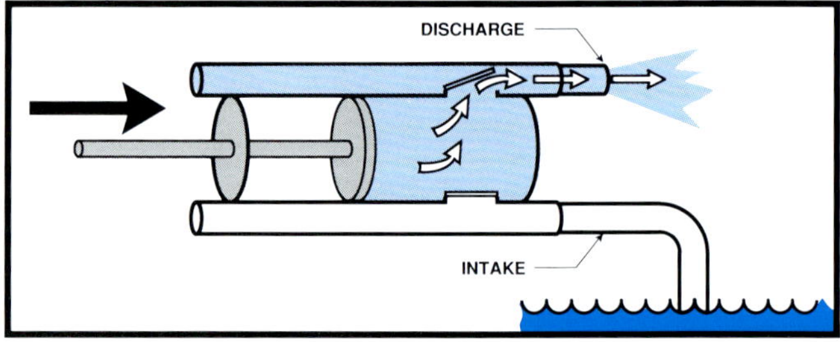

Figure 5.14 The higher pressure inside the pump causes the discharge valve to open, allowing air to escape through the discharge lines.

action continues until the piston completes its travel on the forward stroke and stops. At that point, pressures equalize and the discharge valve closes. As the piston begins the return stroke, the area in the cylinder behind the piston increases and the pressure decreases, creating a partial vacuum. At this time, the intake valve opens, allowing some of the air from the suction hose to enter the pump (Figure 5.15).

As the air from the suction hose is evacuated and enters the cylinder, the pressure within the hose and the intake area of the pump is reduced.

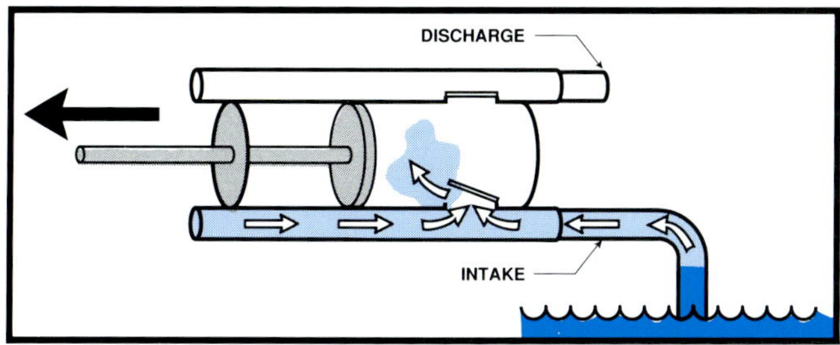

Figure 5.15 The partial vacuum created as the piston begins the return stroke causes the intake valve to open. This allows air from the suction hose to enter the pump.

Atmospheric pressure forces the water to rise within the hose until the piston completes its travel and the intake valve closes. As the forward stroke is repeated, the air is again forced out of the discharge. On the return stroke, more of the air in the intake section is removed and the column of water in the suction hose is raised. This action is repeated until all the air has been removed and the intake stroke results in water being introduced into the cylinder. The pump is now considered to be primed, and further strokes cause water to be forced into the discharge instead of air (Figure 5.16).

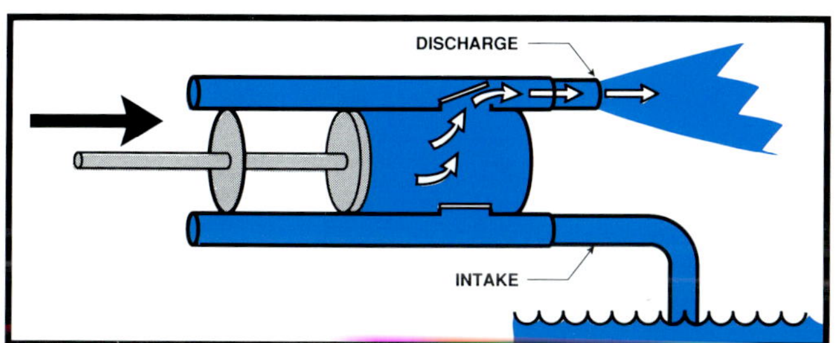

Figure 5.16 Once all the air has been evacuated, only water will be pushed through the pump.

The forward stroke causes water to be discharged and the return stroke causes the pump to fill itself with water again. This is known as a single-acting piston pump. Obviously, this would not produce a usable fire stream, since no water would be flowing during the return stroke, and the discharge would be a series of surges of water followed by an equal length of time with no water. A more constant stream can be produced with the addition of two additional valves. This is known as a double-acting piston pump because it both receives and discharges water on each stroke of the piston (Figure 5.17). Even with the double-acting pump, the output is a series of pressure surges

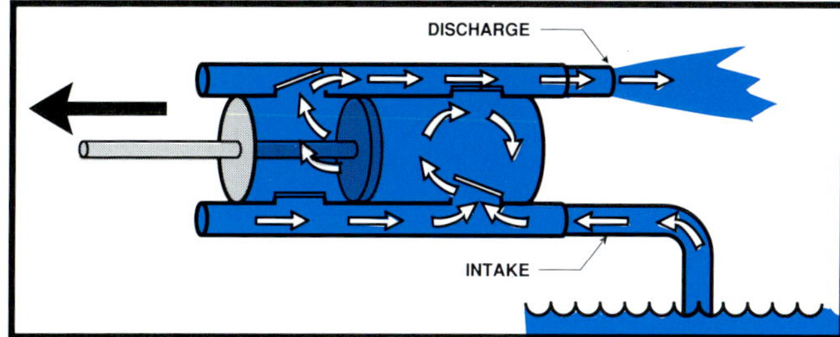

Figure 5.17 A double-action piston pump pushes water on both the forward and return strokes.

with two periods of no flow. These occur when the piston ends its travel in each direction.

Since the pump cylinder contains a definite amount of water, just that amount will be delivered by each stroke of the piston. The output capacity of the pump is determined by the size of the cylinder. In practice, it is more practical to build multicylinder pumps than one large single-cylinder pump. Multiple cylinders are more flexible and efficient, since some of them can be disengaged when the pump's full capacity is not needed. Multicylinder pumps also provide a more uniform discharge since the cylinders are arranged to reach their peak flows at different parts of the cycle.

Older large capacity piston pumps are equipped with a pressure dome or air chamber on the discharge to even out the pulses (Figure 5.18). During peak pressures, water is forced into the air chamber, compressing the air in the chamber until the air pressure trapped in the dome is equal to the water pressure being applied. When the pres-

Figure 5.18 The pressure dome on a 1926 Ahrens-Fox pumper. *Courtesy of OSU Fire Protection Society.*

sure in the pump discharge drops, the air pressure stored in the chamber forces the water into the line, thus counteracting the drop in pressure from the pump.

Piston pumps have not been used as a major fire pump in pumpers for many years. In addition to the problems of the pulsating fire stream, they were very susceptible to wear. The efficiency of the piston pump depends upon a close tolerance between the piston and the walls of the cylinder. As the piston wears, water can escape from the discharge side of the pump to the intake, and the capacity of the pump is reduced accordingly. Wear also takes place on the walls of the cylinder. Lastly, pumping contaminated water can quickly damage the pump. For all these reasons, the piston pump has been phased out of the fire service as a high capacity pump.

Although piston pumps are no longer used as high capacity pumps, they are still used in high pressure fire fighting (Figure 5.19). One manufacturer uses a multicylinder, PTO-driven pump to provide pressures up to 1,000 psi (7 000 kPa) for high pressure fog lines. This manufacturer overcomes the wear problem by using rubber pistons, which are easily replaced, and ceramic walls in the cylinder, which are very resistant to wear. This application also requires a dependable relief valve, since pressures would quickly build to dangerous levels if the discharge flow was interrupted.

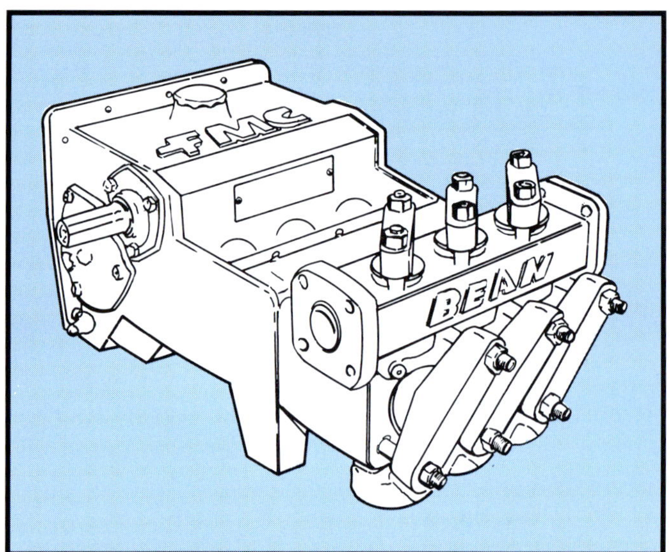

Figure 5.19 Piston pumps are now used only for small capacity high pressure pumping applications.

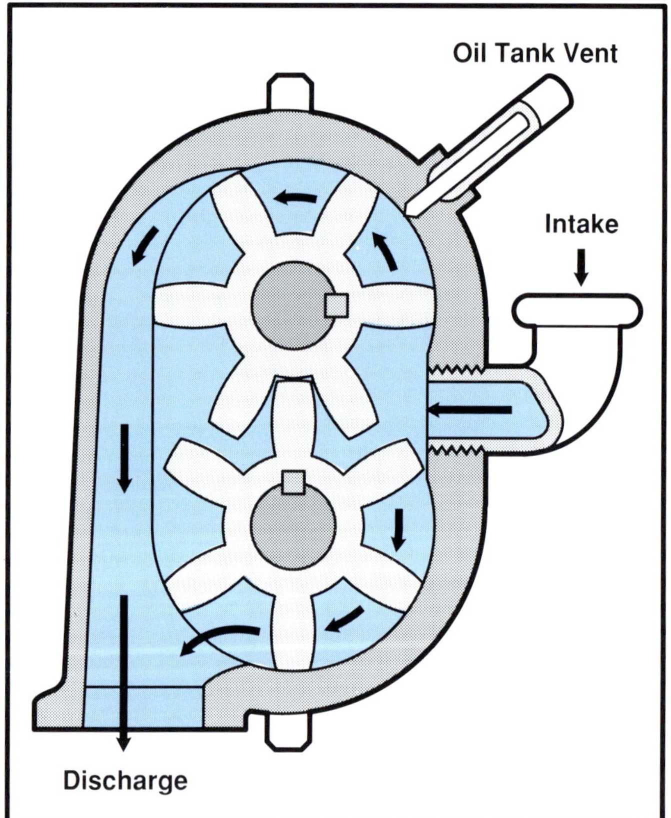

Figure 5.20 Another style of rotary gear priming pump.

ROTARY PUMPS

Rotary-type pumps are the simplest of all fire apparatus pumps from the standpoint of design. Older fire apparatus used these pumps extensively as the major pump, but in recent years, their use has been confined to small capacity booster-type pumps and priming pumps. Most of the rotary-type pumps in use today are either of the rotary gear or rotary vane construction.

Rotary Gear Pump. The rotary gear pump consists of two gears that rotate in a tightly meshed pattern inside a watertight case. The gears are constructed so they contact each other and are in close proximity to the case (Figure 5.20). With this arrangement, watertight and airtight pockets are formed by the gears within the case as they turn from the intake to the outlet. As each gear tooth reaches the discharge chamber, the air or water contained in that pocket is forced out of the pump. As the tooth returns to the intake side of the pump, the gears are meshed tightly enough to prevent the water or air that has been discharged from returning to the intake.

The total amount of water that can be pumped by a rotary gear pump depends upon the size of the pockets in the gears and the speed of rotation. The rotary gear pump is a positive displacement pump. This is because each pocket in the gears contains a definite amount of water, and because this water is forced out of the pump each time the gears turn with a positive action.

If the pump is trying to move more water than the discharge lines in use can take away, pressure builds up. An adequate pressure relief device must be provided to handle any excess pressure that may result.

Like the piston pump, the rotary gear pump is very susceptible to damage from normal wear and from pumping contaminated water. To prevent damage to the casings, most gear pumps use bronze or another soft metal in the gears, which can easily be replaced, and a strong alloy, such as cast iron, in the casing.

Some rotary gear pumps have power delivered to one gear, which then drives the other gears. In this case, the drive gear is usually made of steel with inserts of bronze to provide the strength needed to handle the torque that will be developed. Other designs use steel pilot gears mounted outside the case connected to the shaft that drives the pump gears.

Rotary Vane Pump. The rotary vane pump is constructed with movable elements that automatically compensate for wear and maintain a tighter fit with closer clearances as the pump is used (Figure 5.21). In this type of pump, the rotor is mounted off-center inside the housing (Figure 5.22). The distance between the rotor and the housing is much greater at the intake than it is at the discharge. The vanes are free to move within the slot where they are mounted. As the rotor turns, the vanes are forced against the housing by centrifugal force. When the surface of the vane that is in contact with the casing becomes worn, centrifugal force causes it to extend further, thus automatically maintaining a tight fit. This self-adjusting feature makes the rotary vane pump much more efficient at pumping air than a standard rotary gear pump.

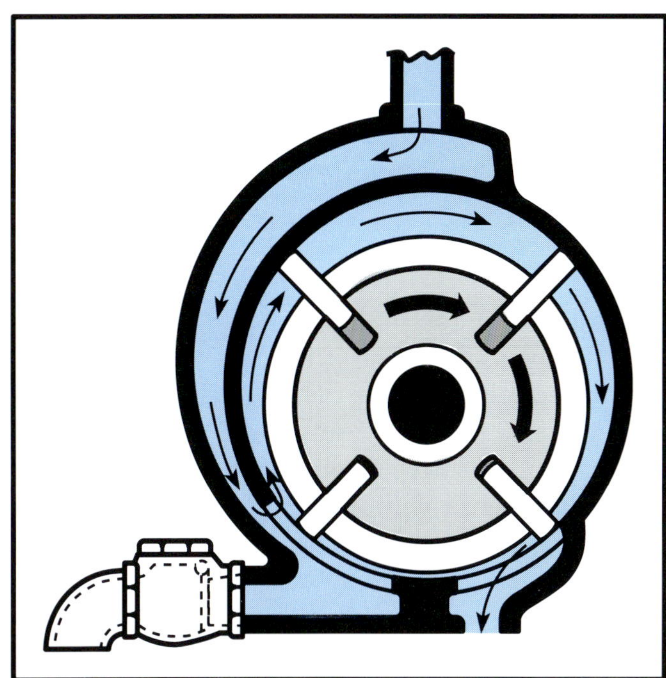

Figure 5.21 Schematic of a rotary vane pump.

Figure 5.22 Interior of a rotary vane pump. Note the eccentric position of the rotor.

As the rotor turns, air is trapped between the rotor and the casing in the pockets formed by adjacent vanes. As the vanes turn, this pocket becomes smaller, which compresses the air and causes pressure to build up. This pocket becomes even smaller as the vanes progress toward the discharge opening. At this point, the pressure reaches its maximum level, forcing the trapped air out of the pump. The air or water is prevented from returning to the intake by the close spacing of the rotor at that point. As in the rotary gear pump, the air being evacuated from the intake side causes a reduced pressure (similar to a vacuum) and water is forced into the pump by atmospheric pressure until the pump fills with water. At this point, the pump is primed and will force water out of the discharge in the same manner as air was forced out.

Centrifugal Pumps

Nearly all fire apparatus being constructed at the present time utilize the centrifugal pump as their major pump. The centrifugal pump is classified as a nonpositive displacement pump since it does not pump a definite amount of water with each revolution. Rather, it imparts velocity to the water and converts it to pressure within the pump itself. This gives the pump a flexibility and versatility that has made it popular with the fire service. It has also virtually eliminated the positive displacement pump as a major fire pump in fire apparatus.

PRINCIPLES OF OPERATION AND CONSTRUCTION OF THE PUMP

In theory, the operation of a centrifugal pump is based on the principle that a rapidly revolving disk tends to throw water introduced at its center toward the outer edge of the disk. The faster the disk is turned, the farther the water is thrown, or the more velocity the water has. If the water is contained at the edge of the disk, the water at the center of the container will begin to move outward. The velocity created by the spinning disk is converted to pressure by confining the water within the container. The water is limited in its movement by the walls of the container and will move upward since it has no other place to go. This shows that pressure has been created in the water. The height to which it rises, or the extent to which it overcomes the force of gravity, depends upon the speed of rotation.

Fundamentally, the centrifugal pump consists of two parts: an impeller and a casing. The impeller transmits energy in the form of velocity to the water. The casing collects the water and confines it to convert the velocity to pressure. Then it directs the water to the discharge of the pump (Figure 5.23).

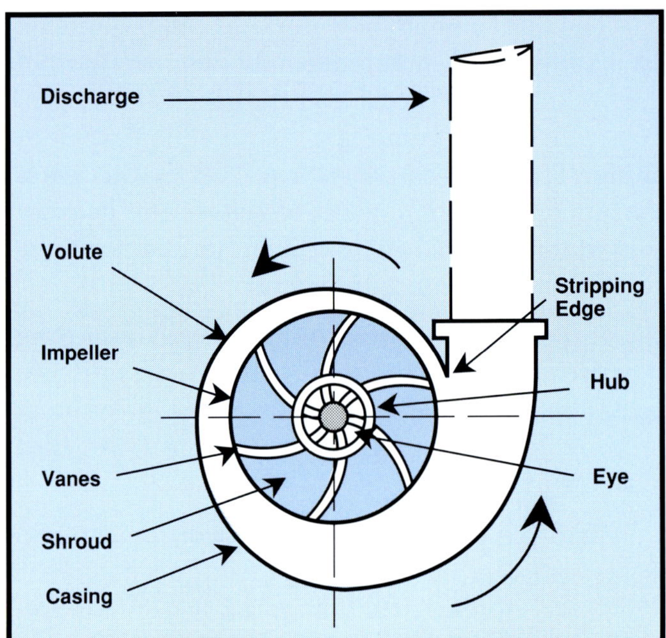

Figure 5.23 This diagram highlights the major parts of a centrifugal pump.

The impeller rotates very rapidly within the casing. Water is introduced from the intake into the eye of the impeller (Figure 5.24). As the water comes into contact with the vanes of the impeller, it is thrown by centrifugal force to the outside of the impeller (Figure 5.25). This water is confined in its travel by the shrouds of the impeller, and increases in velocity with the speed of the impeller.

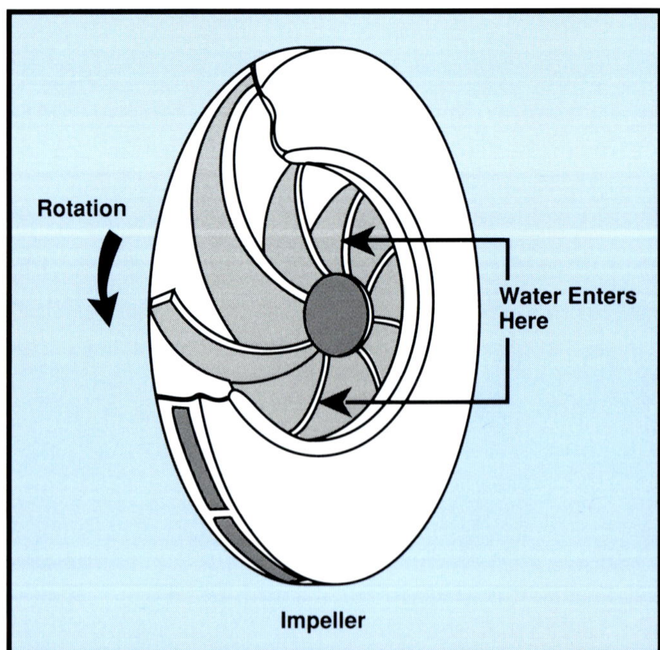

Figure 5.24 Cutaway diagram showing where water enters a centrifugal pump.

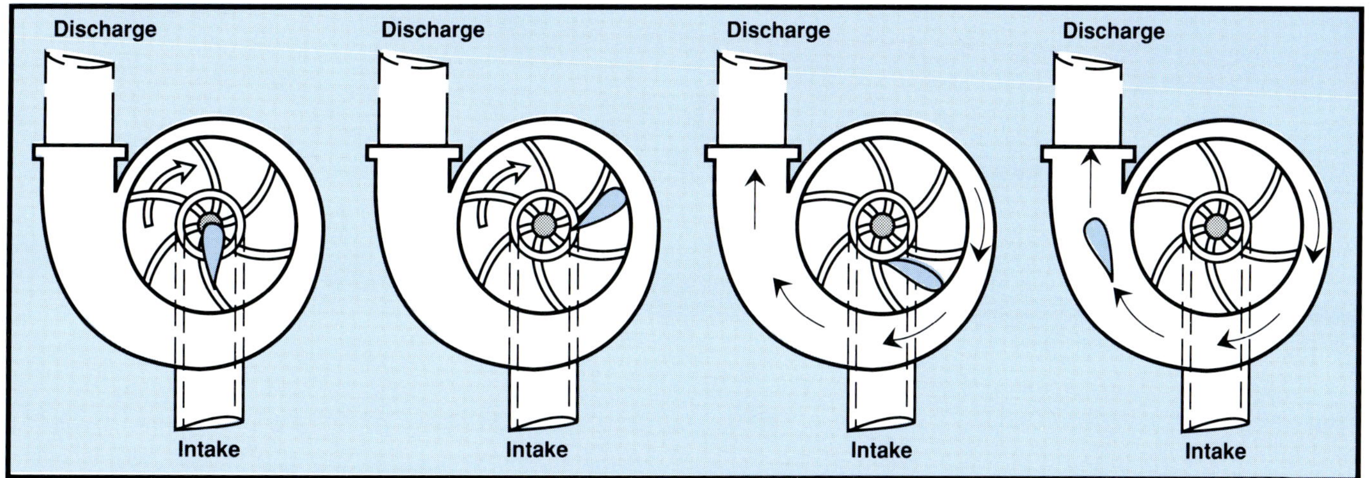

Figure 5.25 Schematic showing the path a single drop of water takes through a centrifugal pump.

The impeller is mounted off-center in the casing. This placement creates a water passage that gradually increases in cross-sectional area as it nears the discharge outlet of the pump. This section of the pump is known as the volute. The increasing size of the volute is necessary because the velocity of water passing through the volute increases as it approaches the discharge outlet. The gradually increasing size of the waterway reduces the velocity of the water, thus enabling the pressure to build up proportionately.

There are three main factors that greatly influence a centrifugal fire pump's discharge pressure: the amount of water being discharged, the speed at which the impeller is turning, and the pressure that the water has when it enters the pump.

If the discharge outlet is large enough in diameter to allow the water to escape as it is thrown from the impeller and collected in the volute, this pressure buildup will be very small. If the discharge outlet is closed off, a very high pressure buildup will result. With all other factors remaining constant, the amount of output pressure that a pump may develop is directly dependent upon the volume of water it is discharging. Simply stated, the greater the volume of water being flowed, the lower the discharge pressure will be.

Since the discharge pressure in the centrifugal pump results from the velocity and the amount of water that is moving, the speed of the impeller is an important factor in determining the pressure to be developed. The greater the speed of the pump, the greater the pressure that will be developed. This increase is approximately equal to the square of the change in impeller speed. For example, doubling the speed of the impeller results in four times as much pressure if all other factors remain constant.

A third factor that influences the discharge pressure is the intake pressure. By its nature, there is no positive mechanical blockage between the intake and the discharge outlet of a centrifugal pump. Isolation of discharge pressure from intake pressure in the impeller occurs entirely by the velocity of the moving water. Water will flow through a centrifugal pump even if the impeller is not turning. When water is supplied to the eye of the impeller under pressure, it moves through the impeller by itself. Any movement of the impeller increases both the velocity of the water and the corresponding pressure buildup in the volute. Since the incoming pressure adds directly to the pressure being developed by the pump, incoming pressure changes will be reflected in the discharge pressure (Figure 5.26 on next page).

Since the centrifugal pump depends on the velocity of the water to move water through the pump, it is unable to pump air and is not self-priming. In order for a centrifugal pump to draft water, some type of external primer must be supplied to remove the air and allow atmospheric pressure to force the water into the pump.

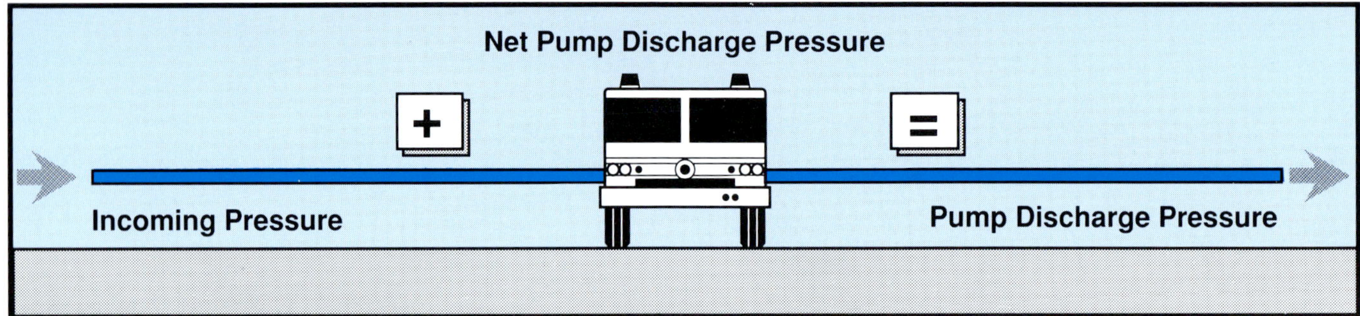

Figure 5.26 The pressure coming into a pump, plus the pressure created by the pump itself, equals the pump discharge pressure.

PUMP WEAR RINGS AND PACKING

Although there is no positive mechanical isolation of the discharge area of the pump from the intake in the impeller, the velocity of the water moving through the impeller prevents any water in the discharge from escaping back into the intake. Since the pressure in the volute is much higher than that in the intake side of the pump at the eye of the impeller, a very close tolerance must be maintained between the pump casing and the hub of the impeller. This opening is usually limited to .01 inch (0.25 mm) or less. Any increase in the opening lessens the pump's effectiveness.

Impurities, sediment, and dirt are present in every water supply. As these pass through the pump, they cause wear when they come in contact with the impeller, which is turning nearly 4,000 rpm when the pump approaches its capacity. This process is greatly accelerated when it is necessary to pump water with a high sand content. The sand particles passing between the impeller and the pump casing act like sandpaper in wearing down the metal surfaces. As the gap increases, greater amounts of water are allowed to slip back into the intake discharge and are not available at the discharge. Eventually, the pump is no longer able to supply its rated capacity. The first indication that wear is becoming a problem is when increased engine rpm is required to pump the rated capacity in pump tests.

To restore the capacity of the pump without replacing the pump itself, replaceable wear rings or clearance rings are provided in the pump casing to maintain the desired spacing between the hub of the impeller and the casing (Figure 5.27). If the hub of the impeller is also worn down, it is possible to install smaller wear rings to compensate for the smaller size and maintain the proper clearance.

The centrifugal pump differs from the positive displacement in that no harm will result from shutting off all discharges for short periods of time. When discharges are shut off, the energy being supplied to the impeller is dissipated in the form of heat as the water within the pump is allowed to churn. If this situation is allowed to continue for extended periods of time, the water in the pump becomes quite hot and the metal parts tend to expand. If the wear rings and the impeller expand too much, they may come in contact with each other and the friction of the two surfaces rubbing together may cause even more heat. In extreme cases, the wear rings may seize, causing serious pump damage. The best insurance against this happening is to ensure that some water is moving through the pump at all times.

Figure 5.27 Wear rings, which are replaceable, are vital in maintaining the proper spacing between the hub of the impeller and the casing.

The impellers are fastened to a shaft that connects to a gear box. The gear box transfers the necessary energy to spin the impellers at a very high rate of speed. At the point where the shaft passes through the pump casing, a tight seal must be maintained to prevent air leaks that could interfere with a drafting operation. Packing rings are used to make this seal in most fire pumps. The most common type of packing is a material made of rope fibers impregnated with graphite. The material is pushed into a stuffing box by a packing gland driven by a packing adjustment mechanism.

As packing rings wear with the operation of the shaft, the packing gland can be tightened and the leak controlled. Where the packing rings come into contact with the shaft, heat is developed as a result of friction against the packing. To overcome this, a lantern ring (spacer) is supplied with the packing to provide cooling and lubrication. A small amount of water will leak out around the packing and prevent excessive heat buildup. If the packing is too tight, water is not allowed to flow between the packing and the shaft, and excessive heat buildup results. A scored shaft will prevent even new packing material from making a good seal.

If the packing is too loose, as indicated by an excessive amount of water leaking from the pump during operation, air leaks will adversely affect the pump's ability to draft. The adjustment of this packing must be made carefully and according to the manufacturer's instructions. In general, however, when the pump is operating under pressure, water should drip from the packing glands but it should not run in a steady stream.

The packing only receives the needed water for lubrication if the pump is full of water and operating under pressure.

WARNING

If the pump is operated with no water in it for any length of time, damage to the shaft will likely result.

Some fire departments make a practice of keeping the pump drained between fire calls. If the pump is not used for extended periods of time, the packing may dry out and excessive leakage will result. Adjustment of the packing after a period of dryness should not be made until the pump has operated for a period of time under pressure and the packing has had a chance to seal properly.

Pump Mounting and Drive Arrangements

There are many types of mounting arrangements and pump drive systems to consider in building a fire department pumper. Cost, appearance, space required, ease of maintenance, and tradition all enter into the decision, but the most important consideration is the use that the system will receive. Each system has certain characteristics that make it more or less adaptable to a particular fire department's needs.

AUXILIARY ENGINE DRIVEN PUMPS

Separate engines are used to drive fire pumps in various specialized applications, such as some airport crash trucks, but the most common in the fire service is as a portable pump. They are also used to supply pressure to hose reels or units designed to fight fires in fields, woods, and brush (Figure 5.28).

Figure 5.28 An auxiliary engine driven pump used on brush apparatus. *Courtesy of MARCO, Inc.*

Auxiliary engine driven pumps offer the maximum amount of flexibility. With a separate engine, the pump can be mounted anywhere on the apparatus it will fit. The pump pressure is independent of the apparatus drive system, which makes it ideal for pump and roll operations. (Pump and roll refers to pumping water while the apparatus is in motion.) It can even operate away from the fire apparatus by being used as a portable unit, skid, or trailer mounted pump and engine. Maintenance of the pump is simplified by its accessibility. Since it is built as a unit with most of the accessories built in, this type of pump is generally less expensive to build and purchase than some of the others.

A typical auxiliary engine driven pump used as a portable or tank mounted unit has an engine 5-89 that limits the capacity to less than 500 gpm (2 000 L/min). This size makes it unsuitable for use as the primary fire pump on a fire department pumper. Higher capacities can be obtained by using a larger engine, but this results in a drastic increase in weight. The heavier weight usually requires going to some type of trailer mounting.

CROSS-MOUNTED ENGINE DRIVE

A separate engine-driven pump rated at 500 to 750 gpm (2 000 L/min to 3 000 L/min) has been introduced into the fire service in recent years. It is usually mounted midship on the apparatus. This pump has pump-and-roll capability, meets UL and ISO certification as a Class A pumper, and is considered an economical multipurpose unit. It is especially useful for the small volunteer fire department that needs a unit that can pump and roll for wildland fires, yet still supply enough water for rural structural fire fighting (Figure 5.29).

Figure 5.29 A midship-mounted auxiliary engine driven pump. Note radiator below pump panel.

POWER TAKE-OFF DRIVE

In this arrangement, the pump is driven by a transmission mounted power take-off unit through a universal joint shaft (Figure 5.30). Proper mounting of these units is essential for dependable and smooth operation. The pump gear case must be mounted in a location that allows for a minimum of angles in the drive shaft. At the same time, it must not extend far enough below the chassis to be readily damaged as the unit travels on or off the road. Some type of skid plate can be used to provide protection if the pump extends too far below the chassis.

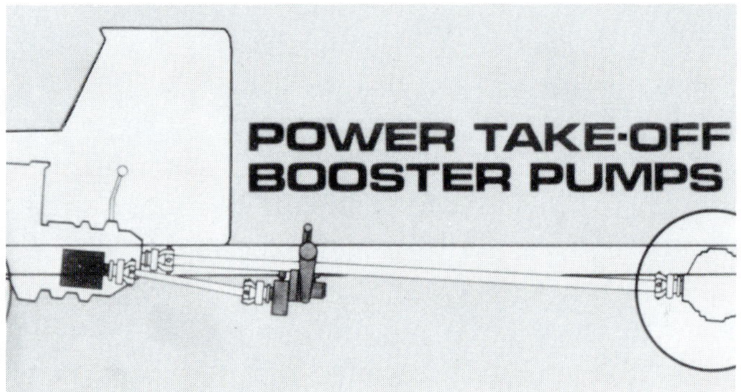

Figure 5.30 A power take-off drive train arrangement. *Courtesy of Hale Fire Pump Company.*

The power take-off unit is powered by an idler gear in the truck transmission. The speed of the shaft is independent of the gear in which the road transmission is operating when the pump is in use, but it is under the control of the clutch. When the operator disengages the clutch to stop momentarily or to change gears, the pump will also stop turning. The PTO pump does permit "pump and roll" operation, but it is not as effective as the separate engine unit.

Since the pressure being developed is determined by the speed of the engine, the pressure will change when the driver changes the vehicle speed. If the unit is designed for pump and roll operation, a pressure gauge should be mounted inside the cab in full view of the driver, and the vehicle should be driven by the pressure gauge instead of the speedometer while hoselines are in operation.

The conventional power take-off unit limits the capacity of the pump to about 500 gpm (2 000

L/min). This is because the PTO unit is mounted on the side of the transmission and the maximum strain this housing can take limits the available horsepower to approximately 70 hp (52 kw). In recent years, some manufacturers are providing "full torque" power take-offs that permit the installation of large-capacity pumps. This type of power take-off is especially common on some of the modern automatic transmissions where the flywheel of the engine drives a PTO unit. This provides enough torque to drive all but the very largest fire pumps.

Since its capacity is severely limited, the conventional PTO unit is used primarily on brush trucks, minipumpers, and on some mobile water supply apparatus. The full torque power take-off is used increasingly on larger pumpers and allows for pump and roll flexibility with high-capacity pumps.

FRONT MOUNT PUMPS

On some trucks, the bumper is extended and a pump is mounted between the bumper and the grill (Figure 5.31). This unit is driven through a gear box and a clutch connected by a universal joint shaft to the front of the crankshaft. The gear box uses a step-up gear ratio, which causes the impeller of the pump to turn faster than the engine does. This ratio is set to match the torque curve of the engine to the rotation speed required for the impeller to deliver the pump's rated capacity. It is usually between 1½ and 2½:1. The maximum capacity that can be used depends on the limitations of the engine driving it, but typically can go as high as 1,000 gpm (4 000 L/min). To use a front

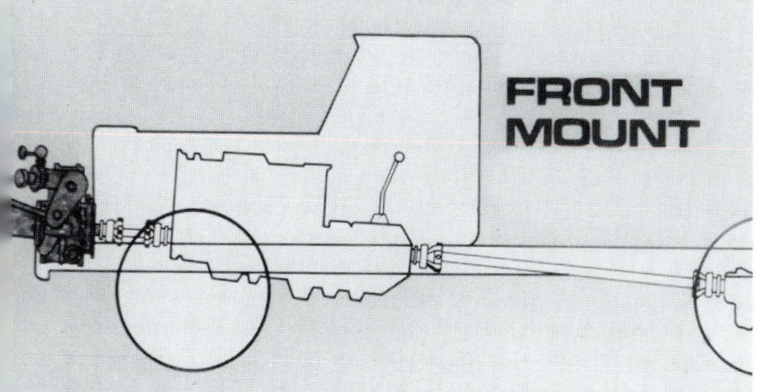

Figure 5.31 A front mount pump drive train arrangement. *Courtesy of Hale Fire Pump Company.*

mount pump, a chassis should have a front mount PTO option that provides a coupling to the front of the crankshaft. The chassis should also have an opening in the radiator for the drive shaft when it is required.

One disadvantage of the front mount pump is that the pump and gauges are susceptible to freezing in cold climates. External lines circulate radiator coolant through the pump body and keep it from freezing. Gauges and connecting pipes, on the other hand, may suffer damage during extremely cold weather. When possible, it is desirable to enclose the gauges and protect them from the weather.

The front mounted pump is in a vulnerable position in the event of a collision. For this reason, it is important that the chassis and frame be modified to provide a solid base of protection for the pump. Where they have been properly installed, front mount pumps have been known to withstand accidents that have done considerable damage to the apparatus, even front-end collisions.

Front mount pumps can be used for pump and roll operations, since the pump is independent of all parts of the vehicle drive system except the engine. Again, if the unit is to be used for pump and roll, a pressure gauge is needed inside the cab so the driver can use the gauge for a reference instead of using the speedometer.

The front mount pump is engaged and controlled from the pump location itself. This puts the pump operator in a vulnerable spot, namely, standing in front of the vehicle while operating the pump. Therefore, it is essential that a lock be provided to prevent the road transmission from being engaged while the pump is operating. This lock is needed with either a manual or an automatic transmission and should always be used when the pump is in service. Since the operating lever that engages the pump is located at the pump, a warning light in the cab should be provided to alert the driver that the pump is engaged. If the vehicle is driven with the pump turning, damage to the pump will result. Damage occurs either to the packing from running the pump dry, or to the pump itself from overheating the water by churning it inside the pump with no water flowing.

MIDSHIP TRANSFER DRIVE

Most fire department pumpers have the fire pump mounted laterally across the frame behind the engine. Power is supplied to the pump through the use of a split-shaft gear case (transfer case) located in the drive line between the transmission and the rear axle (Figure 5.32). By shifting a gear and collar arrangement inside the gear box, power can be diverted from the rear axle and transmitted to the fire pump. The pump is then actually driven by either a series of gears or a drive chain (Figure 5.33).

The gear ratio is set to match the engine torque curve to the speed of rotation required for the impeller to deliver the rated capacity of the pump. This ratio is arranged so that the impeller turns faster than the engine, usually $1\frac{1}{2}$ to $2\frac{1}{2}$ times as fast. The maximum capacity that can be obtained by this system is limited only by the engine horsepower and the size of the pump. Most fire pumps can operate over a range of capacities. These capacities are determined by the piping arrangement and gear ratio used. For example, a particular two-stage centrifugal pump, popular in the fire

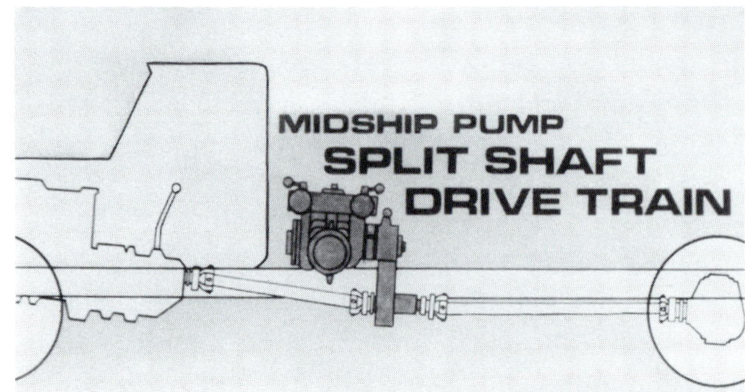

Figure 5.32 A midship pump drive train arrangement. *Courtesy of Hale Fire Pump Company.*

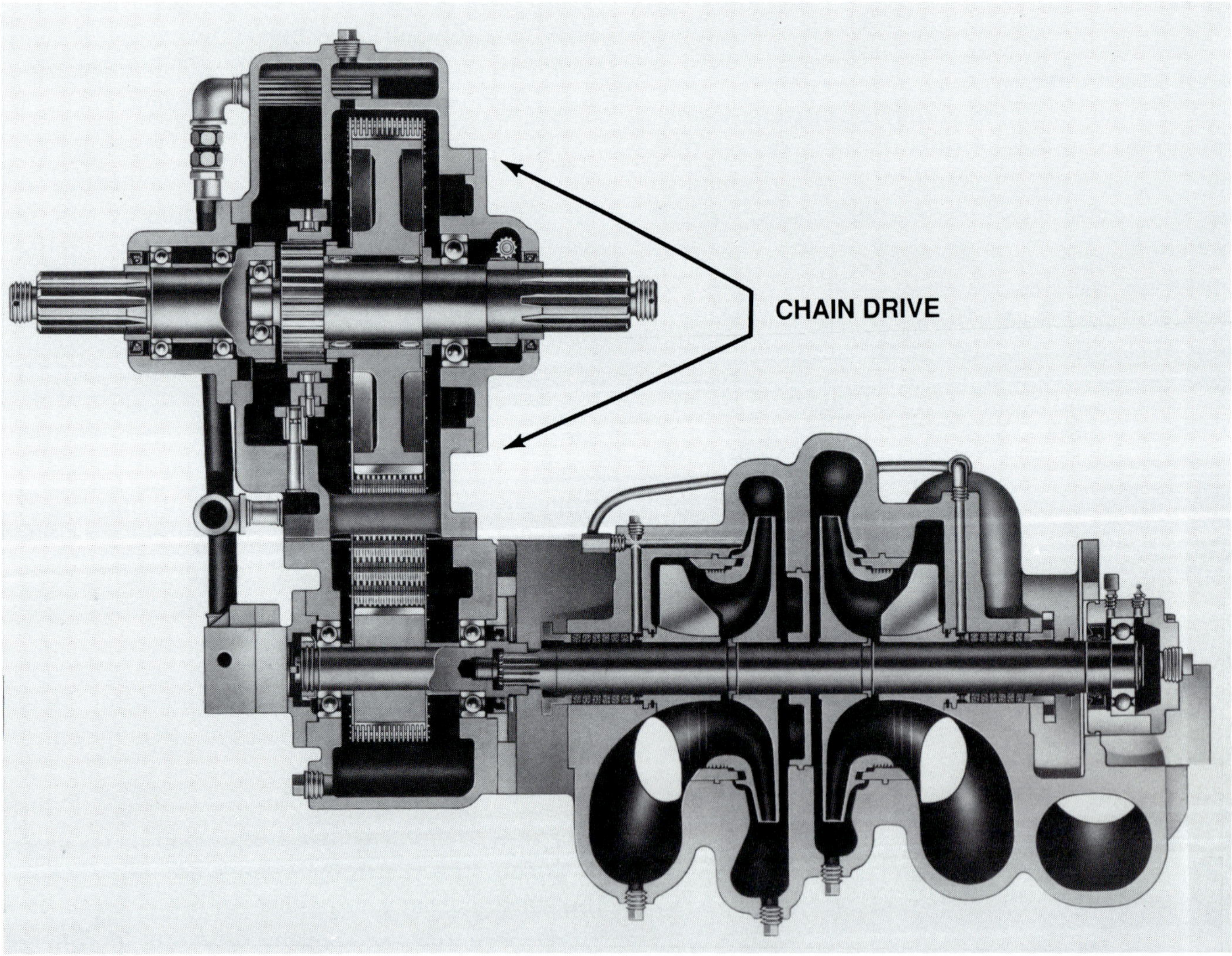

CHAIN DRIVE

Figure 5.33 Cutaway showing construction of chain drive for a fire pump. *Courtesy of Waterous Co. (Photo inverted for emphasis.)*

5-89

service, can be rated anywhere from 500 to 1,500 gpm (2 000 L/min to 6 000 L/min) with no major changes to the pump itself.

The normal arrangement is for the transfer case to be controlled from inside the cab of the apparatus. This can be done by mechanical linkage, or electrical, hydraulic, or air operated controls (Figure 5.34). If the pumper is so equipped, the operator should engage the pump and put the road transmission in the proper gear before leaving the cab. If the road transmission is not placed in the correct gear, the pump will not turn at the needed rpm to operate effectively. Since the transmission will be turning, most apparatus will register the engine speed in road miles per hour on the speedometer when the pump is operating. Check to see that the transmission is in the correct gear by observing the speedometer reading after the pump is engaged. With the engine idling and the pump engaged, most speedometers will read between 10 to 15 mph (16 km/h to 24 km/h), depending on the apparatus.

Figure 5.34 Some transfer controls have lights that indicate whether the transfer is complete. 5-89

To prevent damage to the gears, disengage the clutch and place the road transmission in neutral while the power transfer is made. Some power transfer arrangements are provided with a manual override in case of difficulty with the power unit. Operators should practice using this manual override frequently to be prepared for an emergency.

Since engaging the pump involves transferring the power from the rear axle coupling shaft to the pump drive gears, it is not possible to transmit power to the rear axle while the pump is engaged. This eliminates the possibility of a pump and roll operation with a conventional gearing arrangement. One manufacturer has added additional gears to the gear box to permit movement of the apparatus while the pump is operating at a reduced capacity.

To prevent an automatic transmission gear selector from moving during a pumping operation, or a manual transmission from slipping out of gear, a lock is provided on the transmission or shift lever to hold it in the proper gear for pumping. It is also possible to operate the pump shift control and not have the gears complete their travel. If this happens, the vehicle may begin to move as the engine rpm is increased to build up the pressure. To prevent movement, some apparatus are equipped with an indicator light on the dash, and the pumping operation should not begin until the green indicator light goes on. Later model automatic transmissions for fire apparatus should be equipped with a top gear lockup to prevent automatic shifting while the pump is engaged.

Pump Piping

Since priming the pump involves removing all or most of the air from the pump, any air trapped in the pump during the priming operation can prevent a successful drafting operation. For this reason, all intake lines to a centrifugal pump are normally located below the eye of the impeller, and no part of the piping will be above this point. The single exception to this may be the tank-to-pump line where the water will be moving under the natural pressure of gravity. The primer is tapped into the pump at a high point and a valve (priming valve) is used.

INTAKE PIPING

The majority of all fires are fought with the water carried in the fire apparatus tank. Therefore, it is important that the piping from the tank to the pump be large enough to allow for adequate streams to make an effective initial attack. NFPA 1901, *Standard for Automotive Fire Apparatus*, states that piping should be sized so that pumpers with a capacity of 750 gpm (3 000 L/min) or less should be capable of flowing 250 gpm (1 000 L/min) from their booster tank. Pumpers with capacities greater than 750 gpm (3 000 L/min) should be able to flow at least 500 gpm (2 000 L/min).

Some older pumpers had piping as small as 2 inches (50 mm) from the tank to the pump and were hard pressed to supply two 1½-inch (38 mm) attack lines, much less larger lines. Many pumpers today are equipped with tank-to-pump lines as large as 4 inches (100 mm) in diameter. Mobile water supply units may have multiple tank-to-pump lines. Many pumps are equipped with check valves in these lines. The check valves prevent damage to the tank if the tank-to-pump valve inadvertently opens when water is being supplied to the pump under pressure such as during a relay. If this line has a check valve in it, it will be impossible to fill the tank through the pump by opening the tank-to-pump valve. The tank-to-pump valve must be maintained in good condition. If it leaks, priming the pump when the tank is empty will be impossible since air from the tank will be drawn into the pump and no vacuum will be established. It is also possible for the water to be drawn out of the tank during a drafting operation, and prime will be lost when the tank is emptied.

Additional intake lines, usually gated, are provided for use in relay operations or anytime water is being received through supply lines. Most of these intake openings have threads to use 2½-inch (65 mm) hose couplings (Figure 5.35). The amount of flow that can be obtained is determined by the pipe from the valve to the pump inlet. If the pipe is not larger than 2½ inches (65 mm) and if it contains 90-degree bends or T-fittings, the friction loss may be enough to limit the flow through these intakes to 250 gpm (1 000 L/min). If 3-inch (77 mm) pipe is used and care is given to the fittings, it may be possible to flow as much as 450 gpm (2 000 L/min) through one of these intake openings.

Figure 5.35 The gated intake is most commonly found to the side and slightly below the large intake. Some gated intakes are located on the front or the rear of the vehicle.

DISCHARGE PIPING

According to NFPA 1901, every Class A pumper must have at least one 2½-inch (65 mm) discharge for every 250 gpm (1 000 L/min) capacity of the pump (Figure 5.36). These lines are usually equipped with a locking ball valve, and they should always be kept locked when they are open to prevent movement. This is especially true when it has been necessary to “gate down” a line to reduce the discharge pressure being supplied from it.

Figure 5.36 A typical 2½-inch (65 mm) discharge.

When multiple attack lines are in use with different pressures required, the only way to supply them is to set the pump discharge pressure to the highest pressure needed. Then each of the other lines should have the valve partially closed until the friction loss in the valve is sufficient to provide the desired pressure at the hoseline. Individual line pressure gauges or flowmeters are essential to do this properly. Without individual line gauges, providing good fire streams becomes a matter of guesswork or constant feedback from the nozzle.

Most pumpers carry additional discharge lines over and above those that are required. They are used frequently for various preconnected attack lines of different sizes and lengths. A master stream device is often carried preconnected.

A tank fill line should be provided from the discharge side of the pump. This allows the tank to be filled without making any additional connections when the pump is drafting. It also provides a means of replenishing the water carried in the tank after the initial attack has been made from the water tank on the apparatus. In some multiple-stage fire pumps, this line is taken from the first stage, which provides for reduced pressures to the tank as a measure of safety.

The tank fill line can be used to circulate water through the pump to prevent overheating when no lines are flowing. In a two-stage pump with the fill line coming off the first stage, overheating can still result in the second-stage portion of the pump. A circulator valve is more effective for preventing overheating. The circulator valve is connected to the discharge side of the pump that enables water to be dumped into the tank or outside the tank on the ground. One valve in common use has a tank position and spill position to indicate what happens to the water, but there is no closed position designated on the valve. Either position allows air to be introduced into the pump, which will make drafting difficult if not impossible. Although it is not so designated, this particular valve is closed when the handle is in the center position, and it should be in that position when drafting.

Some pumpers have a booster line cooling valve that serves the same function as the circulator valve by diverting a portion of the discharge water into the tank. While either of these arrangements serves the purpose for normal pumping operations, the piping is done with small copper tubing, and the flow is limited to approximately 10 to 20 gpm (40 L/min to 80 L/min) (Figure 5.37). During prolonged operations with intermittent flows, or when operating at very high pressures, this may not be enough water to keep the pump cool. To do this, it may be necessary to discharge water through some type of waste line.

Figure 5.37 The circulator line circulates water through the pump to keep it cool when little or no other water is being discharged.

PUMP DRAINS

Due to the danger of freezing in cold weather, a drain must be supplied at the lowest point on the pump and at the lowest point on each line connected to it. A number of these drains are connected to a common master drain valve. When the control handle is operated, all the drains will be opened simultaneously and the pump can be drained with one operation. This drain valve should not be opened when the pump is operating, either with pressure or vacuum on the intake. An "O" ring is often used as a gasket to maintain an airtight connection when the valve is closed. If the valve is operated with pressure or vacuum on it, damage to the "O" ring inside the valve could result. Some additional drain valves may be supplied if it is not convenient to connect all of the drains to the master valve. It is important that all of them be opened for the pump to drain completely.

Single-Stage Centrifugal Fire Pumps

Many centrifugal pumps used in the fire service are constructed with a single impeller and are referred to as single-stage centrifugal pumps (Figure 5.38). Front mount pumps, power take-off, separate engine driven, and some midship split power train driven pumps use a single intake impeller and a simple casing to provide capacities up to 2,000 gpm (8 000 L/min). Improvements in this basic design have been made as more powerful engines have become available and the design capacity has increased.

Figure 5.38 Single-stage impeller, with double intake, mounted on its shaft. *Courtesy of Hale Fire Pump Company.*

High-capacity pumps require large impellers with waterways that present a minimum of opposition to the movement of water. A law of physics states that for every action there is an equal and opposite reaction. The water in motion in the pump creates stress on the pump, its bearings, other moving parts, and the casings mounted to the truck frame.

To minimize the lateral thrust of large quantities of water entering the eye of the impeller, a double suction impeller was designed. The double suction impeller takes water in from both sides; the reaction being equal and opposite cancels the lateral thrust. It also provides a larger waterway for the movement of water through the impeller. Because the impeller turns at a very high rate of speed, a radial thrust is developed as the water is delivered to the discharge outlet. Stripping edges in the opposed discharge volutes divert the water 180 degrees apart. Since the water is being removed at two places and traveling in opposite directions, the radial thrust is also canceled. This design provides a hydraulically balanced pump, which lessens stress on the pump and chassis, and helps to lengthen the useful life of the pump and the apparatus.

Multistage Centrifugal Fire Pumps

5-89

A common type of fire pump in service today is the midship mounted, two-stage centrifugal pump. This pump has two impellers mounted within a single housing. The two impellers are usually mounted on a single shaft driven by a single drive train (Figure 5.39). The two impellers generally are identical and have the same capacity. What gives the two-stage pump its versatility and efficiency is its capability of connecting these two stages in SERIES for maximum pressure or in PARALLEL for maximum volume by use of a transfer valve (Figure 5.40 and 5.41).

PUMPING IN THE VOLUME POSITION

Figure 5.42 shows the pump in the PARALLEL or VOLUME position. In this configuration, each of the impellers takes water from a source and delivers it to the discharge. Since each of the impellers is capable of delivering its rated pressure while flowing 50 percent of the rated capacity, the total amount of water the pump can deliver is equal to the sum of each of the stages. If the pump depicted was rated at 1,000 gpm (4 000 L/min), each of the impellers would supply a stream of water that carried 500 gpm (2 000 L/min) to the discharge outlet. There the two streams would combine, so the total amount available would be 1,000 gpm (4 000 L/min) at a net pump pressure of 150 psi (1 050 kPa). To discharge water at a higher pressure, changing

Figure 5.39 Two-stage impellers mounted on their shaft. *Courtesy of Hale Fire Pump Company.*

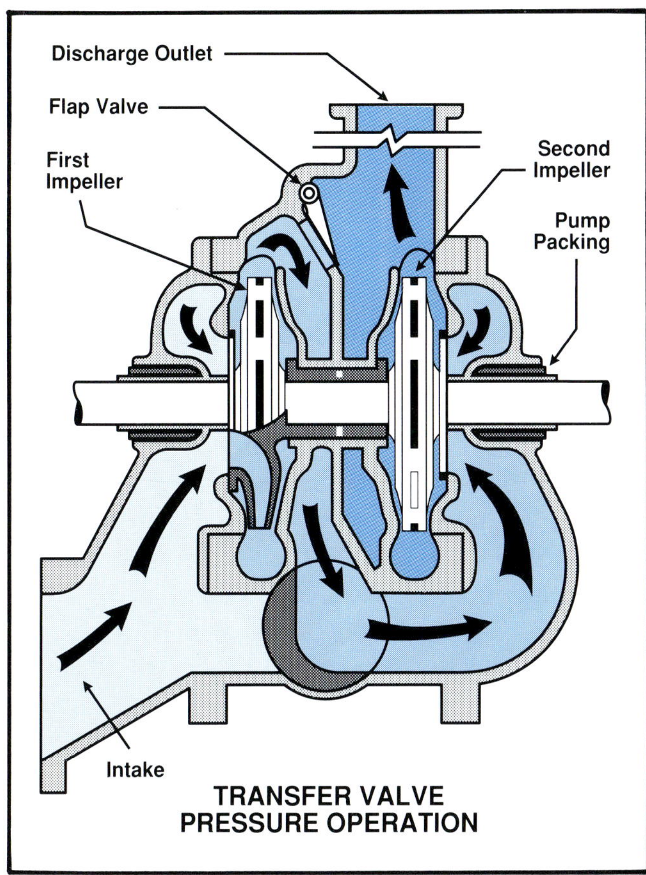

Figure 5.40 The routing of water when a two-stage pump is in the PRESSURE (SERIES) position.

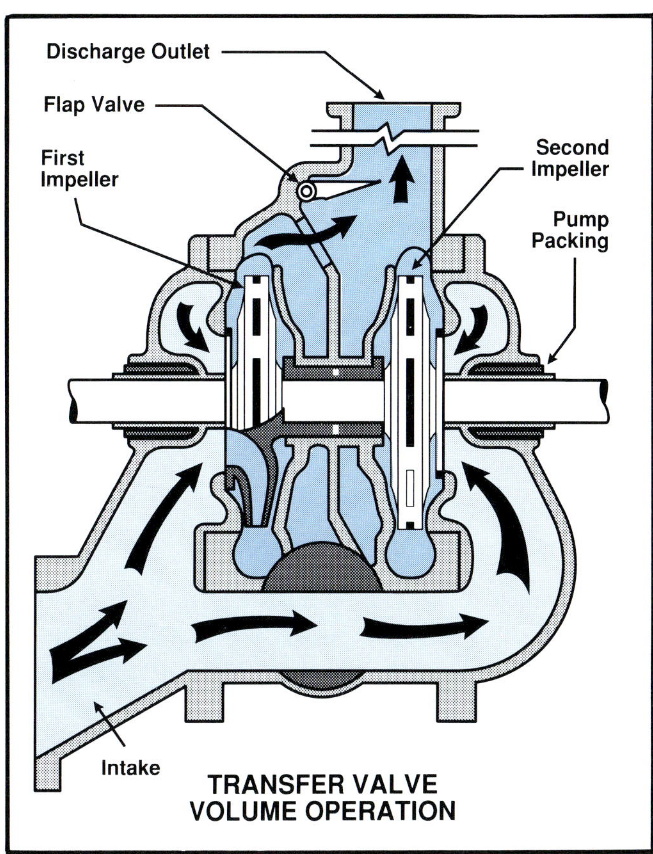

Figure 5.41 The routing of water when a two-stage pump is in the VOLUME (PARALLEL) position.

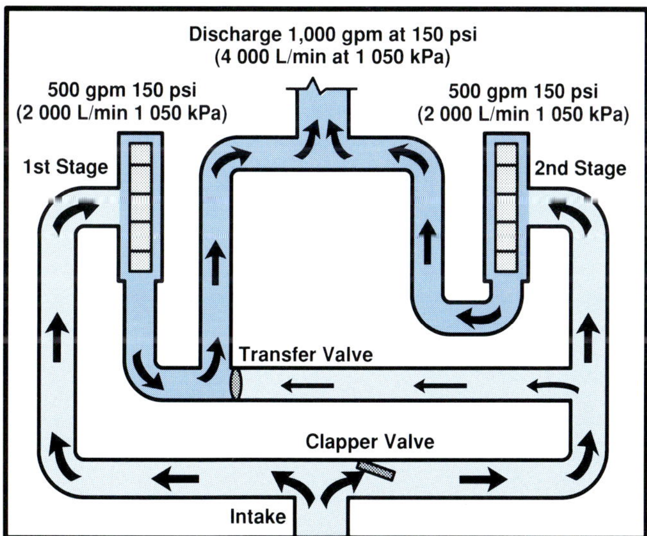

Figure 5.42 During VOLUME operation, each of the impellers takes water from a source and delivers it to the discharge.

the transfer valve to the SERIES or PRESSURE position will greatly increase the maximum pressure attainable. Increasing the pressure, however, will result in a corresponding reduction in capacity.

In Figure 5.43 the water comes into the intake as before. The first stage increases the pressure and discharges up to 50 percent of the capacity into the transfer valve. The transfer valve directs

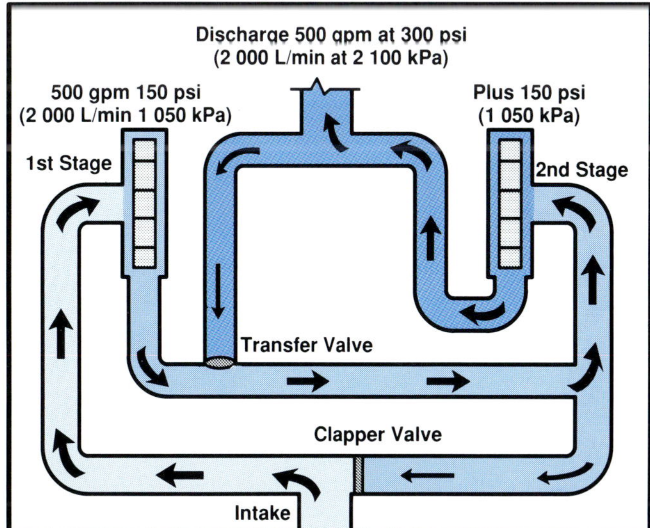

Figure 5.43 During PRESSURE operation, water does not recirculate through the pump; rather, it is forced into the eye of the impeller of the second stage, where pressure is increased.

the water into the intake of the second stage. The closed clapper valve isolates the second stage from the intake manifold and prevents the water from recirculating through the pump. Instead, the water is forced into the eye of the impeller of the second stage. The second stage increases the pressure and delivers the water at the higher pressure into the discharge outlet. At this point, the pressure is much higher than in the PARALLEL (VOLUME) position since the same stream of water has passed through two impellers, with each adding to the pressure. There is only one impeller delivering water to the discharge outlet instead of two, as in the PARALLEL (VOLUME) position, so the total available water is limited to the amount that one impeller can supply.

Since the pump is designed for each stage to deliver half its rated capacity, the optimum operation of the transfer valve is in the PARALLEL (VOLUME) position anytime more than half the capacity will be used. At lower flow rates, operating in the SERIES (PRESSURE) position reduces the load and the required rpm of the engine.

WHENEVER THE PUMP IS USED AT MORE THAN ONE-HALF ITS RATED CAPACITY, THE PARALLEL POSITION SHOULD BE USED.

With the use of variable gallonage and automatic nozzles and modern fire fighting techniques involving intermittent operation of attack lines, it is difficult for the pump operator to determine how much water is being used, unless the apparatus is equipped with a reliable flowmeter. If the apparatus is not equipped with a flowmeter, the driver/operator will be required to estimate the current flow of water. One method of estimating the amount of water flowing is to keep track of the number of hoselines being supplied by the apparatus.

Remember that NFPA 1901 requires that one 2½-inch (65 mm) outlet be provided for every 250 gpm (1 000 L/min) of capacity. Therefore, a good rule of thumb is that anytime more than half the large outlets are in use, the pump should be in the PARALLEL (VOLUME) position. This means that whenever more than one large line is in use from a 750 gpm (3 000 L/min) pump or more than two lines from a 1,000 gpm (4 000 L/min) pump, it should be arranged for PARALLEL (VOLUME) operation.

When the transfer valve is operated on a two-stage pump, sudden changes in pressure will occur as the water changes its direction of flow. This pressure can cause damage to hoselines and fire pumps, as well as injury to firefighters on the hoselines if it takes place suddenly with excessive pressure on the pump. From the standpoint of safety to firefighters and equipment, this changeover should not take place at a net pump pressure of more than 50 psi (350 kPa). It will be necessary to shut down the fireground operation to make this change since effective fire streams cannot be maintained at the reduced pressure. It is important that the changeover be coordinated with attack crews so that lines are not shut down at times that crews are in a precarious position.

The operator should attempt to anticipate the requirements that will be placed on the pumper as the fire fighting operation progresses and have the pump in the proper position. If there is any question as to the proper operation of the transfer valve, it is better to be in PARALLEL (VOLUME) than in SERIES (PRESSURE). While the PARALLEL (VOLUME) position may make it difficult to attain the desired pressure, it will be able to supply the amount of water needed. The SERIES (PRESSURE) mode of operation may make it impossible to maintain the necessary amount of water to the attack lines.

Operation of the transfer valve is done manually on many two-stage pumps (Figure 5.44). If this is the case, a built-in safeguard makes it physically impossible to accomplish the transfer while the pump is operating at high pressures. Other manufacturers provide a power-operated transfer valve (Figure 5.45). This transfer valve can be activated by using electricity, air pressure, vacuum from the engine intake manifold, or even by using the water pressure itself to accomplish the transfer.

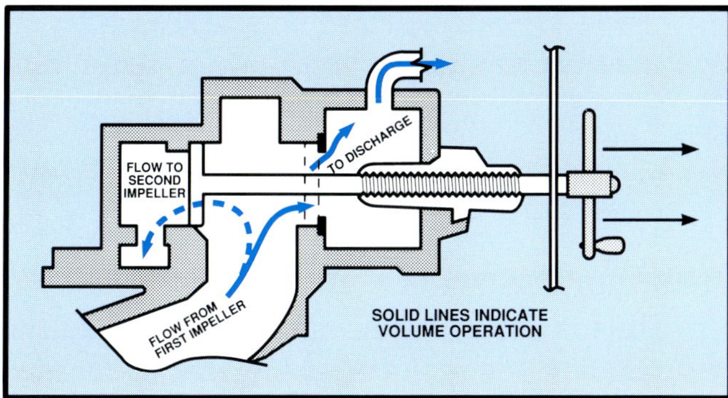

Figure 5.44 A manually operated transfer valve. The solid lines indicate VOLUME operation. The dotted line indicates SERIES operation.

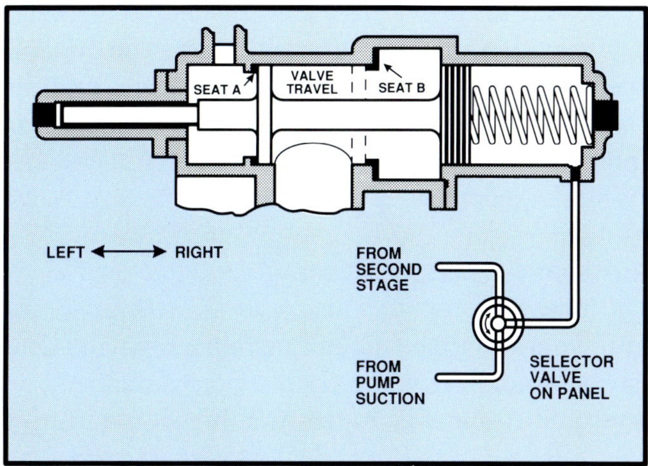

Figure 5.45 An American LaFrance hydraulic transfer valve. Solid lines indicate VOLUME position.

Many power-operated transfer valves will operate at pressures as high as 200 psi (1 379 kPa). These pressures can present extreme danger to personnel and equipment, so special care must be used with these types of controls. If it is available, the power control should have some type of manual override to allow the transfer to be operated if the power equipment should fail. It is important for all operators to be familiar with this manual override and to practice with it frequently.

The clapper valves are essential in a two-stage pump. If they should stick open or get debris caught in them, the pump will not operate properly in the SERIES position. When the transfer valve is operated, the clapper valve would allow the water to escape back into the intake, and it would just churn through the pump instead of building up pressure. The operation of this valve can be inspected by removing the strainer from the big intake openings, reaching into the pump with some type of rod, and ensuring that the valve swings freely.

The routing of water for VOLUME and PRESSURE operations is illustrated in Figures 5.46 and 5.47, using sectional views of two popular centrifugal pumps. In these illustrations, light blue indicates water supply, medium blue indicates intermediate pressure, and dark blue indicates discharge pressure.

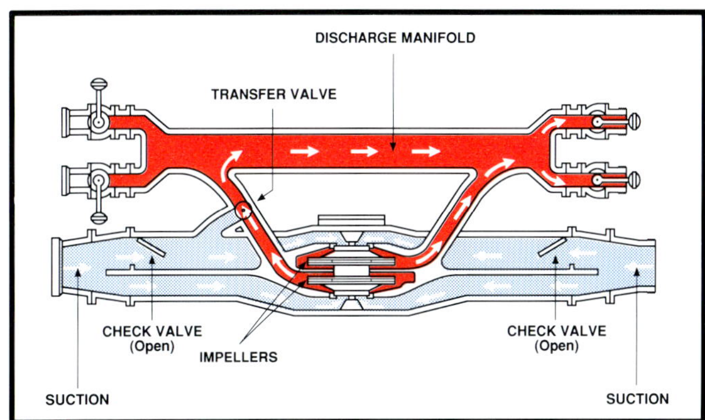

Figure 5.46 The routing of water for VOLUME operation. *Courtesy of Hale Fire Pump Company.*

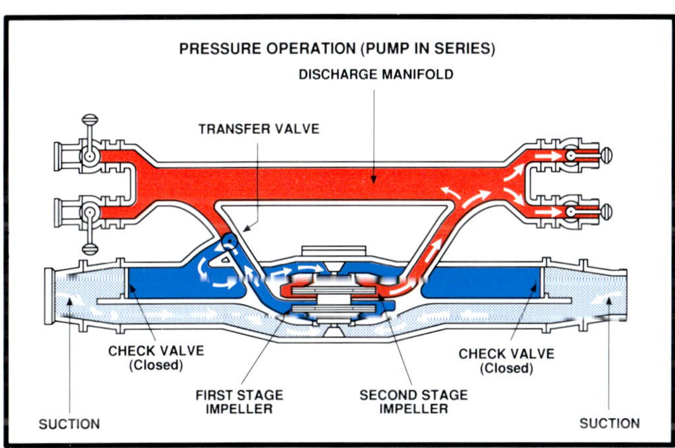

Figure 5.47 The routing of water for PRESSURE operation. *Courtesy of Hale Fire Pump Company.*

High-Pressure Pumps

Some manufacturers have used as many as four impellers connected in series to develop pressures up to 1,000 psi (7 000 kPa) for high-pressure fog fire fighting. Another approach is to supply an auxiliary high pressure centrifugal pump mounted on a conventional midship pump with a separate drive system. When pressures higher than 250 psi (1 750 kPa) are needed, the separate

5-89

5-89 pump can be engaged and used to increase the pressure to a much higher level (Figure 5.48).

Figure 5.48 A high pressure pump. *Courtesy of Hale Fire Pump Company.*

Characteristics of Centrifugal Fire Pumps

The major characteristics of centrifugal fire pumps are as follows:

- Centrifugal pumps will not prime themselves and require an external priming device.
- The impeller turns at a very high speed and can be damaged by pumping contaminated water or cavitating.
- Discharges can be shut down for short periods of time while the pump is turning with no damage, but operating with insufficient flow for an extended time will cause damage to the pump from overheating.
- There is no positive blockage through the pump and it can take advantage of incoming pressures. The pump will also be affected by changes in intake pressure.
- Changes in the amount of water being discharged will change the discharge pressure of the pump.

PUMP ACCESSORIES

The previous sections mentioned some of the various additions to the fire pump that are needed to make the pump operator's job of supplying water more effective. While most fire pumps are of basically similar construction, each manufacturer's approach to providing the needed accessories varies widely. This section will cover the general operating principles of each type of device and the type of accessories offered in some detail. For more information about specific pumps, see Appendix A through Appendix G.

Automatic Pressure Control

The volume of water moving through the pump may change suddenly when a nozzle is shut down rapidly, or when the setting is changed on a variable gallonage nozzle. While modern nozzles can tolerate less-than-ideal pressures and still maintain an effective fire stream, the firefighter on the nozzle cannot tolerate any sudden changes in pressure. During the critical stages of the attack, a sudden change in pressure and the accompanying change in fog pattern or stream reach can be disastrous. When a pump is supplying multiple attack lines, any sudden flow change in one line can cause a pressure surge on the other. Even an alert operator will find it impossible to compensate for these sudden changes in time to protect the nozzleman on the other lines. Some type of automatic pressure regulation is essential to ensure the safety of personnel operating the hoselines.

Relief Valve

A pressure relief valve is provided on most fire pumps (Figure 5.49). The main feature of a relief valve is its sensitivity to pressure change and its ability to relieve excessive pressure within the pump discharge. An adjustable spring-loaded pilot valve actuates the relief valve to bypass water from the discharge to the intake chamber of the pump. Although only a small quantity of

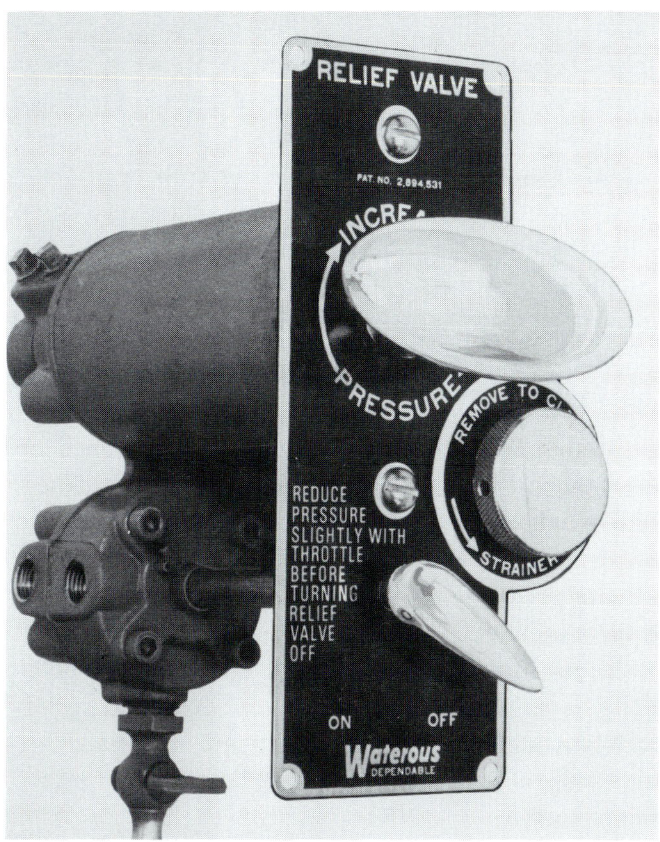

Figure 5.49 A pressure relief valve is designed to relieve excessive pressure within the pump discharge. Shown is the pressure relief valve controller. *Courtesy of Waterous Co.*

water is rerouted, the rerouting permits the pump to continue in operation when pressure rises above the working, or set, pressure.

There are many types of relief valves available. Most of them operate similar to the schematic drawings in Figures 5.50 and 5.51. Figure 5.50 illustrates what happens when the pump discharge pressure is lower than allowed by the pilot valve setting. Water flows from the pump discharge to the pressure chamber of the pilot valve. A diaphragm separates this pressure chamber from the pilot valve. Tension against the diaphragm is regulated by adjusting the handle of the pilot valve against the spring. As long as the hydraulic force in this chamber is less than the force of the spring, the pilot valve will stay closed. The water is then transmitted back to the main valve chamber and the main valve stays closed.

Figure 5.51 illustrates what happens when the pump discharge pressure rises higher than allowed by the pilot valve setting. When the discharge pressure gets high enough to compress the spring in the pilot valve, the needle valve moves to the left, which permits water to dump back into

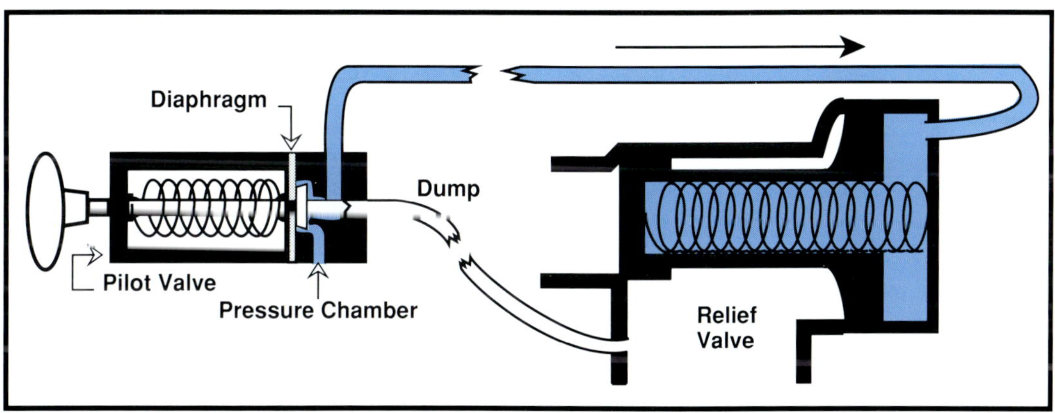

Figure 5.50 Schematic of water flow when pump discharge pressure is lower than allowed by pilot valve setting. *Courtesy of Waterous Co.*

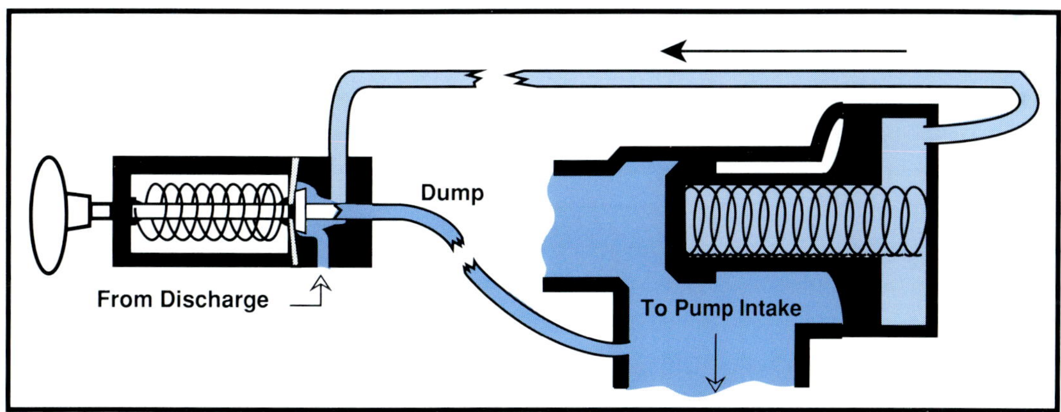

Figure 5.51 Schematic of water flow when pump discharge pressure is higher than allowed by pilot valve setting. *Courtesy of Waterous Co.*

the pump intake. This action in turn reduces the pressure in the tube and behind the main valve. The hydraulic force on the small end of the main valve is now greater than that behind the main valve. Consequently, the main valve opens and permits part of the water to return to the intake side, thus reducing discharge pressure. When the discharge pressure drops below the set pressure in the pilot valve, the pilot valve reseats, pressure increases behind the large end of the main valve, and it closes.

Figure 5.52 shows a simplified diagram of another automatic relief valve, with its components as they would appear when the discharge pressure is below the set pressure. When the pressure goes above the set pressure, the pilot valve moves, compressing its spring until the opening in the pilot valve housing is uncovered. Water flows through this opening and through the bleed line and also into the pump intake.

This flow reduces the pressure on the pilot valve side of the churn valve, allowing the higher pressure on the discharge side to force the churn valve open. Water flows from the discharge into the intake, relieving the excessive pressure.

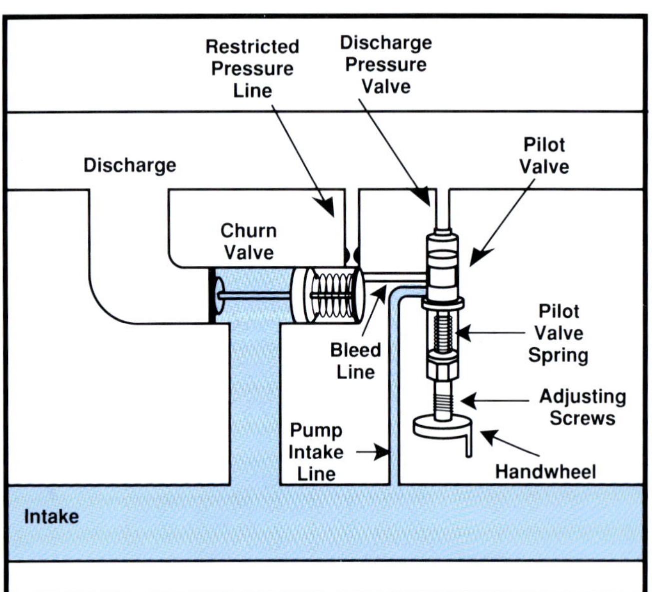

Figure 5.52 Another type of relief valve shown as it would operate when pump discharge pressure is below set pressure.

Pressure Governor

Pressure can also be regulated on centrifugal pumps by a governor that is pressure actuated to adjust the engine throttle. The main feature of a pressure governor is that it regulates the power output of the engine to match set pump discharge requirements. When the pressure in the discharge chambers of the pump exceeds the pressure necessary to maintain safe fire streams, the excessive pressure must be reduced. Since the speed of the impellers determines the pressure, and the engine speed determines the speed of the impellers, it is only necessary to reduce the engine speed to reduce the pressure.

Excessive pressures are generally caused by shutting down one or more operating hoselines. When excessive pressure builds up, a tube from the discharge side of the pump transmits the resulting pressure rise to a governing device, which then cuts back the throttle. The device varies with each manufacturer's design; it may be attached to either a regular or an auxiliary throttle. A pressure governor device can be used in connection with a throttle control, engine throttle, and pump discharge (Figure 5.53).

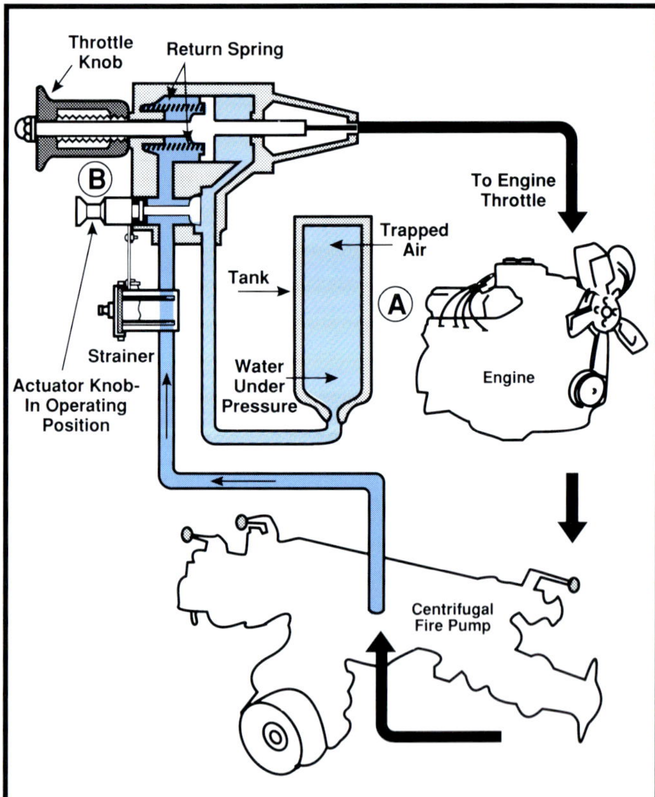

Figure 5.53 When activated, this governor uses the trapped air and water in area A as a reference pressure for the piston that controls the engine throttle. A change in pump pressure affecting area B moves the piston adjusting the engine speed. *Courtesy of Hale Fire Pump Company.*

One type of pressure control governor, found on older apparatus, operates as follows: When the control handle is pulled out, it lifts the split nut from the serrated piston shaft (Figure 5.54). The engine throttle is then advanced until the desired pump pressure is reached. The water control valve from the discharge side of the pump is then opened to permit water from the discharge or pressure side of the pump to enter the cylinder. This hydraulic pressure causes the piston to compress the control and move the serrated shaft. This movement actuates the carburetor linkage and butterfly valve to control engine speed.

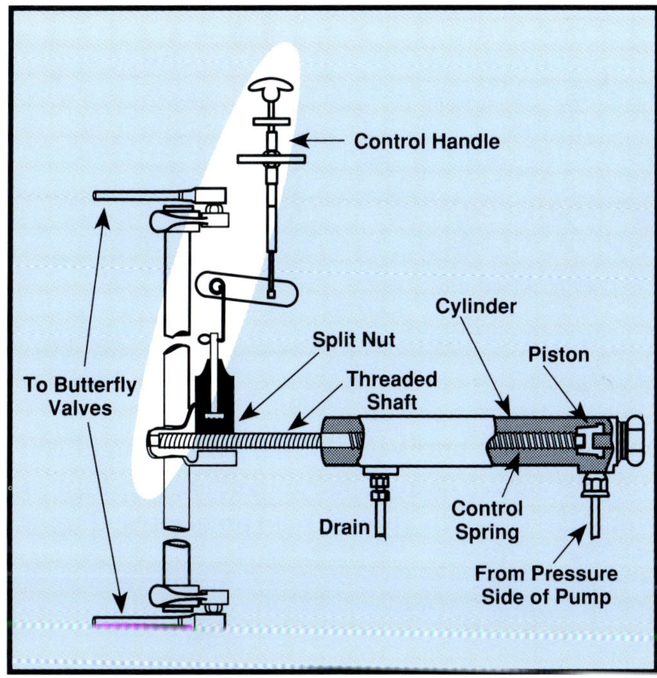

Figure 5.54 The control handle is pulled out to disengage the serrated nut from the threaded shaft.

The control handle must then be pushed in to lower the serrated split nut against the serrated piston shaft (Figure 5.55). Engine speed is then governed by the water pressure in the pump discharge. If a hoseline is shut off, the increase in pressure travels through the pipe from the pump discharge into the cylinder, moving the piston from its preset position. This movement causes the carburetor control linkage to adjust the butterfly valve and reduce the engine speed. When a discharge line is reopened, the reverse occurs and the piston moves back to increase engine speed and returns to the preset pump pressure position.

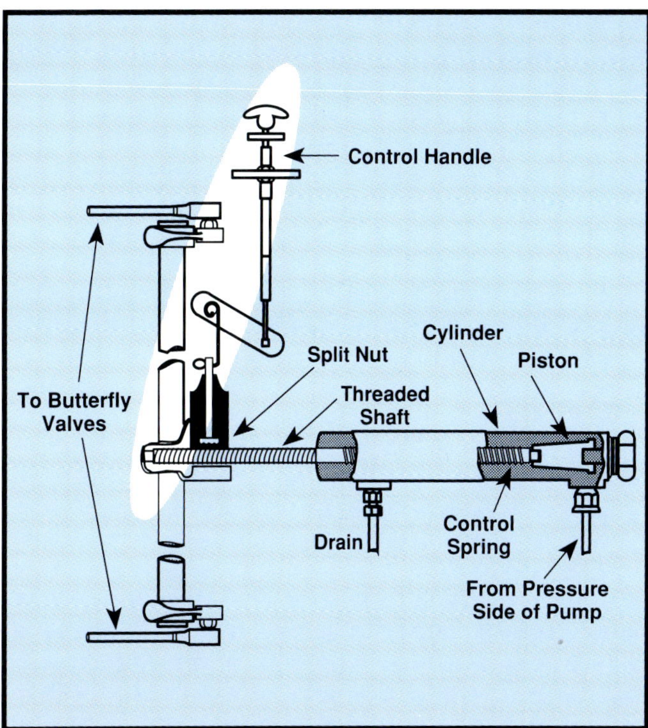

Figure 5.55 After the desired pressure is reached, the control handle is pushed back in, setting the governor for that pressure.

Another type of governor is the piston assembly governor (Figure 5.56 on next page). The speed of the engine is controlled by a governor assembly. This assembly fits onto the carburetor and reduces or increases the engine speed under the control of a rod connected to a piston in a water chamber. The governor chamber contains a flexible diaphragm to isolate the piston and spring assembly from the water while exposing the face of the piston to the water pressure. Pressure comes into the water chamber through a strainer from the pressure side of the pump. Water returning from the water chamber goes through the governor valve assembly to a two-piston governor return valve. This valve allows the water from the pump to be discharged on the ground or return to the intake side of the pump.

When drafting or pumping from the tank, the DRAFT (or CLOSED) position is used; therefore, the water returns to the intake side of the pump. When pumping from a hydrant (or external pressure source), the HYDRANT (or OPEN) position is used so the water is discharged to the ground. This hydrant drainline to the ground can be run through a check valve to the tank to prevent water from continuously running on the ground.

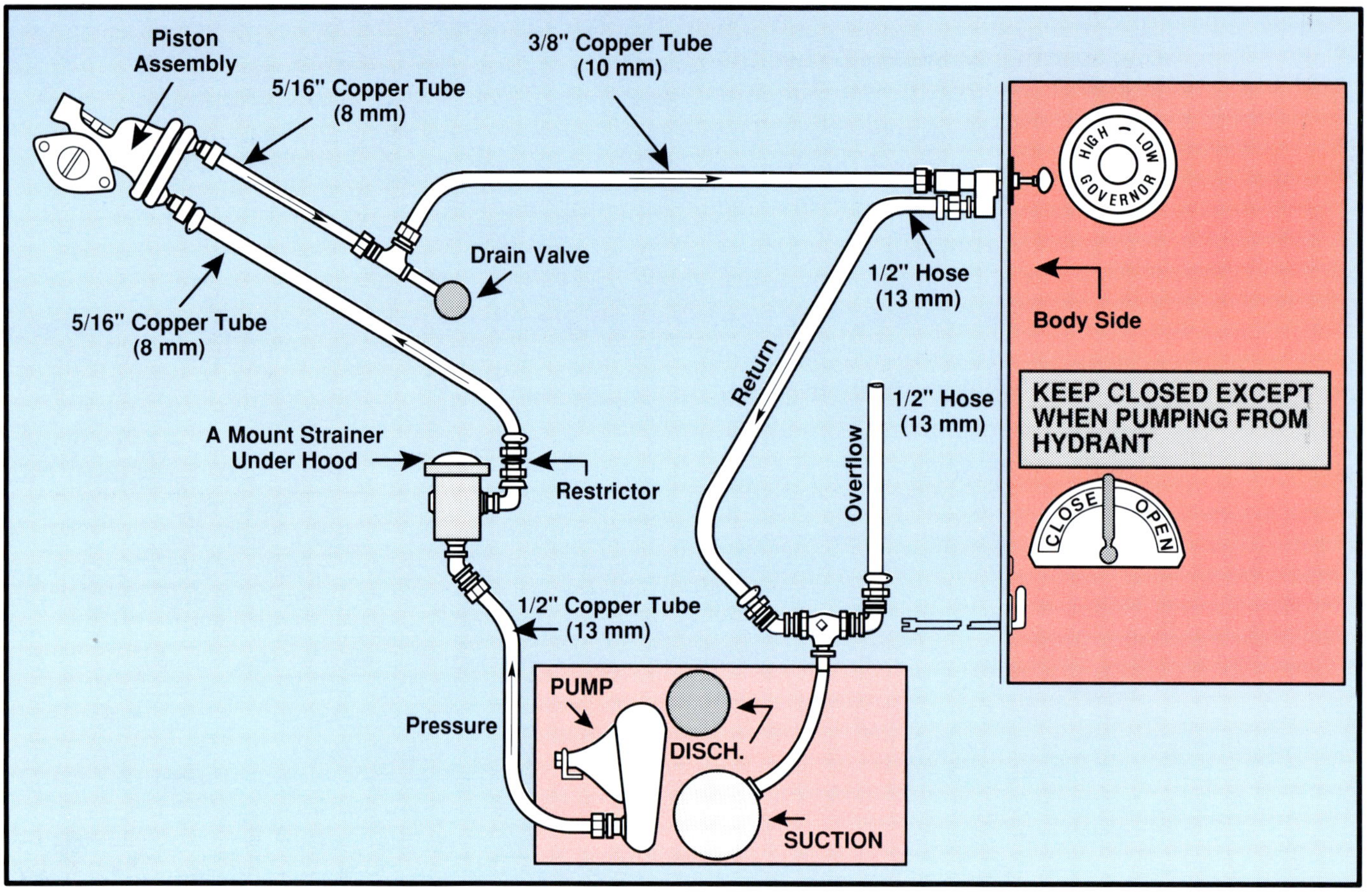

Figure 5.56 The piston assembly of this governor is mounted on the carburetor.

The electronic governor depends entirely on electricity for its operation (Figure 5.57). A pressure sensing element is connected through a supply line to the pump discharge. This element controls the action of an electronic amplifier. First, it compares the pump pressure to an electrical refer-

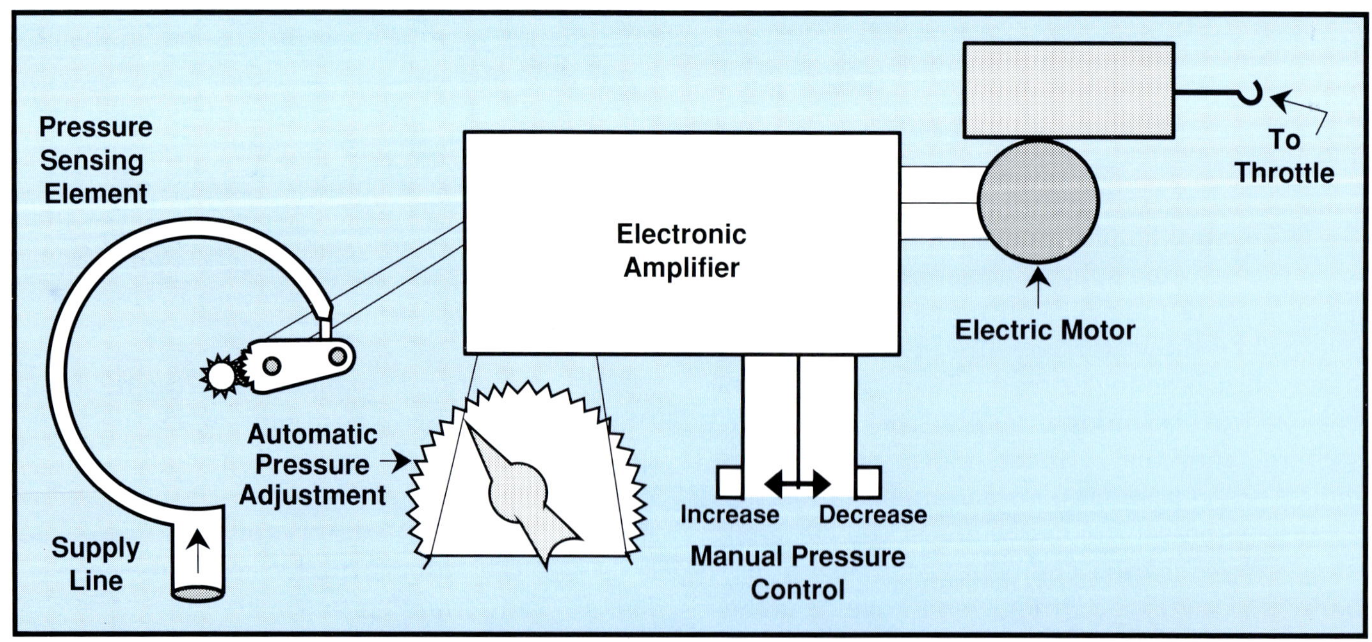

Figure 5.57 In an electronic pressure governor, a pressure sensing element drives an electric motor to change the throttle setting. In this way, the desired operating pressure is reached.

ence point. Then the element drives an electric motor to change the throttle setting, thus bringing the pump to the desired operating pressure. This governor will maintain any pressure that is set on the control knob above 50 psi (350 kPa). It will also return the engine to idle speed anytime the pressure drops below 50 psi (350 kPa).

PRIMING METHODS AND DEVICES

Successfully drafting water depends upon the ability to create a partial vacuum within the pump and the intake hose. A positive displacement pump can pump air as the pump operates; therefore, it is self-priming. On the other hand, the centrifugal pump depends on the movement of the water within the housing to develop pressure. It has no ability to pump air. The centrifugal pump requires an external priming device to operate from draft. Primers fall into three categories: positive displacment pumps, exhaust primers, and vacuum primers.

Positive Displacement Primers

The rotary vane primer is used extensively by many manufacturers. It requires a relatively high rpm as compared to a rotary gear primer, and can be driven by either mechanical means from the pump transfer gear case or by an electric motor. Since a relatively small amount of horsepower is required to drive this pump, the electric motor is the most popular for this type of primer. Use of an electric motor as the driver allows for maximum flexibility in mounting and application to different chassis and pumps (Figure 5.58).

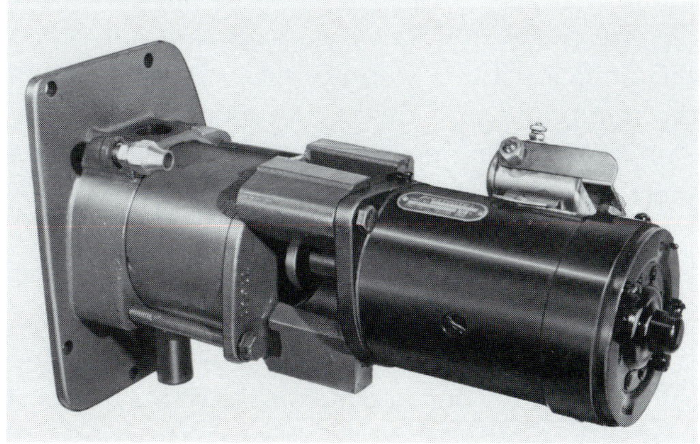

Figure 5.58 A rotary vane priming pump with electric motor drive. *Courtesy of Waterous Co.*

The inlet is connected to a priming valve that provides a connection to the fire pump. If there are high spots in the way the pump is constructed or mounted, the line from the priming pump may be connected to the pump in several different locations. If the pump is driven electrically, the priming valve usually incorporates a switch into the valve. In this way, only one operation is necessary to prime the pump.

Most primers use an oil supply (Figure 5.59). The oil serves two purposes. As the pump wears, the clearances between the gears and the case increase, and the pump loses efficiency. A thin film of oil is drawn into the pump and seals the gaps between the gears and the case. The oil fills any irregularities in the housing caused by pumping contaminated water and improves the efficiency of the primer. The oil also acts as a preservative and minimizes deterioration of the metal parts by inhibiting corrosion when the pump is not in use. To derive the maximum benefit from the oil, it is necessary to operate the pump periodically so that a coating of oil will form on all the metal parts.

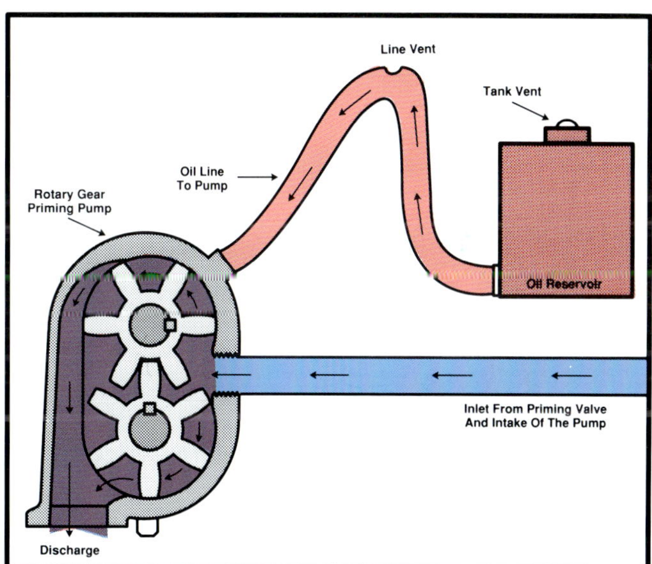

Figure 5.59 The oil supply in a rotary gear primer increases the efficiency of the primer and acts as a preservative.

There is a vent in the oil line from the reservoir to the priming pump. Since the tank is frequently mounted higher than the priming pump, a siphon action would drain the tank after the primer has been used. This vent breaks the siphon. It must be large enough to serve the purpose, but small enough to allow the priming pump to draft the oil out of the tank when the primer is in use.

The described engine rpm depends on the construction of the primer, the gear ratio of the transfer case, and other features that are unique to a particular installation. The instruction manual for the pumper should specify the current rpm for priming, but in general, should operate with an engine rpm around 1,000 to 1,200.

Exhaust Primers

The exhaust primer operates on the same principle as a foam eductor. The exhaust gases are prevented from escaping to the atmosphere by the exhaust deflector. The gases are diverted to a chamber where the velocity of the gases passing through a venturi creates a vacuum. This chamber is connected through a line and a priming valve to the intake of the pump. There the air is evacuated into the venturi chamber and then discharged along with the exhaust gases into the atmosphere.

Creating a vacuum by using exhaust requires high engine rpm to operate the maximum velocity of the exhaust gases. At best this primer is not very efficient. It is used primarily on portable pumps where price and weight are of prime importance.

The exhaust primer requires a great deal of maintenance because the exhaust gases moving through the primer itself tend to leave carbon deposits. These deposits decrease the efficiency of the device even further. Use of an exhaust primer requires that any air leaks in the pump be kept to an absolute minimum, and that the suction hose and gaskets be kept in good condition (Figures 5.60 and 5.61).

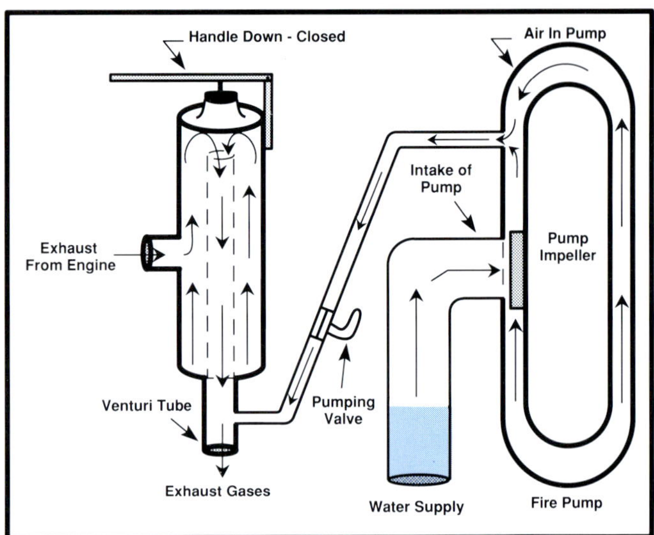

Figure 5.60 Schematic diagram of an exhaust primer with the primer handle in the CLOSED position. *Courtesy of Bennie Spauling.*

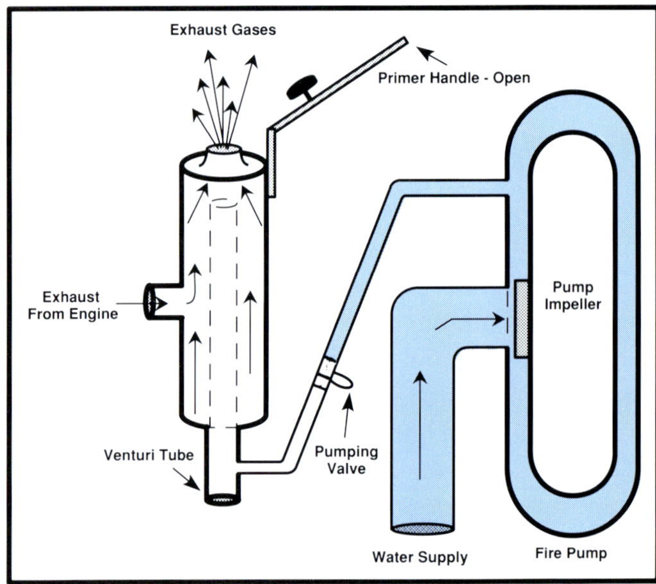

Figure 5.61 The same exhaust priming system with the primer handle in the OPEN position. *Courtesy of Bennie Spauling.*

Vacuum Primers

Probably the simplest type of primer makes use of the vacuum already present in the intake manifold of any gasoline engine. To prime the pump, it is necessary only to connect a line from the intake manifold of the engine to the intake of the fire pump with a valve connected in the line to control it. This connection creates two dangerous conditions, however. First, the intake manifold contains explosive gases that could be drawn into the fire pump and cause damage. Second, when the pump is primed, water could be drawn through the pump and into the intake manifold, causing damage to the engine. One approach to overcoming these problems is illustrated in Figure 5.62.

The line from the intake manifold contains a check valve. When this valve is under pressure, such as occurs during a backfire, it closes, cutting off the pump from the engine. Under a vacuum, the valve stays open and allows the pump to prime normally. As the air comes through the line from the intake side of the pump, the pump is primed

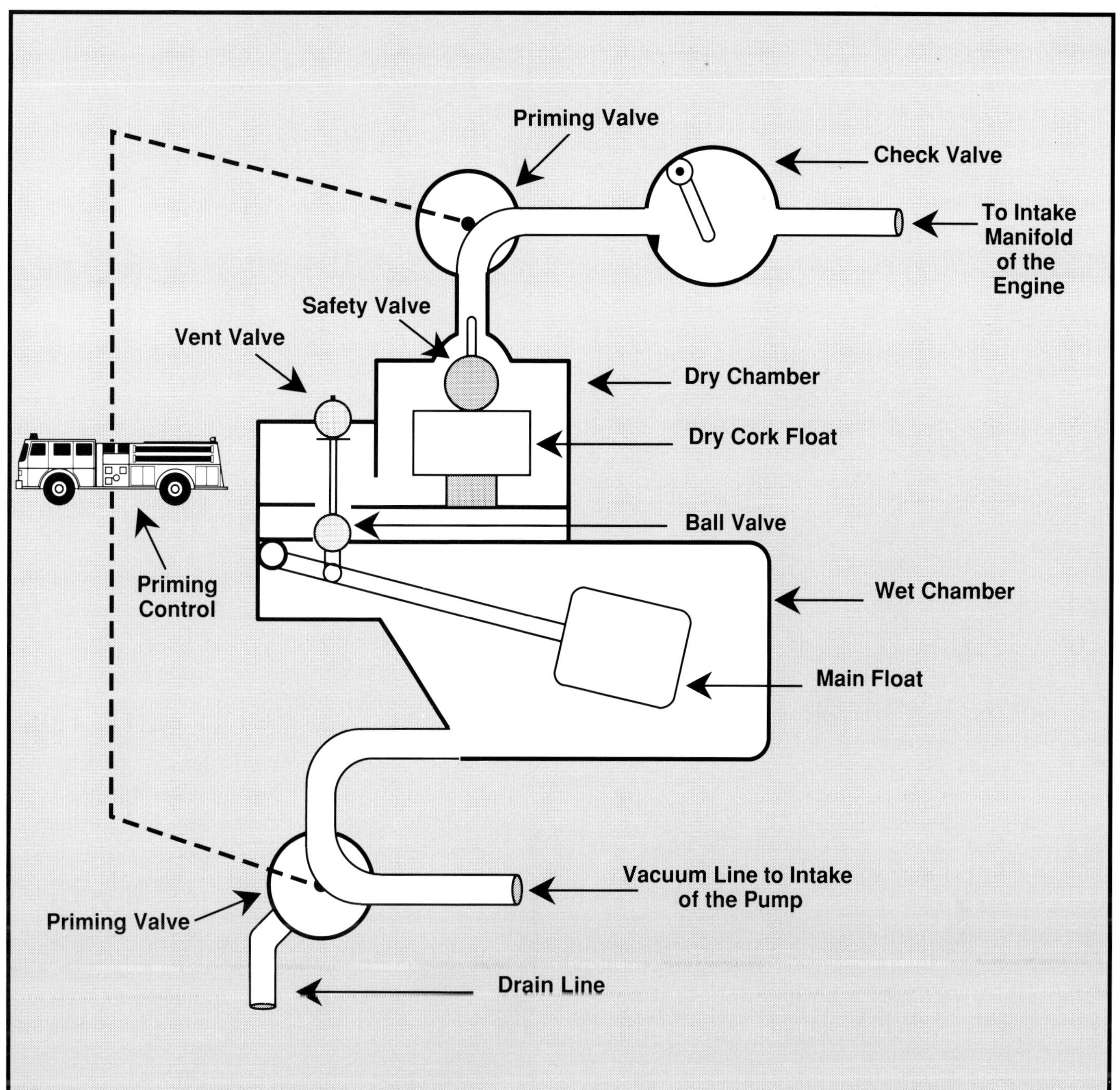

Figure 5.62 Schematic diagram of a vacuum priming system.

and fills with water. When the water enters the wet chamber of the primer, the float operates and closes the ball valve. This action opens the vent valve and allows the vacuum line to take air from outside the pump. If the float does not operate, the water continues up into the dry chamber, where a cork float rises and closes the safety valve. When the priming valve is closed, the drain opens and allows the water from the primer to drain, and the primer is ready to operate again.

Since engine vacuum is maximum near an idle, the primer works best at low engine rpm. It is only necessary to speed the engine enough to enable it to keep from stalling when the vacuum line is opened to prime the pump.

DISCHARGE AND INTAKE VALVES

Most of the intake and discharge lines from the pump are controlled by valves. These valves are usually of ball-type construction and permit full flow through the lines with a minimum of friction loss (Figure 5.63). When the pump is new, valves must be airtight. The initial dry prime tests are made with the discharge and intake lines uncapped, depending on the valves to prevent air leaks. Even though the valves are constructed to resist wear (in some cases, they are self-adjusting), they do require repair as they age and are subjected to frequent use.

Figure 5.63 Shown is the relation of the ball valve to the actual discharge orifice.

The major discharges are equipped with 2½-inch (65 mm) valves on the typical pumper, but the use of larger hose sizes has led some manufacturers to use larger valves. Preconnected master stream devices may also require larger or multiple valves to realize their potential for large fire flows. Pumpers are equipped with a variety of preconnected hoselines, and these may use 2½-, 1½-, or even 1-inch (65 mm, 38 mm, or 25 mm) valves depending on the size hoselines.

Valves in common use are either ball valves or butterfly valves. The actuation of these valves can be accomplished by any one of a series of operating controls. One common operating device is the push-pull rod. The push-pull rod is connected to a series of mechanical linkage or a gear-tooth rack arrangement which in turn operate the valve (Figure 5.64). The gear-tooth arrangement provides a mechanical advantage that makes the valve easier to operate under pressure. Another operating device that is common is the twist-lock lever. This lever can be installed in either a vertical or horizontal orientation. The lever can be locked in any position by twisting the handle to the LOCK position.

Figure 5.64 Typical push/pull valves. Valve number 3 is in the OPEN position, while valve number 4 is CLOSED.

The butterfly valve has a simpler mechanical linkage with the handle mounted directly on the valve stem. The valve is opened or closed by a 90-degree movement of the handle (Figure 5.65). Some of the older quarter-turn valves are locked in position by raising and lowering the handle, but the more modern versions lock by rotating the handle in a clockwise direction. One manufacturer has recently designed a valve that locks automatically when the handle is released, but the majority of valves in use require a positive action to lock the valve in position anytime the line is in use.

Figure 5.65 A quarter-turn valve used on the rear dump of a tanker. *Courtesy of Tuolumne County (California) Fire Department.*

DRAIN OR BLEEDER VALVES

Most connections to the pump are equipped with drain valves on the line side of the main valve. On the discharge fittings, these drain valves provide a way to relieve the pressure from the hoselines after the discharge valve has been closed. This is useful when one or more attack lines have been shut down while the pump is still supplying other lines. If there were no way to get rid of the pressure trapped in the line, it would be very difficult to break the connection. By use of the drain valve, the unused line can be drained and the hose disconnected while the pump is still in service.

The bleeder line on the gated intake serves another purpose. If a supply line is connected to the gated intake of a pump while the attack lines are being supplied from the tank, it is desirable to make the changeover to the supply line without interrupting the fire streams. Since the hoseline is full of air when it is dry, the water coming through the line would force air into the pump, probably causing it to lose its prime. At the very least, it would cause fluctuations in the nozzle pressure. By opening the bleeder valve on the line side of the intake valve, the air can be forced out of the hose through the bleeder as the line fills with water. When all the air has been evacuated from the line and the bleeder is discharging a steady stream of water, the drain valve can be closed, the intake valve opened, the tank-to-pump valve closed. These actions occur on a coordinated basis, and the transition can be made with no interruption to the flow.

PUMP GAUGES

Every fire department pumper must be supplied with at least two gauges: a vacuum gauge and a pressure gauge. The vacuum gauge must be a compound gauge connected to the intake side of the pump capable of measuring either positive or negative pressure (Figure 5.66). The gauge is usually calibrated from 0 to 600 psi (0 kPa to 4 200 kPa) positive pressure and from 0 to 30 inches (0 mm to 762 mm) of vacuum on the negative side. This gauge will provide an indication of the vacuum present at the intake of the pump during priming or when the pump is operating from draft.

Figure 5.66 A typical compound gauge.

It also provides an indication of the residual pressure when the pump is operating from a hydrant or is receiving water through a supply line from another pump. Since the physical size of the vacuum scale is so small, even a slight error in the zero setting of the gauge can result in a large error in measuring the vacuum on the intake side of the pump. If the gauges used are not of a type safe for use in a vacuum, they can be damaged. Refer to NFPA 1901 for further requirements of the vacuum gauge.

Another use of the vacuum gauge is to indicate the vacuum being developed while the pump is operating from draft. This provides an approximation of how much pump capacity is not being used. As the flow from the pump increases, the vacuum reading increases as more negative pressure is required to overcome the friction loss in the suction hose. As the vacuum reading approaches 20 inches (508 m), the pump is near its maximum capacity and will not be able to supply any additional lines.

A pressure gauge is also required on a fire department pumper. It must be calibrated to measure 600 psi (4 200 kPa) unless the pumper is equipped to supply high-pressure fog streams, in which case the gauge may be calibrated up to 1,000 psi (7 000 kPa). The pressure gauge registers the pump discharge pressure (Figure 5.67). On a centrifugal pump, this gauge may also be a compound gauge, since both the intake and the discharge of the centrifugal pump will register negative pressure during priming.

Figure 5.67 A typical pressure gauge.

External connections to these gauges must be made to allow installation of calibrated gauges when service tests on the apparatus are performed (Figure 5.68). These connections are also used for standard gauges when the initial acceptance tests of the pump are made. These gauges are equipped with valves in the pressure line from the pump.

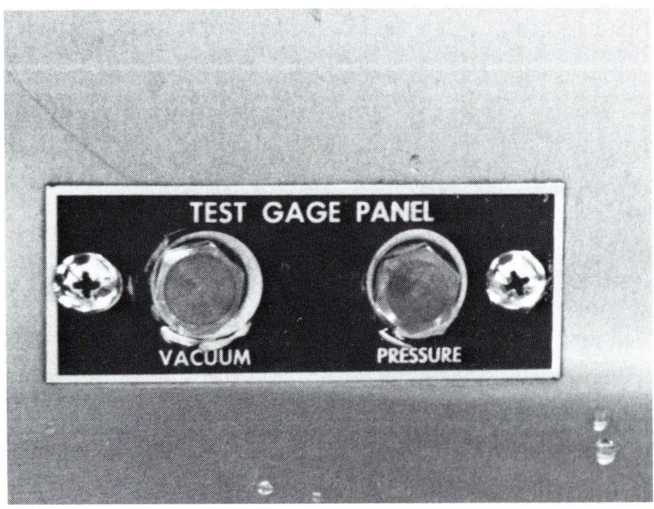

Figure 5.68 External gauge connections are used for pump testing or calibrating the fixed gauges.

Individual Line Discharge Gauges

Pressure gauges can be connected to the individual discharge fittings of the pump. These gauges must be connected to the line side of the discharge valve so that the pressure actually being applied to the hoselines after the gate valve is adjusted to reduce pressure can be determined accurately. The gate valve must be readjusted each time the flow at the nozzle being supplied is changed. Readjustment is necessary because the pressure loss in the valve is determined by the amount of water flowing through it. If the nozzle is shut down on the hoseline being supplied, the individual pressure gauge will read the same as the master pressure gauge, since there is no flow through the valve to reduce the pressure. No attempt should be made to readjust the gate valve until water is flowing again.

Individual pressure gauges may also be supplied with master stream devices or the lines that are supplying them. These gauges are very important since it is impossible to maintain effective master streams unless the proper pressure is being supplied to the appliance. With the large flows required, friction loss is high in the supply lines. An individual line gauge is the only way to be certain that the pump has been adjusted properly.

FLOWMETERS

Determining fireground hydraulics has long been an inexact science. Disastrous results can come from inaccurate fireground computations. The use of pressure gauges to determine water flow can further contribute to this situation.

Flowmeters virtually eliminate the possibility of operator error because they provide an accurate reading of water flow in gallons per minute (liters per minute). This is the term used when determining how much water is needed to extinguish the fire. The number displayed on the flowmeter requires no further calculation because it reflects how much water is moving through the discharge valve. This quantity of water will not diminish before it reaches the nozzle unless there is a leak or break in the hoseline.

Flow readouts on the pump panel are the only way to accurately determine how much water is being discharged at any given time. With the advent of the automatic nozzles, flowmeters are becoming more common. With the automatic nozzle, nozzle pressure is automatically maintained at a predetermined level. As the pump pressure increases, the nozzle opens wider to increase the flow. This increases the friction loss in the line to keep the discharge pressure of the nozzle constant. The amount of water discharged by the nozzles is determined by the pressure being applied to the supply line by the pump.

NFPA 1901 allows flowmeters to be used instead of pressure gauges on all 2½-inch (65 mm) or larger outlets (Figure 5.69). Flowmeters consist of a self-averaging primary sensor and a linear flow-rate indicator. Good flowmeters can make it possible for pump operators to pump (within the limits of the pump) the correct volume through nozzles at fires, without having to know the length of hoseline, the amount of friction loss, or whether the nozzles are above or below the pump.

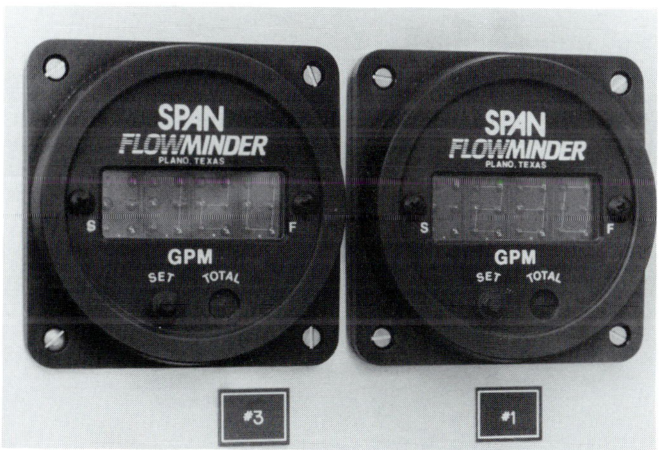

Figure 5.69 Some pumping apparatus are equipped with flowmeters for each discharge. *Courtesy of Span Instruments.*

This relieves the pump operator from having to make calculations that are, at best, only close approximations of the amount of water reaching the nozzle. Relieved of this responsibility, the operator is better able to monitor pump and apparatus functions, communicate with fire fighting crews, and distribute fire fighting equipment.

All flowmeters are designed to "read" water flow, but all flowmeters are not alike. Several designs are available, some of which are more reliable than others. There are four basic types of flowmeter sensors:

- Differential
- Turbine
- Spring Probe
- Paddlewheel

The differential sensor utilizes a pair of small tubes, each placed in opposite directions within the flow. Each tube measures pressure; the difference in pressures between the two orifices is translated electronically into a gallons per minute (liters per minute) readout. The primary problem with this type of sensor is clogging due to silt and mineral deposits. Because the tube orifices are very small, they require frequent cleaning. Differential sensors are used successfully in the chemical industry for measuring flows in such chemicals as alcohol or solvents, where contaminants are seldom a problem. This type of sensor, however, is not practical for fire service use.

The turbine sensor, common in the farming industry where only low-volume flows are required, is an accurate device by design. It is mounted directly in the center of the flow to rotate as water passes through its blades. Turbine revolutions are translated mechanically into flow rates expressed in gpm (L/min). The primary problem with this sensor is that it impedes the flow so significantly that high-volume streams are difficult to push through the turbine blades. The blades are also vulnerable to damage by stones and other debris that may be pulled into the pump during drafting operations. The units tend to be bulky and are not easily installed in fire apparatus plumbing, where space is limited.

One of the newest flowmeter designs to be introduced to the fire service is the spring-type probe. This flowmeter uses a stainless steel, spring-type probe to sense water movement in the discharge piping (Figure 5.70 on next page). The greater the flow of water through the piping, the more the spring probe will be forced to bend. This will send a corresponding electrical charge to the digital display unit. Because the spring probe is the only moving part of the system, these devices are relatively maintenance free.

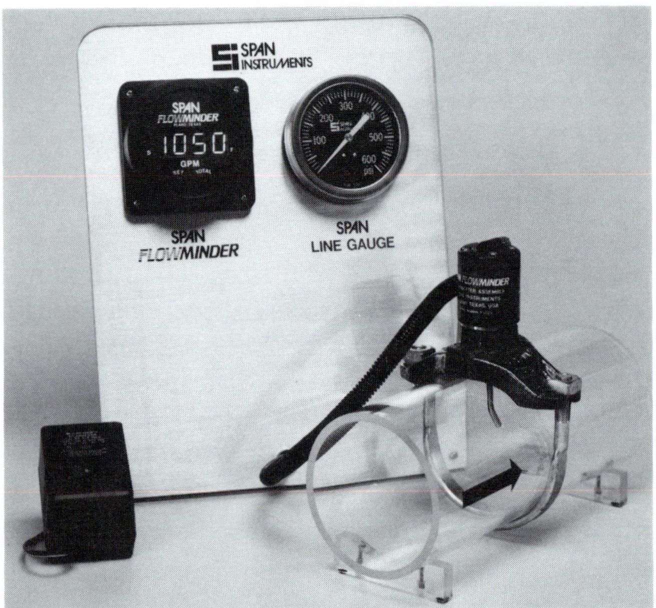

Figure 5.70 This flowmeter features a spring-type probe that senses water movement in the discharge piping. *Courtesy of Span Instruments.*

Of the two types commonly found, the paddlewheel sensor has been in use in the fire service for a longer period. Operating on the same principle as the turbine sensor (translating rpms in gpm [L/min]), the paddlewheel is mounted in the top of a pipe in such a manner that very little of the device extends into the flow (Figure 5.71). This placement eliminates the problems of impeded flow and damage by debris. Because the paddlewheel is located at the top of the pipe, sediment will not deposit on the paddlewheel.

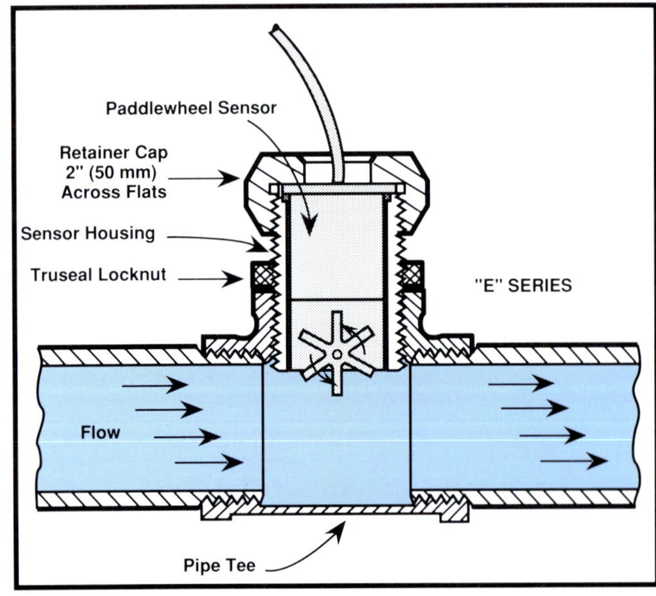

Figure 5.71 The paddlewheel sensor is another type of sensor used in flowmeters. *Courtesy of Fire Research Corp.*

Flowmeter Accuracy

When properly calibrated and in good working condition, flowmeters should be accurate to a tolerance of ±3 percent. This means that the readout should not be more than 3 gallons (12 L) high or low for every 100 gallons (400 L) flowing. This kind of accuracy may not be possible with pressure gauges, as they are often calibrated to read in 5 to 10 psi (35 kPa to 70 kPa) increments.

Flowmeters can be installed in a new pump at the factory or can be retrofitted later by the fire department. The sensors must be mounted in straight pipe runs in the apparatus plumbing to ensure accuracy. These runs should be free of turbulence so that a uniform flow of water passes over the sensor. Turbulence is caused by fittings such as elbows, valves, and tees.

If installed properly, flowmeters require little maintenance and are highly dependable. Sensors can be remotely located within pumping apparatus plumbing because the devices operate electronically. They are connected by watertight wiring and circuitry to analog gauges or digital LCD displays mounted in the pump panel.

The Auto Scan Unit

One manufacturer provides an automatic scanning unit to give the pump operator a number of useful options (Figure 5.72). The unit will function in the manual mode, which allows the operator to check individual discharges. The automatic mode displays each discharge valve flow

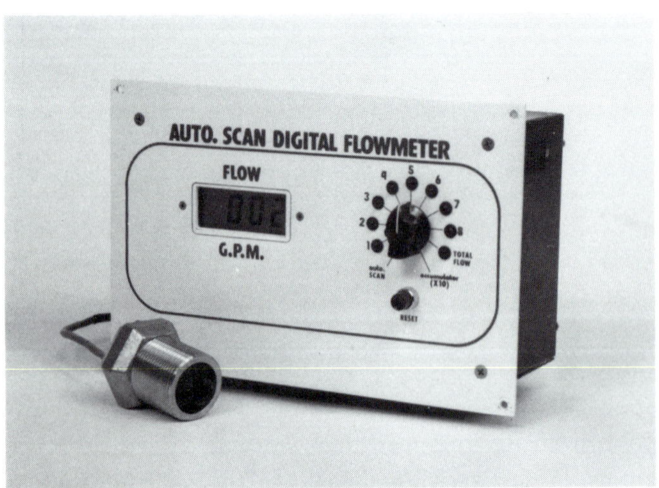

Figure 5.72 An automatic scanning flowmeter displays each discharge valve flow for four seconds, then moves to the next flow when in automatic mode. *Courtesy of Fire Research Corp.*

for four seconds before automatically moving to the next flow.

Some types of flowmeters allow the operator to receive an instant readout of the total flow being delivered at any moment during the pumping operation. This is important when calculating how much pump capacity remains for supplying additional water needs. The unit will also compute the total amount of water that has been pumped since the pumping operation started (accumulated flow). This feature is useful when pumping from a single water source such as a 5,000-gallon (20 000 L) tanker. It gives the operator the ability to monitor the limited water supply so that it does not become depleted unexpectedly. Another use of this feature is to monitor the amount of water that has been used on the fire since the operator started. This figure is especially important when firefighters calculate how much weight has been added to the structure by water. At about 8⅓ pounds per gallon (1 kg/L), it does not take long to increase the interior load on structural members by several tons.

Flowmeter Applications

DIAGNOSING WATERFLOW PROBLEMS

In the hands of an experienced pump operator, the flowmeter is an excellent diagnostic tool. If the flow does not increase when the operator increases pressure, several problems are likely. The hose may be kinked or a midline valve (such as a gated wye) may be partially closed. If a firefighter communicates that water volume at the nozzle has suddenly diminished, it can be assumed that a hose has burst if there is no reduction in the flowmeter reading.

RELAY PUMPING

The use of a flowmeter during relay pumping will greatly simplify the operation. Since a flowmeter provides the gallons (liters) per minute, there is immediate knowledge of how much is being discharged to the receiving pumper. If more water is needed on the fireground, the throttle can be increased to the needed flow. However, if the supply is insufficient as the engine speed is increased, the flowmeter reading will no longer increase. This indicates pumping at the maximum volume available and near the point of cavitation. The throttle should be adjusted down slightly. If more water is required, a better supply and/or additional hoselines will be necessary between the relay pumpers.

STANDPIPE OPERATIONS

When pumping to standpipes, it is difficult to determine where hoselines and nozzles are being placed in a multistory building. Pressure losses due to elevation must be accounted for when calculating discharge pressures. This problem is compounded when automatic nozzles are used because the nozzles compensate for inadequate nozzle pressure by providing a stream that may not be of adequate volume to control the fire.

When a flowmeter is used, the problem can be solved by determining the number and type of constant flow nozzles connected to the standpipe, adding their maximum rated flows, then pumping the volume of water that matches this figure. An example is when three hose packs are taken to three different floors. One 2½-inch (65 mm) hose pack is equipped with a 300 gpm (1 200 L/min) fog nozzle, a second pack with a 1⅛-inch (29 mm) straight tip nozzle (250 gpm [1 000 L/min]), and a third 1½-inch (38 mm) hose pack with a 125 gpm (500 L/min) fog nozzle. In this case, the total maximum flow for the three nozzles is 675 gpm (2 700 L/min). The operator increases the flow until a volume of 675 gpm (2 700 L/min) is flowing. When using automatic nozzles, a reasonable flow rate should be selected for each nozzle.

It is important for the pump operator to establish communication with firefighters on the nozzles to determine that the fire streams are adequate and effective because nozzles placed several floors apart may receive pressures somewhat greater or lesser than optimum. This is a problem and communication is the best way to make adjustments to correct the problem.

IN-LINE FOAM EDUCTORS

Another common problem overcome by using a flowmeter is that of pumping the correct flow through an in-line foam eductor. Eductor design requires that a certain eductor inlet pressure be maintained to create a venturi effect within the eductor. The venturi effect causes foam concentrate to move through the device and into the water stream, thus creating correctly proportioned foam for fire fighting use.

With flowmeters all that is required when setting up a foam eductor operation is to increase the throttle until the flowmeter reaches the flow rate of the eductor and the nozzle. If the eductor is set at the correct percentage, the foam produced will be in the correct proportion.

Flowmeters are also excellent tools when used in conjunction with fixed foam delivery systems on apparatus. Use of the flowmeter allows the pump operator to maintain the proper amount of flow for the equipment being used. It also allows the pump operator to estimate the rate at which foam concentrate is being expanded.

OTHER GAUGES

In addition to pump gauges, there are a number of other gauges that will be found on any apparatus.

Engine Gauges

Adequate dashboard instrumentation is important to ensure that the engine is operating correctly (Figure 5.73). These gauges are required when the vehicle is in motion, both to ensure that it operates efficiently and to warn off impending problems. Gauges are also necessary during fireground operations because the engine is subjected to more rigorous conditions while operating on the fireground than on the road. The engine gauges can be duplicated on the pump panel so that the operator can monitor operating conditions during pumping operations. Some of the gauges that are usually provided are as follows:

Figure 5.73 It is important that the driver/operator be familiar with the operation and location of all gauges on the dashboard.

- Speedometer
- Tachometer
- Oil Pressure Gauge
- Ammeter
- Voltmeter
- Air Pressure Gauge
- Water Temperature Gauge
- Fuel Gauge

The speedometer is used primarily on the road, but it can also provide some indication of proper operation of the fire pump. To begin pump operations, the pump is initially engaged and the transmission set for pump operation. Most speedometers will register when the pump starts to turn. By knowing the normal speedometer reading while the pump is turning and the engine is operating at idle, the operator can check that everything is set properly before leaving the driver's seat. The speedometer can also provide significant information for trouble analysis by giving an indication of the speed of the drive shaft when difficulty is experienced in pumping capacity.

The tachometer records the engine speed in revolutions per minute (rpm). It can give an experienced driver the necessary information to operate the vehicle on the road in the most efficient manner. It can also provide valuable information about the condition of the pump. Like the speedometer, the tachometer is useful as a means of trouble analysis when difficulty with the pump is encountered.

When the initial acceptance tests of the pump are made, the rpm required to pump its rated capacity is determined, and the information recorded on an identification plate on the pump panel. A gradual increase in the amount of rpm required to pump the rated capacity indicates wear in the pump and a need for repairs.

The oil pressure gauge indicates that an adequate supply of oil is being delivered to the critical areas of the engine. It is not a measure of the oil level in the crankcase, but if the oil level in the crankcase drops too low, it will be impossible for the oil pump to maintain the required amount of pressure. While normal operating oil pressures are specified in the maintenance manual, the pressure for a particular engine will vary, so the operator needs to be familiar with the expected reading. Any significant deviation from the normal oil pressure reading is an indication of pending problems.

The ammeter indicates the amount of current being drawn from the battery to operate the electrical equipment or the amount of current being supplied to the battery to charge it. When the engine is running, the electrical system should have sufficient capacity to supply all the current needed by the lights and other electrical equipment. The system should also have sufficient reserve to replace the current drawn from the battery to start the engine. The reading on the ammeter does not include the current being supplied directly from the generator or alternator to the various devices, but only indicates the amount of current flowing into the battery or being taken from it. When a pumper is equipped with an electrically driven primer, it too requires a very large starting current. If the batteries are not maintained in good condition and at full charge, these accessories may not operate.

The voltmeter is an optional gauge. It provides a relative indication of battery condition by showing the amount of drop in voltage that is measured as some of the more demanding electrical accessories are used. It will indicate the top voltage available when the battery is fully charged.

All vehicles equipped with air brakes have an air pressure gauge to indicate whether there is sufficient air pressure to operate the brakes. If the system is in good condition, the vehicle can sit for extended periods without a decrease in air pressure in the system, but most systems have minor leaks that are very difficult to eliminate. These leaks will cause the pressure to decrease to inoperable levels over a period of inactivity.

The water temperature gauge indicates the temperature in degrees of the coolant in the engine. The operating temperature of the engine is important: an engine that operates too cool will not be efficient and an operating temperature that is too high may cause damage to the mechanical parts. Auxiliary cooling devices are provided to compensate for the lack of movement and can be used to maintain the engine temperature within tolerable limits when the pump is in operation.

The fuel gauge displays the level of fuel in the apparatus tank. One gauge should be located in the cab and one on the pump panel. The pump panel fuel gauge will be important during extended pumping operations when fuel may run low.

AUXILIARY COOLERS

The primary function of auxiliary coolers is to control the temperature of cooling water in the apparatus engine during pumping operations. There are two types of auxiliary coolers in common use today: the marine type and the immersion type. The marine type is inserted in one of the hoses used in the engine cooling system in such a way that the engine coolant must travel through it as it circulates through the system (Figure 5.74).

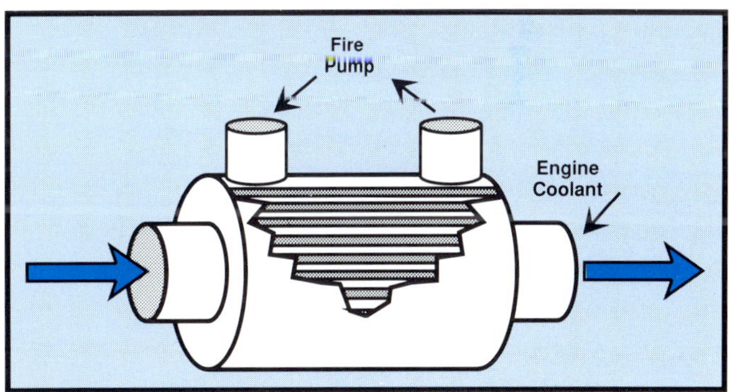

Figure 5.74 A marine-type cooler. As the coolant from the radiator passes through the tubes in the cooler, the colder water from the pump reduces the temperature of the coolant.

The cooler itself contains a number of small tubes similar to the flues in a steam boiler. Surrounding the tubes is a water jacket that is connected to the discharge of the fire pump. When the fire pump is operating, water from the pump can be circulated through the water jacket of the auxiliary cooler. As the coolant from the radiator passes through the tubes in the cooler, the colder water from the fire pump comes in contact with the metal tubes. The colder water conducts heat away from the tubes, thus reducing the temperature of the coolant flowing through them. A valve is provided on the pump panel to control the amount of water being supplied to the auxiliary cooler from the pump panel. This valve is needed because it is not desirable to lower the engine temperature below the recommended level.

The immersion-type auxiliary cooler is mounted in a similar manner to the marine type, with the radiator coolant passing through the body of the cooler. In this case, the water being supplied by the fire pump passes through a coil or some type of tubing mounted inside the cooler so that it is immersed in the coolant. As the cooler water from the fire pump passes through a tubing, some of the heat from the coolant is absorbed by the tubing and dissipated in the water from the fire pump. This type of cooler is also controlled by a valve on the pump operator's panel to regulate the degree of cooling desired (Figure 5.75).

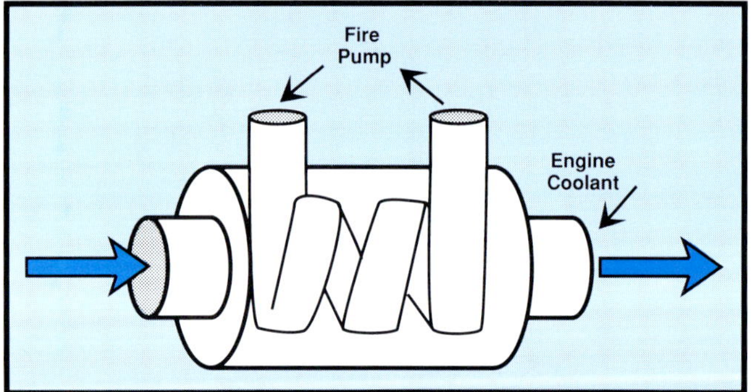

Figure 5.75 An immersion cooler. Tubing immersed in the coolant absorbs heat and dissipates it in the water from the fire pump.

Both types of auxiliary coolers are constructed so that the coolant in the radiator does not come in contact with the cooling water from the fire pump. Thus, the radiator cooler can be used without contaminating the engine coolant.

Some manufacturers also supply a radiator fill valve. This valve can be used to refill the radiator if the coolant level drops too low for effective cooling during a pumping operation. The radiator fill valve should only be used in an emergency, since whatever type of water is being supplied to the fire pump will get into the radiator and be circulated through the engine and the entire cooling system. If antifreeze is installed in the radiator, it will be diluted by the water from the fire pump and may become ineffective.

If it is absolutely necessary to use the radiator fill valve to replenish the coolant in the radiator, the line from the radiator fill valve should be restricted sufficiently to reduce the pressure to the radiator to safe levels. It is still possible, however, that dangerous pressures could build up in the cooling system faster than the overflow line can relieve them. The radiator fill valve should then be closed a very small amount and the radiator closely watched to keep the amount of contaminated water pumped into the system to an absolute minimum.

As soon as possible after the pumping operation, the cooling system should be serviced. It should be thoroughly flushed and the system refilled with the correct amount of clean, fresh antifreeze solution.

Chapter 5 Review

Answers on page 360

TRUE-FALSE: Mark each statement true or false. If false, explain why.

1. One of the basic differences between liquids and gases is that gases have a specific volume, while liquids do not.

   ☐ T ☐ F ____

   ____

2. Centrifugal pumps are self-priming.

   ☐ T ☐ F ____

   ____

3. All the discharges on a centrifugal pump may be shut down for short periods of time with no ill effects to the pump.

   ☐ T ☐ F ____

   ____

4. The compound gauge measures the pressure on the intake.

   ☐ T ☐ F ____

   ____

MULTIPLE CHOICE: Circle the correct answer.

5. Which of the following is *not* an advantage of water as an extinguishing agent?
   A. It freezes at 32°F (0°C)
   B. It is capable of absorbing large amounts of heat energy
   C. Its liquid to steam expansion ratio is 1 to 1700
   D. It is nontoxic
6. How many 2½-inch (65 mm) discharges must a 1,500 gpm (6 000 L/min) pumper be equipped with?
   A. 4
   B. 5
   C. 6
   D. 8

MATCHING: Write the correct letter in the space provided.

7. Match terms associated with types of pressure to their descriptions.

_____ Atmospheric Pressure	A. Pressure exerted in all directions at a point in fluids at rest.
_____ Residual Pressure	B. The pressure remaining in a system while fluid is flowing.
_____ Static Pressure	C. The pressure exerted by the atmosphere at the surface of the earth due to the weight of air.
_____ Head Pressure	D. Pressure found on a water system during normal consumption demands.
_____ Normal Operating Pressure	E. Pressure created when the water source is higher than the discharge.

LISTING

8. List the three ways to create pressure for a water supply.

   A. ______

   B. ______

   C. ______

9. List at least four characteristics of positive displacement fire pumps.

   ______

   ______

   ______

   ______

   ______

   ______

10. List the three factors that influence discharge pressure.

    A. ______

    B. ______

    C. ______

11. List at least three methods of power transfer from a vehicle engine to the pump.

    ______

    ______

    ______

12. List the three categories of primers used in conjunction with centrifugal pumps.

    A. ______

    B. ______

    C. ______

13. Name the two main gauges that every pumper should be equipped with.

    A. ______

    B. ______

14. List the four principal types of flowmeters.

    A. ______

    B. ______

    C. ______

    D. ______

10-91

15. List at least five types of gauges or indicators that are found on the dashboard of any piece of fire apparatus.

____

____

____

____

____

FILL IN THE BLANK: Fill in the blank with the correct response.

16. The two main parts of a centrifugal fire pump are the ________ and the ________.

17. The number of impellers a pump has determines what ________ pump it is called.

18. The parallel pumping position of a two-stage centrifugal pump is also called the ________ position.

19. The series pumping position of a two-stage centrifugal pump is also called the ________ position.

SHORT ANSWER: Answer each item briefly.

20. What property of water makes it potentially damaging when it freezes?

____

21. Define water hammer.

____

____

____

22. Explain the basic operating principle of a positive displacement pump.

____

____

____

____

23. Explain the basic operating principle of a centrifugal pump.

____

____

____

24. What is the rule of thumb for using the transfer valve to switch from PARALLEL to SERIES position?

____

____

____

25. Briefly explain the operating principles of a two-stage pump when used in the SERIES and the PARALLEL positions.

26. Briefly outline at least three characteristics of centrifugal pumps.

27. Describe the main feature of a pressure relief valve.

28. What is the basic operating principle of a pressure governor?

29. Briefly explain the function of drain valves.

30. Describe the primary function of the bleeder valve.

31. What is the purpose of the master pressure gauge?

32. Briefly explain the primary function of an auxiliary cooler.

____

____

____

33. Which type of flowmeter is most common in fire service applications?

____

6

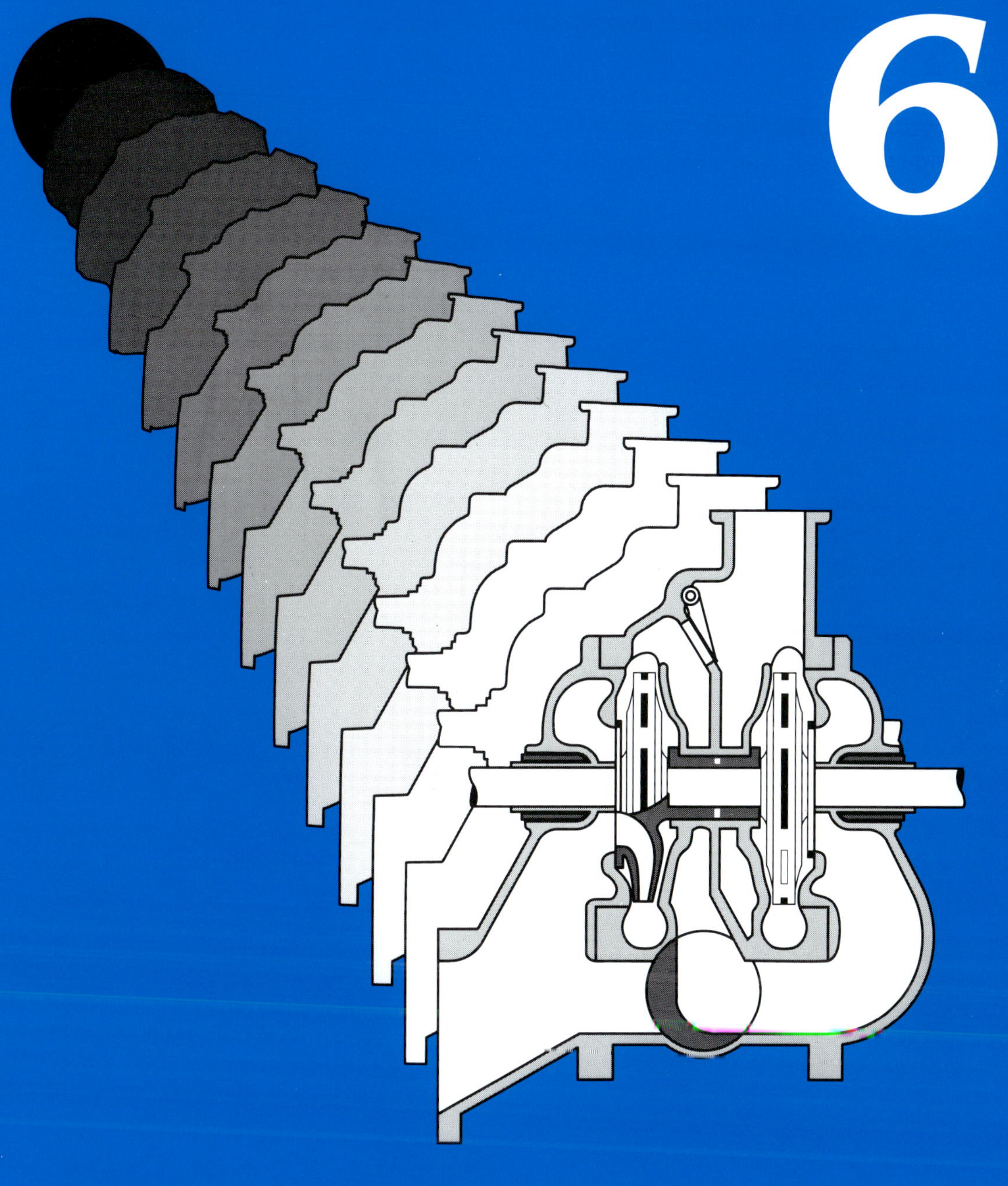

Operating Fire Pumps

NFPA STANDARD 1002
Fire Apparatus Driver/Operator Professional Qualifications

3-6 Operations.

3-6.1 The fire apparatus driver/operator, given a fire department pumper used by the authority having jurisdiction, shall demonstrate the method(s) of power transfer from vehicle engine to pump.

3-6.2 The fire apparatus driver/operator, given a fire department pumper and a series of fireground situations, shall produce effective hand and master streams specified by the authority having jurisdiction.

3-6.3 The fire apparatus driver/operator, given a fire department pumper, shall draft water, and demonstrate a systems check when the pumper will not draft.

3-6.5 The fire apparatus driver/operator, given a fire department pumper, shall properly position, set up the apparatus, and perform the following operations:

(a) pump at maximum delivery rate from the apparatus water tank
(b) pump at maximum rated capacity from a hydrant
(c) pump at maximum rated capacity from draft
(d) pump in a relay operation
(e) pump in a tandem pumping operation
(f) pump in a dual pumping operation.

3-6.7 The fire apparatus driver/operator, given a fire department pumper with a multiple-stage pump, shall demonstrate the operation of the volume/pressure transfer valve under actual pumping conditions.

3-6.11 The fire apparatus driver/operator, given a fire department pumper, shall demonstrate the operation of the pumper pressure relief system, or the pressure control governor, or both.

5-89

3-6.12 The fire apparatus driver/operator, given a fire department pumper, shall demonstrate the operation of the auxiliary cooling system.

3-6.13 The fire apparatus driver/operator, given a series of fireground situations, shall identify the capabilities and limitations of the water supply operation.

Chapter 6
Operating Fire Pumps

Once the driver/operator understands how the pump works, the next step is to become proficient at operating the pump under a variety of conditions. A knowledge of the operational theory of fire pumps is of no use on the fireground if the driver/operator is unable to put that knowledge to practical use. In this chapter, the driver/operator will be instructed in methods for operating the pump under a number of different circumstances.

There are several operating situations commonly encountered during fire suppression activities. The most common of these occurs when the pump gets its water supply from a tank mounted on the apparatus. When more water is required than is available from the tank, municipal fire departments usually obtain it from fire hydrants. Rural departments more often must depend on taking water from a static source by drafting it. Finally, any fire department may find that the source of supply is far enough away from the fire scene that a relay operation is required. Also covered in this chapter are sprinkler and standpipe support operations and operating mobile water supply apparatus.

WATER SUPPLY TO THE FIRE PUMP

Water can be supplied to a fire pump from three different sources: the water tank mounted on the apparatus, a hydrant or relay source, and a static source. All three are used from time to time in any type of fire department operation.

The first means of supply is the water tank mounted on the apparatus. Most fire apparatus has a tank that will provide water for the initial attack. The most common size on a Class A pumper is 500 gallons (2 000 L). Some modern attack pumpers have as much as 2,000 gallons (8 000 L) on board and some older pumpers have as little as 100 gallons (400 L) on board.

The water tank is normally mounted on the chassis higher than the intake of the fire pump (Figure 6.1). When the tank-to-pump valve is

Figure 6.1 An apparatus water tank is mounted directly on top of the main frame of the chassis. Shown above is an 1,800 gallon (7 200 L) tanker. *Courtesy of Saulsbury Fire Equipment Corporation.*

opened, the force of gravity causes the pump to fill with water. If the pump is normally drained when it is not in service, air may be trapped inside it. This air will prevent the pump from filling with water until a discharge is opened and air is allowed to escape. While priming is not normally required to get the water from the tank to the pump, operating the primer may speed the operation.

Hydrant System or Relay Operation

A hydrant system or a relay operation may be used to supply water under pressure (Figure 6.2). When utilizing either of these two sources, water enters the pump under pressure from the system. As the discharge increases, however, the incoming pressure will drop due to friction loss in the water system. If the flow is increased too much, the pressure in the system may be reduced below 0 psi (0 kPa). Operation at a negative pressure is dangerous because it increases the possibility of damage to the system due to water hammer if the flow is cut off suddenly. Water heaters or other domestic appliances may also be damaged by a negative pressure.

Figure 6.2 One method of taking water under pressure is by hooking directly to a hydrant.

Water may be supplied under pressure from another fire pump in a relay operation (Figure 6.3). If the volume of water being discharged is increased too much, the friction loss in the supply line may cause the residual pressure to drop below 0 psi (0 kPa). If this happens, the supply may collapse, resulting in an interruption in the water supply. Operating at or near this point can also cause damage to the pump from cavitation.

Operation at near-zero residual pressures requires close attention to the compound gauge on

Figure 6.3 Relays involve at least two engines, but may involve many more if the distance requires it. *Courtesy of Bill Eckman.*

the intake and prompt action before the danger point is reached. As a standard practice, it is not desirable to reduce the incoming pressure from a hydrant or relay below 20 psi (140 kPa).

Static Water Supply

All fire department pumpers should be capable of pumping water from a static water supply (Figure 6.4). In many cases, the supply will be located lower than the fire pump. Since one drop of water will not stick to another, it is not possible to pull water from above. It is possible to evacuate some of the air inside the fire pump. This creates a pressure differential, which allows atmospheric pressure acting on the surface of the water to force

Figure 6.4 All pumpers should be able to draft from a static water supply source.

water into the fire pump. In order to accomplish this, an airtight, noncollapsible waterway is needed between the fire pump and the body of water to be used.

In Figure 6.5, enough air has been evacuated to reduce the atmospheric pressure inside the fire pump and the intake hose to 12.7 psi (88 kPa). A negative pressure of -2 psi (-14 kPa) is measured on the compound gauge as 4 inches (102 mm) of vacuum. This vacuum will cause water to rise 4.6

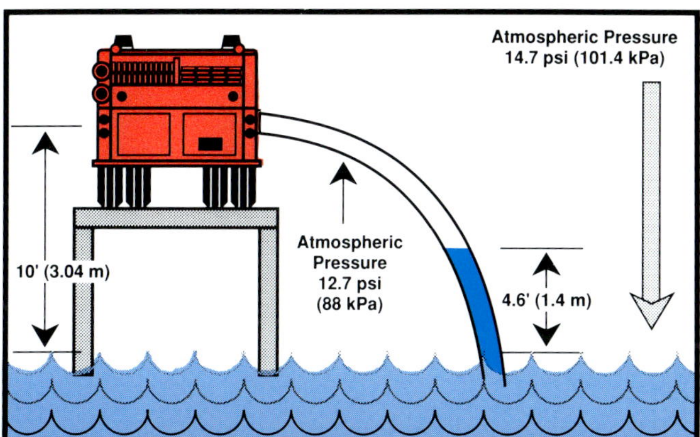

Figure 6.5 In this example, enough air has been evacuated to reduce the atmospheric pressure by 2 psi (14 kPa). The resulting vacuum causes the water to rise 4.6 feet (1.4 m) into the intake hose.

feet (1.4 m) into the intake hose from the surface of the water. The weight of the water, combined with the reduced air pressure acting on its surface, will create a balance (Figure 6.6).

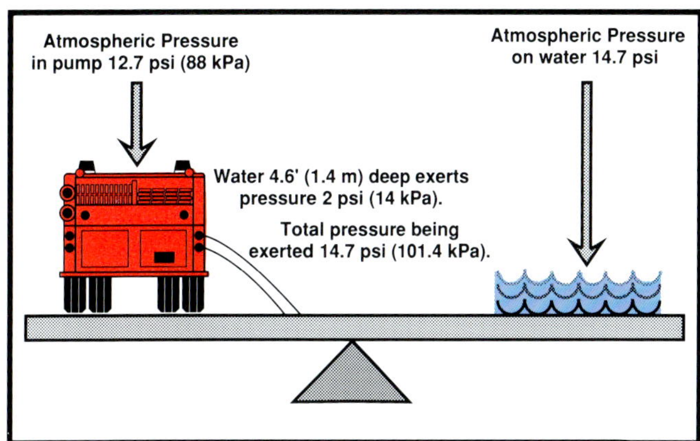

Figure 6.6 As the pressure inside the intake hose changes, the level will go up or down to maintain a balance at all times.

As the water begins to move through the pump, additional pressure losses are encountered. Any type of fire hose, strainer, or appliance creates a certain amount of friction loss. The amount of friction loss will be proportional to the amount of water moving through it.

The inertia of water is an additional factor in pressure loss. As the water begins to move through the pump, a certain amount of energy is consumed in getting the water at rest to begin to move and increase its velocity sufficiently to supply the amount of water needed.

The amount of friction loss in the hard suction hose is dependent upon the diameter of the hose. The smaller the size of the hose, the greater the friction loss will be, and accordingly, less water will be able to be supplied to the pump. To allow for this, the size of the suction hose is increased for larger capacity pumps. For example, a pumper rated at 750 gpm (3 000 L/min) is normally supplied with a 4½-inch (115 mm) suction hose, which will allow the pumper to reach its capacity. By equipping the same pumper with a 5-inch (125 mm) hose, however, the capacity can be increased to 820 gpm (3 280 L/min) if all other factors remain the same.

The total pressure available to overcome all of these pressure losses is limited to the atmospheric pressure at sea level (14.7 psi or 101 kPa).

(Atmospheric pressure decreases .5 psi (3.4 kPa) for each 1,000 feet (305 m) of elevation.) Since the same 14.7 psi (101 kPa) of atmospheric pressure must overcome the elevation pressure as well as the friction loss, increasing the height of the lift also decreases total pump capacity. If the lift were increased from 10 to 16 feet (3 m to 5 m), the vacuum needed to handle the increased lift would increase from 9 to 14 inches (229 mm to 356 mm) of mercury, which leaves 5 inches (125 mm) less to overcome friction loss. Using the previous example of the 750 gpm (3 000 L/min) pumper, increasing the lift to 16 feet (5 m) with the 4½-inch (115 mm) suction hose would reduce the pump capacity to 585 gpm (2 340 L/min).

While the pump is moving water, the vacuum reading on the compound gauge provides an indication of the remaining pump capacity. The maximum amount of vacuum that most pumps will develop is approximately 22 inches of mercury (hg) (560 mm/hg). A reading approaching this figure is a warning to the pump operator that the pump is getting close to the limit of its ability to deliver. Any further requirements will necessitate some other means of supply. If an attempt is made to increase the discharge from the pump beyond the point of maximum vacuum on the intake, cavitation will result.

Cavitation can be described as that condition where water is being discharged from the pump faster than it is coming in. Cavitation is sometimes expressed as "the pump running away from the water." Cavitation occurs when cavities are created in the pump or bubbles pass through the pump. They move from the point of highest vacuum into the pressurized section, where they collapse or fill with fluid.

The high velocity of the water filling these cavities causes a severe shock to the pump. In extreme cases or over prolonged usage, this shock can result in damage to the pump. A more scientific and accurate explanation is as follows: As the pressure drops below atmospheric, the boiling point of water drops to the point that the liquid changes to a vapor and creates a bubble of water vapor or steam. As the vapor passes through the impeller of the pump, the pressure increases, the vapor condenses, and water rushes in to fill the void. It is evident that the temperature of the water, the height of the lift, and the amount of water being discharged affect the point at which cavitation begins.

There are a number of indications that a pump is cavitating. The hose streams will fluctuate, as will the pressure gauge on the pump. A distinctive sound described as a popping or sputtering sound may be heard as the water leaves the nozzle. In cases of severe cavitation, the pump itself will be noisy, sounding rather like gravel is passing through it. The best indication of cavitation, however, is the lack of reaction on the pressure gauge to changes in the setting of the throttle.

When a pump reaches the point of cavitation, it is discharging all of the water that the atmospheric pressure or other pressure source can force into the intake. Since discharge pressure builds when water is supplied from the pump faster than it can be taken away, increasing pump rpm will not increase discharge pressure when there is no more water available to be supplied (Figures 6.7 and 6.8).

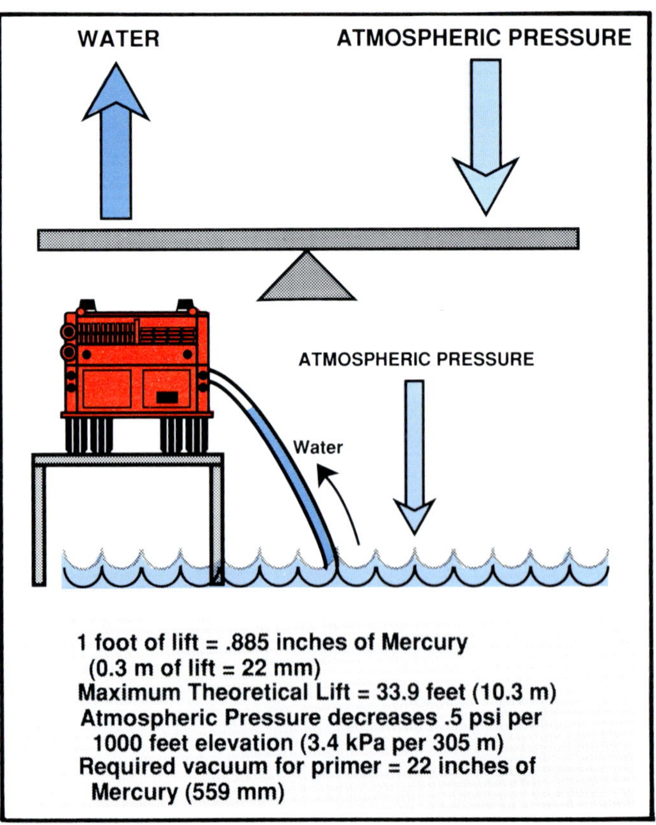

Figure 6.7 Principles of lift.

LIFT WATER FEET (meters)	VACUUM MERCURY INCHES (mm)	AIR PRESSURE PSI (kPa)
33.9' (10.3 m)	30.00" (762 mm)	14.7 psi (101 kPa)
32.2' (9.8)	29.04" (738)	14 psi (96.5)
29.9' (9.1)	26.96" (685)	13 psi (89.6)
27.6' (8.4)	24.88" (632)	12 psi (82.7)
25.3' (7.7)	22.8" (579)	11 psi (75.8)
23.0' (7.0)	20.72" (526)	10 psi (69.0)
20.7' (6.3)	18.64" (473)	9 psi (62.1)
18.4' (5.6)	16.56" (420)	8 psi (55.2)
16.1' (4.9)	14.48" (367)	7 psi (48.3)
13.8' (4.2)	12.4" (314)	6 psi (41.4)
11.5' (3.5)	10.32" (262)	5 psi (34.5)
9.2' (2.8)	8.24" (209)	4 psi (27.6)
6.9' (2.1)	6.12" (155)	3 psi (20.7)
4.6' (1.4)	4.08" (104)	2 psi (13.8)
2.3' (.7)	2.04" (52)	1 psi (6.90)

Figure 6.8 The chart shown above compares the ratios between lift, vacuum in mercury, and air pressure.

Cavitation often results when a pump has been equipped with inadequate piping from the water tank. Since a pump operates most often from its own tank, the pump can be severely damaged from habitual attempts to pump more water from the tank than the piping will allow to flow into the pump. Such damage can be especially severe when the pump is operating at relatively high pressures supplying long attack lines. Cavitation can occur while operating from a poor hydrant system but it most often occurs during drafting operations.

Successfully operating a fire department pumper from draft is one of the most challenging tasks most pump operators will face. It requires a thorough knowledge of the principles involved in drafting as well as a familiarity with the apparatus. All driver/operators should master drafting even if it is unlikely that they will be required to do so in an actual emergency operation.

Operating from the Tank

During the majority of fire calls, the initial attack is made by using the water carried in the tank on the attack pumper. Many fires can be controlled with this amount of water, so an operator must be able to operate the pumper competently from its tank. Even this type of operation can develop many problems that require a thorough knowledge of operating principles and practices to correct. If the fire progresses and an extended water supply operation is needed, the pump operator must be able to make the transition from the apparatus tank to an incoming water supply smoothly, with no disruption of the fireground operation.

OPERATING THE FIRE PUMP

Putting the Pump into Gear

After spotting the pumper in the desired position, the driver should set the parking brake. The sequence of operation for putting a fire pump into gear and getting water into the pump will vary with local department policy, design of apparatus, and type of priming device. On some types of apparatus, the pump must be put into gear before the driver leaves the cab. The steps required to put a fire pump into gear depends largely upon the location and type of the pump.

Front mounted pumps are provided with a clutch between the pump and motor. This requires that the operator leave the seat of the cab and go to the front of the apparatus to engage the pump.

PTO pumps have shifting levers that may be mounted on the floorboard or instrument panel in the cab. The operations usually required for placing a PTO drive into gear are as follows:

For Pump and Roll Operation:

Step 1: Bring the apparatus to a full stop.

Step 2: Disengage the clutch.

Step 3: Operate the PTO shifting lever to mesh the gears.

Step 4: Place the transmission in the proper gear.

Step 5: Engage the clutch slowly.

For Stationary Pumping:

Step 1: Bring the apparatus to full stop and set the brakes.

Step 2: Disengage the clutch.

Step 3: Operate the PTO shift lever to mesh the gears.

Step 4: Place the transmission in neutral.

Step 5: Engage the clutch slowly. The PTO pump should now be ready for operation.

Most midship mounted pumps require that both the pump and truck transmission be in gear for operation. The engine power is then transferred to the pump instead of to the drive wheels. The drive transmission must be placed in the proper gear to efficiently use the engine power for the pump. The gear used should be the one recommended in the operator's manual. The steps for putting most midship pumps into gear are as follows:

Step 1: Bring the apparatus to a full stop and set the brake. Allow the engine to idle.

Step 2: Shift the drive transmission into neutral. This applies for both standard and automatic transmissions.

Step 3: Operate the pump shift lever to transfer power from the drive axle to the pump drive.

Step 4: Shift the truck transmission into the proper gear for pumping and lock the lever in place.

Step 5: If the apparatus is equipped with a standard transmission, engage the clutch slowly. In apparatus equipped with automatic transmission, depress the accelerator to ensure the shift has been completed and the apparatus will not "drive away."

Putting the Pump into Operation

If the pump is a standard midship drive, it will be turning when the operator reaches the pump panel. If it is a front mount or one that is engaged from the pump panel, the driver/operator should first engage the pump drive system. The tank-to-pump valve must be opened fully (Figure 6.9). If a locking arrangement has been supplied, the valve must be locked in the OPEN position to prevent vibration or inadvertent contact from closing it and interrupting the flow of water to the pump. When the valve is locked open, the pump will fill with water. If the pump is a multistage one, the transfer valve should be set to the proper position before pressure builds up in the pump.

Figure 6.9 Once the pump is in gear, the water tank-to-pump valve should be opened fully.

In most cases, the pump should be in a SERIES (PRESSURE) position if the operation is from the tank since maximum flow is limited by the size of the pump piping. If it appears that the pump will be required to furnish more than 50 percent of its rated capacity, it should be set to the PARALLEL (VOLUME) position from the start. This will prevent having to shut down later on to make the changeover from the SERIES position.

The operator should then increase engine rpm by using the hand throttle, at the same time observing the pressure gauge (Figure 6.10). If the pump is normally full of water, the pressure gauge should start to rise as soon as the rpm is increased.

Figure 6.10 Once the pump is full of water, the discharge pressure can be built up by increasing the hand throttle.

If the pump has been drained, it will be full of air. The air will have to be forced out of the pump by the water as it enters before discharge pressure can build up. If this is the case, at least one of the discharge valves must be open before the air can escape and the pump can fill with water. Operating the primer will speed the removal of air from the pump.

If none of the attack lines are ready to be charged by the time the pump pressure has built up, the pump-to-tank line (tank fill valve) can be partially opened to allow water to circulate. Any air that has been trapped in the pump or piping will be removed and a stable pressure can be established. If the hoselines are ready for water when the pressure is set, the discharge valve can be opened slowly, locked into position, and the fire flow begun. The automatic pressure regulating device should be set for the operating pressure. Information on setting pressure regulating devices for specific makes of pumps is contained in the appendices to this manual.

While the pump is in operation, the operator should carefully observe all the gauges, including engine gauges and those gauges associated with operation of the pump. The operator should be ready to take whatever action is required to remedy any abnormal reading.

During the initial attack, hoselines operate intermittently. If no water is being used for an extended time while the discharge pressure is maintained at relatively high levels, the pump will overheat. To prevent this, some means of maintaining water movement through the pump must be found. If the pump is equipped with a circulator valve or a booster cooling valve, it can be opened and set to the tank position. Water will circulate through the pump and back into the tank, thus providing some cooling without wasting any water.

The pump operator must be conscious of the water level in the tank. As the level drops, the operator should warn the officer in charge of the amount of water remaining in the tank. The operator should also be able to give an estimate of the amount of time the water will last at the present rate of consumption.

Transition to an External Water Supply

As the water in the tank becomes depleted, an external supply may be provided from the tank on another apparatus, from a relay operation, or from a supply line connected to a hydrant. When this happens, it will be necessary to change over from the water tank to the new supply. This transition must be made without disrupting the fire attack.

The supply line should be connected into a gated intake fitting of the pump while the gate valve remains closed (Figure 6.11). There will be

Figure 6.11 Gated intakes are most often used to accept water from another apparatus.

air in the empty hoseline that will cause problems if it enters the pump while the pump is supplying attack lines under pressure. To purge this air, open the bleeder valve on the gated intake line so the air can escape as the supply line fills with water. When all the air has been forced from the line and a steady stream of water comes from the bleeder valve, the bleeder valve should be closed. The water supply is now at the pump, ready for the transition.

To begin the transition, the operator should open the intake valve from the supply line very slowly while observing the pressure on the intake and discharge gauges (Figure 6.12). At the same time the intake valve is opened, the tank-to-pump valve can be closed. Many apparatus are designed so that the pump-to-tank line may be cracked open to refill the water tank without interrupting other pumping operations.

MAKE SURE ALL AIR IS BLED FROM INTAKE HOSES PRIOR TO OPENING THE INTAKE VALVE.

Figure 6.12 Once water is being used from the external source, the driver/operator may have to adjust the throttle to correct for any increase in pressure that may have resulted.

After the changeover is made, the water for the attack lines will come from the external supply. If the supply is sufficient, an effort should be made to fill the tank on the apparatus as well. This should be done as quickly as possible so that any interruption to the water supply will leave the attack pumper with a full tank. Only excess water supply not needed for the attack lines should be used to fill the tank.

OPERATING PROCEDURE FOR TRANSITION TO AN EXTERNAL WATER SUPPLY

The steps for making the transition from the pumper's water tank to an external water supply are as follows:

Step 1: Position the apparatus in a safe position and immobilize it by setting the parking brake and blocking the wheels.

Step 2: Engage the pump and select the proper gear in the road transmission. Lock the gear in place.

Step 3: Open the tank-to-pump valve.

Step 4: Set the transfer valve to the SERIES position, if necessary.

Step 5: Increase the throttle setting to obtain the desired pressure, priming if necessary.

Step 6: Set the relief valve or pressure governor.

Step 7: Open the circulator valve or partially open the tank fill valve.

Step 8: When an external supply becomes available, open the intake valve while closing the tank-to-pump valve. As soon as an adequate supply of water is available, divert enough of it through the tank fill line to replenish the supply in the tank.

OPERATING FROM A HYDRANT

Except in rural areas, taking water from hydrants will be the second most common water supply that driver/operators will use. Whether the pumper directly supplies attack lines or supplies another pumper in a relay, the operation of the pumper on the hydrant is basically the same.

Choosing a Hydrant

The first consideration in selecting a hydrant is to determine which hydrant is most appropriate in terms of fire fighting and safety needs. The hydrant closest to the fire is not always the best choice. Due to limitations of the water system, the closest hydrant may not be capable of supplying the necessary amount of water. Connecting to such a hydrant would mean that supply lines would have to be laid to another hydrant with a larger capacity. In addition, the closest hydrant may be too close to the fire; connecting to it would put the safety of personnel and equipment in jeopardy.

In order to select the best hydrant, a thorough knowledge of the water system is required (Figure 6.13). The best hydrants are located on large water mains that are arranged in a grid pattern so they can receive water from several directions at the same time (Figure 6.14). The worst hydrants are those located on "dead-end mains." Water mains that are interconnected into a grid and that have

Figure 6.13 The only way a driver/operator can become familiar with a water system is to study maps and other information about the system.

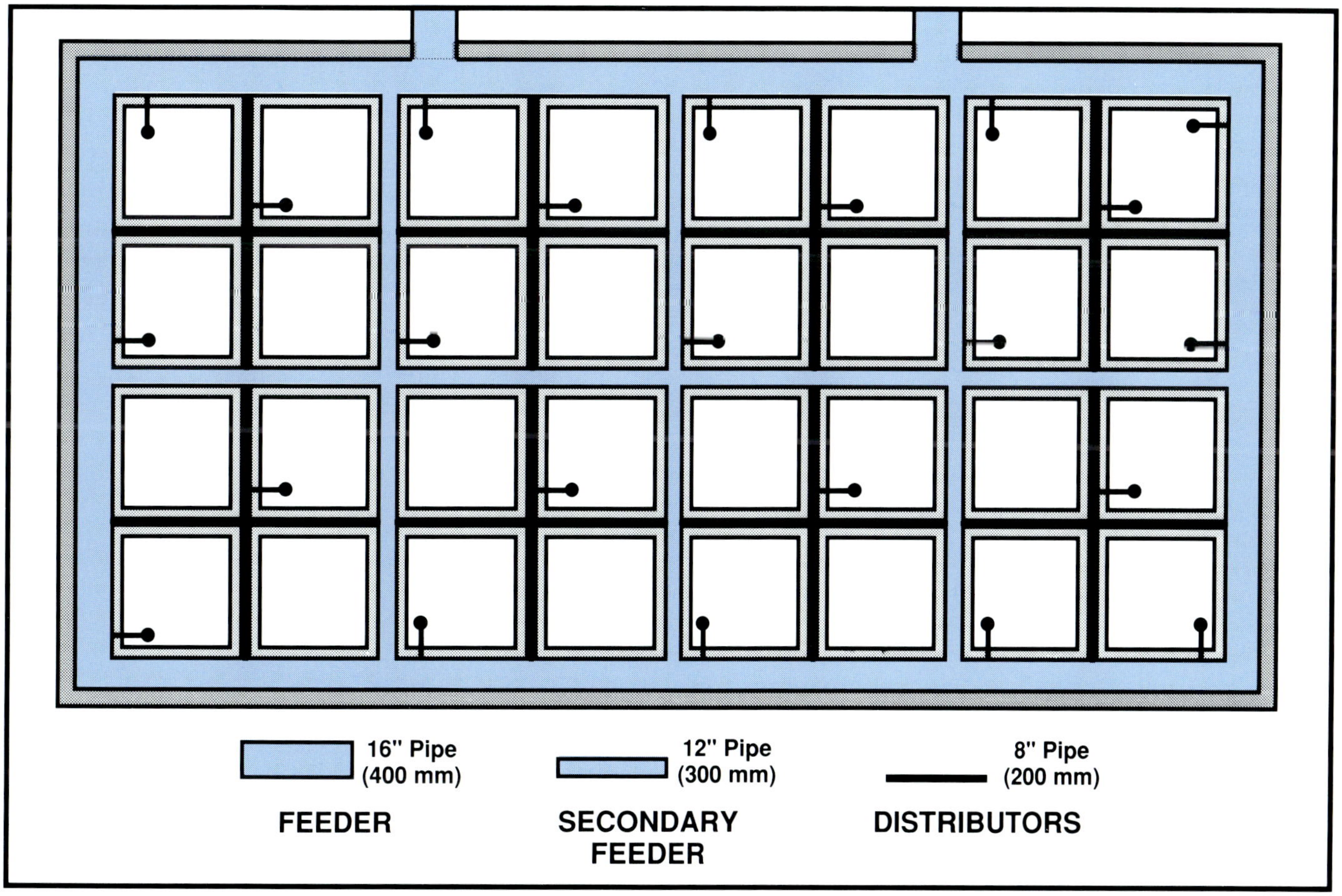

16" Pipe (400 mm)
FEEDER

12" Pipe (300 mm)
SECONDARY FEEDER

8" Pipe (200 mm)
DISTRIBUTORS

Figure 6.14 A well-gridded water distribution system consists of primary and secondary feeders and distribution mains.

relatively high amount of circulation are usually free of sedimentation and deterioration. Single lines used to supply relatively small amounts of water may be partially clogged, which further reduces their capacity.

Once a hydrant has been selected, the pumper should be hooked to it as quickly as possible. The driver/operator and the pumper crew should be well trained and practiced at making this connection as smoothly as possible. For detailed instructions on making hydrant connections using hard suction, soft sleeve, and short sections of 2½- or 3-inch (65 mm or 77 mm) hose, see Chapter Four, Positioning Apparatus.

Use of Valves or Unused Outlets

In order to receive the maximum amount of water possible, the pumper is usually connected to the hydrant's 4½-inch (115 mm) discharge outlet. Some hydrants have the ability to flow even more water than can be taken through this connection. For this reason, some departments make it a standard practice to attach one or two gate valves to the unused 2½-inch (65 mm) connections of the hydrant. This would allow additional lines to be connected to the original pumper, or for one or two supply lines to be run to a second pumper without having to shut down the hydrant to make the connection.

Forward Lays

Many departments make it standard practice to drop and lay lines directly from a hydrant to the fire scene. In some cases, those lines will merely lay there until a second pumper comes along to hook up to the hydrant and supply the line(s). Other departments connect those lines directly to the 2½-inch (65 mm) discharges of the hydrant and begin to work off hydrant pressure. If this is done, gate valves should be attached to the unused discharges so that other connections may be made to the hydrant without having to interrupt the flow in the original line(s).

When using 2½- or 3-inch (65 mm or 77 mm) hoselines to supply the pumper directly off hydrant pressure, it is not recommended that the lines be longer than 300 feet (90 m). Beyond this point, the amount of friction loss in the lines is too great to provide an adequate supply of water. This distance may be lengthened somewhat if the hydrant is on a special high-pressure main. The 300-foot (90 m) maximum also does not apply when using large diameter supply hose (4-inch [100 mm] or larger). These lines may be operated off hydrant pressure for significantly longer distances than smaller hose. The exact distances will vary with the size and manufacturer of the hose, the required flow, and the strength of the water system supplying the hydrant. Whenever using a forward lay, it is desirable to use a four-way valve to make the connection so the hydrant can be used to its fullest extent.

Four-Way Hydrant Valves

Several manufacturers make a special hydrant valve that allows the supply line to be charged initially from hydrant pressure. The valve may have a 4½- or 5-inch (115 mm or 125 mm) shutoff that enables a pumper to be connected to the outlet without interrupting the flow to the supply line. A second intake connection on the valve then allows the original supply line, which was first charged from the hydrant, to be supplied from the pump at increased pressure.

Figures 6.15 through 6.17 illustrate the operation of one type of four-way valve. In Figure 6.15, the valve is connected to a hydrant. The 5-inch (125 mm) supply line to the attack pumper has been connected to the 5-inch (125 mm) outlet. The hydrant has been opened, and the clapper valve has operated to allow the water to flow into the supply line. The pumper then connects to the 5-

Figure 6.15 The quint has laid a supply line to the fire and is operating off hydrant pressure. *Courtesy of George Braun, Gainesville (Florida) Fire-Rescue.*

inch (125 mm) pumper intake connection on the four-way valve (Figure 6.16). The pumper intake shutoff is opened, and the water from the hydrant fills the pump without interfering with the flow through the original supply line. The pump operator connects one of the pumper discharge outlets to the second intake of the hydrant valve (Figure 6.17).

As the pressure in the pump builds up, it overcomes the pressure from the hydrant that keeps the clapper valve open. When the pump pressure gets high enough, the clapper valve closes. When this happens, the supply line is fed from the pump, not directly from the hydrant. Figures 6.18a and b give an overview of the entire operation. The Humat valve used is one of the more sophisticated devices of this type. There are many variations on this principle available, including the use of simple valve and siamese fittings to do the same job.

Figure 6.16 The engine arrives at the hydrant and connects its intake hose to the hydrant valve discharge. *Courtesy of George Braun, Gainesville (Florida) Fire-Rescue.*

Figure 6.17 The engine boosts the pressure of the original supply line laid by the quint. This gives the attack apparatus the ability to supply more lines on the fire scene. *Courtesy of George Braun, Gainesville (Florida) Fire-Rescue.*

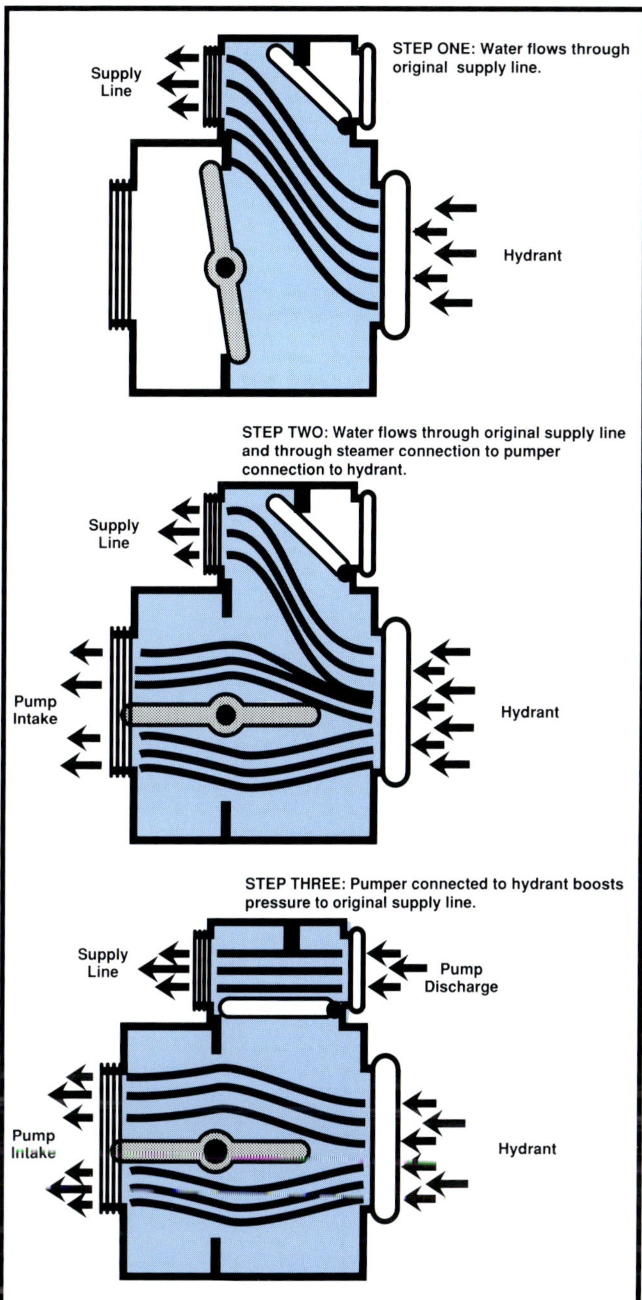

Figure 6.18a This drawing highlights the flow of water when using a Humat valve.

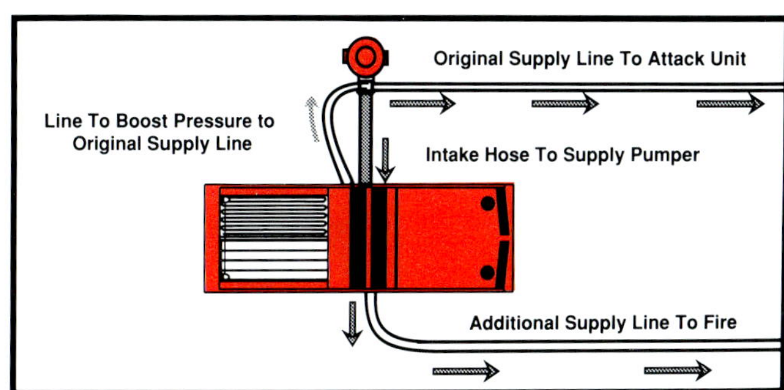

Figure 6.18b This illustration shows the flow of water in a completed four-way valve.

Getting Water into the Pump

After the connections to the hydrant are made, certain preliminary checks should be made before opening the hydrant. The tank-to-pump valve must be closed. Many of the newer pumps have check valves in the tank-to-pump line. These check valves prevent water from entering the tank under pressure from the pump intake. Many older pumps do not have this type of protection. If water is allowed to enter the tank under pressure, the venting may not be adequate to allow the pressure to dissipate and damage to the tank may result. Whether or not a particular apparatus has a check valve in the tank-to-pump line, the driver/operator should always make sure the tank-to-pump valve is closed when switching to an external water supply.

The hydrant must be opened all the way. If it is not opened fully, the drain valve at the base of the hydrant may be open at the same time water is coming in from the main. This flow of water (coming from the system and leaving through the drain valve) will wash away the gravel that is supporting the body of the hydrant. Loss of support will, of course, eventually damage the hydrant.

After the pump is full of water and the pressure in the system has stabilized, a reading of the pressure on the intake gauge will indicate the static pressure in the system. This reading is very important for estimating the remaining capacity of the hydrant as the water begins to move. Therefore, it may be desirable for the pump operator to record the static pressure before the operation begins. Without knowing the static pressure, it will be impossible to establish a reference for determining whether additional hoselines can be supplied.

Putting the Pump in Service

When operating from a hydrant, it is not advisable to engage the pump drive system before leaving the cab. This is because there may be a long time when the pump would be running dry with no lubrication or cooling while the connections are being made. Once the connections have been made, the pump may be put into gear as previously outlined in this chapter.

Before building up pressure, set the transfer valve to the proper position. The proper position to use is dictated by the amount of water that will be supplied. Even if only one line is initially put in service, it may be necessary to supply additional lines as the fire progresses. If the pump is placed in the SERIES position, it may not be possible to do this. Since the pump on the hydrant is usually called on to supply large amounts of water, it is good operating practice to set the transfer valve to the VOLUME position. An exception to this rule would be if very high pressures are required for very long supply lines.

Open discharge valves slowly. Since the lines are empty, they must fill before pressure can be established. While observing both the intake and discharge gauges, increase the engine rpm. If the intake gauge drops below 20 psi (140 kPa), do not increase the throttle further because the pump is in danger of cavitation. If there is adequate supply, then advance the throttle until the desired pressure and flow have been established. The relief valve or governor can now be set.

Since the pump operator at the hydrant may be some distance from the actual fireground operation, it may be difficult to determine how much water is being used or even if any of the lines are flowing. Because of the intermittent use of water in fire fighting operations, it is necessary to keep the pump from overheating by assuring a continued minimum flow. Some methods for preventing overheating are

- Open a booster line
- Open a discharge drain valve
- Partially open the tank fill valve
- Use a bypass or circulator valve if so equipped

The flow should be sufficient to cool the pump without seriously diminishing the amount of water available for fire suppression. If an open booster line is used, the operator must secure the line properly to protect against injury from a loose hoseline. One way to do this is to place the line into the top fill opening of the booster tank and tie off the hoseline with a short piece of rope. Use recirculating valves when available on the apparatus.

Shutting Down the Hydrant Operation

Due to the danger of water hammer and pressure surges on water systems, all changes in flow should be made smoothly, with no sudden actions. This is especially true when shutting down hydrant operations. Hydrants have been broken and pushed out of the ground by water hammer when a large flow has been stopped suddenly. To shut down the pump, first take the pressure governor out of service if the vehicle is so equipped. Then gradually slow the engine rpm to reduce the discharge pressure and slowly and smoothly close the discharge valves. Disengage the pump and put the road transmission in neutral.

Close the hydrant slowly and completely to prevent water hammer. The last turn of the operating nut opens the drain holes in the base of the hydrant, allowing the water to drain from it (Figure 6.19). This is particularly important in areas subject to freezing temperatures. In these areas, water left in the hydrant barrel will freeze, making the hydrant unusable. Frozen water in the hydrant may also cause permanent damage to the hydrant. After a hydrant is closed and before the caps are put in place, the operator can visually check to be sure the water is draining from the hydrant.

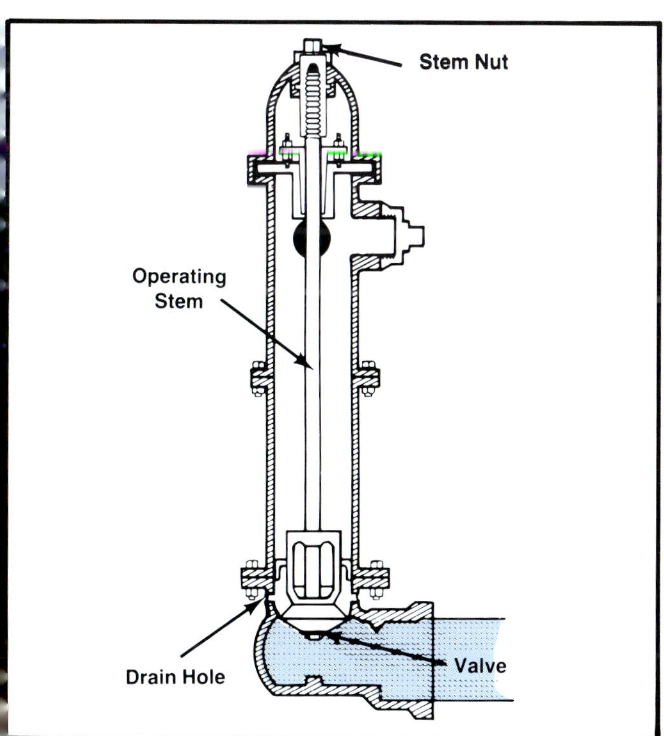

Figure 6.19 Hydrant drain holes allow water to drain from dry-barrel hydrants.

OPERATING THE PUMPER FROM DRAFT

The first consideration in establishing a successful drafting operation is selecting the site. If the purpose of the operation is to supply water to a fireground directly or through a relay, there may not be much choice as to where to set up. If the draft is being established to supply tankers, there may be several choices. The choice is dictated by the following factors:

- Amount of water
- Type of water
- Accessibility of water

The most important factor in the choice of the draft site is the amount of water available. If the draft is from a static body of water, such as a pool or lake, the size of the body becomes significant. A backyard swimming pool holding only 12,000 gallons (48 000 L) of water will not last long if large streams are in use. A small pond may not have a large capacity, but the rate of replenishment may make up for lack of volume. Even a rather small stream can provide a good supply if the water is moving rapidly. Here, the problem becomes more one of accessibility.

In order for a pumper to approach its rated capacity, there must be a minimum of 24 inches (610 mm) of water over the strainer. It is also desirable to have 24 inches (610 mm) of water all around the strainer. This will help to ensure maximum capacity and to avoid drawing foreign objects, such as sand and gravel, into the pump (Figure 6.20). If there is not 24 inches (610 mm)

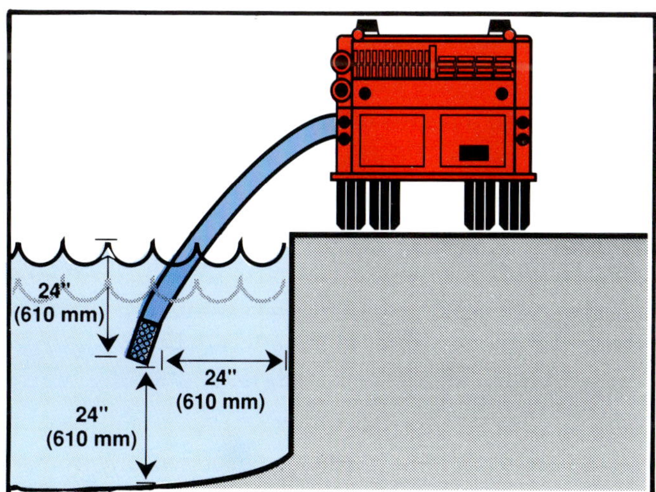

Figure 6.20 It is best to have a minimum of 24 inches (610 mm) of water all around a strainer for maximum water intake.

above the strainer, the rapid movement of the water into the suction strainer will cause a whirlpool to form. In extreme cases, this whirlpool can allow air to get into the suction hose and cause the pump to lose its prime. A wooden board or beach ball may be placed above the strainer to prevent a whirlpool effect (Figure 6.21).

Figure 6.21 A beach ball placed over a strainer will prevent a whirlpool from forming.

To use a swiftly moving shallow stream, a dam can be constructed from available material to raise the level, the bottom can be dug out to form a pool of greater depth, or a combination of the two can be used (Figure 6.22). A better solution is the use of a "floating" strainer. With this type of strainer, the end of the suction hose floats on the surface and the water is drawn into the suction hose through a series of holes on the bottom of the strainer (Figure 6.23). Since the water has to enter the hose from the bottom and come around the float, there is no way that a whirlpool can develop. In order for a floating strainer to work properly, it must float free in the water and not be constrained by the rigidity of the suction hose. Figures 6.24 and 6.25 illustrate the right and wrong ways to use a floating strainer.

In an emergency, any type of water can be used for fire fighting; however, pumping nonpotable water can be harmful to the pump. In some

Figure 6.22 Constructing a makeshift dam can enable a normally insufficient water source to be used for fire fighting operations.

Figure 6.23 A whirlpool cannot develop with a floating dock strainer because water enters the hose from the bottom.

areas, most of the available water is salt water. Salt water will cause corrosion and a gradual deterioration of the pump if it is not thoroughly flushed out after each use. Sulfur water is common in the vicinity of coal mines. Other chemicals may be present near industrial plants.

In addition to flushing the pump with fresh water each time it is used to pump nonpotable water, other precautions must be taken. It is not a good idea to fill the tank with nonpotable water, since it will be very difficult to completely remove it when fresh water is available.

Drafting operations can be seriously affected by foreign material and debris clogging the

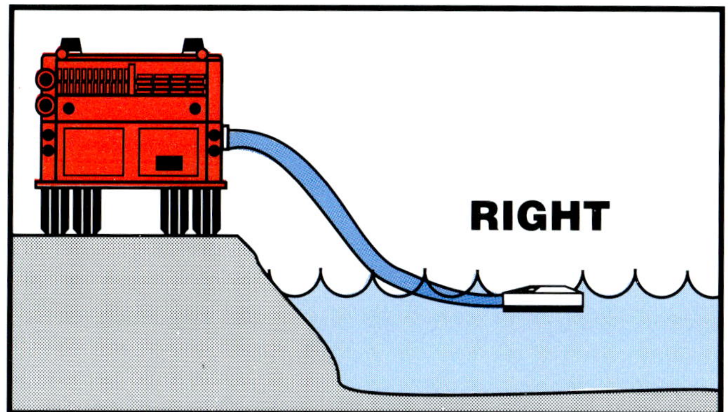

Figure 6.24 When used properly, the floating dock strainer should be perfectly parallel to the surface of the water.

Figure 6.25 The floating dock strainer should not be at an angle when in use.

strainer. An open body of water may contain a high concentration of algae or other marine vegetation that will block the strainer and prevent effective operation. Leaves or other debris floating on the water can also block the strainer. Using a floating dock strainer or covering the end of the strainer with a bucket will help prevent this. However, if high flow rates are attempted, the water source capacity may be limited by unavoidable blockages.

The most common type of contamination, and possibly the most damaging, is dirty or sandy water. If dirt particles are too small to be caught in the strainer, they pass into the pump where they can cause serious deterioration. As the water passes into the eye of the impeller, the sand acts as an abrasive in the area between the clearance rings and the hub of the impeller. Such abrasion will quickly increase the spacing to an intolerable amount, causing increased slippage from the discharge back into the suction and reducing the capacity of the pump.

As the dirty water passes through the pump, it is forced into the packing by the discharge pressure. The packing is contaminated by the dirt and can no longer be adjusted to form a good seal. This will cause air leaks at the intake of the pump and prevent it from being able to draft.

When operating from draft, be sure to maintain water in the booster tank. A leaking tank valve can cause the water to be drawn out of the pump, then let air into the pump, causing loss of prime.

Accessibility to the water source is another important consideration in selecting a drafting site. Since drafting is accomplished by evacuating the air from the pump and allowing the atmospheric pressure to push water into it, a maximum of 14.7 psi (101.3 kPa) is available. This 14.7 psi (101.3 kPa) has to overcome elevation pressure and friction loss in the suction hose. The higher the lift, the greater the elevation pressure, the less friction loss that can be overcome, and the more the capacity of the pump is decreased.

All Class A pumpers are rated to pump their capacity at 10 feet (3 m) of lift. If the lift is lower, the capacity is higher; if the lift is greater, the capacity decreases. While a pumper in good condition can lift water approximately 23 or 24 feet (7 m or 7.3 m), all of the available atmospheric pressure is required to overcome the lift. As a result, the tolerable amount of friction loss is so low that the capacity is too small to be of practical value. For effective operation, the maximum lift is about 20 feet (6 m) and the amount of water that can be supplied is only about 60 percent of the rated capacity of the pump. In selecting a drafting site, it is important to keep the lift as low as possible. It would be better to use an extra 100 feet (30 m) of supply line to set up the draft in a location where the lift will be lower and more water can be supplied.

Other considerations in selecting a drafting site are the stability of the ground, the convenience of connecting hoselines, and the safety of the operator. A well-trained operator will be familiar with the potential drafting sites in the area as well as the limitations and special considerations needed in using each site. It is a good practice to do some advance preparation at the most strategic drafting sites to improve their accessibility and usability before they are actually needed.

Connecting to the Pump and Preliminary Actions

After positioning the pumper, immobilize it by setting the parking brake and placing the wheel chocks. If the pumper is close to a road, leave the warning lights operating. Since there will be some delay before getting water into the pump, it should not be engaged until all connections are made and it is ready to put into operation.

Before connecting sections of hard suction hose, check the gaskets to make sure they are in place and that no dirt or gravel has accumulated inside the couplings (Figure 6.26). If a floating strainer is available, it should be used at all times. If a conventional barrel-type strainer is used, fasten a rope to it to aid in handling the hard suction hose after it is connected and in positioning it properly.

Figure 6.26 Always check hard suction hose for dirt in the couplings or other foreign objects before use.

The strainer and the necessary sections of hard suction hose must be coupled and made airtight. The key to doing this is getting each section of hose in line with the other before the coupling is turned. If the gaskets are in good condition and the coupling is done properly, it should be possible to get an airtight connection when the couplings are hand tight. If necessary, use a rubber mallet to make the connections tighter. It is important to have enough help to connect the hard suction hose without putting it on the ground since dirt may get into the coupling if it is laid in the dirt.

Once the sections of hard suction hose are connected with the strainer attached, it is usually easier to connect the hose to the intake fittings by first putting the strainer in the water and pulling the apparatus into position. If a keystone valve is normally connected to the intake, it should be removed before connecting the suction hose since the butterfly valve obstructs the waterway and will reduce the capacity. The rope that was tied to the strainer can be used to suspend the strainer above the bottom by tying it to the front or back of the pumper or to a tree or other natural object (Figure 6.27). If the bottom slopes away steeply from the edge, it may be possible to put a roof ladder in the water and rest the suction hose on it (Figure 6.28). Under extremely adverse conditions, a shovel or some other piece of flat metal can be placed on the bottom to protect the strainer.

Inside the cab, the pump drive system shift control must then be operated, the road transmission placed in the proper gear, and the clutch engaged. The pump will begin to turn. Again, the speedometer reading provides an indication that everything is properly set. The indicator light that confirms the disengagement of the rear axle is especially important in a drafting operation because the pumper is in a vulnerable position in case of any unwanted movement.

Figure 6.27a Shown above is a correct method for attaching a rope to a barrel strainer.

Figure 6.27b The rope may be used to maneuver the hose into the water as well as to keep the strainer off the bottom.

Figure 6.28 In some cases, it may be desirable to use a ladder to support the hard suction hose and keep the strainer from resting on the bottom.

During priming, the transfer valve on a two-stage pump should be in the PARALLEL position. When the two stages are connected in SERIES, air may be trapped inside the pump and it will be more difficult to remove. Before beginning priming, make sure that all drains and valves are closed and all unused openings capped to make the pump as airtight as possible.

Priming the Pump and Beginning Operation

If the primer is a positive displacement pump that is driven by the transfer case, the engine rpm should be set according to the manufacturer's instructions. Most pumps prime well between 1,000 and 1,200 rpm. If the priming pump is driven by an electric motor, the exact rpm is not critical, but 1,000 to 1,200 rpm should be sufficient to keep the alternator charging and prevent loss of prime once the pump fills with water. If the apparatus uses a vacuum-type primer, the engine rpm should be kept as low as possible without causing the engine to stall out.

After the pump has been made airtight and the engine rpm set, operate the primer control. As the air is evacuated from the pump, the compound gauge on the intake will register a vacuum reading. This reading should be 1 inch (25 mm) for each 1 foot (0.3 m) of lift. This vacuum, which is measured from the surface of the water to the eye of the impeller of the pump, is required to overcome the elevation pressure and cause the water to enter the pump.

As the primer operates, the vacuum reading increases and water is forced into the hose. As the hose fills with water, the weight of the water will cause the hard suction hose to drop. When the body of the pump fills with water, the primer will discharge water on the ground under the apparatus. At first, there will not be a steady stream as the water entering the pump mixes with the remaining air that is being removed.

The priming action should not be stopped until all the air has been removed and the primer is discharging a steady stream of water. As the pump fills with water, there will be a pressure indication on the discharge gauge.

The entire priming action typically requires 10 to 15 seconds from start to finish, but should not take more than 30 seconds (45 seconds in pumps larger than 1,250 gpm [5 000 L/min]) to accomplish. If water has not been obtained in 30 seconds, stop priming and check to find out what the problem is. The most common cause of inability to prime is an air leak that prevents the primer from developing enough vacuum to successfully draft water.

The first area to check is that the couplings on the hard suction hose are airtight. Open drains are also common. They should all be closed during priming and drafting operations. If the pump is equipped with a circulator valve or booster cooling valve, air can leak if it is not turned off completely. An intake relief valve is another frequent offender. An intake relief valve should always be equipped with a shutoff valve or an outlet fitting that can be capped. Cap the outlet before attempting to prime the pump again.

After the pump has been successfully primed, increase the setting of the throttle before attempting to open any of the discharges. Increasing the setting of the throttle is needed to boost the pressure to somewhere between 50 and 100 psi (350 kPa and 700 kPa). Open a discharge valve slowly while observing the discharge pressure. If the presssure drops below 50 psi (350 kPa), pause for a moment to allow it to stabilize before opening the valve further. If the valve is opened too quickly, air may enter the pump and cause it to lose its prime.

If the pressure continues to drop, momentarily operating the primer may eliminate the air still trapped in the pump and restore the pressure to the original value. Do not try to set the discharge pressure to the desired value until water is flowing. If the hoselines are not ready for water, a discharge can be opened, allowing water to discharge back into the source.

Even though the pump has been primed, there may be small amounts of air trapped inside the pump or in the piping that will not allow the pressure to increase normally. It is not until water moves through the pump that it becomes fully primed.

Constant movement of water through the pump is necessary to prevent overheating, but there is a more important reason to maintain flow at draft. The vacuum to initially prime the pump comes from the primer, but once the pump begins to operate, the primer is no longer used. The vacuum is now maintained by the movement of water through the pump. When no water is being discharged, even the smallest air leak may result in a loss of vacuum. This can be prevented by utilizing the same methods for preventing overheating of the pump discussed earlier. Once the desired pressure has been established and hoselines are in service, set the relief valve or presssure governor.

Operating the Pump from Draft

Operating from draft is the most demanding type of pump operation, both from the standpoint of the apparatus and the operator. It demands a careful monitoring of the gauges associated with the motor as well as those associated with the pump. The engine temperature gauge must be kept at the normal operating temperature by use of the auxiliary cooler. Only under the most critical conditions should the radiator fill valve be used while operating at draft since it is impossible to be certain of the quality of the water being pumped and damage to the engine could result. Other engine gauges are not as critical, but any deviation from normal engine temperature is a signal that another pumper should be used for drafting.

The problems that can occur while operating from draft fall into one of three categories:

- Air leak on the intake side of the pump
- Whirlpool allowing air to enter the pump
- Air leakage due to defective packing in the pump

Air leaks on the intake side of the pump are the most common problems. If the discharge pressure gauge begins to fluctuate with a corresponding loss of vacuum on the intake gauge, air is probably coming into the pump along with the water. The first place to check is the suction hose. Even if the connections were tight enough initially, vibration may have caused them to leak. If the water tank is empty, the tank-to-pump line may be a source of air leakage. In addition, check all drains and intake openings for leakage.

If there is not enough water over the strainer, a whirlpool may be allowing air to enter the pump. This can be corrected by placing a beach ball or other floating object above the strainer. A floating dock strainer can also be used in shallow water where whirlpooling is a problem.

Defective packing in the pump may cause air leakage. If there is an excessive amount of water leaking from the packing, that is, if it is leaking a steady stream instead of dripping, the packing is probably the cause. If this is the case, nothing can be done about it while the pump is operating. If the problem is severe enough to cause the pump to lose its prime, the pumper will have to be replaced and put out of service for repairs.

While the pump is operating, a gradual increase in the vacuum reading may be noted with no change in the flow rate. This is an indication that a blockage is developing. Blockage most often occurs after the pump has been operating satisfactorily for some time at a high rate of discharge, which creates a maximum velocity of the water that is entering the strainer. In extreme cases, the pump may go into cavitation, resulting in a fluctuation and a gradual decrease of the discharge pressure. An immediate cure is to decrease the engine rpm until the pressure drops. A drop in pressure indicates that the flow has decreased below the point of cavitation.

Since the reduced pressure may not be adequate to maintain the desired flow, try to clear the blockage and restore the operation to normal. The most common place for blockage to develop is at the strainer. If leaves or debris are present in the water being pumped, an accumulation against the strainer may be partially blocking the flow. This can be cleared by physically cleaning it if the water is shallow enough to allow access to the strainer. An alternate method is to charge the booster line under pressure and attempt to wash away the material blocking the strainer (Figure 6.29).

Blockage may also occur against the strainer that is located in the intake of the pump. This blockage cannot be cleared without interrupting the water flow because it is necessary to remove the suction hose to clear this strainer. Blockage can occur for other reasons. Since the suction hose is operating under a vacuum, the inner liner can become detached from the hose itself and collapse, thus blocking a portion of the waterway. This may not be immediately apparent since the vacuum is removed when the hose is disconnected and the liner may return to the proper position.

Figure 6.29 Turbulence caused by the stream from a charged booster line may be used to clear foreign debris that has been sucked against the outside of a strainer.

Exceeding the capacity of the pump will also lead to cavitation. This condition is also accompanied by a high reading of the vacuum gauge. Cavitation can occur at a flow well below the rated capacity of the pump if there is a high lift involved. The actual capacity of the pump in a given situation is determined by the size and length of the suction hose as well as the lift.

Shutting Down the Operation

When shutting down after a drafting operation, slowly decrease the engine speed to an idle, take the pump out of gear, and allow the pump to drain. Stabilize the engine temperature in the same manner as after a hard road run. After the pump is drained and the connections have been removed, operate the mechanical primer for a few seconds until oil comes out of the discharge from the priming pump. This will lubricate the parts of the primer and help to preserve them in good condition. Unless the pump has been pumping very clean uncontaminated water, it should be thoroughly flushed when a supply of fresh water is available.

OPERATING A PUMPER IN A RELAY

A relay operation is required anytime the source of water is located more than a few hundred feet from the fire. A relay can be set up by a two-piece engine company, with an attack pumper being supplied by a pumper at a hydrant. In rural areas, water may have to be supplied over distances as great as 1 mile (1.6 km) or more by the use of many pumpers and multiple hoselines. No matter how simple or complicated a relay may be, the principle is the same. Fire pumps are connected to the supply lines at intervals to counteract the pressure losses that occur as the water moves through the fire hose.

Designing a Relay

A relay operation is based on the amount of water needed and the distance the fire scene is from the water source. There are two main types of pressure loss that a relay operation must overcome: elevation pressure and friction loss. Elevation pressure is not affected by the amount of water to be moved, only by the topography. A relay operation must also overcome friction loss. Friction loss is directly affected by the amount of water being flowed, the size of the hose being used, and the distance between pumpers.

WATER RELAY OPERATIONS ARE BASED ON FIRE FLOW NEEDS AND DISTANCE.

There are certain limitations to any relay operation. Pumpers are rated for maximum capacity with 150 psi (1 000 kPa) net pump discharge pressure. If higher pressures are needed, the capacity of the pumper is reduced proportionally. Pressures are also limited by the fact that most fire hose is tested annually at a pressure of 250 psi (1 700 kPa). A maximum working pressure of 200 psi (1 350 kPa) in a relay allows a 50 psi (345 kPa) safety factor to provide for a certain amount of deterioration in condition of the hose and for pressure variations during operation. Table 6.1 (U.S. and metric) lists the distances various flows can be pumped at a pump discharge pressure of 200 psi (1 350 kPa). A minimum of 20 psi (140 kPa) remains as the intake pressure at the next pumper.

The number of pumpers needed to relay a given amount of water can be determined by using the following formula:

$$\frac{\textit{Relay Distance}}{\text{Distance From Table 6.1}} + 1 = \text{Total Number of Pumpers Needed}$$

Example 6.1 (U.S.)

If a single line of 3-inch hose is used, how many pumpers will be needed to supply a 1,000 gpm master stream device that is 1,000 feet from the water source (Figure 6.30)?

Answer

From Table 6.1 (U.S.), it can be seen that the maximum distance water will flow at 1,000 gpm through 3-inch hose is 225 feet. Divide this figure into the distance to the fire; then add 1 for the attack pumper:

$$\frac{1{,}000}{225} = 4.4 + 1 = 5.4, \text{ or 6 pumpers needed}$$

Example 6.1 (metric)

If a single line of 77 mm hose is used, how many pumpers will be needed to supply a 4 000 L/min master stream device from a source 300 meters away (Figure 6.30)?

Answer

From Table 6.1, it can be seen that the maximum distance water will flow at 4 000 L/min through 77 mm hose is 61 meters. Divide this figure into the distance to the fire; then add 1 for the attack pumper:

$$\frac{300}{61} = 4.9 + 1 = 5.9 \text{ or 6 pumpers needed}$$

In determining the desired flow from a relay, the maximum setting of the nozzle should be used rather than the amount of water actually being discharged. If the nozzle operator decides to increase the setting of the nozzle, the water must be available to supply it without changing the pressures in the relay. It would be impossible to change the relay every time an attack line was shut down or the nozzle setting changed, and if more water is needed it must be there.

TABLE 6.1 (U.S.) MAXIMUM DISTANCE OF WATER FLOW AT 180 psi (In Feet)					
Flow	Hose Diameter (inches)				
GPM	2½	3	4	5	6
100	9,000	22,500	90,000	225,000	360,000
200	2,250	5,625	22,500	56,250	90,000
250	1,440	3,600	14,400	36,000	57,600
300	1,000	2,500	10,000	25,000	40,000
400	563	1,406	5,625	14,060	22,500
500	360	900	3,600	9,000	14,400
750	160	400	1,600	4,000	6,400
1,000	90	225	900	2,250	3,600

Example 6.2 (U.S.)

If two lines of 3-inch hose are used, how many pumpers will be needed to supply a 1,000 gpm master stream device that is 2,000 feet from the water source (Figure 6.31)?

Answer

First divide the flow in half (500). Find the distance in Table 6.1 for 3-inch hose at 500 gpm (900 feet). Then divide this distance into the total distance.

$$\frac{2{,}000}{900} = 2.2 + 1 = 3.1, \text{or 4 pumpers needed}$$

TABLE 6.1 (metric) MAXIMUM DISTANCE OF WATER FLOW AT 1 260 kPa (In Meters)					
Flow	Hose Diameter (mm)				
L/min	65	77	100	125	150
400	2 445	6 102	25 410	56 159	93 373
800	611	1 526	6 352	14 040	23 343
1 000	391	976	4 066	8 986	14 940
1 200	272	678	2 823	6 240	10 375
1 500	174	434	1 807	3 994	6 640
2 000	98	244	1 016	2246	3 735
3 000	43	108	452	998	1 660
4 000	24	61	254	562	940

Example 6.2 (metric)

If two lines of 77 mm hose are used, how many pumpers will be needed to supply a 4 000 L/min master stream device that is 600 meters away from the water source (Figure 6.31)?

Answer

First divide the flow in half (2 000 L/min). Find the distance in Table 6.1 for 77 mm hose at 2 000 L/min (244 meters). Then divide this distance into the total distance.

$$\frac{600}{244} = 2.5 + 1 = 3.5, \text{or 4 pumpers needed}$$

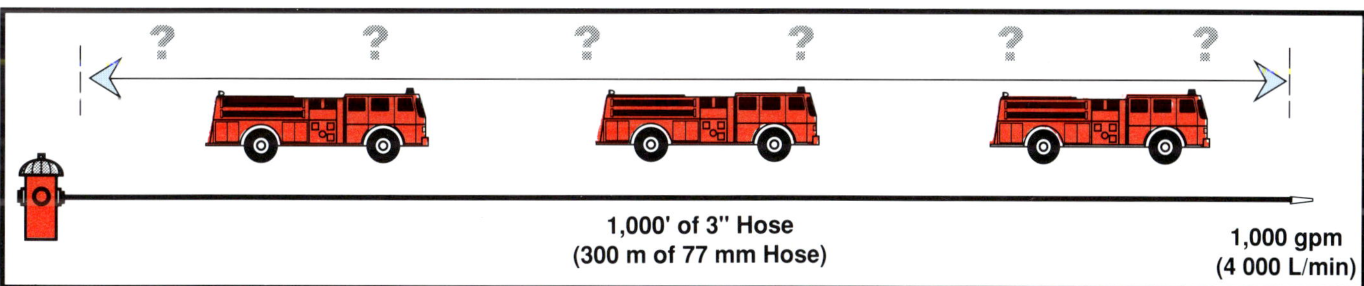

Figure 6.30 Determine the number of pumpers it will take to supply the required amount of water in the example shown above.

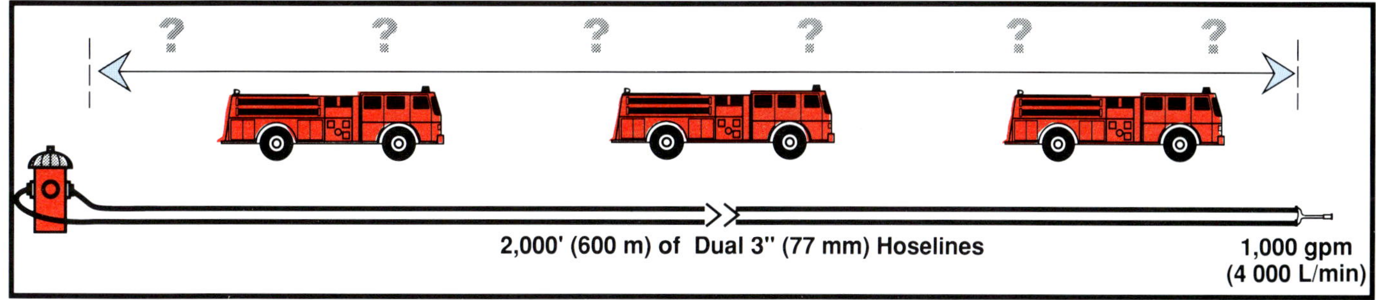

Figure 6.31 Determine the number of pumpers it will take to supply the required amount of water in the example shown above.

The number of pumpers necessary for the preceding examples is adequate if the pumpers have the ability to deliver 1,000 gpm (4 000 L/min) at 180 psi (1 260 kPa). The amount of water a pumper can deliver decreases with each increase in pressure. Put another way, an increase in volume decreases the pressure that the pump can develop. Table 6.2 shows this relationship between volume and pressure.

TABLE 6.2
MAXIMUM PRESSURE/VOLUME OF CLASS A PUMPERS

Percent of Capacity	Pressure (psi)	Pressure (kPa)
100	150	1 000
70	200	1 350
50	250	1 700

Source: American Insurance Association.

From Table 6.2, one finds that a pumper rated at 1,000 gpm (4 000 L/min) will deliver 1,000 gpm at 150 psi (4 000 L/min at 1 000 kPa), 700 gpm at 200 psi (2 800 L/min at 1 350 kPa), and 500 gpm at 250 psi (2 000 L/min at 1 700 kPa). The pumpers for Examples 6.1 and 6.2 then, would have to be rated at 1,500 gpm (6 000 L/min) to deliver 1,000 gpm at 180 psi (4 000 L/min at 1 240 kPa). If 1,500 gpm (6 000 L/min) pumpers were not available, more pumpers would be needed.

As a rule, the largest capacity pumper should be used at the source. If the relay is being supplied from draft, the source pumper will have to develop a higher net pump discharge pressure. This higher net pump discharge pressure is needed since the relay pumpers will have a residual pressure at the intake to reduce the amount of pressure needed from the pump. It is important to remember that the maximum capacity of the relay will be determined by the capacity of the smallest pump and the smallest hoseline used within the relay.

Another relay method that works well when pumpers lay hose long distances is the *constant pressure relay*. This relay method establishes the maximum flow available from a particular relay setup by utilizing a constant pressure in the system.

The source pumper and the relay pumpers operate at 175 psi (1 225 kPa) while the attack pumper maintains a constant flow through attack lines and uses an open uncapped discharge to handle excess flow. This allows the attack pumper driver/operator to throttle the discharge pressure to the correct level for the attack lines while still reserving the remainder of the total flow in case additional lines are needed. When fire situations dictate the use of a relay, the total available flow may already be committed.

The constant pressure may be modified for varying situations to increase or decrease the available flow but driver/operators must remember the maximum working pressure of hoselines (Figure 6.32). Source and relay operators should keep correcting to 175 psi (1 225 kPa) during the relay unless the following situations occur:

- Intake pressure from pressurized sources drops to 20 psi (140 kPa).
- Increasing the engine rpm does not result in an increase in discharge pressure (pump capacity reached).
- Operating the hand throttle does not result in an increase in rpm (engine has reached governed speed).

Putting a Relay Into Operation

A relay operation always begins with the source pumper. Once the water supply has been established, the source pumper opens an uncapped discharge or allows water to waste through a dump line until the first relay pumper is ready for water. The discharge pressure is built up to the calculated value by increasing the throttle. While doing this, the dump line discharge valve should be slowly closed to keep from wasting all the water from the water supply.

The relay pumper should be waiting for water with the dump line or discharge open and the pump out of gear. When both the source pumper and the relay pumper are ready, the discharge on the source pumper is opened while the valve on the

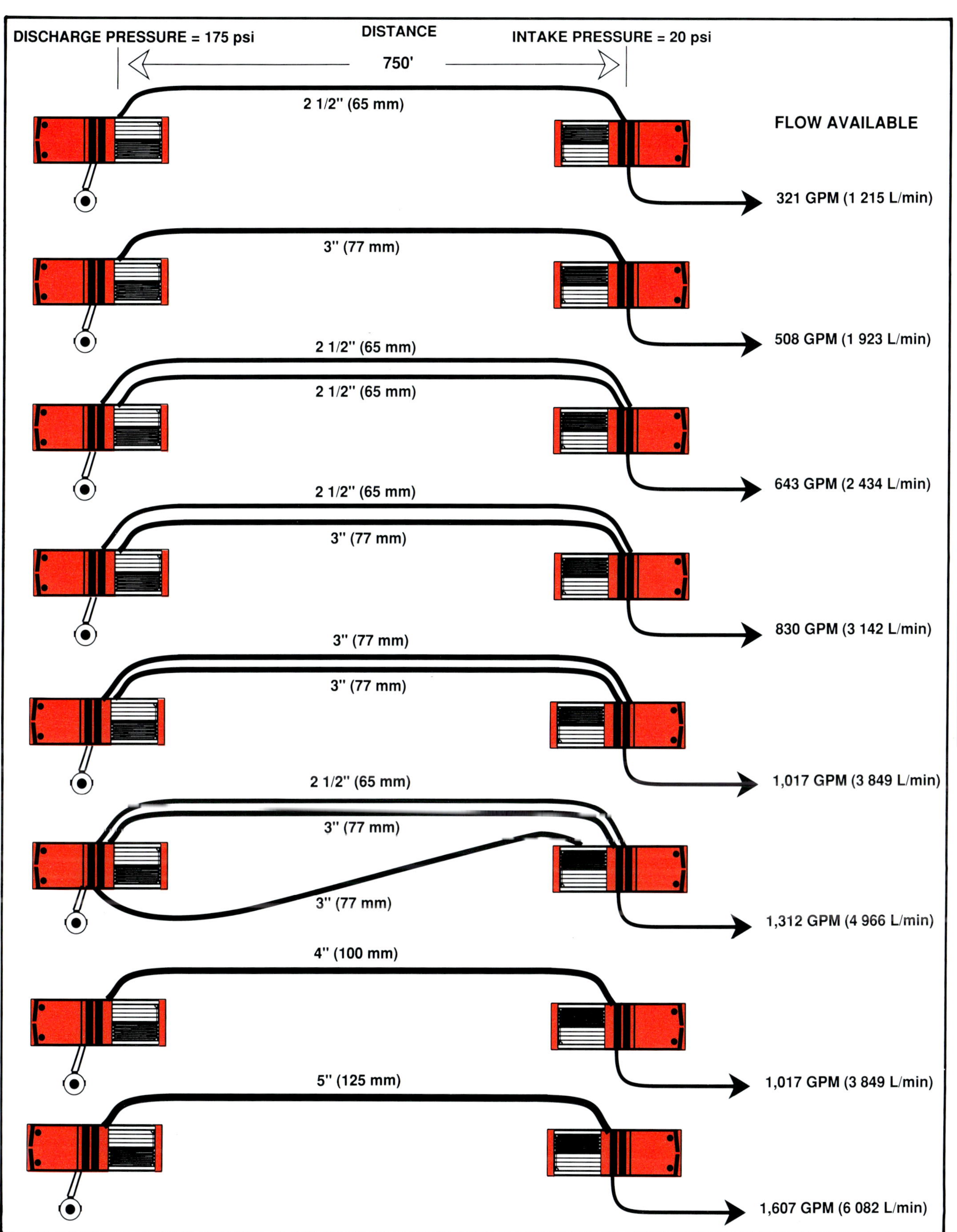

Figure 6.32 Example of available flows with maximum layout conditions using a standard pump discharge pressure of 175 psi (1 225 kPa).

5-89

dump line is closed with a coordinated action. The water then begins to move from the source pumper to the relay pumper. The discharge to the supply line must be opened slowly to prevent a sudden discharge into the empty hoseline. As the water fills the line, the air will be forced through the pump and out the open dump line of the relay pumper. When water comes out of the dump line, the pump on the relay pumper can be engaged.

If the relay pumper is receiving an intake pressure greater than 50 psi (350 kPa), the valve to the dump line on the relay pumper must be adjusted to limit the residual to the 50 psi (350 kPa) maximum. The pump discharge pressure increases as the throttle setting on the relay pumper is increased; therefore, the discharge valve to the dump line will have to be gated down to maintain the 50 psi (350 kPa) residual pressure. If the dump line is allowed to flow unrestricted, the friction loss would increase in the hoseline from the source pumper to the point that the pump would go into cavitation.

Once the pump discharge pressure on the relay pumper has reached the calculated value with the water being discharged, this portion of the relay has been established and no further adjustments should be necessary. When the next relay pumper is ready for water, the same procedure will be followed. The first relay pumper opens the discharge valve supplying the next pumper while closing the dump line on a coordinated basis. The operator does this while carefully observing the intake gauge to maintain the 50 psi (350 kPa) residual pressure. The next relay pumper allows the water to waste through the dump line and follows the same procedure used by the first relay pumper in receiving water from the source pumper. This action can go on until the relay is complete, building even the most complex relay section by section with a minimum of delay.

When the water reaches the attack pumper, the operator should bleed air out of the supply line by opening the bleeder valve on the intake being used. The intake valve on the attack pumper can then be opened and a water supply established through the relay. There is a need for a dump line on the attack pumper as well. When one of the attack lines is shut down, an alert operator can open the dump line to allow water to flow, thus preventing a dangerous pressure buildup in the relay.

Operating the Relay

Once the relay is in operation and water is moving, all pump operators set the relief valve or pressure governor. The use of automatic pressure control devices is essential when operating in a relay due to the cumulative nature of pressure increases when changes in flow occur. The auxiliary cooler can be adjusted as necessary to maintain the proper operating temperature over the extended operating periods that are often necessary during a relay.

If the pumper is equipped with an intake relief valve, it should be put in service by taking off any caps on the outlet or by opening any valves associated with it (Figure 6.33). If the valve is readily adjustable, it should be set for 100 psi (700 kPa) or more on the relay units. At this setting, the valve will not open and cause excessive fluctuations when minor changes in flow occur. If the attack pumper is equipped with a readily adjustable intake relief valve, set it between 50 and 75 psi (350 kPa and 525 kPa) to establish a stable operating condition for the attack pumper. If an attack line is shut down or the amount of discharge

Figure 6.33 An intake relief valve operating to discharge excess pressure.

changes, the friction loss in the supply line will decrease and the residual pressure will increase. The intake relief valve will open, allowing water to dump out of the intake. When this happens, the flow through the supply line and the entire relay increases and pressures return to their original settings. Additional flow requirements will reduce the residual pressure and cause the valve to close. The dumping action will stop, allowing the pressure to again return to the original setting.

There is a tendency to overreact when operating in relay. Small variations in pressure are not significant and no attempt should be made to maintain exact pressures. As long as the intake pressure does not drop below 20 psi (140 kPa) or increase above 100 psi (700 kPa), no action should be required. Changing the pressure at any of the pumpers in a relay operation has some effect on the others. Excessive changes can result in constantly varying pressure throughout the relay.

Shutting Down the Relay

Relay operations should be shut down from the fire scene first. If the source pumper is shut down while the rest of the relay is still operating, the pumpers will run out of water and cavitation can result. Starting with the terminal pumper, each operator should slowly decrease the throttle, open the waste line, and take the pump out of gear.

Communications

Effective relay operations, like all other fireground operations, require good communications to be effective. Each unit in the relay must be aware of the actions of other units so that operations can be coordinated properly. Radios are an obvious means of communication but must be used cautiously. With all the activity on the fire scene, too much radio usage in establishing a relay can hamper not only fire fighting activities, but also efforts to establish the water supply. When pumpers are within sight of each other, hand signals can be used. In extreme cases, messengers on foot can still be effective.

Where additional radio frequencies are available, one channel should be dedicated to coordination of the water supply operation. Once the water is moving, however, a minimum of communication should be required. When units involved in the relay are equipped with radios that are incompatible, portable radios may be useful (Figure 6.34). Ambulances or utility units that are radio equipped and not otherwise occupied can be used to establish communications throughout the relay.

Figure 6.34 Portable radios can be used in relay situations where all apparatus do not have the same radio frequencies.

Summary of Relay Operating Procedures

The following is a summary of the previously outlined methods for establishing a relay operation:

- As soon as the source pumper gets water, advance the throttle until the pressure reaches the calculated value. At the same time, allow water to waste through the dump line.
- When the first relay pumper is ready for water, the operator of the source pumper opens the valve to the supply line, closes the dump line, and starts the water moving toward the fire scene.

- When the water reaches the first relay pumper, it is allowed to flow through the dump line until all of the air is forced out. The pump is then engaged and the throttle advanced to establish the specified pressure while adjusting the valve on the dump line to maintain 50 psi (350 kPa) on the intake.
- Each pumper in turn duplicates the previous two steps until the water arrives at the fire.
- The pumper at the fire establishes the pump discharge pressure needed to provide the required nozzle pressure on all lines.
- All pumpers set relief valves or pressure governors.
- Any adjustment of pressure at relay pumpers is made only when gauge readings clearly indicate that changes are necessary.
- The pumper at the fire is shut down first by slowly decreasing the pressure until the pump can be disengaged. Water is allowed to waste through a dump line or uncapped discharge.
- Each pumper in turn reduces pressure gradually and allows the water to waste after the pump is disengaged.

OPERATING A MOBILE WATER SUPPLY UNIT

Operating a mobile water supply unit from the tank is similar to operating any other pump from a tank, but there are a few basic differences. In this case, the objective is not to supply attack lines, but rather to transfer the water in the tank to some type of reservoir as rapidly as possible (Figure 6.35). High discharge pressures are not required to do this. What is required are hoselines large enough to handle the desired discharge rate without causing excessive friction loss. A good operating practice is to use one 2½-inch (65 mm) line for each 350 gpm (1 400 L/min) to be discharged or one 3-inch (77 mm) line for each 500 gpm (2 000 L/min) to be supplied. If these lines are kept short (less than 100 feet [30 m]), 50 psi (350 kPa) will be adequate to maintain the desired rate of flow. If the lines are longer than 100 feet (30 m), an additional 25 psi (175 kPa) should be added for each 100 feet (30 m) of hose.

Figure 6.35 For maximum speed, use large dump valves when dumping water into portable tanks.

Limiting the discharge pressure makes the hoselines easier to handle if they are discharging into a folding tank and prevents damage to the tank if the lines are supplying a nurse tanker. Limiting the discharge pressure also keeps the intake pressure of the attack pumper at a desirable level if a direct supply is being used.

If an attack pumper is being supplied directly through a gated intake, the flow will vary with the fire attack rate. In fact, no water at all may be moving for extended periods of time. Pressure should be maintained on the supply line to the attack pumper at all times so that the water supply will be available when needed. Give special attention to the possibility of overheating the pump when it is being used in this manner; the circulating water should be adequate to prevent overheating.

Mobile water supply units may be equipped with large dump valves to enable rapid unloading during water shuttle operations. These outlets can be used only if some type of open reservoir, such as a folding tank, is available to dump into by the force of gravity and the head pressure of the water in the apparatus tank. When a gravity dump is being used, total unloading time can be shortened by using the fire pump to discharge through a hoseline at the same time. This procedure is useful even if the pump is of small capacity.

The venturi principle may be used to create a jet dump. This principle utilizes a pressure line supplied by a small pump to discharge a stream of water into the large dump line, thus increasing the velocity of the water and the discharge rate. First the pump is put into operation in the usual manner. The pump discharge pressure is set to the amount specified by the manufacturer or by departmental policy. Next, the operator opens the valve on the dump line, which allows the water to flow by gravity. Then the operator opens the valve to the pressure line (Figure 6.36). By opening and closing this valve, and by changing the pump discharge pressure, the flow from the dump outlet can be changed by the operator.

Figure 6.36 A jet dump increases the rates at which water can be discharged. *Courtesy of Towers Fire Apparatus.*

SPRINKLER AND STANDPIPE SUPPORT

Sprinklers

Properly installed and maintained fire sprinklers have a long history of providing reliable, automatic protection to all types of occupancies. The water supply for sprinkler systems is designed to supply only a fraction of the sprinklers actually installed on the system. If a large fire should occur, or if a pipe breaks, the sprinkler system will need an outside source of water and pressure in order to do its job effectively.

The pump operator should be prepared to support sprinkler systems by attaching lines to the fire department sprinkler connection (Figure 6.37). Fire department sprinkler connections consist of a siamese with at least two 2½-inch (65 mm) female connections or one large diameter sexless connection that is connected to a clappered inlet. As water flows into the system, it first passes through a check valve (Figure 6.38). This valve prevents water flow from the sprinkler system back into the fire department connection. It does allow water from the fire department connection to flow into the sprinkler system.

Figure 6.37 A typical fire department sprinkler connection.

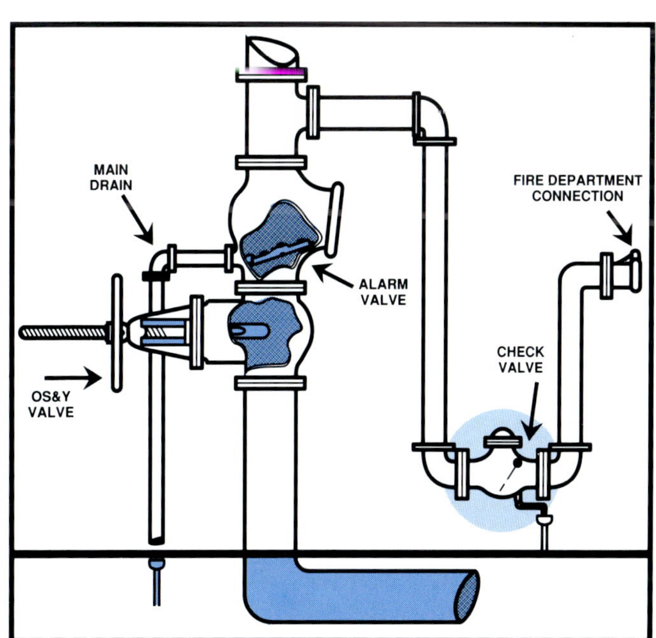

Figure 6.38 The fire department connection check valve prevents water flow back into the fire department connection.

At least one line 2½-inches (65 mm) or larger should be attached to the sprinkler connection when operating at a fire in a sprinklered occupancy. The sprinklered occupancies in your response area should be identified in planning area surveys or by looking for water motor gongs attached to the outside walls of a building (Figure 6.39). Additional lines should be attached to the sprinkler connection if the volume of fire in the building indicates that more water will be needed to adequately supply all operating sprinkler heads. Another indication that additional sprinkler supply lines are needed is when interior attack crews report that operating sprinklers do not have full discharge patterns.

Figure 6.39 A water motor gong indicates the presence of an automatic sprinkler system.

Sprinkler connections should be supplied with water from high-capacity pumpers (1,000 gpm [4 000 L/min] and greater). Each sprinkler system is different, but a good general rule is to supply the sprinkler connection with 150 psi (1 050 kPa) pump discharge pressure. Pump discharge pressures in excess of 200 psi (1 400 kPa) may damage the sprinkler piping and result in breaks.

Standpipes

Standpipes are used to speed fire attack in multistory or single-story buildings with large floor areas. Fire attack teams attach lines to 2½- or 1½-inch (65 mm or 38 mm) connections provided on each floor (Figure 6.40). House or standpipe lines should not be used by fire personnel unless a hose testing program similar to the one used to test all fire department hose is implemented. These lines are often unlined, single-jacket cotton hose that have not been tested or removed since they were installed. Fire attack crews should bring attack lines with them when they enter a property protected with standpipes to ensure an attack without the fear of house lines bursting (Figure 6.41).

Standpipes may be wet or dry, depending upon owner preference or local code requirements. Wet-pipe systems contain water under pressure and are ready to be used as soon as lines can be attached to the outlet. Dry-pipe systems must be supplied with water from a pumper that attaches to a standpipe connection outside the building. Standpipe connections should be clearly identified

Figure 6.40 Fire department standpipe connections may be located in or just outside stairwells, or in hose or extinguisher cabinets.

Figure 6.41 Firefighters entering a building in which a fire alarm is sounding should carry fire fighting hose and nozzles, forcible entry tools, and flashlights. They should be wearing full personal protective equipment, including SCBAs.

to prevent confusion between sprinkler and standpipe connections. Wet standpipes also incorporate a fire department connection that should be supplied with water under pressure to supplement the system's water supply.

The pump discharge pressure needed to supply standpipes depends upon the following factors:

- Height of the fire floor
- Friction loss in the standpipe
- Size and number of fire streams in operation
- The pressure desired at the nozzle of each fire stream

Approximately 5 psi (35 kPa) should be added to the desired nozzle pressure for each floor above the standpipe connection that will have operating fire streams. Friction loss for the attack line(s), standpipe piping, and layout from the pumper to the standpipe connection should also be included. These are lengthy calculations that cannot be made each time a standpipe line is needed. The fire department should have a planned pump discharge pressure or develop a rule of thumb for each building in the area equipped with a standpipe. Pump discharge pressures in excess of 200 psi (1 400 kPa) are not encouraged unless the standpipe system has been designed to withstand higher pressures.

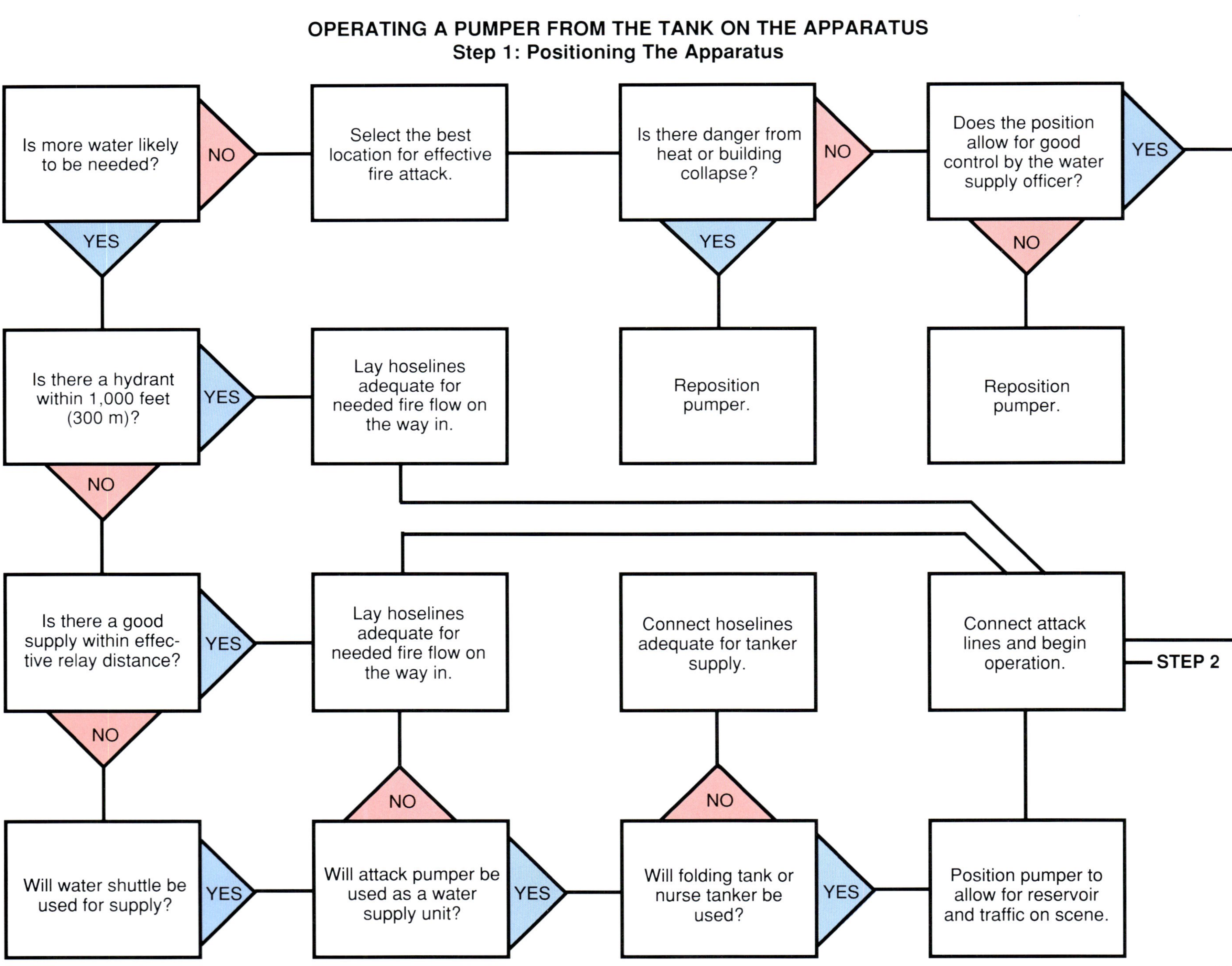
OPERATING A PUMPER FROM THE TANK ON THE APPARATUS
Step 1: Positioning The Apparatus
Is more water likely to be needed?
NO
Select the best location for effective fire attack.
Is there danger from heat or building collapse?
NO
Does the position allow for good control by the water supply officer?
YES
YES
YES
Reposition pumper.
NO
Reposition pumper.
Is there a hydrant within 1,000 feet (300 m)?
YES
Lay hoselines adequate for needed fire flow on the way in.
NO
Is there a good supply within effective relay distance?
YES
Lay hoselines adequate for needed fire flow on the way in.
Connect hoselines adequate for tanker supply.
Connect attack lines and begin operation.
STEP 2
NO
NO
NO
Will water shuttle be used for supply?
YES
Will attack pumper be used as a water supply unit?
YES
Will folding tank or nurse tanker be used?
YES
Position pumper to allow for reservoir and traffic on scene.

OPERATING A PUMPER FROM THE TANK ON THE APPARATUS
Step 2: Put Pump In Service And Establish Operating Pressure

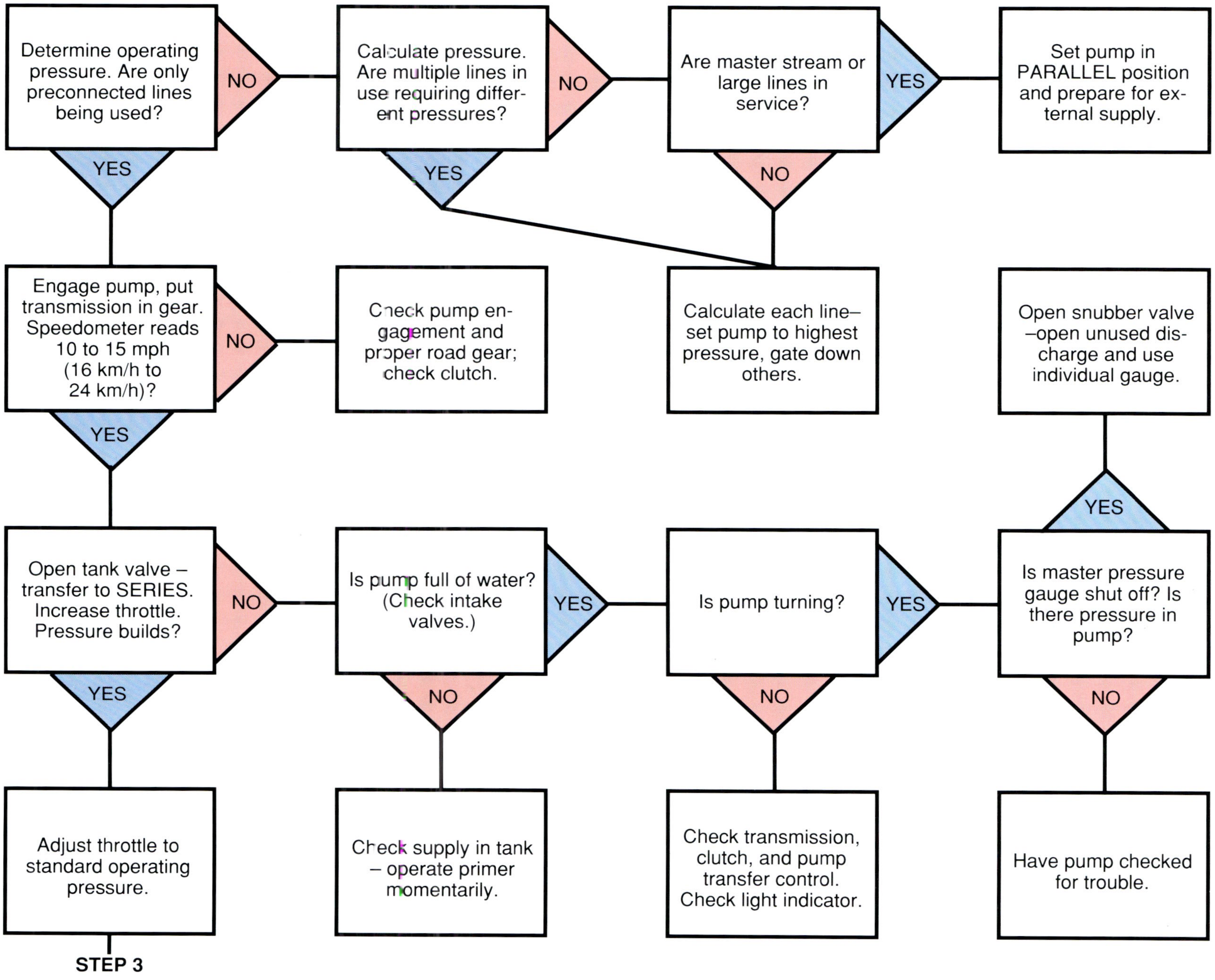

Determine operating pressure. Are only preconnected lines being used?

NO

Calculate pressure. Are multiple lines in use requiring different pressures?

NO

Are master stream or large lines in service?

YES

Set pump in PARALLEL position and prepare for external supply.

YES

NO

YES

NO

Engage pump, put transmission in gear. Speedometer reads 10 to 15 mph (16 km/h to 24 km/h)?

NO

Check pump engagement and proper road gear; check clutch.

Calculate each line—set pump to highest pressure, gate down others.

Open snubber valve—open unused discharge and use individual gauge.

YES

YES

Open tank valve – transfer to SERIES. Increase throttle. Pressure builds?

NO

Is pump full of water? (Check intake valves.)

YES

Is pump turning?

YES

Is master pressure gauge shut off? Is there pressure in pump?

YES

NO

NO

NO

Adjust throttle to standard operating pressure.

Check supply in tank – operate primer momentarily.

Check transmission, clutch, and pump transfer control. Check light indicator.

Have pump checked for trouble.

STEP 3

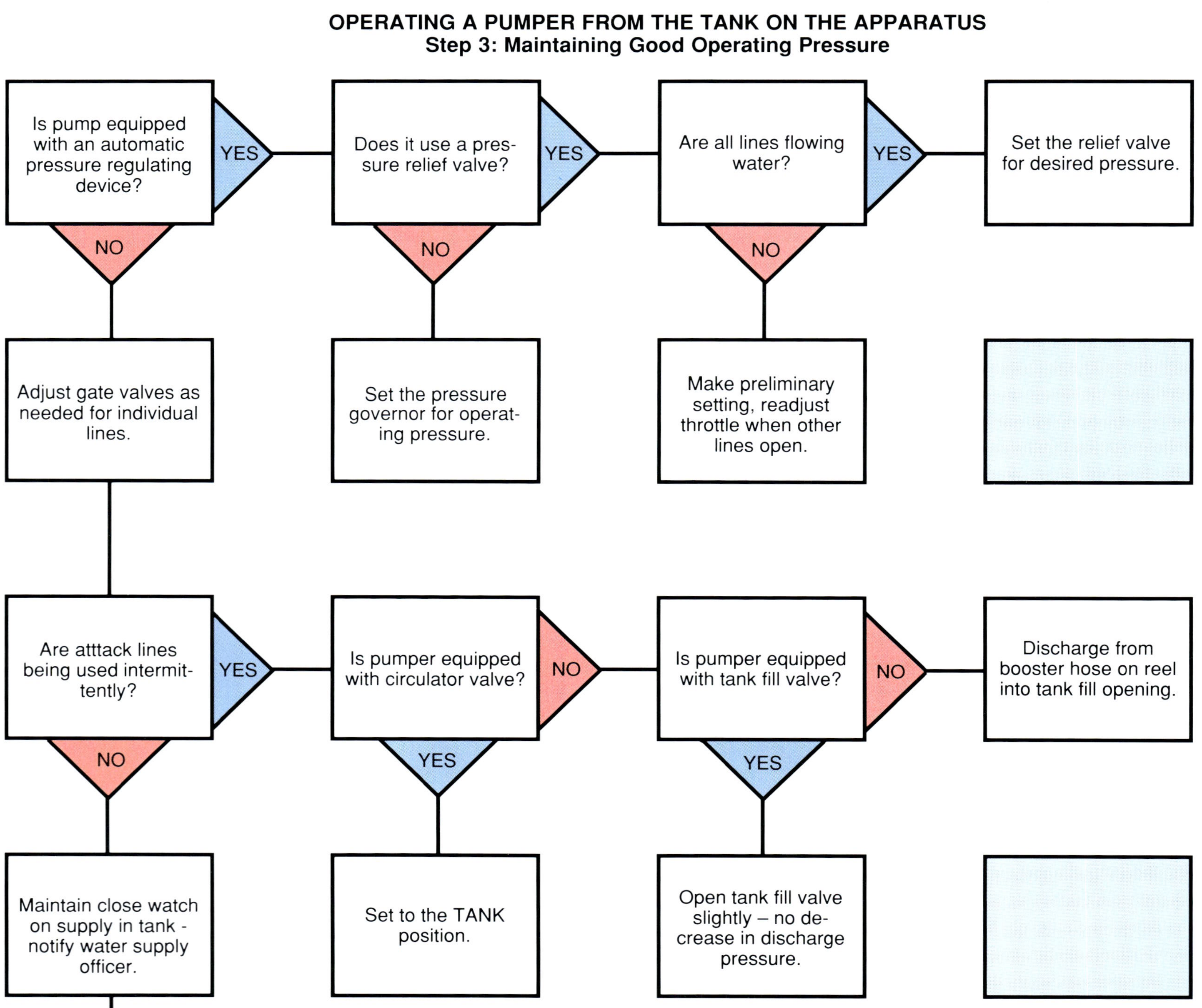
OPERATING A PUMPER FROM THE TANK ON THE APPARATUS
Step 3: Maintaining Good Operating Pressure
Is pump equipped with an automatic pressure regulating device?
YES
NO
Adjust gate valves as needed for individual lines.
Does it use a pressure relief valve?
YES
NO
Set the pressure governor for operating pressure.
Are all lines flowing water?
YES
NO
Make preliminary setting, readjust throttle when other lines open.
Set the relief valve for desired pressure.
Are atttack lines being used intermittently?
YES
NO
Maintain close watch on supply in tank - notify water supply officer.
Is pumper equipped with circulator valve?
NO
YES
Set to the TANK position.
Is pumper equipped with tank fill valve?
NO
YES
Open tank fill valve slightly – no decrease in discharge pressure.
Discharge from booster hose on reel into tank fill opening.

OPERATING A PUMPER FROM THE TANK ON THE APPARATUS
Step 4: Making The Transition To An External Supply

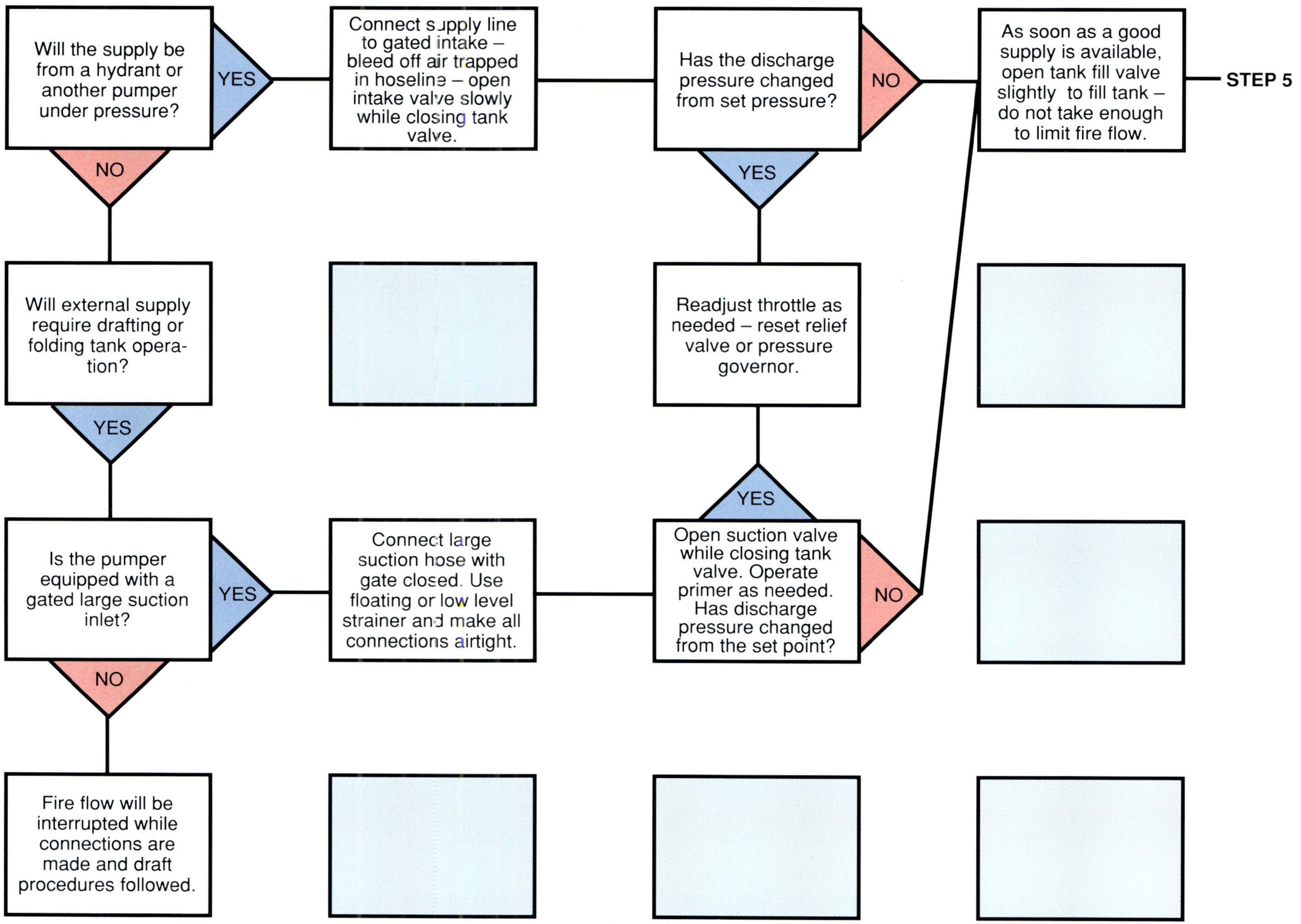

Will the supply be from a hydrant or another pumper under pressure?

YES

Connect supply line to gated intake – bleed off air trapped in hoseline – open intake valve slowly while closing tank valve.

Has the discharge pressure changed from set pressure?

NO

As soon as a good supply is available, open tank fill valve slightly to fill tank – do not take enough to limit fire flow.

STEP 5

YES

Readjust throttle as needed – reset relief valve or pressure governor.

NO

Will external supply require drafting or folding tank operation?

YES

Is the pumper equipped with a gated large suction inlet?

YES

Connect large suction hose with gate closed. Use floating or low level strainer and make all connections airtight.

Open suction valve while closing tank valve. Operate primer as needed. Has discharge pressure changed from the set point?

YES

NO

NO

Fire flow will be interrupted while connections are made and draft procedures followed.

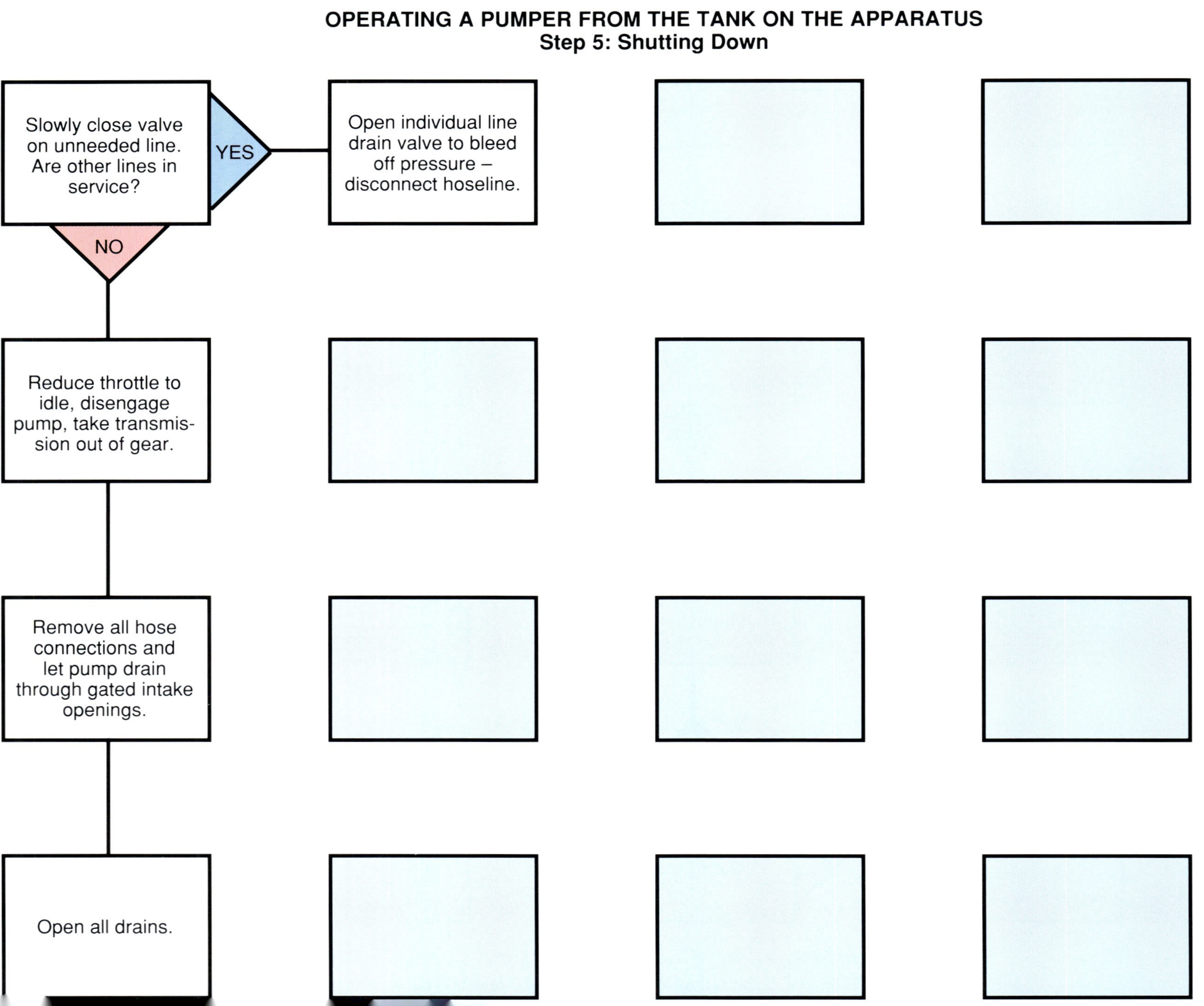

OPERATING A PUMPER FROM THE TANK ON THE APPARATUS
Step 5: Shutting Down
Slowly close valve on unneeded line. Are other lines in service?
YES
Open individual line drain valve to bleed off pressure – disconnect hoseline.
NO
Reduce throttle to idle, disengage pump, take transmission out of gear.
Remove all hose connections and let pump drain through gated intake openings.
Open all drains.

OPERATING FROM A HYDRANT
Step 1: Positioning The Pumper And Connecting To The Hydrant

Is there a hydrant within 200 feet (60 m) of the fire location?

NO

Select the most accessible hydrant for the fire location.

Is the selected hydrant within 1,000 feet (300 m) of the fire location?

NO

Set up a relay to get the water from the hydrant to fire location.

YES

YES

Will the closest hydrant supply the needed fire flow?

NO

Arrange for a supplemental water supply.

Lay supply line on arrival. Use 4-way hydrant valve and gate valve to provide supported supply line.

Connect soft sleeve or supply hose from steamer connection on hydrant to pumper.

YES

YES

Is the selected hydrant a safe distance from the fire?

YES

Will the attack pumper also be used as the water supply pumper?

YES

Position and immobilize the pumper – will suction hose reach to steamer connection on hydrant?

Open hydrant, note static pressure and begin operation.

STEP 2

NO

NO

NO

Select another hydrant – set up a relay if needed.

Lay sufficient hoselines from the water supply pumper to the attack pumper for needed fire flow.

Connect short section of 3-inch (77 mm) hose from hydrant to suction siamese or gated inlet.

OPERATING FROM A HYDRANT
Step 2: Putting The Pumper In Service

Open the hydrant. Note static pressure. Does the compound gauge register at least 35 psi (245 kPa)?

NO

Is hydrant fully opened?

YES

Is suction hose clear of blockage, no kinks?

YES

Is system pressure low due to excess flow from other hydrants in system?

YES

Find a hydrant with more pressure. Set up for relay or supplemental pumping.

NO

Check valve – make sure it has been opened enough turns.

NO

Check suction hose – rearrange to eliminate kinks.

YES

Engage pump. Put road transmission in pump gear. Does speedometer register between 10 and 15 mph (16 km/h to 24 km/h)?

NO

Is transmission lock in place? Is road transmission in pump gear? Verify settings.

NO

Is pump turning? Verify by sound.

YES

Is pump full of water?

NO

Check hydrant and intake valves on the pumper.

YES

Open unused and uncapped discharge. Is there pressure in the pump?

YES

Open snubber valve on master gauge – compare with pressure readings on individual line gauges.

YES

Increase throttle setting – does pressure build up on discharge gauge?

NO

Is pump turning? Verify by sound.

YES

Adjust throttle to set operating pressure.

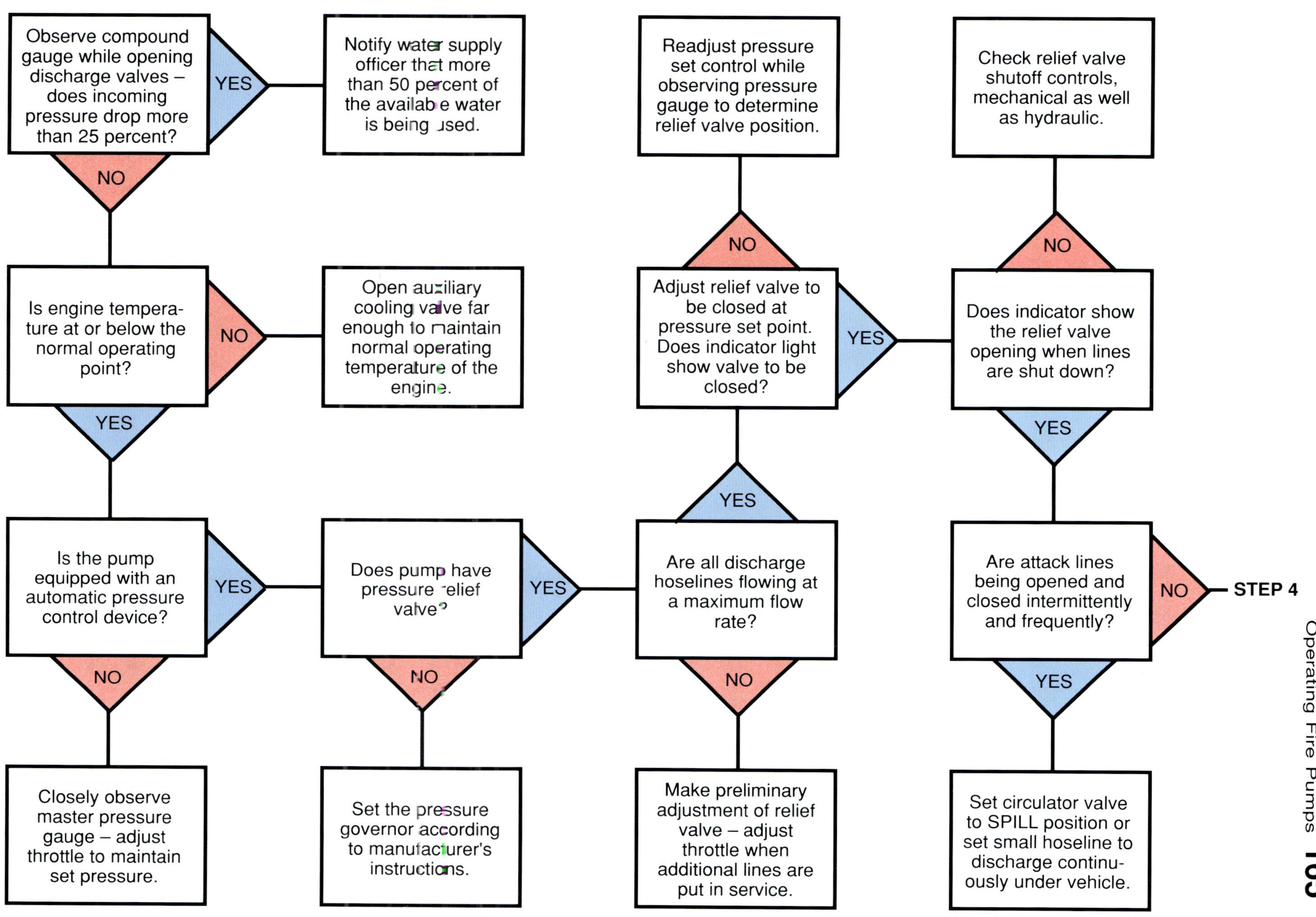
OPERATING FROM A HYDRANT
Step 3: Maintain Operating Pressure
Observe compound gauge while opening discharge valves – does incoming pressure drop more than 25 percent?
YES
Notify water supply officer that more than 50 percent of the available water is being used.
NO
Is engine temperature at or below the normal operating point?
NO
Open auxiliary cooling valve far enough to maintain normal operating temperature of the engine.
YES
Is the pump equipped with an automatic pressure control device?
NO
Closely observe master pressure gauge – adjust throttle to maintain set pressure.
YES
Does pump have pressure relief valve?
NO
Set the pressure governor according to manufacturer's instructions.
YES
Are all discharge hoselines flowing at a maximum flow rate?
NO
Make preliminary adjustment of relief valve – adjust throttle when additional lines are put in service.
YES
Adjust relief valve to be closed at pressure set point. Does indicator light show valve to be closed?
NO
Readjust pressure set control while observing pressure gauge to determine relief valve position.
YES
Does indicator show the relief valve opening when lines are shut down?
NO
Check relief valve shutoff controls, mechanical as well as hydraulic.
YES
Are attack lines being opened and closed intermittently and frequently?
NO
STEP 4
YES
Set circulator valve to SPILL position or set small hoseline to discharge continuously under vehicle.

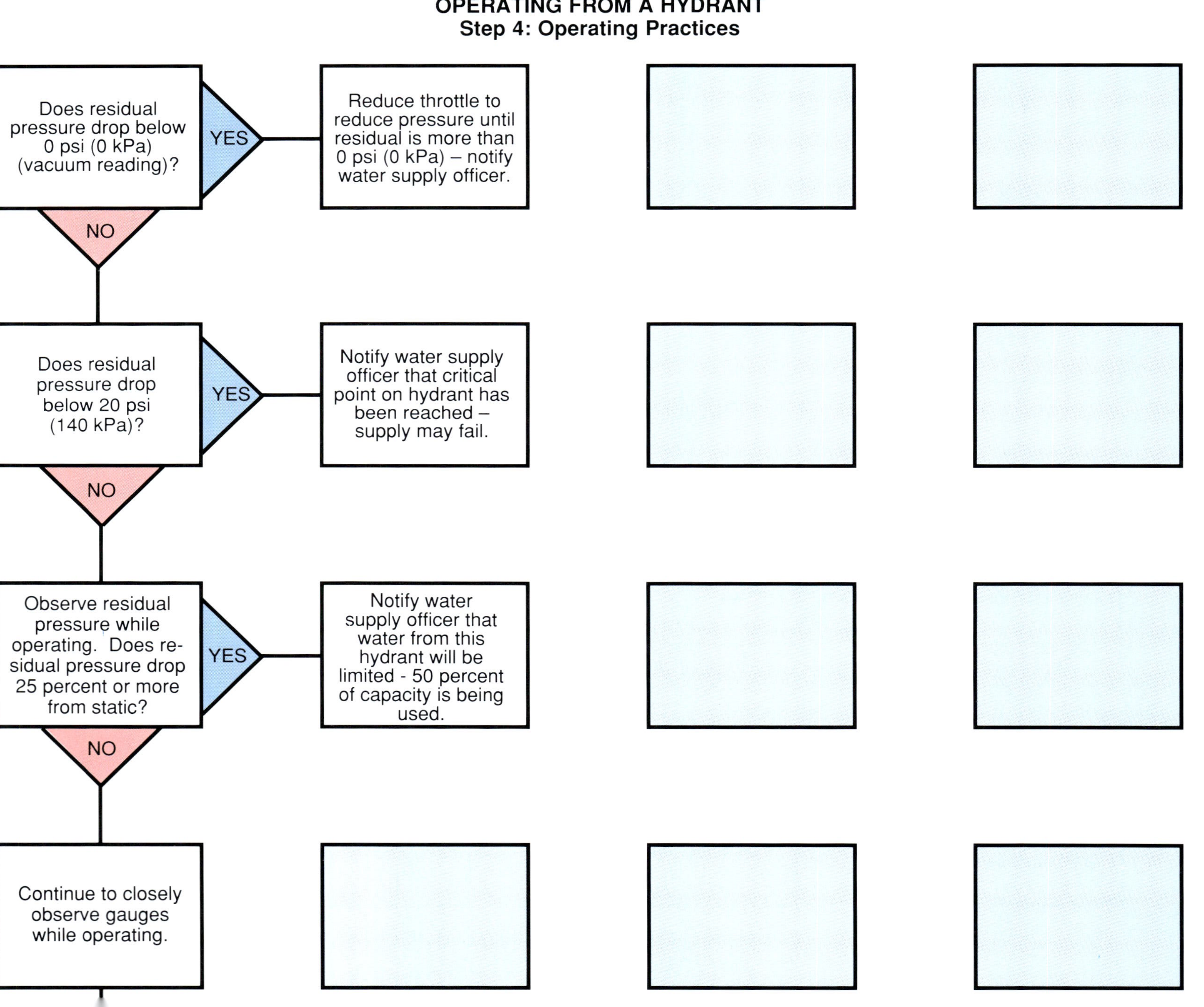
OPERATING FROM A HYDRANT
Step 4: Operating Practices
Does residual pressure drop below 0 psi (0 kPa) (vacuum reading)?
YES
Reduce throttle to reduce pressure until residual is more than 0 psi (0 kPa) – notify water supply officer.
NO
Does residual pressure drop below 20 psi (140 kPa)?
YES
Notify water supply officer that critical point on hydrant has been reached – supply may fail.
NO
Observe residual pressure while operating. Does residual pressure drop 25 percent or more from static?
YES
Notify water supply officer that water from this hydrant will be limited - 50 percent of capacity is being used.
NO
Continue to closely observe gauges while operating.

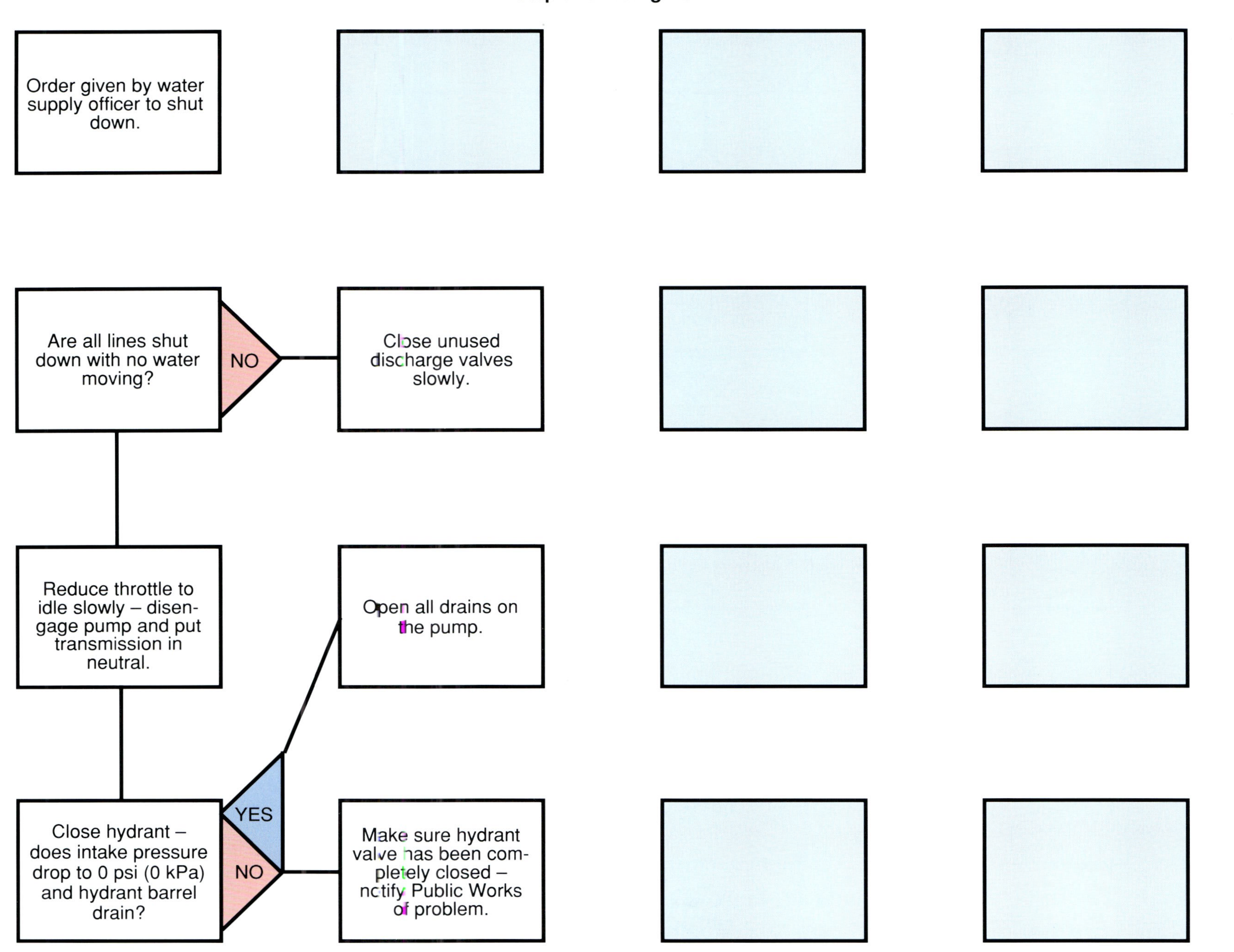
OPERATING FROM A HYDRANT
Step 5: Shutting Down
Order given by water supply officer to shut down.
Are all lines shut down with no water moving?
NO
Close unused discharge valves slowly.
Reduce throttle to idle slowly – disengage pump and put transmission in neutral.
Close hydrant – does intake pressure drop to 0 psi (0 kPa) and hydrant barrel drain?
YES
Open all drains on the pump.
NO
Make sure hydrant valve has been completely closed – notify Public Works of problem.

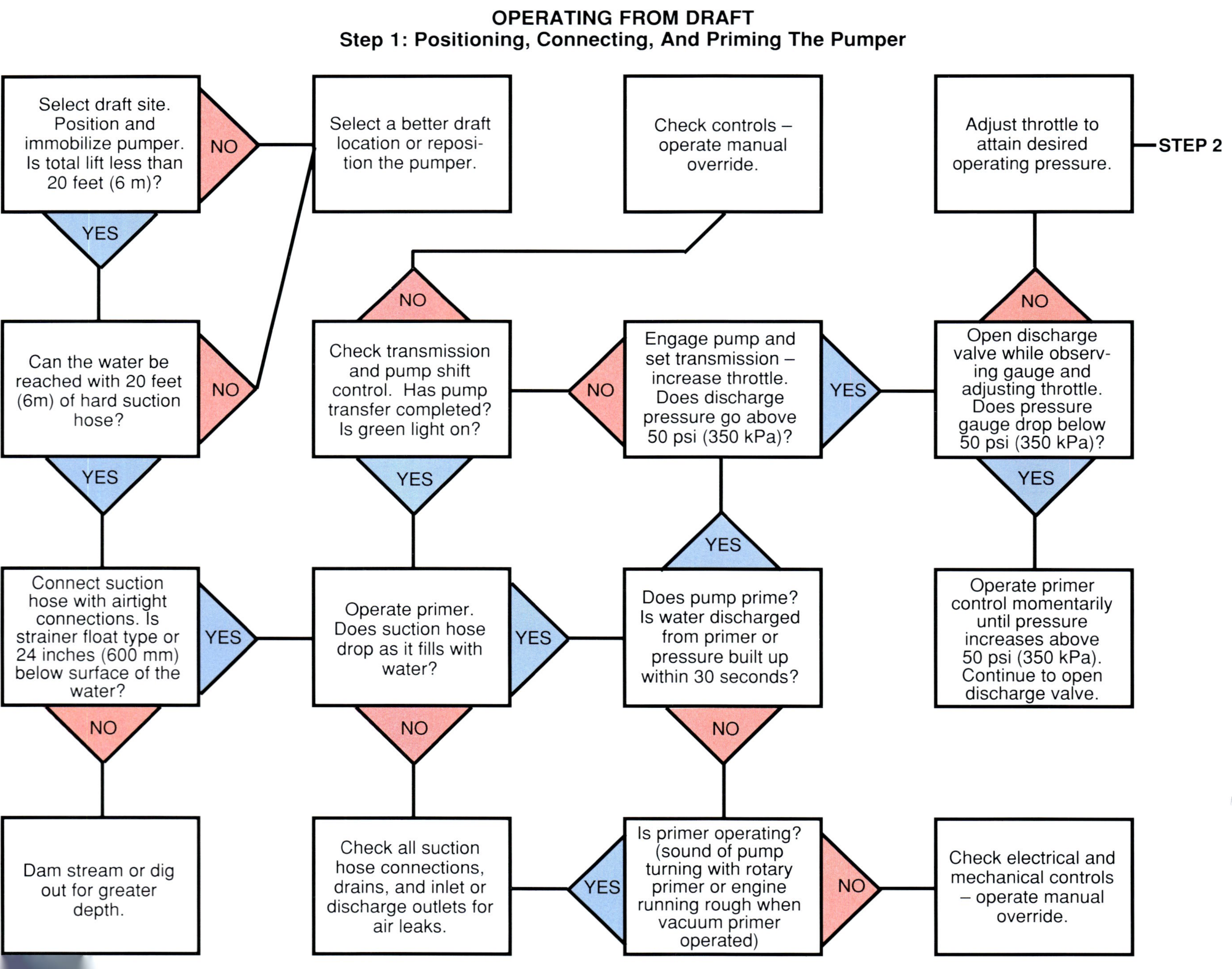
OPERATING FROM DRAFT
Step 1: Positioning, Connecting, And Priming The Pumper
Select draft site. Position and immobilize pumper. Is total lift less than 20 feet (6 m)?
NO
Select a better draft location or reposition the pumper.
Check controls – operate manual override.
Adjust throttle to attain desired operating pressure.
STEP 2
YES
NO
NO
Can the water be reached with 20 feet (6m) of hard suction hose?
NO
Check transmission and pump shift control. Has pump transfer completed? Is green light on?
NO
Engage pump and set transmission – increase throttle. Does discharge pressure go above 50 psi (350 kPa)?
YES
Open discharge valve while observing gauge and adjusting throttle. Does pressure gauge drop below 50 psi (350 kPa)?
YES
YES
YES
YES
Connect suction hose with airtight connections. Is strainer float type or 24 inches (600 mm) below surface of the water?
YES
Operate primer. Does suction hose drop as it fills with water?
YES
Does pump prime? Is water discharged from primer or pressure built up within 30 seconds?
Operate primer control momentarily until pressure increases above 50 psi (350 kPa). Continue to open discharge valve.
NO
NO
NO
Dam stream or dig out for greater depth.
Check all suction hose connections, drains, and inlet or discharge outlets for air leaks.
YES
Is primer operating? (sound of pump turning with rotary primer or engine running rough when vacuum primer operated)
NO
Check electrical and mechanical controls – operate manual override.

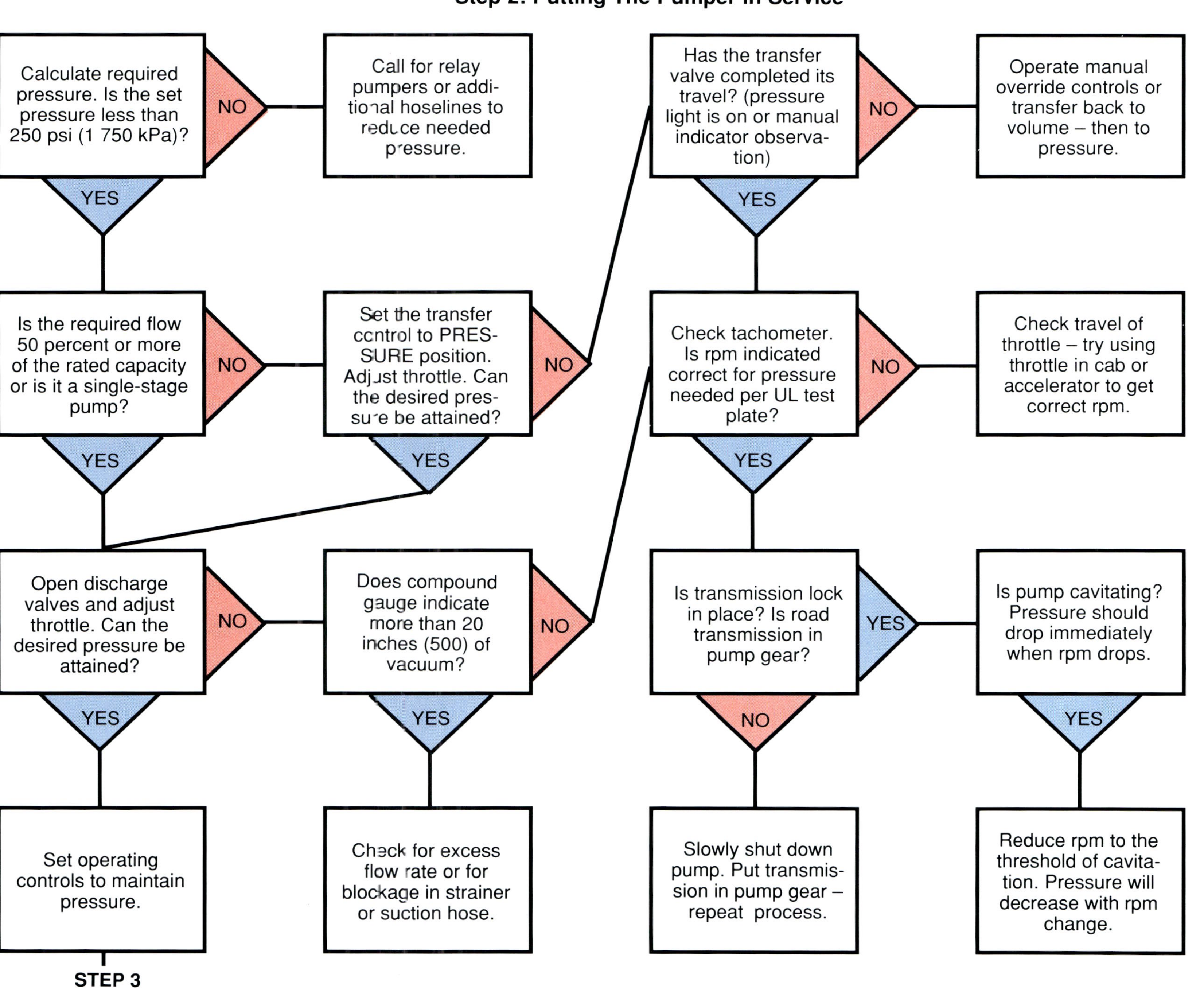
OPERATING FROM DRAFT
Step 2: Putting The Pumper In Service
Calculate required pressure. Is the set pressure less than 250 psi (1 750 kPa)?
NO
Call for relay pumpers or additional hoselines to reduce needed pressure.
YES
Is the required flow 50 percent or more of the rated capacity or is it a single-stage pump?
NO
Set the transfer control to PRESSURE position. Adjust throttle. Can the desired pressure be attained?
NO
YES
YES
Open discharge valves and adjust throttle. Can the desired pressure be attained?
NO
Does compound gauge indicate more than 20 inches (500) of vacuum?
NO
YES
YES
Set operating controls to maintain pressure.
STEP 3
Check for excess flow rate or for blockage in strainer or suction hose.
Has the transfer valve completed its travel? (pressure light is on or manual indicator observation)
NO
Operate manual override controls or transfer back to volume – then to pressure.
YES
Check tachometer. Is rpm indicated correct for pressure needed per UL test plate?
NO
Check travel of throttle – try using throttle in cab or accelerator to get correct rpm.
YES
Is transmission lock in place? Is road transmission in pump gear?
YES
Is pump cavitating? Pressure should drop immediately when rpm drops.
NO
YES
Slowly shut down pump. Put transmission in pump gear – repeat process.
Reduce rpm to the threshold of cavitation. Pressure will decrease with rpm change.

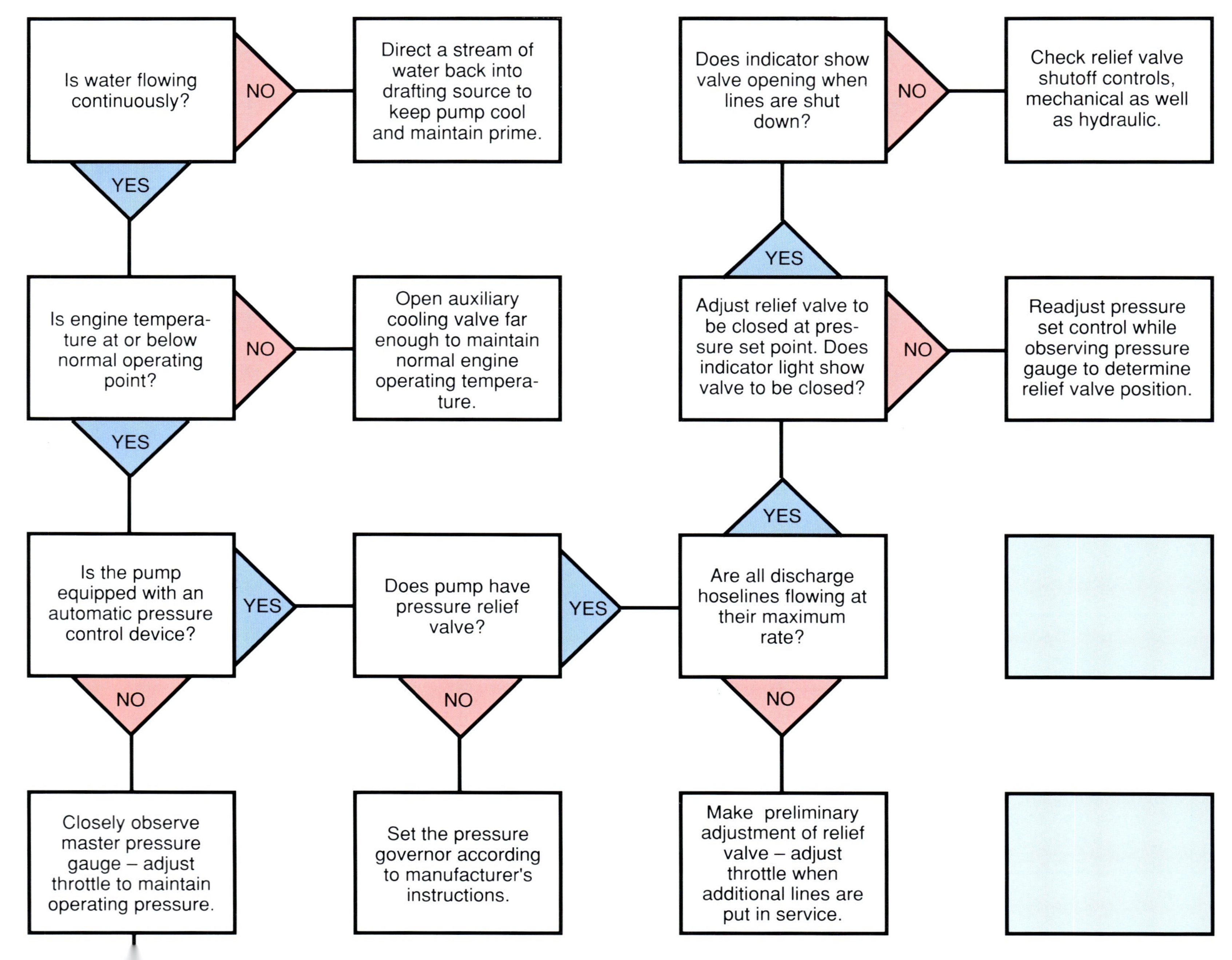

OPERATING FROM DRAFT
Step 3: Maintain Operating Pressure
Is water flowing continuously?
NO
Direct a stream of water back into drafting source to keep pump cool and maintain prime.
YES
Is engine temperature at or below normal operating point?
NO
Open auxiliary cooling valve far enough to maintain normal engine operating temperature.
YES
Is the pump equipped with an automatic pressure control device?
NO
Closely observe master pressure gauge – adjust throttle to maintain operating pressure.
YES
Does pump have pressure relief valve?
NO
Set the pressure governor according to manufacturer's instructions.
YES
Are all discharge hoselines flowing at their maximum rate?
NO
Make preliminary adjustment of relief valve – adjust throttle when additional lines are put in service.
YES
Adjust relief valve to be closed at pressure set point. Does indicator light show valve to be closed?
NO
Readjust pressure set control while observing pressure gauge to determine relief valve position.
YES
Does indicator show valve opening when lines are shut down?
NO
Check relief valve shutoff controls, mechanical as well as hydraulic.

OPERATING FROM DRAFT
Step 4: Shutting Down

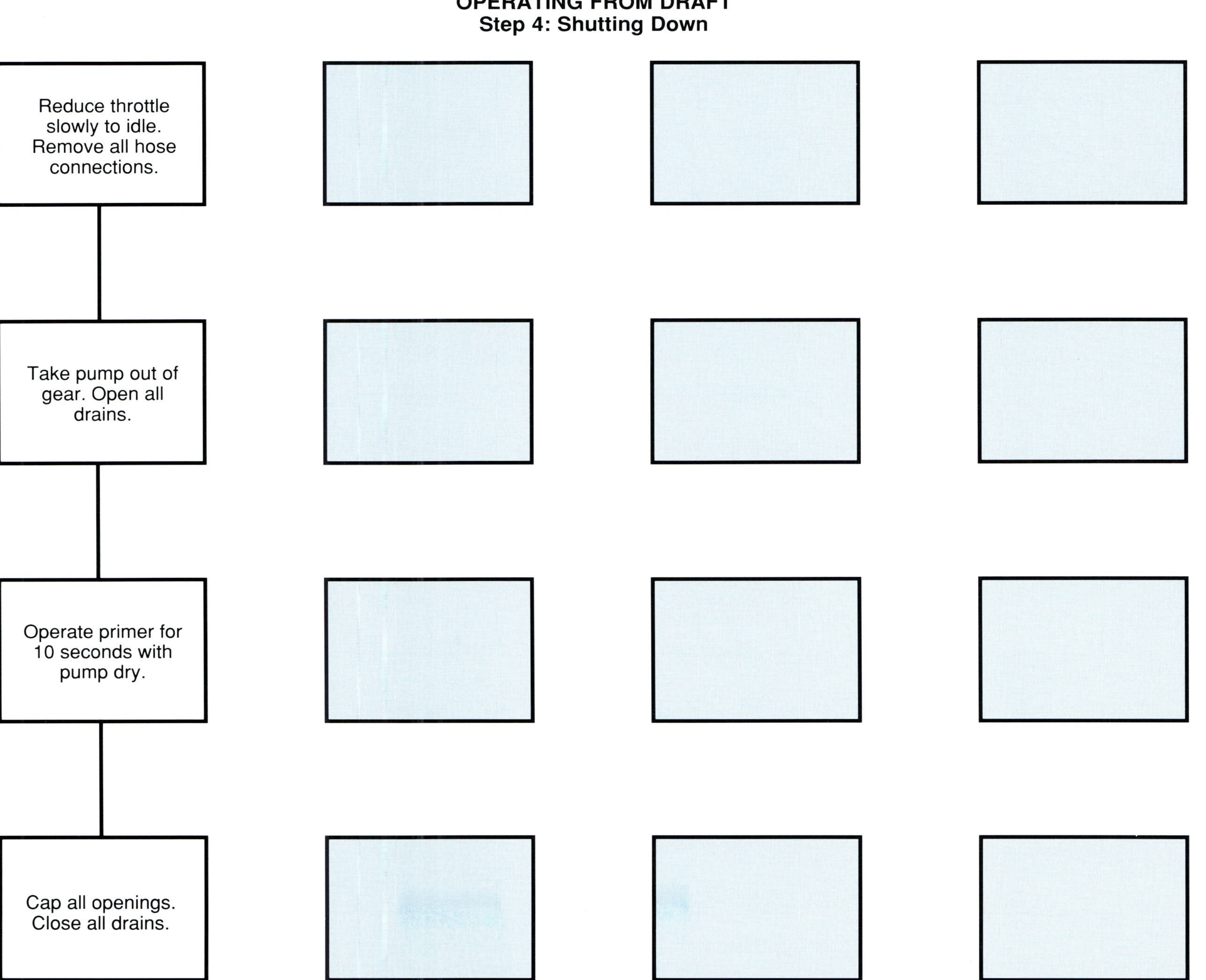

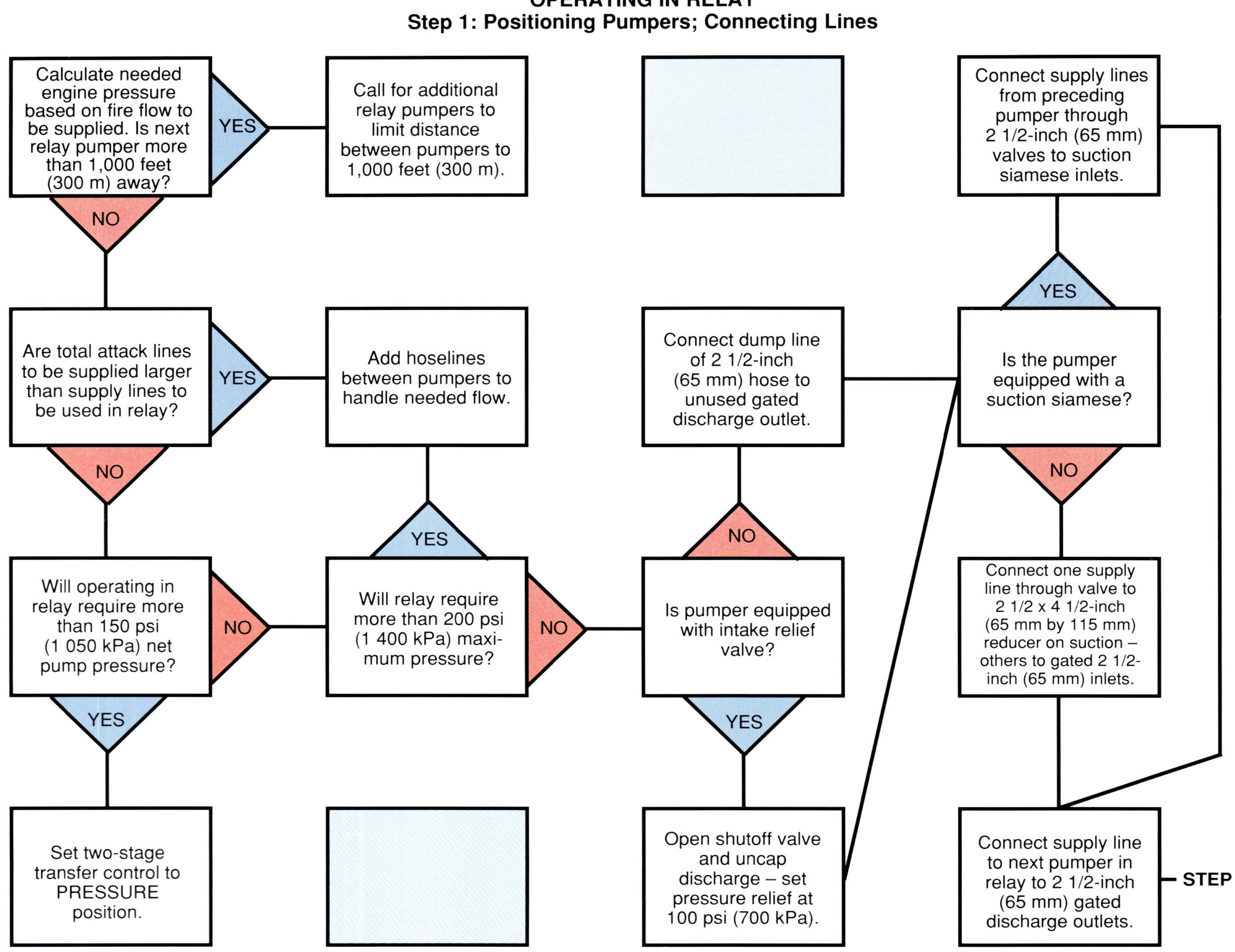
OPERATING IN RELAY
Step 1: Positioning Pumpers; Connecting Lines
Calculate needed engine pressure based on fire flow to be supplied. Is next relay pumper more than 1,000 feet (300 m) away?
YES
Call for additional relay pumpers to limit distance between pumpers to 1,000 feet (300 m).
NO
Are total attack lines to be supplied larger than supply lines to be used in relay?
YES
Add hoselines between pumpers to handle needed flow.
NO
Will operating in relay require more than 150 psi (1 050 kPa) net pump pressure?
YES
Set two-stage transfer control to PRESSURE position.
NO
Will relay require more than 200 psi (1 400 kPa) maximum pressure?
YES
NO
Is pumper equipped with intake relief valve?
YES
Open shutoff valve and uncap discharge – set pressure relief at 100 psi (700 kPa).
NO
Connect dump line of 2 1/2-inch (65 mm) hose to unused gated discharge outlet.
Is the pumper equipped with a suction siamese?
YES
Connect supply lines from preceding pumper through 2 1/2-inch (65 mm) valves to suction siamese inlets.
NO
Connect one supply line through valve to 2 1/2 x 4 1/2-inch (65 mm by 115 mm) reducer on suction – others to gated 2 1/2-inch (65 mm) inlets.
Connect supply line to next pumper in relay to 2 1/2-inch (65 mm) gated discharge outlets.
STEP 2

OPERATING IN RELAY
Step 2: Putting The Relay In Operation

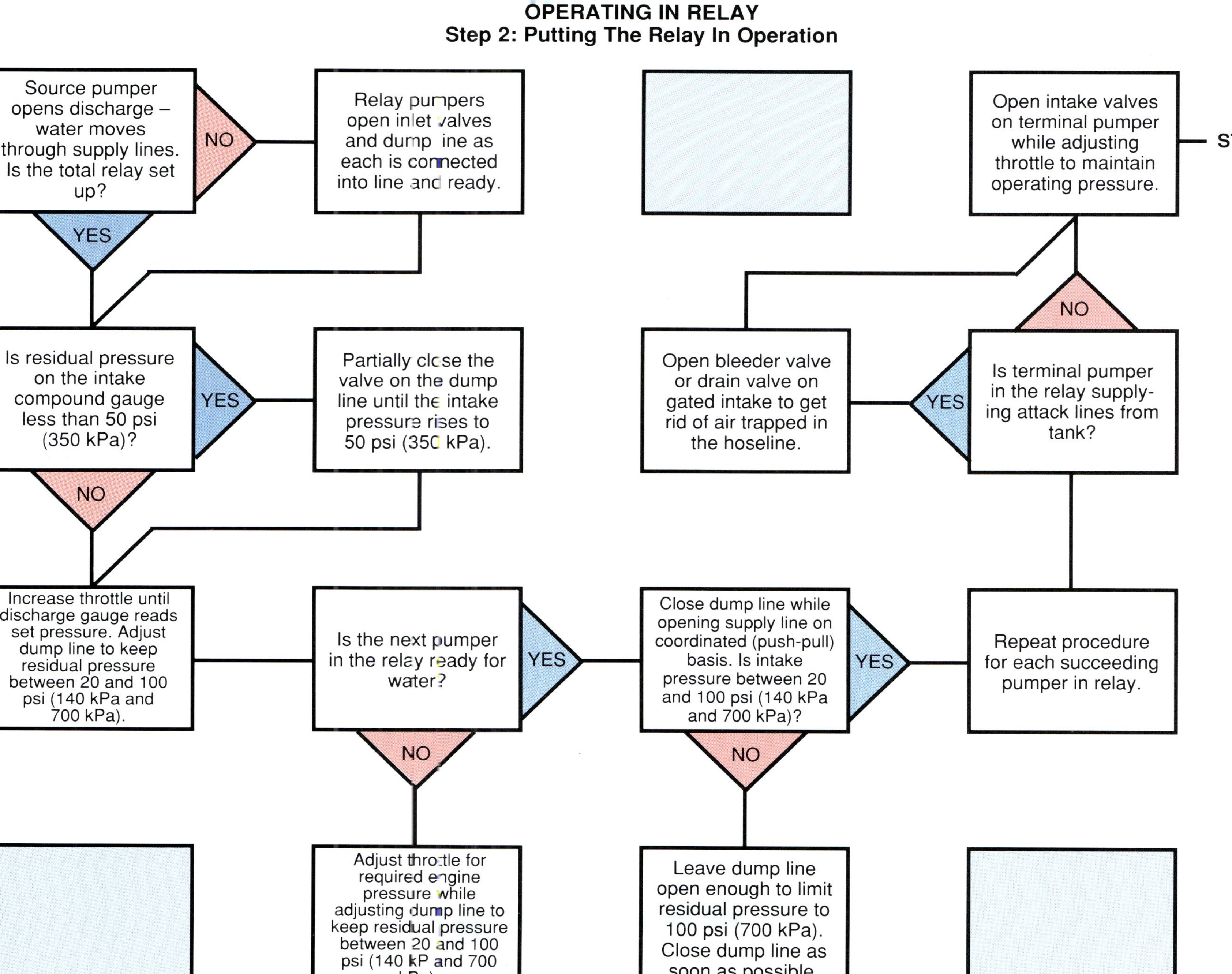

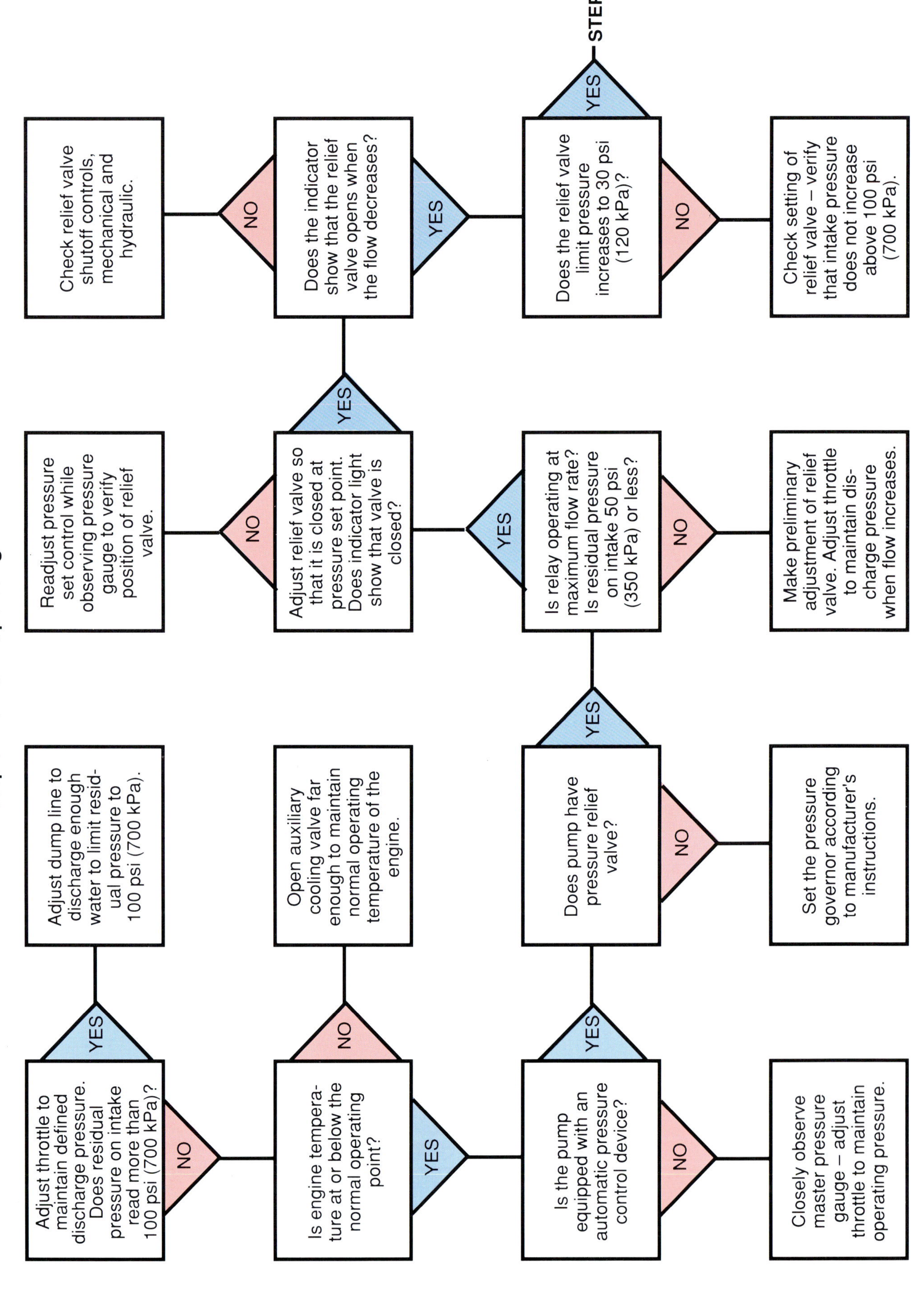
OPERATING IN RELAY
Step 3: Maintain Operating Pressure
Adjust throttle to maintain defined discharge pressure. Does residual pressure on intake read more than 100 psi (700 kPa)?
YES
NO
Adjust dump line to discharge enough water to limit residual pressure to 100 psi (700 kPa).
Is engine temperature at or below the normal operating point?
NO
YES
Open auxiliary cooling valve far enough to maintain normal operating temperature of the engine.
Is the pump equipped with an automatic pressure control device?
YES
NO
Closely observe master pressure gauge – adjust throttle to maintain operating pressure.
Does pump have pressure relief valve?
YES
NO
Set the pressure governor accorder's to manufacturer's instructions.
Is relay operating at maximum flow rate? Is residual pressure on intake 50 psi (350 kPa) or less?
YES
NO
Make preliminary adjustment of relief valve. Adjust throttle to maintain discharge pressure when flow increases.
Adjust relief valve so that it is closed at pressure set point. Does indicator light show that valve is closed?
NO
YES
Readjust pressure set control while observing pressure gauge to verify position of relief valve.
Does the indicator show that the relief valve opens when the flow decreases?
NO
YES
Check relief valve shutoff controls, mechanical and hydraulic.
Does the relief valve limit pressure increases to 30 psi (120 kPa)?
YES
NO
STEP 4
Check setting of relief valve – verify that intake pressure does not increase above 100 psi (700 kPa).

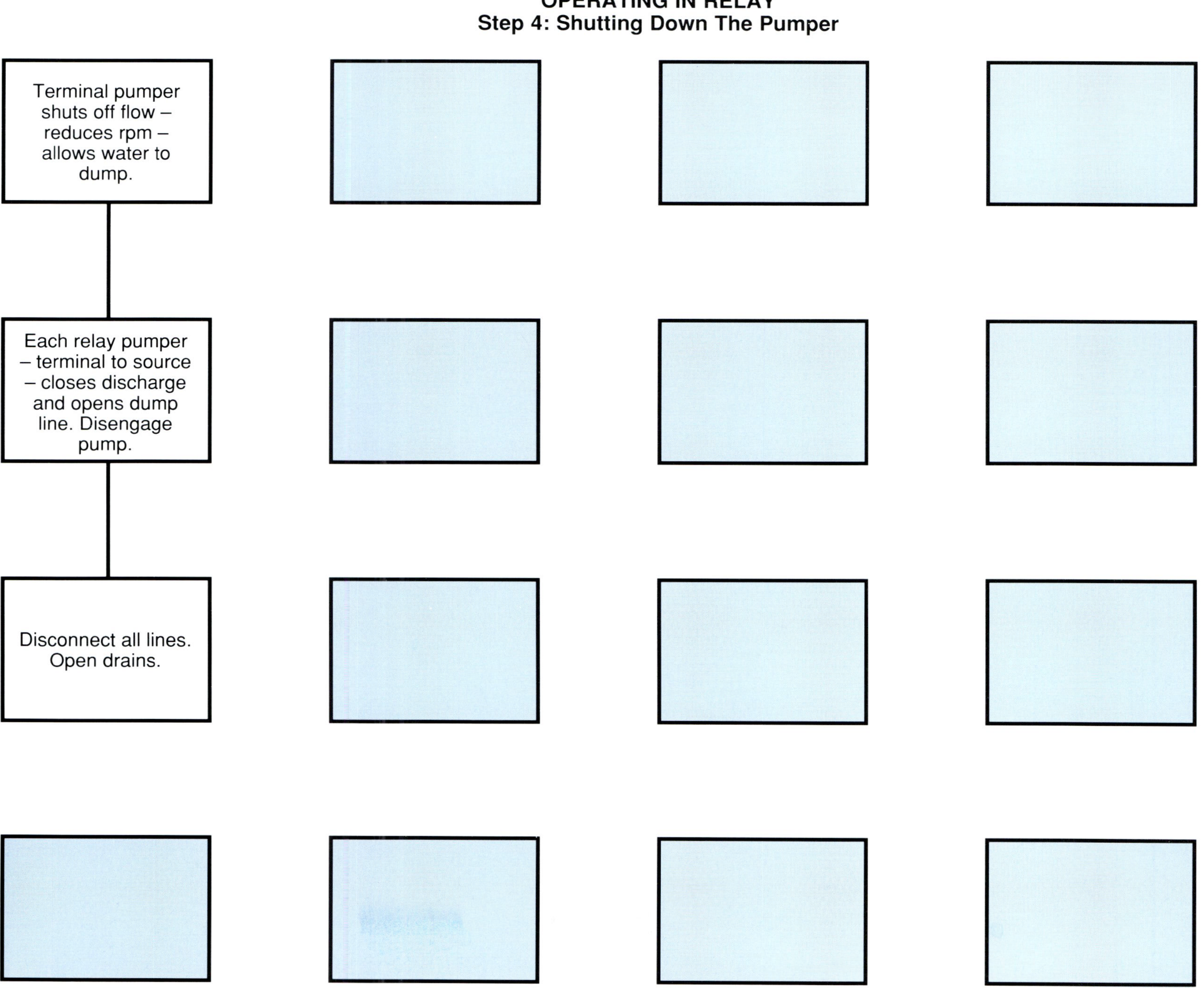

These charts show the proper sequences for operation and describe possible variables in the process. *Flowcharts courtesy of Bill Eckman.*

TROUBLESHOOTING DURING PUMPING OPERATIONS

PROBLEMS COMMON TO ALL TYPES OF OPERATION			
PROBLEM	**SYMPTOMS**	**PROBABLE CAUSE**	**POSSIBLE SOLUTIONS**
Unable to get a reading on the pressure gauge when the pump is put in service.	Green light indicating that the pump shift transfer is complete is not illuminated.	Pump drive system is not fully engaged.	Check the position of the shift transfer control. If it is in the proper position, release the transfer control, allow the gears to turn, then operate it again. Automatic transmission: repeat pump shifting procedure to ensure power transfer to pump operation.
	Green light is on, no mph (km/h) reading registers on the speedometer.	Vehicle clutch is not engaged.	Check the remote clutch control on the pump operator's panel.
		Road transmission is not in the proper gear. Automatic transmission selector is in wrong position.	Check the shift lever: lock it in the proper position for pump operation.
	Speedometer reading is normal for pump operation. All indications are correct and rpm reading is as specified.	There is no water in the pump.	Check the water supply. Ensure that all applicable valves are open. Primer pump may need to be operated to eliminate air in the main pump.
		Gauge is defective. Pressure may be there but not reading on the master pressure gauge.	Check the shutoff (snubber) dampening valve associated with the gauge. Open the gate valve to one of the capped discharge outlets and look for pressure reading on the individual line gauge. Open an uncapped discharge outlet from the pump to see if water is discharged under pressure.
Pump will not develop sufficient pressure.	Rpm reading on the tachometer is normal when compared with the UL plate.	Two-stage pump: • Pump transfer valve is in the wrong position.	The transfer valve should be in the SERIES or PRESSURE position anytime more than 200 psi (1 400 kPa) is needed.
		• Swing check valve may be leaking if the pump is in the SERIES position.	Set the discharge pressure to 50 psi (350 kPa), change from PARALLEL to SERIES, and listen for the metallic sound of the valve operating. If it is blocked, remove the strainer from the large intake and attempt to clear the valve seat of any debris.
		• The transfer valve has not completed its travel and is only partially operated.	Check all mechanical and electrical indications as well as observing pressure gauge readings as the valve is operated. Use the manual override controls to complete the operation of the valve. It may be possible to assist the action of the valve mechanically if the power transfer mechanism is faulty.
		Automatic transmission: transmission not staying in pumping gear lockup and is downshifting as load increases.	Remove pump from service and repair or adjust so apparatus will operate in correct pumping gear.

PROBLEMS COMMON TO ALL TYPES OF OPERATION			
PROBLEM	**SYMPTOMS**	**PROBABLE CAUSE**	**POSSIBLE SOLUTIONS**
Pump will not develop sufficient pressure.	Rpm reading on the tachometer is normal when compared with the UL plate.	Wear on the clearance rings inside the pump may cause excessive slippage.	Take the pumper out of service until it can be repaired.
	Relief valve is operating and the indicator light is on.	Relief valve pressure adjustment is set too low.	Increase the operating pressure of the relief valve by turning the adjustment control clockwise until it closes.
	The indicator light shows that the relief valve is closed.	The relief valve may be stuck open or may not be properly seated, allowing water to bypass back to the intake.	Turn the relief valve operating control to the OFF position.
			Increase the setting of the adjustment control to maximum clockwise position.
			Exercise the valve by rapidly turning the control valve on and off or alternately increasing and decreasing the pressure adjustment control to cause the valve to operate.
	Engine rpm cannot be raised to the value required as determined by the UL plate, even at full throttle. Tachometer reading is low, pressure gauge reading is too low.	Flow requirements may be exceeding the capacity of the pump.	The capacity of the pump in the SERIES or PRESSURE position is limited to 50 percent of its rated capacity at 250 psi (1 700 kPa) and 70 percent at 200 psi (1 400 kPa) net pump pressure.
		Throttle linkage may have slipped or be stuck.	Check the action of the linkage. It may be possible to override the action of the throttle by using the accelerator or hand throttle in the cab or by manually operating the linkage at the carburetor.
		Severe engine overheating can reduce the power available to drive the pump.	Check the engine temperature gauge. Adjust the auxiliary cooling valve to maintain the proper operating temperature.
			Check the level of the coolant in the radiator. If it is too low and the water that is being pumped is clean and pure, open the radiator fill valve to bring the coolant up to the proper level.
		Reduced engine power.	If additional pressure is essential, use another pumper. Take the unit out of service until it can be repaired.
The pump is unable to supply its rated capacity.	The rpm reading on the tachometer is normal when compared to the UL plate.	Two-stage pump: • Transfer valve is in the wrong position.	The transfer valve should be in the PARALLEL or VOLUME position anytime 50 percent or more of the capacity of the pump is needed.
		• The swing check valve is not opening completely.	Remove the strainer from the large intake opening on each side of the pump. Make sure that the valve swings freely by inserting a rod and pushing on the face of the valve.

PROBLEMS COMMON TO ALL TYPES OF OPERATION			
PROBLEM	**SYMPTOMS**	**PROBABLE CAUSE**	**POSSIBLE SOLUTIONS**
The pump is unable to supply its rated capacity.	The rpm reading on the tachometer is normal when compared to the UL plate.	Two-stage pump: • The swing check valve is not opening completely.	Remove the strainer from the large intake opening on each side of the pump. Make sure that the valve swings freely by inserting a rod and pushing on the face of valve.
	The intake gauge registers 0 or has a positive pressure indicated.	Blockage in the waterways of the pump. An object lodged inside the impeller can reduce the capacity.	Thoroughly back-flush the pump by connecting a supply line to the highest discharge outlet and opening the large intake fittings.
		Wear in the pump, usually the clearance rings or relief valve allowing slippage from the discharge back to the intake.	Remove the pumper from service until it can be repaired.
	The intake compound gauge is registering a high vacuum, and the discharge pressure gauge is fluctuating (cavitation).	Blockage of the strainer at the intake fitting of the pump.	Disconnect the intake line and clean any accumulated debris from the strainer.
		Inadequate water supply or supply lines.	Connect an additional supply line to the intake of the pump.
			Reduce the amount of discharge lines being supplied.
			Lower the discharge pressure to reduce the amount of water flowing through the lines.
	Unable to develop enough engine rpm at full throttle to supply the rated capacity.	See items listed under "Pump will not develop sufficient pressure with low engine rpm."	
Pump overheating while in operation.	Pump overheating warning light or physical observation.	Inadequate flow through the pump while operating under pressure.	Open the booster cooling valve or set the circulator valve to TANK or SPILL position as appropriate.
			Open the tank fill valve if it is connected to the discharge side of the pump.
			Use the booster line to maintain a minimum flow of water while pumping.
		Excessive throttle on relief valve-equipped pumps.	Reduce engine rpm.
Relief valve is inoperative or slow acting.	Pressure surges are excessive when individual hoselines are shut down.	Strainer in the pressure line to the pilot valve is dirty.	Open the flush line while pumping clean water to back-flush the in-line strainer.
			Remove the strainer element and wash in clear water.
		Relief valve is corroded or dirty and slow acting.	If the relief valve is equipped with a shut-off, set the discharge pressure on the pump to 150 psi (1 050 kPa), set the relief valve adjustment control to minimum and alternately turn the valve off and on for 60 seconds. If no control valve is provided, set the pump pressure to 150 psi (1 050 kPa) and cause the relief valve to operate by cranking the adjustment control in and out rapidly.

PROBLEM	SYMPTOMS	PROBABLE CAUSE	POSSIBLE SOLUTIONS
PROBLEMS COMMON TO ALL TYPES OF OPERATION			
Relief valve is inoperative or slow acting.	Pressure surges are excessive when individual hoselines are shut down.	Defective relief valve.	Dismantle the valve and clean all working parts, replacing all defective items as needed.
OPERATING FROM THE TANK			
Unable to establish an adequate operating pressure or a loss of pressure occurs when the first discharge valve is opened.	Pressure increases with the engine rpm up to a point, then holds steady or fluctuates.	Some air is trapped in the pump as it fills with water.	Operate the primer until all of the air is removed from the pump. Transfer the pump from SERIES to PARALLEL and back several times while flowing water.
			If only small lines are being supplied, opening the pump to tank valve and increasing the velocity of the water through the pump for a short time may remove the rest of the air.
		Automatic transmission: • Intermittent transmission slippage.	Adjust transmission.
		• Low transmission fluid level.	Increase fluid level.
		• Transmission not remaining in correct pumping gear.	Adjust transmission.
Fluctuation of the pressure gauge and a reduction of discharge pressure when additional lines are put in service.	High vacuum reading on the intake compound gauge.	The tank-to-pump valve may not have been fully opened or has vibrated to a partially CLOSED position.	Check the tank-to-pump valve and lock it in the OPEN position if the valve has a locking mechanism.
		Tank-to-pump piping may be too small to supply the amount of water required by the hoselines in service.	Shut down one of the hoselines if fireground conditions permit.
			Reduce the discharge pressure until fluctuations stop and pressure gauge begins to drop.
While pumping, the discharge pressure drops to a very low value and water supply is interrupted.	Compound gauge on the intake reads 0 or fluctuates. Engine speed increases.	An air leak in the pump.	Check all caps and valves on the intake side of the pump.
	Water gauge reads empty.	The water supply from the tank is exhausted or nearly so.	Reduce the pressure until the gauge becomes steady and flow resumes.
			Arrange for another water supply as soon as possible.
OPERATING FROM A HYDRANT			
The suction line collapses when the discharge valve to a hoseline is opened.	The intake pressure drops to less than 0. Discharge pressure also drops.	Kinks in suction lines can cause excessive friction loss when the water begins to move.	Rearrange the suction line to eliminate any kinks or restrictions.

OPERATING FROM A HYDRANT			
PROBLEM	**SYMPTOMS**	**PROBABLE CAUSE**	**POSSIBLE SOLUTIONS**
The suction line collapses when the discharge valve to a hoseline is opened.	The intake pressure drops to less than 0. Discharge pressure also drops.	Suction line may be too small for the amount of flow needed.	3-inch (77 mm) suction hose can supply approximately 500 gpm (2 000 L/min), but gated 2½-inch (65 mm) intake fittings may limit the flow to less than 300 gpm (1 200 L/min). If 2½- or 3-inch (65 mm or 77 mm) suction line is used, it should be brought into the large intake, either through a suction siamese or bell reducer.
	Water coming out of the ground around the barrel of the hydrant.	Hydrant not fully opened.	Turn the hydrant wrench in a counter-clockwise direction until it reaches the limit of its travel.
While supplying water, the suction line collapses and the pump begins to cavitate.	Intake pressure drops to less than 0. Discharge pressure fluctuates and decreases.	Additional water being demanded by the hoselines that are being supplied.	Reduce the number of hoselines in service or the flow settings on the nozzles.
			Reduce the discharge pressure on the pump by reducing the throttle setting until cavitation stops.
		Additional demands on the water system may have reduced the residual pressure in the system.	Decrease the amount of water being used by changing the nozzle settings or taking attack lines out of service.
			Obtain a supplementary supply from another hydrant, a relay, or from a water shuttle.
OPERATING FROM DRAFT			
Pump will not prime.	Unable to get water into the pump through the hard suction hose. No vacuum reading is registered on the intake compound gauge.	Drain valve left open.	Check the master drain valve to make sure it is fully closed. Check the individual drain valves for governor, auxiliary cooler, and so on.
		Intake valves left open or caps not airtight.	Tighten caps on large suction fittings that are not being used.
			Make sure the tank-to-pump valve is closed if the tank is empty.
		Booster line cooler valve or circulator valve left in OPEN position (Barton American pump).	Shut the booster line cooling valve.
			Put the circulator valve in the VERTICAL position, between the TANK and SPILL settings.
		Intake relief valve may be leaking.	If the relief valve is equipped with a shut-off valve, close it. If it has fire hose threads on the discharge opening, put a cap on it.
		Suction hose connections are not airtight.	Listen for air leaks at each connection. Tighten with a rubber mallet if any connection appears to be leaking.

OPERATING FROM DRAFT			
PROBLEM	**SYMPTOMS**	**PROBABLE CAUSE**	**POSSIBLE SOLUTIONS**
Pump will not prime.	Unable to get water into the pump through the hard suction hose. No vacuum reading is registered on the intake compound gauge.	Suction hose connection to the floating strainer not airtight.	Tighten the coupling with a rubber mallet or a spanner wrench.
		Pump packing is too loose and leaking air.	Take pumper out of service until the packing can be adjusted or the pump repacked.
		Not operating the primer long enough to get rid of all of the air.	A typical fire pump requires 15 seconds to prime through 20 feet (6 m) of suction hose but may take as long as 30 seconds to prime completely.
		Tank-to-pump valve not sealing with an empty tank.	Partially fill tank (temporary). Repair valve (permanent).
			Engage the pump, build up 100 psi (700 kPa), and discharge water from the booster line. While keeping the end of the suction hose submerged, take the cap off the end and install the strainer. Slowly close the tank-to-pump valve while continuing to flow water from the booster line. The pump is now primed.
		Engine rpm is too low.	The engine rpm should be as specified in the manufacturer's instructions but is usually 1,000 to 1,200 rpm.
		No oil in the reservoir for the priming pump.	A supply of oil should be carried on any pumper that uses oil for priming. If this is the case, oil can be put in the reservoir to replenish the supply.
			If the regular priming oil is not available, putting water in the reservoir may enable the primer to operate, but it should be thoroughly cleaned out and a supply of oil put in at the first opportunity.
	The electric motor will not operate to drive the primer.	A bad battery or a poor connection between the cable and battery terminals.	Inspect the battery terminals. If they are corroded, it may be possible to loosen the fastening and clean the terminal sufficiently to run.
			If a mechanical override is provided, open the priming valve and engage the mechanical drive system.
	Very little air is being discharged from the primer.	Defective primer.	If the primer will not operate for mechanical reasons, the pump can be primed by connecting the hard suction hose to the pump and installing the cap on the end of the hose, then submerging the hose under water. Open the tank-to-pump valve, allowing the suction hose to fill with water as well as the pump.
The pump loses its prime when the first discharge valve is opened and water begins to flow.	The discharge pressure gauge drops sharply.	Valve may have been opened too rapidly.	Carefully observe pressure gauge while slowly opening the discharge valve. If the pressure begins to drop suddenly, pause and allow it to stabilize before continuing.

OPERATING FROM DRAFT			
PROBLEM	**SYMPTOMS**	**PROBABLE CAUSE**	**POSSIBLE SOLUTIONS**
The pump loses its prime when the first discharge valve is opened and water begins to flow.	The discharge pressure gauge drops sharply.	The pump was not completely primed and still had some air trapped in it when the primer was released.	Allow the primer to continue to operate until a steady stream of water is discharging from it before closing the priming valve.
			If pressure drops suddenly while opening a discharge valve, operate the primer momentarily to remove any remaining air from the pump.
		The pump may not be turning fast enough to sustain the prime when water begins to flow.	Adjust the discharge pressure to 75 psi (525 kPa) or more before opening the discharge valve.
		The priming valve can stick causing air to leak into the pump.	Check the priming valve control to see that it is in the CLOSED position. Exercise the valve to clear any debris from the valve seat.
		Rock or debris in impeller.	Take pumper out of service until it can be repaired.
	The reading on the pressure gauge drops sharply and the intake gauge returns to the 0 reading.	A high spot in the suction line has trapped a quantity of air when the pump was primed. This air has been drawn into the pump when the water began to move through the suction hose.	Attempt to eliminate the high spot by moving the suction hose and prime the pump again.
			If the high spot cannot be eliminated, it may be possible to scavenge the air from the suction hose by operating the primer again each time the pressure begins to drop.
The pump loses its prime during the course of a pumping operation.	The pump loses its prime when all nozzles are closed and no water is flowing.	An air leak on the intake side of the pump.	Check all connections to see that they are airtight.
		The packing may be mis-adjusted, allowing air to leak into the pump around the shaft.	Open a discharge outlet and allow water to flow at all times to maintain a higher vacuum reading in the intake side of the pump. If the problem becomes acute, take the pumper out of service until it can be repaired.
	The pump loses its prime when it is operating near its maximum capacity. The vacuum reading on the intake gauge is near 0 and is fluctuating.	A whirlpool over the strainer is allowing air to get into the pump through the suction hose.	Put a board or other object over the whirlpool to break up the swirling motion and stop the air from getting into the suction hose.
		Air leak on the intake side of the pump.	Check all connections.
The pump goes into cavitation when the flow increases.	The intake gauge registers more than 22 inches (559 mm) of vacuum, the pressure gauge fluctuates and decreases reading.	The flow exceeds the capacity of the pump at the lift that is required.	The capacity of the pump decreases as the lift increases. At a 20-foot (6 m)lift, the capacity of the pump is only 60 percent of what it would be with a 10-foot (3 m) lift.

OPERATING FROM DRAFT			
PROBLEM	**SYMPTOMS**	**PROBABLE CAUSE**	**POSSIBLE SOLUTIONS**
The pump goes into cavitation when the flow increases.	The intake gauge registers more than 22 inches (559 mm) of vacuum, the pressure gauge fluctuates and decreases reading.	The suction line may be partially blocked.	Debris blocking the strainer on the suction hose. Clean the strainer manually.
			Debris is trapped in the strainer at the intake of the pump. Shut down the pump, remove the suction hose and clear the strainer.
		Inner rubber liner of suction hose has become separated from hose. The result is restriction caused by the "bubble" of inner liner.	The suction hose can collapse. It will have to be replaced if the reduced capacity is unacceptable.
			Replace suction hose.
OPERATING IN RELAY			
The intake supply line collapses when the throttle setting is increased to establish the initial discharge pressure as required.	The intake pressure gauge reading is negative, that is, reading vacuum instead of pressure.	The dump line or uncapped discharge used to waste water while establishing the relay may still be open.	Close the valve to any uncapped discharge or dump line.
			If the pumper is operating as the terminal unit in a relay, adjust the valve on the dump line to bring the residual pressure at the intake of the pump to 50 psi (350 kPa).
		The terminal pumper may be attempting to take more water from the relay than it can supply.	Notify the water supply officer that the relay is unable to supply the amount of water being called for so that additional supply lines can be put in service.
			Reduce the discharge pressure until the intake gauge registers a positive reading.
While the relay is operating, the intake pressure increases above 50 psi (350 kPa).	The intake pressure gauge is reading above 50 psi (350 kPa) and the discharge pressure also increases accordingly.	Changes in flow of the attack line cause the friction loss to decrease and the residual pressure to increase.	No action is necessary unless the pressure increase becomes dangerous. Minor variations are to be expected in a relay and frequent adjustments are undesirable.
While the relay is operating, the intake pressure increases dangerously.	Intake pressure gauge is reading above 150 psi (1 050 kPa), the discharge pressure is above 200 psi (1 050 kPa).	Hoselines have been shut down with no corresponding dumping of excess water.	Open the uncapped discharge or dump line until the intake residual pressure returns to 50 psi (350 kPa).

Courtesy of Bill Eckman.

Chapter 6 Review

Answers on page 361

TRUE-FALSE: Mark each statement true or false. If false, explain why.

1. Ideally, the pumper with the largest capacity should be positioned at the fire scene end of the relay.

   ☐ T ☐ F ____________________

2. A 2½-inch (65 mm) line is the smallest line that should be connected to a fire department sprinkler connection.

   ☐ T ☐ F ____________________

MULTIPLE CHOICE: Circle the correct answer.

3. When placing a PTO pump into gear for a stationary operation, in what gear should the transmission be placed after operating the PTO shift lever?
   A. First gear
   B. Second gear
   C. Reverse
   D. Neutral

4. When pumping from the tank and the vacuum reading exceeds ______ inches (______ mm) of mercury, the pump is in danger of cavitating.
   A. 10 inches (250 mm)
   B. 15 inches (375 mm)
   C. 20 inches (500 mm)
   D. 25 inches (625 mm)

5. Which type of hydrant is generally considered to provide the least reliable amount of water for fire fighting operations?
   A. Dead-end hydrant
   B. Looped hydrant
   C. Dry-barrel hydrant
   D. Wet-barrel hydrant

6. For effective operation, the maximum amount of lift for a pumper is about 20 feet (6 m). What amount of the rated capacity can be supplied from this amount of lift?
   A. 50 percent
   B. 60 percent
   C. 80 percent
   D. 100 percent

7. Which of the following is generally *not* considered to be a major factor influencing the pump discharge pressure required to supply a fire department standpipe system connection?
   A. Height of the building
   B. Friction loss in the standpipe
   C. Occupancy classification
   D. Size of hose streams in operation

8. When supplying a standpipe system, what is the rule of thumb for determining how much the pump discharge pressure must be increased to compensate for elevation loss?
   A. 1 psi (7 kPa) per floor
   B. 3 psi (21 kPa) per floor
   C. 5 psi (35 kPa) per floor
   D. 8 psi (56 kPa) per floor

LISTING

9. List the three basic types of water supply sources used to supply a fire department pumper.

A. ______________________________

B. ______________________________

C. ______________________________

10. List at least two indications that a pump is cavitating.

____

____

11. In order, list the six steps for putting most midship pumps into gear.

A. ______________________________

____

B. ______________________________

____

C. ______________________________

____

D. ______________________________

____

E. ______________________________

____

F. ______________________________

____

12. List at least two methods of preventing the pump from overheating when pumping from a hydrant.

____

____

13. In addition to maintaining as small a lift as possible, list at least two other considerations when selecting a drafting site.

____

____

14. Name at least three possible sources of an air leak when trying to prime a pump.

____

____

____

15. What are the two critical factors in designing a relay operation?

A. ______________________________

B. ______________________________

FILL IN THE BLANK: Fill in the blanks with the correct response.

16. A two-stage pump will usually be in the __________ position when operating from the vehicle's water tank.

17. When drafting, __________ inches (__________ mm) of water should be above a strainer in order for the pumper to be able to pump its maximum capacity.

18. During priming operations, the transfer valve of a two-stage pump should be in the __________ position.

19. As a general rule, a good pressure for supplying fire department sprinkler connections is __________ psi (__________ kPa).

SHORT ANSWER: Answer each item briefly.

20. Define cavitation.

____

____

____

21. If pump pressure has been built up but there are no attack lines ready to be charged, what should you do?

____

____

____

22. What is the best method of removing air from a supply line coming into a gated intake?

____

____

23. Outline the eight steps, in order, for switching to an extended water supply.

A. ____

____

B. ____

____

C. ____

____

D. ____

____

10-91

E.

F.

G.

H.

24. Excess water flowing from the drain holes on a hydrant may result in damage to that hydrant. How can this be prevented?

10-91

25. During a drafting operation, you notice a gradual increase in the vacuum reading, yet there has been no appreciable increase in the flow rate. What possible situation does this indicate?

26. Under what circumstance is a relay operation required?

27. If a single line of 4-inch (100 mm) supply line is used, how many pumpers are needed to supply 500 gpm (2 000 L/min) to feed the attack lines at the scene 6,000 feet (1 800 m) away?

7

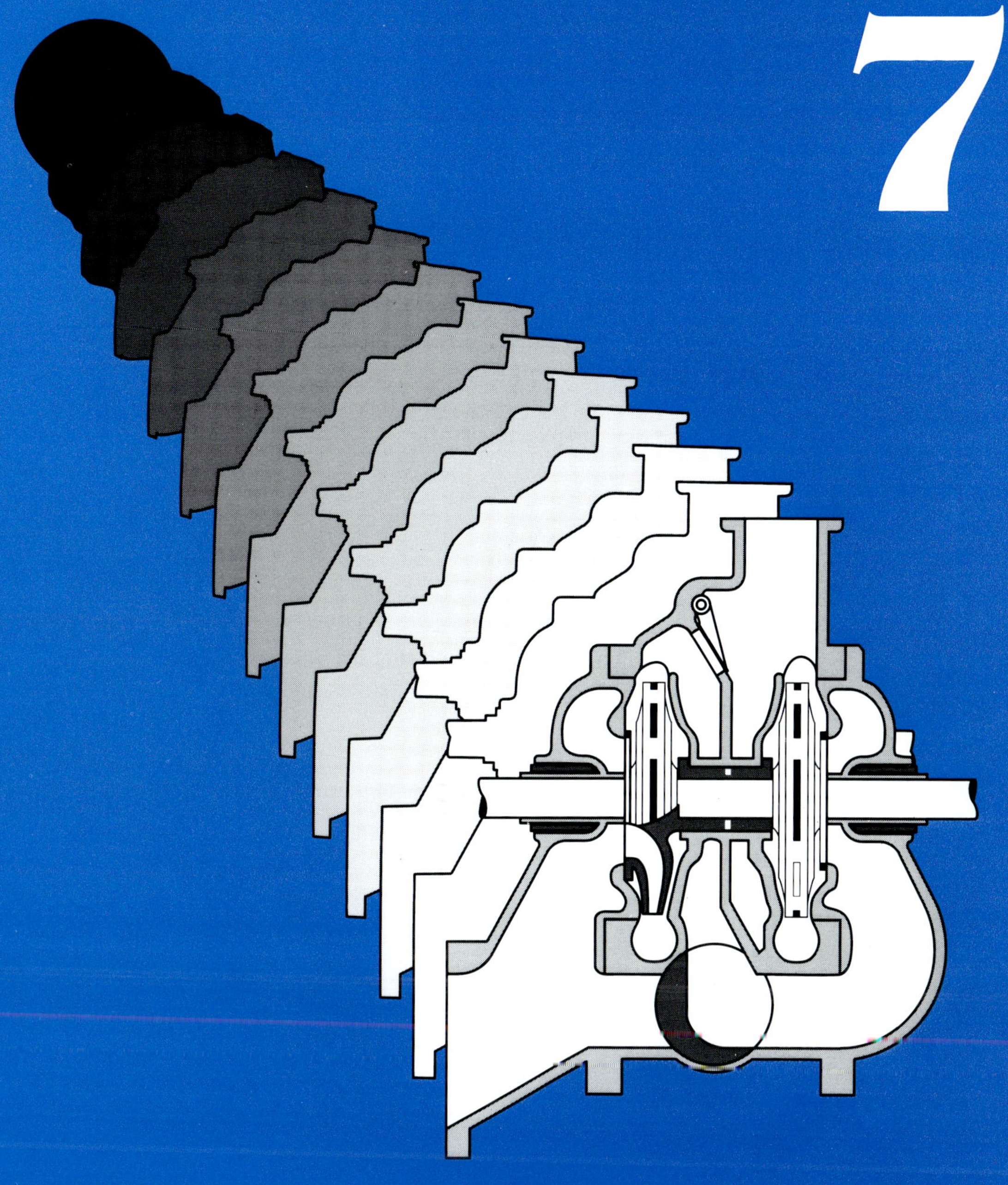

Apparatus Maintenance

NFPA STANDARD 1002
Fire Apparatus Driver/Operator Professional Qualifications

2-1 Preventive Maintenance.

2-1.1* The fire apparatus driver/operator shall demonstrate the performance of routine tests, inspections, and servicing functions required to assure the operational status of fire department vehicles, including:

(a) battery check
(b) braking system
(c) coolant system
(d) electrical system
(e) fueling
(f) hydraulic fluids
(g) lubrication
(h) oil levels
(i) tire care
(j) steering systems
(k) tools, appliances and equipment.

2-1.2 The fire apparatus driver/operator shall demonstrate the recording and reporting, as specified by the authority having jurisdiction, of all servicing functions.

3-1 General.

3-1.1* The fire apparatus driver/operator shall demonstrate the performance of routine tests, inspections, and servicing functions required to assure the operational status of fire department pumpers, including:

(a) battery check
(b) booster tank level (if applicable)
(c) braking system
(d) coolant system
(e) electrical system
(f) hydraulic fluids
(g) fueling
(h) lubrication
(i) oil levels
(j) pumping system
(k) steering systems
(l) tire care
(m) tools, appliances and equipment.

Chapter 7
Apparatus Maintenance

The apparatus driver/operator or person responsible for the upkeep and readiness of the apparatus should establish a system of record keeping for reference and review of the need for preventive maintenance. The specifications of each manufacturer should be reviewed and daily, weekly, monthly, semiannual, and annual inspections established on a simple check sheet. This report serves as a reminder and audit for assurance that the apparatus is ready and operable. The time of an alarm is not a proper occasion for preventive maintenance.

One of the most misunderstood concepts concerning the upkeep of fire apparatus is the difference between maintenance and repair. The word *maintenance,* as used here, means keeping apparatus in a state of usefulness or readiness. *Repair* means to restore or replace that which has become inoperable. Apparatus or equipment that is said to be in a good state of repair has probably been well maintained. Preventive maintenance ensures apparatus reliability, reduces the frequency and cost of repairs, and lessens out-of-service time.

A SYSTEMATIC MAINTENANCE PROGRAM

The need for a planned fire apparatus maintenance program is obvious. Maintenance schedules and records have previously been discussed; these are an integral part of a well-planned maintenance program. All mechanical repairs should be made by a mechanical technician or automotive mechanic. Supervised repair shops are often maintained for this purpose. Even in departments where fire department repair shops are not available, a maintenance program is still necessary.

Although any individual firefighter may be held responsible for the maintenance of the apparatus, this responsibility is usually assigned to the driver/operator. The following section highlights some of the particular things that the driver/operator may be responsible for in maintaining the apparatus.

Cleanliness

Much has been said and written concerning the importance of maintaining good public relations. The public sees a piece of fire apparatus as a unit of protection in which they have invested many thousands of dollars. Permitting apparatus to become marred with dirt, oil, and road grime damages public relations because individuals may feel their investment is not being properly protected (Figure 7.1).

Figure 7.1 Apparatus should be cleaned anytime dirt can be seen on the vehicle.

Although the public relations aspect of cleaning apparatus is important, there are other, more important reasons for keeping fire apparatus clean. A clean engine and clean functional parts permit proper inspection, thus helping to ensure efficient operation as needed. Fire apparatus should be kept clean underneath as well as on top. Oil, moisture, dirt, and grime should not be permitted to collect. Some of the more vulnerable areas are the engine, its wiring, carburetor or fuel injectors, and controls.

Fire Apparatus Records

An accurate and descriptive record should be kept of all items carried on each apparatus. Apparatus records should also show routine in-service preventive maintenance for each apparatus. Apparatus records are not complete unless shop checkup, major and minor repairs, and parts replacement are itemized. Records concerning lubrication and other servicing are more meaningful on a time-frequency basis rather than an engine-miles (engine-kilometers) or road-miles (road-kilometers) basis. Some departments find it desirable to keep a more detailed inventory file, including the apparatus purchase specifications and a parts inventory. Driver/operators are responsible for filling out the appropriate forms for the maintenance functions they perform. These forms should be given to whomever is responsible for seeing that they are properly recorded or filed. (**NOTE:** A number of fire apparatus maintenance forms are included at the end of this chapter.)

Any fire apparatus maintenance or inventory form can also be stored and updated as a computer file. There are a number of software programs designed for inventory and maintenance records that can be used on a microcomputer. These programs are usually referred to as data base programs. Custom programs can be developed to suit the special needs of any fire department maintenance program.

APPARATUS MAINTENANCE

Fire department personnel generally take pride in keeping their apparatus in good operating condition. This process requires performing daily, weekly, and periodic maintenance. If a well-established schedule is followed, operating troubles will be reduced to a minimum. The following schedules are suggested as a guide for establishing a maintenance program.

All maintenance should be performed in accordance with the operator's handbook furnished by the manufacturer. The list under daily maintenance should also be followed before any apparatus that has been out of service is returned to service.

Daily Maintenance

- Check crankcase oil for proper level (Figure 7.2).
- Check coolant level of radiator (Figure 7.3).
- Check all batteries (Figure 7.4).
- Check all visible and audible warning signals.
- Check fuel level.

Figure 7.2 Checking the oil level is a simple task that every driver/operator should be able to perform.

Figure 7.3 The driver/operator should visually check that the level of coolant in the radiator is adequate.

Figure 7.4 Use care when checking batteries. The acid contained in them may injure the driver/operator and damage the apparatus.

- Visually check the water tank water level (Figure 7.5).
- Maintain a full tank at all times.
- Check each tire for cuts, breaks, and proper inflation (Figure 7.6).
- Pressure test all brakes by operating foot pedal.
- Clean apparatus windows. Wash and dry entire vehicle if necessary.
- Operate changeover valves on two-stage pumps.

Figure 7.5 The water level in the booster tank can be visually confirmed by looking down through the tank vent.

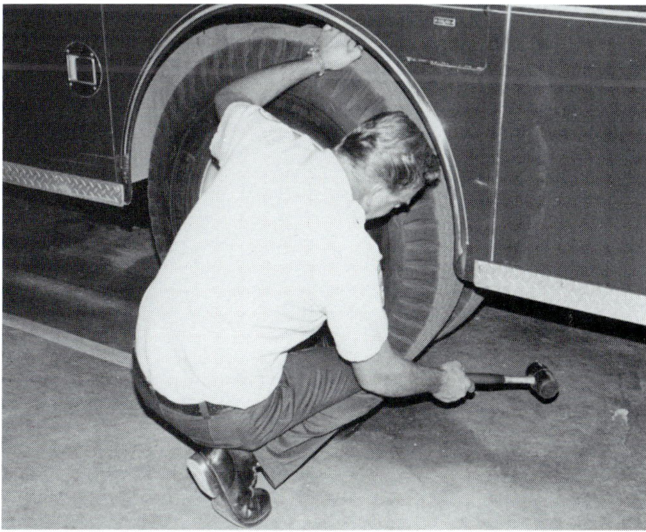

Figure 7.6 The condition of the tires should be checked daily.

Weekly Maintenance

- Check transmission oil level.
- Check differential oil level.
- Check power steering fluid level.
- Check brake system when equipped with hydraulic brakes.
- Check master cylinder brake fluid level and watch for any wheel cylinder or hose leaks.
- On apparatus equipped with air brakes, check air system for leaks and bleed moisture from air tanks.
- Check fan belt and generator or alternator belts (Figure 7.7).
- Check battery terminals and cables.
- Operate valves in auxiliary cooling system.
- Check drains and all hose connections for security (Figure 7.8).
- Check drive shaft and universal joints.
- Start motor and observe oil pressure and idling speed after engine is warm.
- Clean underneath chassis.
- Clean the engine and electrical motors.
- Check for loose nuts, studs, and pins.

Figure 7.7 Check belts to ensure tightness and good physical condition.

Figure 7.8 Check intake connections to make sure they do not contain any foreign objects that could be drawn into the pump.

Periodic Maintenance (Qualified Service Personnel)

- Service air filters as required.
- Drain and refill crankcase twice annually as a minimum or as required by the manufacturer.
- Check fuel pump.
- Perform engine tune-up (annually) for both diesel and gasoline engines.
- Lubricate the chassis, distributor, starting motor, water circulating pump, generator or alternator, and steering gear.
- Service test fire apparatus.

EQUIPMENT MAINTENANCE

Equipment refers to those items carried on the fire apparatus but not permanently attached to or a part of the apparatus. Most removable equipment is common to all fire apparatus. The following lists are recommended guidelines for establishing an equipment maintenance procedure.

The procedures under daily maintenance should also be followed each time the equipment is used.

Daily Maintenance

- Check all portable extinguishers by lifting and checking gauges (Figure 7.9).
- Check hose loads for correct finish (Figure 7.10).
- Inventory all nozzles and appliances. Operate all valves.
- Check air pressure in protective breathing equipment (Figure 7.11).
- Examine regulator.
- Check all handlights and flashlights (Figure 7.12).

Figure 7.9 Examine fire extinguishers for any damage or loss of pressure.

Figure 7.10 Check hose loads to make sure they are ready for rapid deployment when needed.

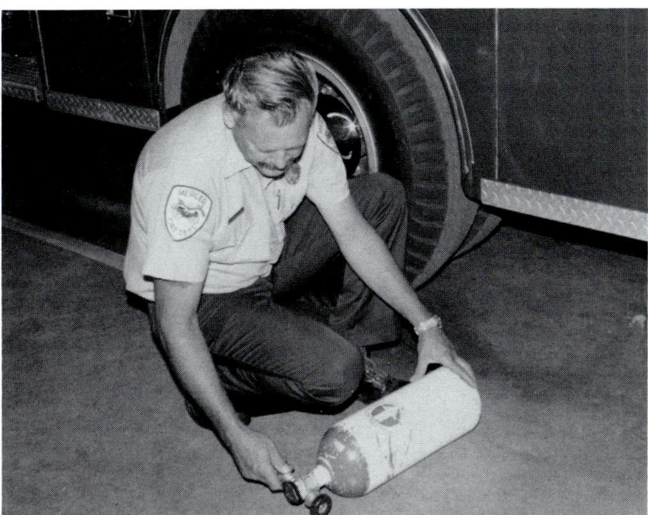

Figure 7.11 SCBAs and spare bottles should be inspected to ensure proper operation.

Figure 7.12 Turn on battery-operated handlights to make sure they are in good working order.

Weekly Maintenance

- Remove all appliances from the apparatus and clean if necessary.
- Start all portable engine or motor-driven equipment, such as saws or generators, and permit it to warm up (Figure 7.13).

Figure 7.13 Motor-driven equipment should be started and run once a week.

Periodic Maintenance

- Remove all hose from fire apparatus, check condition and reload, changing the bends. Service test all fire hose annually.
- Remove all ladders from their racks and clean them (Figure 7.14).

Figure 7.14 In order to clean ladders properly, remove them from the apparatus so they can be hand cleaned.

- Lubricate all motor-, gear-, or hand-operated appliances according to manufacturer's specifications.
- Flush the pump and water tank (monthly).

ELECTRICAL COMPONENTS

There are numerous electrical connections in an apparatus; damage from moisture or corrosion can render an electrical connection inoperative. The following list is suggested as a guide for a thorough inspection of electrical components.

Lighting System

- Headlights and switches
- Dimmer switch
- Clearance, stop, and backup lights
- Compartment lights and switches
- Warning lights and switches (Figure 7.15)
- Floodlights and switches (Figure 7.16)

Figure 7.15 Inspect warning lights and replace any burned-out lamps or flashers.

Figure 7.16 Be careful when inspecting and handling floodlights, as they can become dangerously hot when they have been on for awhile.

Figure 7.17 Keep battery terminals clean to ensure maximum contact between the battery and the wires.

Electrical Motors

- Rotating lights
- Hose reel rewind
- Windshield wipers
- Apparatus controls
- Heater and defroster fan

Battery

- Corrosion of terminals (clean if needed) (Figure 7.17).
- Cell electrolyte level (need for water).
- Specific gravity.
- Test for voltage.
- Recharge if necessary. **NOTE:** Use caution when attaching battery jumper cables to prevent arcs.

Voltage Regulator

- If battery shows full charge and ammeter shows high charge rate under normal circuit load, report to shop mechanic for correction.
- If ammeter shows low charge rate when the battery is below full charge, report to shop mechanic for correction.

Generator or Alternator

- These devices require special maintenance. Repairs should be undertaken by someone trained and equipped with test equipment.

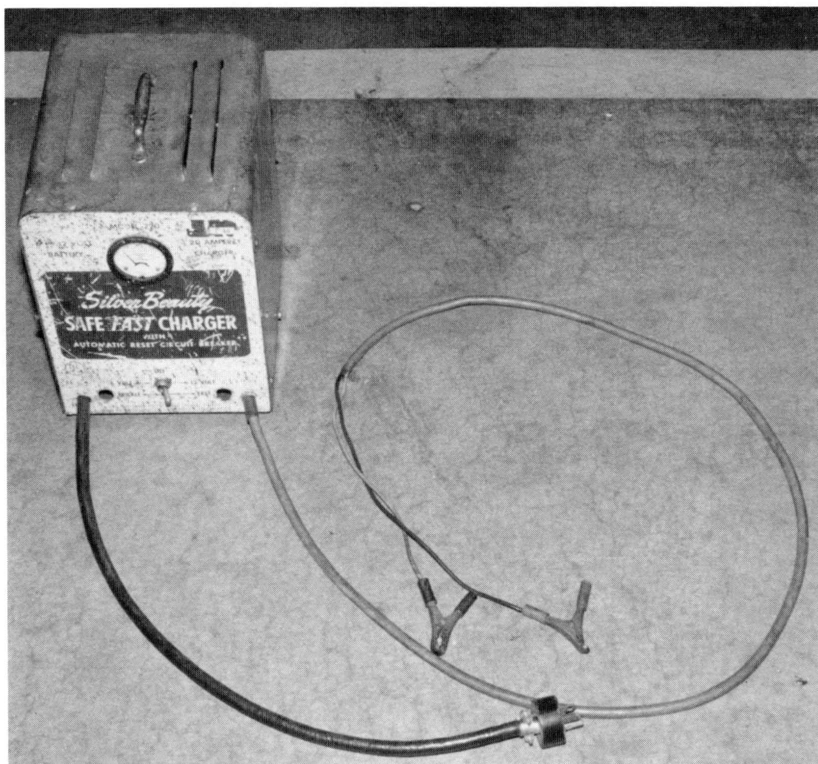

Figure 7.18 It may be necessary to use a battery charger to ensure maximum operation of a battery.

Figure 7.19 Failure to properly connect a battery charger to the battery may result in damage to the battery and could harm the person making the connection.

CHARGING BATTERIES

Battery chargers can be found in almost every fire station (Figure 7.18). Because batteries produce explosive hydrogen gas when being charged, they *must* be used correctly to prevent needless accidents from occurring. Follow the procedure outlined below when attaching battery charger cables to any vehicle battery:

Step 1: Identify the polarity of the battery to be charged (positive or negative ground).

Step 2: Attach the red charger cable to the positive battery post.

Step 3: Attach the black charger cable to the negative battery post (Figure 7.19).

Step 4: Connect the battery charger to a reliable power source (away from gasoline and other flammable vapors).

Step 5: Set the desired battery charging voltage and charging rate if so equipped. **NOTE:** Switches on battery chargers should be in the OFF position when not in use.

Step 6: Reverse the procedure to disconnect the battery charger.

CARE OF ENGINES

House starting of engines simply to warm them up each day is not recommended by engine manufacturers. Starting an engine unnecessarily contaminates the oil, wastes fuel, and causes added engine wear. If the battery is kept fully charged and the apparatus is kept fueled, there should be no need to start the engine until an alarm is sounded or the apparatus is used for drills. These situations generally require using the apparatus at least once a week.

The wear on a cold engine is many times greater than that on a warm engine. As a cold engine becomes warm, condensation is formed in the crankcase. If an engine is started to reposition the apparatus, it is not beneficial to allow the engine to run for any period of time. Running an engine for 10 to 15 minutes in a station under a no-load condition does not evaporate the condensation. Allowing an engine to run does not ensure that the engine will start when the next alarm is received.

Sudden temperature changes are not good for any engine. When an engine is started for nonemergency response, do not run it under full load until the engine has had time to warm up to its operating temperature. Likewise, anytime an engine has been running at full load, follow the engine manufacturer's recommendation for shutdown procedures.

Engine Maintenance

The engine of a fire apparatus is the functional part upon which all other fire fighting devices are dependent. If the engine does not function, everything else stands still. In addition to providing a source of power for transportation, the engine provides power to operate fire pumps, aerial ladders, elevating platforms, and, sometimes, other equipment.

Engine adjustments and repairs should be made by a qualified mechanic. The most important phases of engine maintenance are regular service inspections. These regular inspections are actually engine maintenance routines. Because these inspections are performed regularly, they will often reduce failures to a minimum.

Lubrication

Proper lubrication is one of the prime objectives of good maintenance. Proper lubrication saves maintenance and repair dollars and reduces out-of-service time. Effective lubrication depends upon the use of a proper grade of recommended lubricant, the frequency of lubrication, the amount used, and the method of lubrication. To select the proper lubricant, consideration must be given to the requirements of the unit to be lubricated, the characteristics of lubricants, and the manufacturer's recommendations. The manufacturer's manual will "recommend" the Society of Automotive Engineers (S.A.E.) number for the engine oil. The S.A.E. number indicates only the viscosity. Some other essential characteristics of oil are corrosion protection, foaming, sludging, and carbon accumulation, which may be controlled by the refiner (Figure 7.20 on next page).

SUMMARY OF SERVICE CATEGORIES

AUTOMOTIVE GASOLINE ENGINE SERVICE

Current API Engine Service Categories	Previous API Engine Service Categories	Use and Definition
SA	ML	This service category has no performance requirements and is now considered obsolete.
SB	MM	Provide mild antiscuff capability and resistance to oil oxidation and bearing corrosion. This service category is now considered obsolete.
SC	MS (1964)	Denotes service typical of gasoline engines in 1964 through 1967 models of passenger cars and some trucks. Provide control of high- and low-temperature deposits, wear, rust, and corrosion in gasoline engines.
SD	MS (1968)	Denotes service typical of gasoline engines in 1968 through 1970 models of passenger cars and some trucks. These oils provide more protection against high- and low-temperature engine deposits, wear, rust, and corrosion in gasoline engines than SC-category oils. May be used when API Engine Service Category SC is recommended.
SE	None	Denotes service typical of gasoline engines in passenger cars and some trucks beginning with 1972 and certain 1971 models. Provide more protection against oil oxidation, high-temperature engine deposits, rust, and corrosion in gasoline engines than oils classified as SC or SD. May be used when either of these categories is recommended.
SF	None	Used for gasoline engines beginning with the 1980 model year. Provide greater oxidation stability and improved antiwear performance than oils in Category SE. Also protect against engine deposits, rust, and corrosion. May be used where API Service Categories SE, SD, or SC are recommended.

COMMERCIAL DIESEL ENGINE SERVICE

Current API Engine Service Categories	Previous API Engine Service Categories	Use and Definition
CA	DG	Used with diesel engines operated in mild- to moderate-duty with high-quality fuels. Protect from bearing corrosion and from ring belt deposits in some naturally aspirated diesel engines when using fuels of such quality that they impose no unusual requirements for wear and deposit protection. These oils were widely used in the late 1940s and 1950s but should not be used in any engine unless specifically recommended by the equipment manufacturer.
CB	DM	Used for diesel engines operated in mild- to moderate-duty but with lower-quality fuels, which necessitate more protection from wear and deposits. Protect against bearing corrosion and ring belt deposits in some naturally aspirated diesel engines with higher sulfur fuels. Oils in this category were introduced in 1949 but should not be used in any new engine unless recommended by the manufacturer.
CC	DM	Used for certain naturally aspirated, turbocharged or supercharged diesel engines operated in moderate- to severe-duty service and certain heavy-duty gasoline engines. Protect against high-temperature deposits and bearing corrosion, and low-temperature deposits in gasoline engines. These oils were introduced in 1961.
CD	DS	Typically used for certain naturally aspirated, turbocharged or supercharged diesel engines where highly effective control of wear and deposits is vital or when using fuels of a wide quality range including high sulfur fuels. These oils were introduced in 1955 and provide protection from bearing corrosion and from high-temperature deposits.

Figure 7.20 The API Engine Service Classification System defines and explains classes of service for both diesel and gasoline engine applications. *Courtesy of American Petroleum Institute. For more information about motor oils, contact the Marketing Department, American Petroleum Institute, 1220 L. St. NW, Washington, D.C. 20005.*

CARE OF CHASSIS COMPONENTS

Although all the components that make up the chassis are dependent upon the engine for power, the driveline components behind the engine warrant an equal degree of maintenance and attention. The driver/operator should make a regular check of these components to detect changes, vibrations, noise, and other indications of potential problems. The fact that the drive components are installed in the chassis does not eliminate the possibility of service problems. Regular inspection of the following components is necessary to keep fire apparatus out-of-service time to a minimum:

- Check master cylinder brake fluid reservoir. If vehicle is equipped with air brakes, inspect compressor belts and bleed water from air tanks.
- Check power steering hydraulic fluid reservoir. Check for brake fluid leaks underneath and at wheels.
- Check brake pedal for floorboard clearance. Check brake pedal while driving (hydraulic brakes only).
- Check all tires for proper air pressure using a gauge (Figure 7.21). Inspect tires for breaks and wear.

Figure 7.21 Add air if tire pressure is below the recommended level.

- Check universal joints. Check steering device.
- Check tail pipe and muffler.
- Check parking brake adjustment.

GENERAL FIRE PUMP MAINTENANCE

Although fire pumps are tested at regularly scheduled times, these tests are designed to compare performance to specific standards. In addition, certain preventive maintenance procedures should be regularly carried out to detect pump failure. If it is a department policy to carry the pump full of water, the pump should be periodically flushed with clear water. The only way to check or inspect a fire pump is to operate the controls. It is not necessary to pump full capacity or even to deliver any more water than a booster line will convey. The following maintenance routine should be performed for a good pump inspection:

- Open all pump drains and flush out sediment (weekly).
- Check and clean intake strainers (weekly or after each use) (Figure 7.22).

Figure 7.22 It may be necessary to remove the screens covering pump intakes to clean them properly.

- Check pump gearbox for proper oil level and traces of water.
- Operate pump primer with all pump valves closed.
- Operate changeover valve while pumping from booster tank (weekly).
- Check packing glands for excessive leaks.
- Operate all valves, including relief valve (weekly) (Figure 7.23).
- Check all gauges for proper operation.
- Recalibrate flowmeter according to manufacturer's instructions.
- Refer to manufacturer's recommendations for additional instructions, if any.

FUEL AND WATER SUPPLY TANKS

It is extremely important to maintain a full tank of fuel since it is impossible to predict how long the apparatus engine must run during emergency operations. A full fuel tank is the only way to ensure maximum running time. Even with this precaution, fire apparatus must often be refueled during prolonged operations.

For maximum effectiveness, apparatus must respond with a full water tank as well as a full fuel tank. In terms of maintenance, the water in the tank should be changed if and when rust particles appear. Most water tanks have been treated to make them rust resistant. The water tank and pump should be flushed on a regular schedule to reduce the possibility of rust and stagnation.

Figure 7.23 Valves should be operated weekly.

MAINTENANCE AND RECORD FORMS

When keeping records of fire apparatus maintenance, each fire department should find a system that best meets its needs. The following examples of apparatus forms can be used for a number of purposes. They may be used exactly as is or can be modified as needed (Figures 7.24 – 7.31).

FIRE APPARATUS RECORD

PUMPER No. ________

Make ________ Year ________ Whl. Base ________ Weight ________

Chassis ________ Model ________ Type ________ No. ________

Requisition No. ________ Purchase Price ________ Date in Service ________

Date in Reserve ________ Date Salvaged ________ Salvage Price ________

ENGINE

Make ________ Model ________ Serial No. ________ Motor No. ________

Cylinders ________ Max. Hp. ________ Max. Torque ________ Peak RPM ________ Bore ________

Stroke ________ Displacement ________ Comp. Ratio ________ Carb. Make ________

Transmission ________ Model ________ Serial No. ________

PUMP

Make ________ Model ________ Serial No. ________ Capacity ________

Ratio ________ Suction Size ________ Priming Device ________

Press. Cont. Relief Valve-Type ________ Speed Cont. Valve-Type ________

CAPACITIES

Gas Tank ________ Gals. Radiator ________ Qts. Crank Case ________ Qts.

Booster Tank ________ Gals. Hyd. Reserve Tank ________ Qts.

FIRE APPARATUS RECORD

AERIAL TRUCKS No. ________

Make ________ Year ________ Serial No. ________ Weight ________

Chassis ________ Model ________ Type ________ No. ________ Whl. Base ________

Requisition No. ________ Purchase Price ________ Date in Service ________

Date in Reserve ________ Date Salvaged ________ Salvage Price ________

ENGINE

Make ________ Model ________ Motor No. ________ Cylinders ________

Max. Hp. ________ Max. Torque ________ Peak RPM ________ Bore ________ Stroke ________

Displacement ________ Comp. Ratio ________ Carb. Make ________

Transmission ________ Model ________ Serial No. ________

Aerial Ladders ________ Ft. Ground ________ Ft.

CAPACITIES

Gas Tank ________ Gals. Radiator ________ Qts. Crank Case ________ Qts.

Hyd. Res. Tank ________ Qts.

Figure 7.24 General Fire Apparatus Record for pumpers and trucks.

FIRE APPARATUS RECORD

PASSENGER CARS AND SERVICE TRUCKS No. ______

Assigned To ______

Make ______ Year ______ Model ______ Weight ______

Requisition No. ______ Purchase Price ______ Date in Service ______

Date in Reserve ______ Date Salvaged ______ Salvage Price ______

ENGINE

Make ______ Motor No. ______ Serial No. ______ H.P. ______

Cylinders ______ Carb. Make ______

FIRE DEPARTMENT SHOP ORDER

Order No. ______

Date ______ 19____ Apparatus No. ______

Description	Amount	

INDIVIDUAL FIRE APPARATUS REPAIR COST RECORD

______ No. ______

Date	Order No.	Parts Used	Cost	Labor

Figure 7.25 Fire Apparatus Record, Fire Department Shop Order, and Individual Fire Apparatus Repair Cost Record.

APPARATUS DATA AND HISTORY CARD

Public Property No. **13-**

Make	Type	Year

First Cost	Manufacturer's No.	Title No.	Specification No.

BASIC DATA

Length Ft. In.	Width Ft. In.	Make of Engine	GVW Rate	Cu. In. Displacement	Cylinder

Fuel Tank	Tire Size	Tire Type	Battery	Clutch	Transmission	Wheel Base	Radio 12V 6V

PUMPERS

Pump Make	Pump Type	Pump Volume	Water Tank

LADDERS

Height	Type
Complement of Ladders	

Year	License Plate	Status	Condition of Body	Condition of Pump or Ladder	Motor Hours	Road Mileage	Annual Pump Test Result	Times Out of Service	Down Time	Cost of Repairs Non-Accident		Cost of Repairs Accident	
										Labor	Material	Labor	Material

Figure 7.26 Apparatus Data and History Card.

INDIVIDUAL FIRE APPARATUS
Gasoline, Oil, and Mileage Record

Apparatus No. ______________

Year	________			________			________			________		
Month	Gasoline Gallons	Oil Quarts	Mileage	Gasoline Gallons	Oil Quarts	Mileage	Gasoline Gallons	Oil Quarts	Mileage	Gasoline Gallons	Oil Quarts	Mileage
Jan.												
Feb.												
March												
April												
May												
June												
July												
August												
Sept.												
Oct.												
Nov.												
Dec.												
Total												
Cost												

Figure 7.27 Individual Fire Apparatus Gasoline, Oil, and Mileage Record.

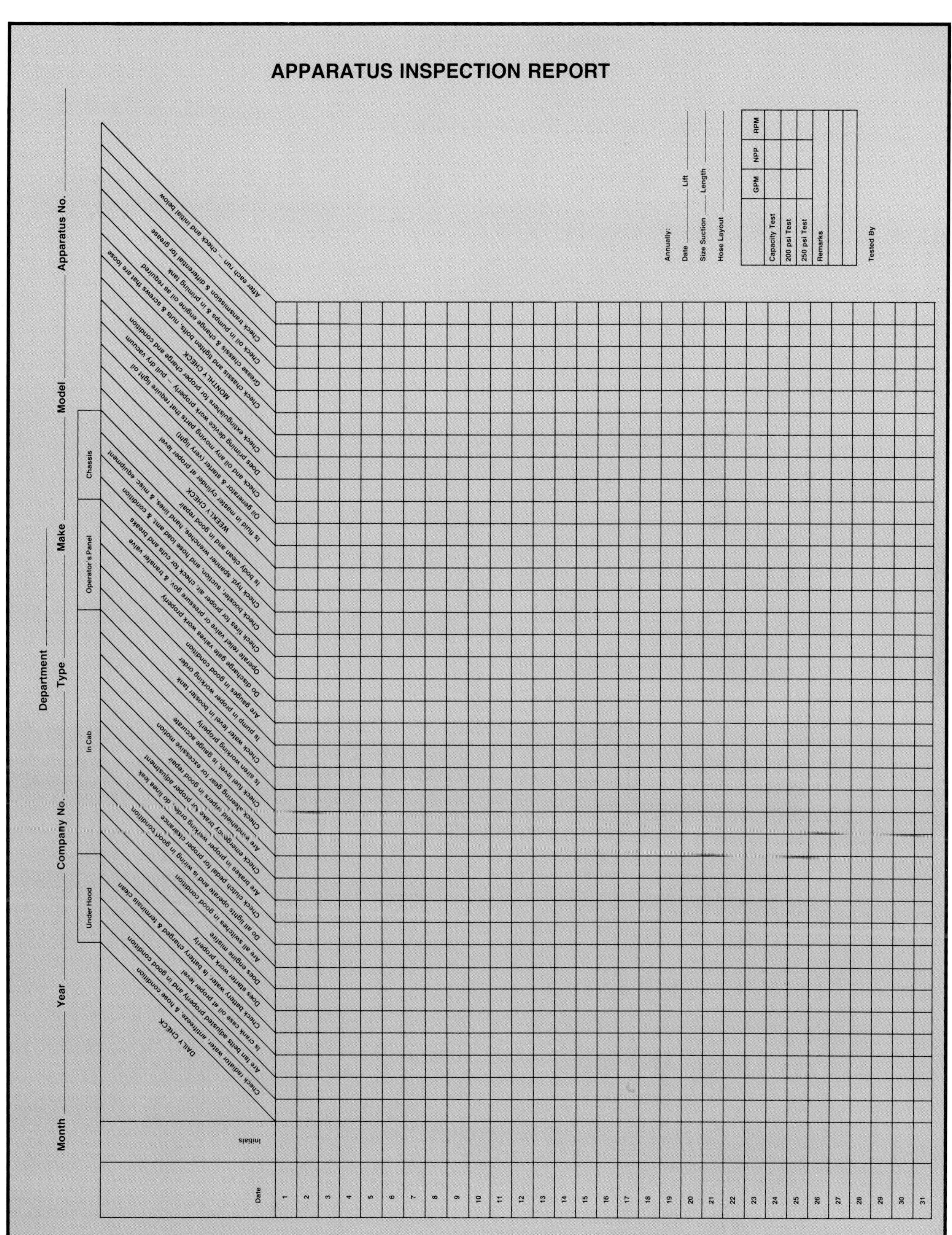

APPARATUS INSPECTION REPORT

Department ______

Month ______ Year ______ Company No. ______ Type ______ Make ______ Model ______ Apparatus No. ______

Under Hood | In Cab | Operator's Panel | Chassis

Date | Initials

DAILY CHECK
Check radiator water, antifreeze, & hose condition
Are fan belts adjusted properly and in good condition
Is crank case oil at proper level
Check battery water. Is battery charged & terminals clean
Does starter work properly
Does engine misfire
Are all switches in good condition
Do all lights operate and is wiring in good condition
Check clutch pedal for proper clearance
Are brakes in proper working order, do lines leak
Check emergency brake for proper adjustment
Are windshield wipers in good repair
Check steering gear for excessive motion
Check fuel level, is gauge accurate
Is siren working properly
Check water level in booster tank
Is pump in proper working order
Are gauges in good condition
Do discharge gate valves work properly
Operate relief valve or pressure gov. & transfer valve
Check tires for proper air, check for cuts and breaks
Check booster, suction, and hose load amt. & condition
Check hyd. spanner wrenches, hand lines, & misc. equipment
Is body clean and in good repair
WEEKLY CHECK
Is fluid in master cylinder at proper level
Oil generator & starter (very light)
Check and oil any moving parts that require light oil
Does priming device work properly — pull dry vacuum
Check extinguishers for proper charge and condition
MONTHLY CHECK
Check chassis and tighten bolts, nuts & screws that are loose
Grease chassis & change engine oil as required
Check oil in pumps & in priming tank
Check transmission & differential for grease
After each run — check and initial below

Date rows: 1, 2, 3, 4, 5, 6, 7, 8, 9, 10, 11, 12, 13, 14, 15, 16, 17, 18, 19, 20, 21, 22, 23, 24, 25, 26, 27, 28, 29, 30, 31

Annually:
Date ______ Lift ______
Size Suction ______ Length ______
Hose Layout ______

	GPM	NPP	RPM
Capacity Test			
200 psi Test			
250 psi Test			
Remarks			

Tested By

Figure 7.28 Apparatus Inspection Report.

ROCKFORD FIRE DEPARTMENT
Preventive Maintenance Work Sheet

Repair Order No. ______

"A" Service — To be performed each 1,000 miles or every 4 month period

"B" Service — To be performed each 4,000 miles or every 12 month period

Date ______ Vehicle No. ______ Company No. ______

"A" Service	"B" Service				
		1. Check driver's report			26. Drain air tank
		2. Fan belt adjustment			27. Springs for broken leaves and alignment
		3. Cooling system hoses			28. Muffler and tailpipe
		4. Radiator core for leaks			29. Tighten spring U-bolts
		5. Starter and generator Commutator and brushes			30. Tires — cuts, pressure, unusual wear
					31. Tighten wheel — lug nuts
		6. Distributor cap — rotor and breaker points			32. Tighten axle flange nuts
					33. Master cylinder fluid level
		7. Generator regulator			
		8. Tighten manifold nuts Check for leaks			34. Fifth wheel
					35. Road and pump transmission level
		9. Engine compression reading 1. ___ 2. ___ 3. ___ 4. ___ 5. ___ 6. ___ 7. ___ 8. ___ 9. ___ 10. ___ 11. ___ 12. ___			36. Rear axle oil level
					37. Drain and refill crankcase
					38. Replace oil filter if necessary
		10. Inspect and regap spark plugs			39. Clean air and fuel filters
		11. Adjust valve lash			40. Oil all linkages
		12. Set ignition timing			41. Wheel bearings — inspect and repack
		13. Adjust carburetor at idle			42. Inspect and adjust brakes
		14. Battery breakdown test			43. Lubricate chassis
		15. Radiator shutters			44. Flush cooling system — add rust inhibitor
		16. Hand brake			45. Door hinges, latches, window glass regulators
		17. Lights			
		18. Siren and bell			46. Check for oil and water leaks
		19. Windshield wipers			47. Cleanliness of apparatus
		20. Instruments and gauges			48. General appearance of paint and metal
		21. Choke and throttle			49. Road test — check brakes, steering, operation of clutch, transmission, engine performance, rattles
		22. Pitman and steering arms drag link and tie rod			
		23. Clutch pedal clearance and linkage			50. Engine oil pressure (hot) Idle — 1,000 rpm
		24. Universal joints and flanges			51. Engine temperature
		25. Hydrovac — inspect and oil			52. Check auxiliary generators, pumps, etc.

Figure 7.29 Preventive Maintenance Work Sheet. *Courtesy of Rockford (Illinois) Fire Department.*

Preventive Maintenance Work Sheet

Each item adjusted shall be explained under "remarks"
Example: (17) Tail light bulb replaced — burned out.

REMARKS: ______________________________

Figure 7.29 Continued.

______________ FIRE DEPARTMENT

"THREE MONTH & ANNUAL"
Maintenance Checklist

Service to be performed:
☐ "A" Three months or 1,000 miles
☐ "B" Annual or 4,000 miles

Date ____________
Vehicle & make ____________
Apparatus no. ____________
Odometer reading ____________
Engine miles ____________
Pump hrs. since last report ____________
Total pump hours ____________

☐ Adjustment Made ☐ Repairs Required

A	B	ITEMS ☐ O.K.	A	B	
		1. Check "Each Week" reports			7. Chassis
		2. Cooling system			• Lubricate chassis
		• Hoses			• Wheel bearing (inspect and repack)
		• Radiator core (leaks)			• Universal joints and flanges
		• Flush (add rust inhibitor)			• Clutch and linkage (automatic transmission)
		3. Oil:			• Oil all linkages
		• Drain and refill crankcase			• Steering mechanism
		• Replace oil filter			• Springs (damage and alignment)
		• Road transmission oil level			• Springs — U-bolts
		• Pump transmission oil level			• Muffler and tailpipe
		• Rear axle oil level			• Axle, flange nuts (tighten)
		4. Brakes			• Wheel lug nuts (tighten)
		• Hand brake			8. Body of Apparatus
		• Inspect and adjust brakes			• Cleanliness, paint
		• Hydrovac — inspect and lubricate			• Doors and windows (hinges, etc.)
		• Drain air tank			9. Lights
		• Master cylinder (fluid level)			10. Siren and bell
		5. Battery breakdown test			11. Windshield wipers (blade, arm, etc.)
		6. Engine			12. Pump mounting bolts (tighten)
		• Spark plugs (clean, etc.)			13. Winterization
		• Distributor cap, rotor, points			14. Auxiliary generators, pumps, etc.
		• Starter & generator			15. Road test
		• Voltage regulator			• Brakes
		• Set ignition timing			• Steering
		• Adjust carburetor at idle			• Clutch
		• Choke and throttle			• Transmission
		• Manifold nuts (tighten, etc.)			• Engine performance
		• Engine compression readings			• Engine oil pressure (hot)
		(1) ______ (2) ______			1) Idle 2) 1,000 rpm
		(3) ______ (4) ______			• Engine temperature
		(5) ______ (6) ______			• Rattles
		(7) ______ (8) ______			
		(9) ______ (10) ______			
		(11) ______ (12) ______			
		• Adjust valve lash			
		• Clean air and fuel filters			
		• Instruments and gauges			
		• Check fuel pumps			

____
Mechanic's Signature

____
Officer in Charge

Figure 7.30 Three-Month and Annual Maintenance Checklist.

FIRE APPARATUS PREVENTIVE MAINTENANCE INSPECTION FORM

UNDER HOOD (Inspection)	1st	2nd
Coolant level and radiator core		
Fan and generator belts		
Hood catches		
Engine mounts		
Oil, fuel, and water leaks		
Carburetor and choke		
Generator, regulator, and alternator		
Motor lights — compartment		
DRIVER COMPARTMENT		
Master cylinder		
Brake pedal		
Clutch pedal		
Steering wheel play		
Siren and/or horn		
Batteries and connections		
Seat upholstery		
Windshield wipers		
Lights warning and others		
Instruments		
Parking brake adjustment		
Vacuum clutch		
Cab doors, windows, etc		
OUTSIDE VEHICLE		
Wheel nuts		
Front end		
General body inspection		
Oil pan and front and rear seals for leaks		
Spring mountings		
Brake system for leaks		
Tail pipe and muffler		
FUEL SYSTEM		
Throttle adjustment		
Carburetor		
Gasoline leaks		
Gasoline tanks		
Gasoline lines		
Fuel pump, electric		
Air cleaner		
FIRE PUMP (Main)		
Packing		
Transfer valve		
Working valves, disch. intake		
Leaks, water, air		
Gaskets		
Shift		
Gauges		
Relief valve or regulator		
Piping		
Priming pump and oil reservoir		
Booster reel, brake, and gland		
Pump transmission		
Pump performance, vacuum, pressure		

AERIALS AND SNORKELS	1st	2nd
Hydraulic system — leaks, relief valve, working pressures, etc.		
Ladders — lubrication, operation, cable, adjustments, etc.		
Turntable — rotation, gears		
Jacks — hydraulic, mechanical, operation		
Control stand — operating levers, gauges, etc.		
Radius rods		
Rear axle oil leaks and level		
Tie rods and knuckle stops		
Transmission oil level		
Drive shafts and universals		
BOOSTER TANK		
Check inside surface		
Water level gauge		
Leaks		
TESTING — ROAD		
Engine performance and acceleration		
Gear shift action		
Overdrive performance		
Brake performance		
Steering action		
TESTING — ROAD (Inspection)		
Front-end action		
General rule		
ENGINE OILING SYSTEM		
Oil pressure		
Oil lines		
Oil filter		
Oil level and condition of oil		
IGNITION SYSTEM		
Spark Plugs		
Wiring		
Timing		
Magneto		
Distributor		

ALL BLANKS ANSWERED BY OK OR IF (✓) SEE REMARKS BELOW

Remarks ____________________

Serviceman ____________________

Date ____________________

Figure 7.31 Fire Apparatus Preventive Maintenance Inspection Form.

Chapter 7 Review

Answers on page 362

TRUE-FALSE: Mark each statement true or false. If false, explain why.

1. Driver/operators are responsible for carrying out periodic apparatus maintenance.

   ☐ T ☐ F

2. Driver/operators are responsible for carrying out daily and weekly apparatus maintenance.

   ☐ T ☐ F

3. It is a good practice to start the apparatus engine every day and let it run for awhile.

   ☐ T ☐ F

LISTING

4. List at least five items that should be checked or performed daily as a part of *apparatus* maintenance.

5. List at least five *apparatus* maintenance activities that should be performed on a weekly basis.

6. State at least three *equipment* maintenance functions that should be performed daily.

7. Name at least two periodic *equipment* maintenance tasks.

10-91

8. Name at least three electrical motors that are part of the apparatus and should be checked periodically.

____

____

____

9. List at least six checks that should be made to ensure proper operation of the fire pump on an apparatus.

____

____

____

____

____

____

FILL IN THE BLANK: Fill in the blank with the correct response.

10. Maintenance should be performed in accordance with the operator's handbook furnished by the ______________________________.

SHORT ANSWER: Answer each item briefly.

11. Explain the difference between repair and maintenance.

____

____

____

12. In addition to the outside of the apparatus, what are some of the other areas that should be regularly cleaned?

____

____

____

13. Effective lubrication will depend on what factors?

____

____

____

14. What is indicated by the SAE number on oil?

____

____

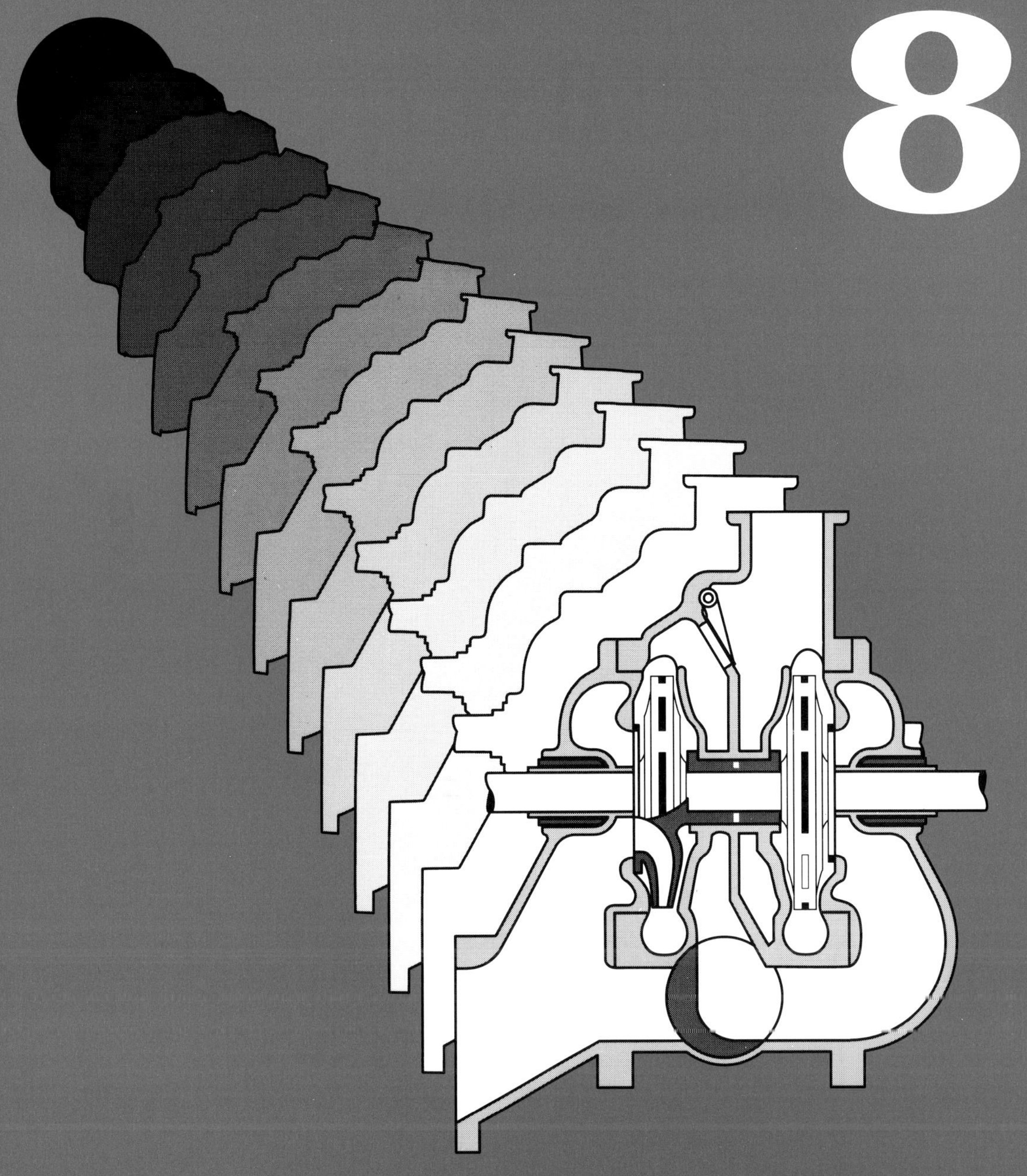

8

Apparatus Testing

NFPA STANDARD 1002
Fire Apparatus Driver/Operator Professional Qualifications

3-1 General.

3-1.5* The fire apparatus driver/operator, given a fire department pumper and the necessary equipment, shall demonstrate an annual pumper service test.

Chapter 8

Apparatus Testing

Fire apparatus is tested regularly to ensure that it will perform properly under emergency conditions. An organized system of apparatus testing, plus regular maintenance, is the best assurance that apparatus will perform within its design limitations. Furthermore, the insurance industry requires that apparatus be tested in order to receive full grading schedule credit. These tests can be grouped into two basic categories: preservice tests and service tests. Preservice tests are conducted before the apparatus is placed in service; service tests are conducted on a regular basis while the apparatus is in service.

This chapter concentrates on both types of tests. Preservice tests include discussions of manufacturer's tests, certification tests, and acceptance tests. Regularly scheduled pumper service tests are discussed next. Important considerations for these tests include correcting net pump discharge pressure for tests, sequence of tests, equipment needed, safety precautions, and possible causes of trouble during tests.

PRESERVICE TESTS

Apparatus equipped with a pump undergo extensive testing before they are put into service. These tests assure the purchaser that the pump and pump components will operate properly under normal use. These tests can be grouped into the categories of manufacturer's tests, certification tests, and acceptance tests. To ensure that they are performed, these tests must be required in the apparatus bid specifications.

NFPA 1901, *Standard for Automotive Fire Apparatus,* is used as a basis for many apparatus bid specifications. When specifications are written for an apparatus that will be equipped with a pump, water tank, and hose bed, a clause such as the following should be included:

> "The apparatus and all equipment herein shall meet the detail specifications of Chapters 1, 2, 3, 4, 6, 10, and 11 of NFPA 1901, *Standard for Automotive Fire Apparatus.* Any equipment that does not meet these specifications shall be unacceptable."

Generally, fire department personnel will not be involved in performing manufacturer's or certification tests. These tests are performed by the manufacturer or Underwriter's Laboratories (UL) personnel. Fire department personnel may be involved with acceptance testing, as this is commonly performed after the apparatus has been delivered to the purchaser. The following section highlights the major points of these preservice tests.

Manufacturer's Tests

If the requirements of NFPA 1901 are included in the apparatus bid specifications, the manufacturer is required to perform the Two-Hour Run-In Test, hydrostatic tests, and other tests.

Two-Hour Run-In Test. This test is designed to ensure that the pump and pump components have been installed correctly. During this test, the pump must be operated at draft for 2 hours. The pump must then be successfully operated as follows:

- The pump must deliver 100 percent of its rated capacity at 150 psi (1 000 kPa) net pump discharge pressure for 1 hour.

- The pump must deliver 70 percent of its rated capacity at 200 psi (1 350 kPa) net pump discharge pressure for 30 minutes.
- The pump must deliver 50 percent of its rated capacity at 250 psi (1 700 kPa) for 30 minutes.

The procedures used to conduct the above tests are similar to the procedures used in service testing. Service testing is discussed later in this chapter.

Hydrostatic Test. This test determines whether the pump and pump piping can withstand pressures normally encountered during fire fighting operations. Single-stage pumps are tested hydrostatically at 350 psi (2 450 kPa) for 10 minutes. Two-stage series-parallel pumps are tested at 500 psi (3 500 kPa) for 10 minutes.

Other Tests. The manufacturer may conduct other tests as prescribed by the purchaser. These tests may include tank-to-pump flow tests, acceleration tests, braking tests, noise level readings, or any other features required by the purchaser.

Certification Tests

Pumper certification tests are conducted by Underwriters Laboratories (UL), the nationally recognized testing organization. A UL technician conducts the tests either at the manufacturer's plant or at the fire department after delivery. These tests assure both the fire department and insurance companies that the apparatus will perform as expected after being placed into service. Certification tests must be required in the apparatus bid specifications, either by referencing NFPA 1901, Chapter 11 or by including a paragraph similar to the following:

> "The manufacturer shall submit the completed pumper for pumping performance certification, including tank-to-pump flow rate tests, by Underwriters Laboratories, Inc. A requirement of acceptance is that a 'Certificate of Inspection for Fire Department Pumper' for the pumper be received by the purchaser from Underwriters Laboratories, Inc."

NFPA 1901 requires the following certification tests:

- Vacuum Test
- Three-Hour Pump Test
- Automatic Pump Pressure Control Tests
- Tank-To-Pump Flow Rate Test

Vacuum Test. This test is designed to ensure that there are no leaks in the pump or pump piping that could cause problems during drafting. During this test, suction inlets are capped and discharge outlets are uncapped (Figure 8.1). The pump is subjected to a vacuum of 22 inches (559 mm) of mercury by means of the pump priming device. The vacuum may not drop more than 10 inches (254 mm) during the 10-minute test period.

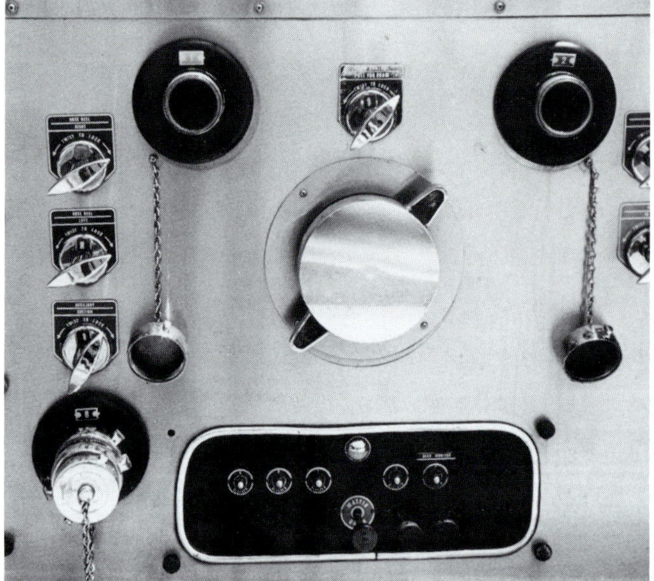

Figure 8.1 Before performing a vacuum test, cap all intakes and uncap all discharges.

Three-Hour Pump Test. This test is similar to the run-in test performed by the manufacturer, with the addition of an overload test. The pump is set up at draft and required to provide the following:

- 100 percent of rated capacity at 150 psi (1 000 kPa) net pump discharge pressure for 2 hours.
- 70 percent of rated capacity at 200 psi (1 350 kPa) net pump discharge pressure for 30 minutes.

- 50 percent of rated capacity at 250 psi (1 700 kPa) net pump discharge pressure for 30 minutes.
- 100 percent of rated capacity at 165 psi (1 150 kPa) net pump discharge pressure for 10 minutes. (This is a 10-minute Overload test.)

Automatic Pump Pressure Control Tests. These tests demonstrate the ability of the pump pressure control device to regulate the pressure created by the pump. If the pressure were unregulated, it could cause damage to pump piping or injury to firefighters operating hoselines. Three tests are conducted. First, the pump controls are adjusted to provide the pump's rated capacity at 150 psi (1 000 kPa). Discharges are then shut down one at a time until they are all closed. The discharge pressure must not increase more than 30 psi (210 kPa).

For the second test, 100 percent of the rated capacity at 150 psi (1 000 kPa) is reestablished by simply reopening the appropriate discharges. Using the throttle, reduce the engine speed (rpm) to lower the pump discharge pressure to 90 psi (630 kPa); set the relief device at that point. Then close each discharge one at a time. When all are closed, the pump discharge pressure must not increase more than 30 psi (210 kPa).

The last pressure relief test is performed with the pump providing 50 percent of its rated capacity at 250 psi (1 700 kPa). After all discharges have been closed, the pump pressure may not increase more than 30 psi (210 kPa).

Tank-To-Pump Flow Rate Test. This test is performed to determine if the tank-to-pump flow rate is sufficient to support an initial attack. The higher required flow rate for apparatus equipped with large water tanks facilitates their use as water supply apparatus. Those apparatus equipped with water tanks ranging in capacity from 300 to 750 gallons (1 200 L to 3 000 L) must provide a tank-to-pump flow rate of 250 gpm (1 000 L/min). Those with tanks larger than 750 gallons (3 000 L) must provide 500 gpm (2 000 L/min). These flows must be maintained for 80 percent of the tank's capacity.

Acceptance Testing

Acceptance tests are conducted to assure the purchaser that the apparatus meets bid specifications. These tests are most often performed after the apparatus is delivered to the purchaser with a representative of the manufacturer present. Departments that have steep hills in their protection area may require that the apparatus be able to climb a certain hill at a given speed. During the acceptance tests, the new apparatus would be driven up the specified hill and its speed would be monitored. If the apparatus failed to perform according to the requirements detailed in the bid specifications, it could be rejected. NFPA 1901 outlines some basic acceptance tests, which are listed below:

- The apparatus must accelerate to 35 mph (56 km/h) from a standing start within 25 seconds. Apparatus equipped with water tanks greater than 800 gallons (3 200 L) capacity are allowed 30 seconds (Figure 8.2).

Figure 8.2 Larger apparatus will accelerate more slowly than smaller apparatus.

- The apparatus must accelerate from 15 to 35 mph (24 km/h to 56 km/h) within 30 seconds without shifting.
- The apparatus must achieve a minimum top speed of 50 mph (80 km/h). This top speed may be lower for apparatus designed for slower speeds.
- The apparatus must come to a full stop from 20 mph (32 km/h) within 30 feet (9 m).

- The apparatus parking brake must control the rear wheels or all wheels, be mechanically actuated, and it must be securely held in position when applied. Lockup devices on service brakes or park positions on automatic transmissions are not acceptable substitutes for a separate parking brake system. The capabilities of this system should be specified by apparatus purchasers.
- The standard sets maximum pressure loss in discharge system.

Beyond these minimums, acceptance tests are centered around the specific needs of each department outlined in their apparatus bid specifications.

PUMPER SERVICE TESTS

A pumper should be given a service test at least once a year or whenever it has undergone extensive repair. These service tests are necessary to ensure that the pumper will perform as it should, and to check for defects that otherwise might go unnoticed until too late. Underwriter's Laboratory's *Guide for Certification of Fire Department Pumpers* and NFPA's *Fire Apparatus Maintenance* by Robert Ely are two additional source books on service tests.

Important Considerations for Pumper Service Tests

- Altitude will affect the pump's performance: lifting ability drops about 1 foot (0.3 m) per 1,000 feet (300 m) increase in altitude; gasoline engines have 3.5 percent less efficiency per 1,000 feet (300 m) increase in altitude.
- Barometric pressure will affect the ability of the pump to draft water. On a rainy day, for example, there will be less ability to draft than on a clear day. A 1-inch (25 mm) drop in barometric pressure reduces a pumper's maximum possible static lift by about 1 foot (0.3 m).
- The engine should not exceed the manufacturer's maximum governed no-load speed during any of the tests.
- The amount of lift is measured from the surface of the water to the center of the intake connection (Figure 8.3).
- The lift must not exceed 10 feet (3 m), and no more than 20 feet (6 m) of hard suction hose should be used (Figure 8.4).
- Ambient temperature may affect the test results if the pumper is powered by a diesel engine. A diesel engine loses horsepower when the temperature is above 90°F (32°C).

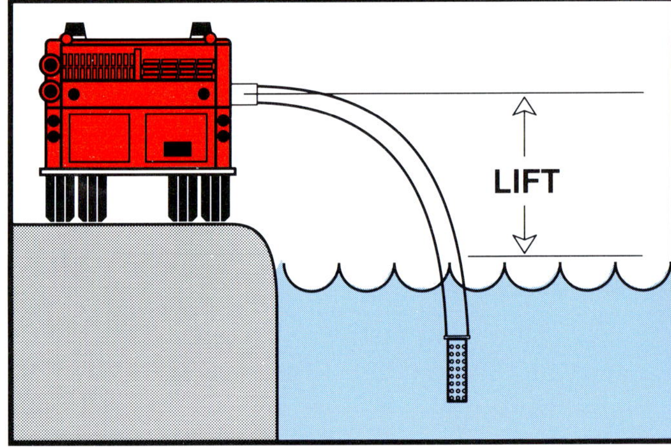

Figure 8.3 Lift is defined as the distance from the middle of the intake to the surface of the water supply source.

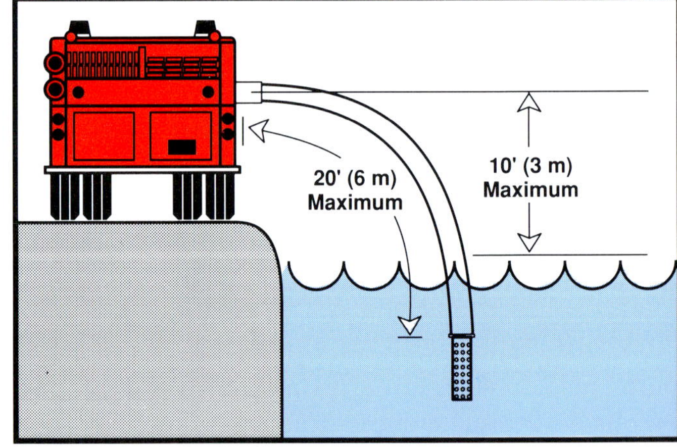

Figure 8.4 During testing, lift must not exceed 10 feet (3 m). No more than 20 feet (6 m) of intake hose should be used.

Correcting Net Pump Discharge Pressure for the Tests

The pump service tests are to be conducted at 150, 165 (overload), 200, and 250 psi (1 000 kPa, 1 150 kPa, 1 350 kPa, and 1 700 kPa) net pump discharge pressure. Net pump discharge pressure is the total work done by the pump to get the water into, through, and out of the pump. When at draft, the net pump discharge pressure is more than the pressure shown on the discharge gauge. Therefore, when the tests are conducted, the allowances for friction loss in the hard suction hose and the height of lift must be taken into account. Table 8.1 gives these allowances.

TABLE 8.1 (U.S.)
FRICTION LOSS ALLOWANCES FOR INTAKE HOSE

Rated Capacity of Pumper (U.S. gpm)	Hose Diameter (inches)	Allowance (feet)	
		For 10 Feet of Hose	Each Additional 10 Feet
500	4	6	+1
	4½	3½	+½
750	4½	7	+1½
	5	4½	+1
1,000	4½	13	+2
	5	8	+1½
	6	4	+½
1,250	5	12½	+2
	6	6½	+½
1,500	6	9	+1

Source: American Insurance Association.

TABLE 8.1 (metric)
FRICTION LOSS ALLOWANCES FOR INTAKE HOSE

Rated Capacity of Pumper (L/min)	Hose Diameter (mm)	Allowance (m)	
		For 3 m of Hose	Each Additional 3 m
2 000	100	1.8	.30
	115	1.1	.15
3 000	115	2.1	.45
	130	1.4	.30
4 000	115	4.0	.61
	130	2.4	.45
	150	1.2	.15
5 000	130	3.8	.61
	150	2.0	.15
6 000	150	2.7	.30

Source: American Insurance Association.

NOTE: The allowance computed in the table should be reduced by 1 psi (7 kPa) for the 200 psi (1 350 kPa) test and by 2 psi (13.8 kPa) for the 250 psi (1 700 kPa) test.

The allowances for friction loss are then used, as in the following example, to determine the correct pump discharge pressure during each test.

Example

Assume that a 1,000 gpm (4 000 L/min) pumper is being tested. The lift is 9 feet (2.7 m) through 20 feet (6 m) of 5-inch (125 mm) hose. Find the necessary pressure correction for this test.

10-91

$$\text{Pressure Correction (U.S.)} = \frac{\text{lift (ft.)} + \text{suction hose friction allowance}}{2.3}$$

$$\text{Pressure Correction (metric)} = \frac{\text{lift (m)} + \text{suction hose friction allowance}}{0.1}$$

From Table 8.1, it can be seen that the friction loss allowance for the first 10 feet (3 m) of intake hose is 8 feet (2.4 m) and for the second 10 feet (3 m) the allowance is 1.5 feet (0.45 m); therefore:

$$\text{Pressure Correction (U.S.)} = \frac{9 + 8 + 1.5}{2.3} = \frac{18.5}{2.3} = 8 \text{ psi}$$

$$\text{Pressure Correction (metric)} = \frac{2.7 + 2.4 + 0.45}{0.1} = \frac{5.55}{0.1} = 55 \text{ kPa}$$

This correction pressure (8 psi [55 kPa]) is subtracted from the desired net pump discharge pressure to determine what the pump's actual discharge pressure can be. For example, 150 psi – 8 = 142 psi (1 000 kPa – 55 = 945 kPa). The same pressure correction is reduced 1 psi (7 kPa) for the 200 psi (1 350 kPa) test and 2 psi (14 kPa) for the 250 psi (1 700 kPa) test. Therefore, the pump discharge pressure should be 193 psi (1 002 kPa) for the 200 psi (1 350 kPa) test and 244 psi (1 659 kPa) for the 250 psi (1 700 kPa) test.

Sequence of Service Tests

The service test is actually a series of related tests. These tests can be completed most efficiently if they are performed in the order given in the following list. Each test is described in greater detail in the sections that follow.

- Dry vacuum test
- Priming test
- Capacity test
- Tachometer and engine rpm check
- Ignition check
- Pressure governor or relief valve test
- Overload test
- 200 psi (1 350 kPa) test
- 250 psi (1 700 kPa) test
- Tank-to-pump flow test
- Other inspections and tests

Equipment Needed for Service Tests

Equipment for performing service tests must be in good shape and be tested regularly. Gauges can best be tested for accuracy with a deadweight tester, which should be available at a local waterworks. The following equipment is needed to perform the service test for pumpers:

- One 0 – 300 psi (0 kPa – 2 100 kPa) test gauge with 5 psi (35 kPa) graduations (Figure 8.5).
- One 0 – 200 psi (0 kPa – 1 400 kPa) test gauge with 1 psi (7 kPa) graduations, marked every 5 or 10 psi (35 kPa or 70 kPa).
- One vacuum gauge or mercury manometer.
- Pitot tube with knife edge and air chamber (Figure 8.6). (**NOTE:** this is not needed if a flowmeter is used.)

Figure 8.5 For testing purposes, the vacuum gauge and both pressure gauges may be mounted on a single test panel.

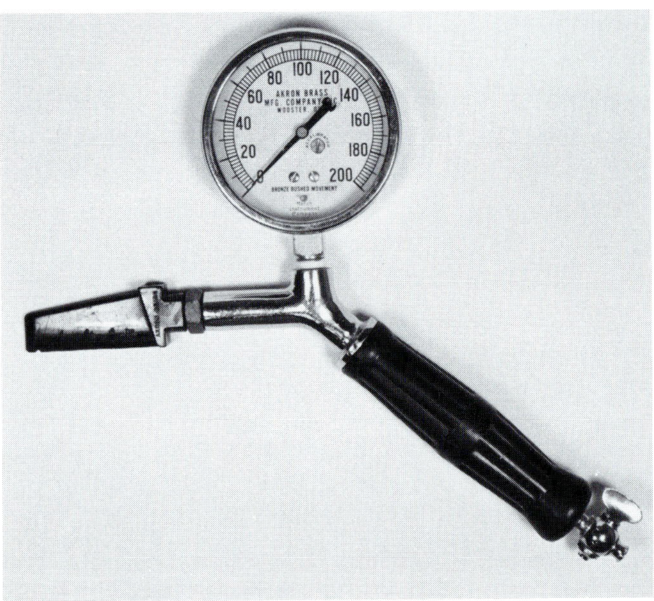

Figure 8.6 A pitot tube is used to measure the velocity pressure of a stream of water. This velocity pressure is used to calculate the amount of water flowing from a discharge outlet.

- Straight stream nozzles of correct sizes to match the volumes pumped for the different tests. (**NOTE:** Only capacity nozzle required with flowmeter).
- Rope or chain for securing test nozzle.
- Revolution counter or hand tachometer.

Recommended, but not necessary, are:

- Two 6-foot (1.8 m) lengths of ¼-inch (7 mm), 300 psi (2 100 kPa) hose with screw fittings. (**NOTE:** these are used to connect the 0-300 psi (0 kPa – 2 100 kPa) gauge and the vacuum gauge to the test gauge fittings at the pump operator's panel.)
- Clamp to hold pitot tube to test nozzle.
- One 50-foot (15 m) ¼-inch (7 mm) hose (used to connect the pitot tube to the 0-200 psi [0 kPa – 1 400 kPa] test gauge).
- Test stand for gauges.
- Thermometer.
- Stop watch or watch with sweep second hand.
- Chokedown valve.

FLOWMETERS

A flowmeter, which reads the flow directly in gallons per minute (L/min), is preferable to a pitot gauge setup (Figure 8.7). Flowmeters allow much more flexibility and help to complete the tests more quickly. When a flowmeter is used, all of the pump tests can be run without shutting down the pump, without changing nozzles, and without having to convert pitot pressure to gpm (L/min).

Figure 8.7 For quicker results, portable flowmeters may be used to measure water flow. *Courtesy of Fire Research Corp.*

CHOKEDOWN VALVE

Although a regular discharge valve may be used to increase or decrease the friction loss when conducting the tests, a special chokedown valve is better (Figure 8.8). The velocity of the water, especially during the 250 psi (1 700 kPa) test, will in time cause the discharge valve seals to leak. A gated-down discharge valve also creates turbulence in the stream, which can damage and loosen hose at the couplings. The chokedown valve has stream straighteners built into it to eliminate this turbulence.

Safety Precautions During Service Tests

- ALL PERSONNEL SHOULD WEAR PROTECTIVE HEADGEAR AND HEARING PROTECTION.
- Prevent water hammer: Open and close all valves and nozzles slowly.
- Do not stand over or straddle hose.
- Manipulate engine throttle slowly. Prevent sudden pressure changes, which can damage equipment and injure personnel.
- Tie down test nozzles and devices *securely*.
- Walk, do not run, and watch where you are walking.
- Cover all open manholes at the test pit.
- Be aware of the location of all personnel in relation to hoselines.

Dry Vacuum Test

The first test, the dry vacuum test, is performed to check the priming device, pump, and hard suction hose for air leaks.

Step 1: Drain the pump.

Step 2: Inspect all gaskets of suction hose and caps.

Step 3: Look for foreign matter in the suction hose. Clean the hose if necessary.

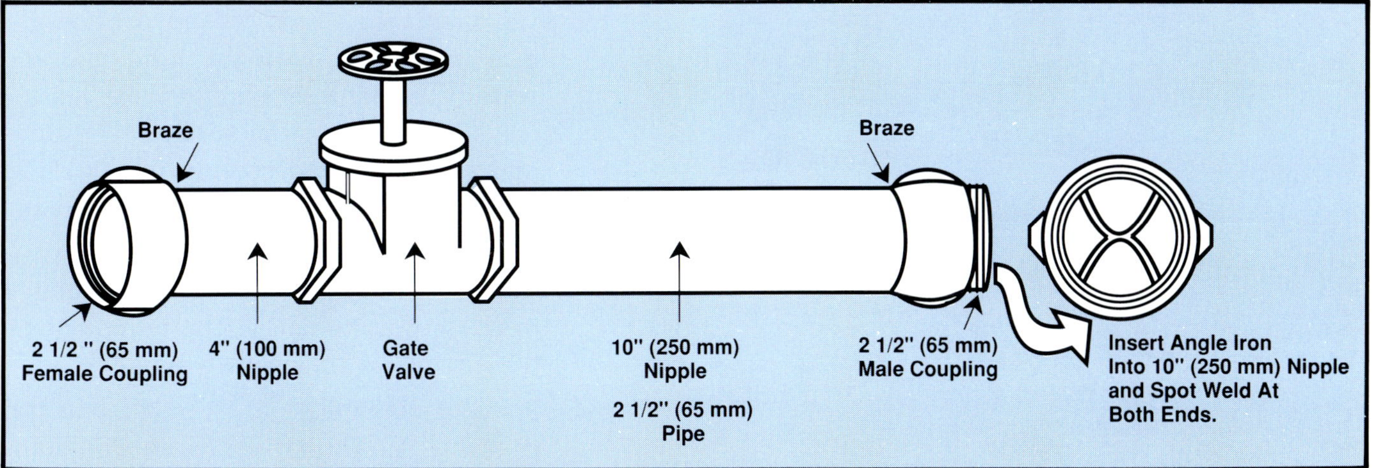

Figure 8.8 Turbulence can be greatly reduced with an easily constructed chokedown valve.

Step 4: Connect 20 feet (6 m) of the correct suction hose to the pump intake connection (check original test records for correct diameter of hose) (Figure 8.9).

Step 5: Cap the free end of the suction hose (Figure 8.10).

Step 6: Close all drain, discharge, and intake valves.

Step 7: Connect an accurate vacuum gauge (or mercury manometer) to the threaded test-gauge connection on the intake side of the pump. **CAUTION:** If the gauge is not connected to the intake side, it will be irreparably damaged.

Figure 8.9 Make hard suction hose connections tight so there are no air leaks that will adversely affect test results.

Figure 8.10 Make sure the free end of the hard suction hose is capped.

Step 8: Check oil level of priming pump reservoir (replenish if necessary).

Step 9: Make pump packing glands accessible for checking (raise floorboards or open compartment doors).

Step 10: Run the priming device until the test gauge shows 22 inches (560 mm) of mercury developed. (**NOTE:** reduce the amount of mercury developed 1 inch [25 mm] for each 1,000 feet [300 m] of altitude.)

Step 11: Compare readings of the compound gauge and test gauge. Record any difference.

Step 12: Shut off the engine. Listen for air leaks. No more than 10 inches (254 mm) of vacuum should be lost in 10 minutes. Excessive leaks will affect the results of subsequent tests and should be located and corrected before performing the rest of the tests.

After the dry vacuum test is completed, prepare the pumper for the remainder of the tests:

Step 1: Open a discharge valve to allow the pressure in the pump to equalize.

Step 2: Replace the cap at the end of the suction hose with the suction strainer (Figure 8.11).

Step 3: Use standard departmental procedure to tie off suction hose in preparation for drafting, then lower the hose into the water. The strainer should be at least 2 feet (0.6 m) below the surface. The sides and bottom of the strainer should also have at least 2 feet (0.6 m) of water surrounding them (Figure 8.12).

Step 4: Connect the 300 psi (2 100 kPa) test gauge to the pressure side of the pump at the test fitting on the operator's panel.

Step 5: For a gasoline engine, connect a vacuum gauge to the intake manifold. The manifold pressure registered on this gauge can

Figure 8.11 The strainer prevents large foreign objects from being drawn into the pump.

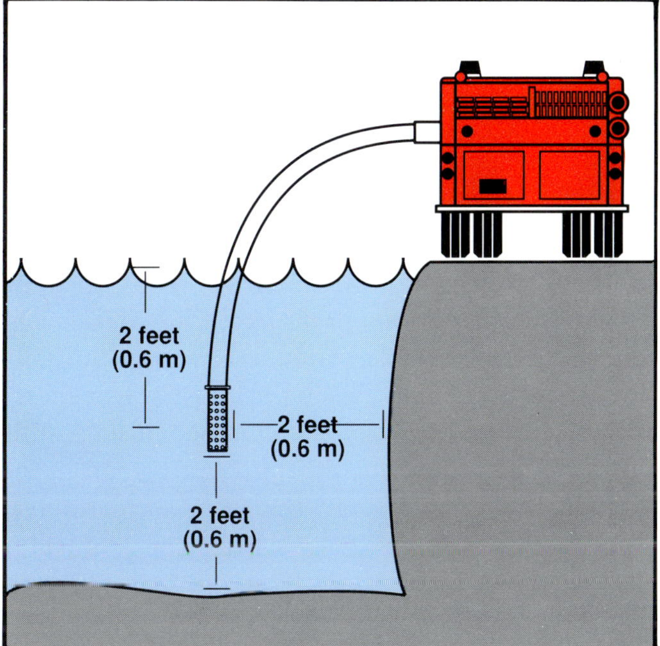

Figure 8.12 For maximum operation, the strainer should be surrounded by 2 feet (0.6 m) of water in all directions.

be used to determine the percentage of available horsepower being used for each test.

Step 6: Connect an adequate number of hoselines to carry the capacity of the pump to the test nozzle. The test nozzle must be the correct size to handle the capacity of the pump (see Table 8.2).

Step 7: Make sure the nozzle is secured so it cannot come loose and injure personnel. **NEVER** hold the test nozzle by hand during a test.

Step 8: Connect the pitot gauge and test stand gauges. It is recommended that the pitot gauge be clamped in position at the nozzle (Figure 8.13).

Figure 8.13 Clamping the pitot tube directly to the test nozzle is safer and provides more accurate results than holding it.

Priming Test

The priming test is timed from the moment the priming device is actuated until water is flowing onto the ground beneath the pump. This time should be no more than 30 seconds for pumps rated less than 1,500 gpm (6 000 L/min) and 45 seconds for pumps rated 1,500 gpm (6 000 L/min) or more. The procedure for performing the priming test is as follows:

Step 1: Close all discharge, drain, and booster tank valves and petcocks. The transfer valve should be in the VOLUME position.

Step 2: Start priming device and timer.

Step 3: Stop time when water discharges onto the ground beneath the pumper. Increase engine rpm to develop pressure.

Step 4: Open discharge valves.

Step 5: Operate the pump at moderate capacity and pressure for a few minutes. This is necessary to warm the engine and transmission for the capacity test (engine temperature should reach at least 160°F [71°C]).

Capacity Test

The capacity test checks the overall condition of the engine and the pump. To obtain the correct engine and nozzle pressures for the capacity test, some adjustments and readjustments will have to be made. All changes must be made slowly to prevent damage to the pump and hose and possible injury to personnel, and to allow time for the resulting pressure changes to register on the test gauges. The procedure for the capacity test is as follows:

Step 1: Gradually speed up the pump until the desired pump pressure is reached (150 psi or 1 000 kPa), adjusted for suction hose friction loss and altitude.

Step 2: Check the flow at the nozzle, using either a pitot gauge or a flowmeter. If the flow is too great, close a valve further. Readjust (lessen) engine speed to correct discharge pressure. If the flow is too low, open a valve further. Readjust (increase) engine speed to correct discharge pressure. (**NOTE:** All these adjustments must be made without the engine speed exceeding 80 percent of its peak.)

Step 3: When both the pump discharge pressure and the volume flowing are satisfactory, the test officially begins. The following readings are made and recorded at the beginning of the test, and at 5-minute intervals thereafter until the 20 minutes for the test are over. (**NOTE:** Fluctuations in pressure will necessitate more frequent readings.)

- Pump discharge pressure
- Nozzle pressure (or flow)
- Engine tachometer
- Rpm using portable rpm counter

Additional points to remember during the capacity test:

- The pitot gauge should be held in the center of the stream and with the tip about one-half the nozzle diameter from the end of the nozzle. If the pitot is too close to the nozzle, the reading will be erroneously increased.
- Short lines of hose are most convenient for a test.
- The engine temperature should be kept within the proper range.
- Check the oil pressure to be sure that proper engine lubrication is maintained.
- Any unusual vibration of pump or engine should be recorded.
- Any other defect in the performance of pump or engine should be recorded. Minor defects should be corrected immediately, if possible.

IGNITION CHECK

If the engine has two ignition systems, use each for a short period of time during the capacity test. Neither the engine speed nor pump discharge pressure should vary greatly.

Pressure Governor or Relief Valve Test

The governor or relief valve should be tested to check the device's capability of operating at capacity and at various pressures. The governor or relief valve should be tested as it was in the original acceptance test. Check the original test sheets and the manufacturer's recommendations. The test should be conducted immediately after the capacity test and without shutting down or marking any changes in layout or pressure. The test is as follows:

Step 1: Set the device.

Step 2: Slowly close all discharge valves.

Step 3: Record the results. The pump pressure must not rise more than 30 psi (210 kPa).

Overload Test

The overload test is conducted to determine whether the pumper has reserve power. The pumper must deliver its capacity at 165 psi (1 150 kPa) net pump pressure, without exceeding maximum no-load governed speed. The same hose layout and nozzles are used for the overload test as were used for the capacity test.

Step 1: Slowly open the throttle to increase engine speed and pump pressure to 165 psi

(1 150 kPa), and at the same time close down a valve enough that capacity is flowing.

Step 2: Record the results (gauge readings).

Step 3: Slowly return the pump speed to idle and disengage.

200 Psi Test

The 200 psi (1 350 kPa) test is similar to the capacity test in that 70 percent of capacity must be delivered at 200 psi (1 350 kPa) pump discharge pressure. See Table 8.2 for correct nozzle sizes.

Check the delivery or acceptance test sheets, or test plate to see whether the 200 psi (1 350 kPa) test should be run with the transfer valve in VOLUME or PRESSURE position.

This test is run for 10 minutes and gauge readings are taken at 5-minute intervals (beginning, middle, and end of test).

250 Psi Test

For the 250 psi (1 700 kPa) test, the transfer valve must be in the PRESSURE position. The nozzle may have to be changed. During this test, the pumper must deliver 50 percent of its rated capacity at 250 psi (1 700 kPa) for 10 minutes. Gauge readings are taken in the same manner as for the 200 psi (1 350 kPa) test.

Tank-To-Pump Flow Test

The tank-to-pump flow test is not a standard service test, but is recommended for delivery tests and after any major piping alterations. The test for tanks with a capacity of from 300 to 750 gallons (1 200 L to 3 000 L) is as follows:

Step 1: Run a hoseline from a hydrant to the top fill opening and keep the tank full while the engine speed is being adjusted.

Step 2: Start the pump and flow 250 gpm (1 000 L/min) from the tank through the test nozzle to assure the flow rate capability of the piping.

Step 3: Make sure the tank is full, then shut off the hydrant supplying the tank. At the same time, start timing the test.

TABLE 8.2 (U.S.)
HOSE AND NOZZLE LAYOUTS

Rated Capacity (gpm)	Pump Pressure (psi)	GPM	Nozzle Size (inches)	2½-Inch Discharge Hose	Pitot Pressure (psi)	Suction Hose (inches)
500	150	500	1½	One 50-foot line	58	4½
	200	350	1¼	One 50-foot line	58	
	250	250	1	One 50-foot line	72	
600	150	600	1½	Two 100-foot lines*	82	4½
	200	420	1⅜	One 100-foot line	56	
	250	300	1⅛	One 100-foot line	66	
750	150	750	1¾	Two 100-foot lines*	68	5
	200	525	1½	One 100-foot line	62	
	250	375	1¼	One 100-foot line	66	
1,000	150	1,000	2	Three 100-foot lines*	72	6
	200	700	1¾	Two 100-foot lines*	60	
	250	500	1½	One 100-foot line	58	

*Siamese these lines into a master stream appliance.

TABLE 8.2 (metric)
HOSE AND NOZZLE LAYOUTS

Rated Capacity (L/min)	Pump Pressure (kPa)	L/min	Nozzle Size (mm)	65 mm Discharge Hose	Pitot Pressure (kPa)	Suction Hose (mm)
2 000	1 000	1 900	38	One 15 m line	400	115
	1 350	1 300	32	One 15 m line	400	
	1 700	940	25	One 15 m line	500	
3 000	1 000	2 800	44	Two 30 m lines*	470	130
	1 350	2 000	38	One 30 m line	425	
	1 700	1 400	32	One 30 m line	455	
4 000	1 000	3 800	50	Three 30 m lines*	500	150
	1 350	2 600	44	Two 30 m lines*	410	
	1 700	1 900	32	One 30 m line	400	

*Siamese these lines into a master stream appliance.

Step 4: When the pump starts "running away" from the water (cavitating) stop timing the test. Shut the discharge valve, reduce engine speed, and disengage the pump transmission.

A flow of 250 gpm (1 000 L/min) is 4.16 gallons (15.7 L) per second. Therefore, multiply the number of seconds the water was flowing before the pump started to cavitate by 4.16 (15.7) to determine how many gallons (Liters) of water flowed through the pump. If the number of gallons (Liters) flowed during the test equals or is greater than 80 percent of the tank's rated capacity, the test is considered to be successful.

For tanks with capacities greater than 750 gallons (3 000 L), the test is the same except that 500 gpm (2 000 L/min) is flowed. The multiplier then is 8.3 (31.4).

Possible Troubles During Service Testing

If the pump fails to meet the requirements of the service test, one or more of the following should be investigated as the probable cause of the problem:

- Transmission in wrong gear
- High gear lockup not functioning (automatic transmission)
- Clutch slipping
- Engine overheating
- Muffler clogged
- Tachometer inaccurate
- Engine governor malfunctioning
- Suction hose too small
- Suction strainer submerged incorrectly (for example, too close to surface, too close to bottom)
- Suction screens clogged
- Lift too high
- Suction hose clogged or inner lining collapsed
- Excessive air leaks at suction side of pump
- Impellers clogged
- Pump or suction hose not fully primed

- Relief valve or pressure governor malfunctioning
- Transfer valve in wrong position
- Inaccurate gauges
- Pitot tube partially clogged
- Nozzle too large
- Seized turbocharger

Every effort should be made to correct any problem that is found. That portion of the test that was unsuccessful should be redone to ensure that the problem has been corrected.

Chapter 8 Review

Answers on page 363

TRUE-FALSE: Mark each statement true or false. If false, explain why.

1. Fire department personnel generally perform manufacturer's and certification tests.

   ☐ T ☐ F ______________________________

   ____

2. A diesel engine loses horsepower when temperature exceeds 90°F (32°C).

   ☐ T ☐ F ______________________________

   ____

MULTIPLE CHOICE: Circle the correct response.

3. Which of the following is *not* one of the three basic categories of preservice tests?
   A. Manufacturer's tests
   B. Hydrostatic tests
   C. Certification tests
   D. Acceptance tests

4. Identify the NFPA standard that is the basis for apparatus design and testing.
   A. NFPA 1001
   B. NFPA 1002
   C. NFPA 1901
   D. NFPA 1971

5. Which of the following is *not* generally considered a manufacturer's test?
   A. Two-Hour Run-In Test
   B. Hydrostatic Test
   C. Acceleration Test
   D. Vacuum Test

6. A pump's ability to lift water drops approximately __________ per every 1,000 feet (300 m) increase in elevation.
   A. 1 foot (0.3 m)
   B. 3 feet (1 m)
   C. 5 feet (1.5 m)
   D. 10 feet (3 m)

7. What is the maximum lift allowance during service testing?
   A. 5 feet (1.5 m)
   B. 10 feet (3 m)
   C. 15 feet (4.5 m)
   D. 20 feet (6 m)

LISTING

8. Name at least three types of certification tests.

____

____

____

9. List at least three acceptance tests recommended by NFPA 1901.

____

____

____

10. State two good sources of information about conducting pumper service tests.

A. ______________________________

B. ______________________________

11. List at least seven problems that may result in a pump failing to pass service tests.

____

____

____

____

____

____

____

12. In order, name the 10 service tests recommended by this book.

A. ______________________________

B. ______________________________

C. ______________________________

D. ______________________________

E. ______________________________

F. ______________________________

G. ______________________________

H. ______________________________

I. ______________________________

J. ______________________________

FILL IN THE BLANK: Fill in the blank with the correct response.

13. The two basic categories of apparatus tests are ______________ tests and ______________ tests.

SHORT ANSWER: Answer each item briefly.

14. What is the purpose of preservice testing?

____

____

____

15. How can a purchaser of apparatus ensure that preservice tests will be performed?

____

____

16. When should pumpers be given service tests?

____

____

____

17. Determine the necesssary correction when testing a 1,500 gpm (6 000 L/min) pumper on a 5-foot (1.5 m) lift through 20 feet (6 m) of 6-inch (150 mm) hose.

9

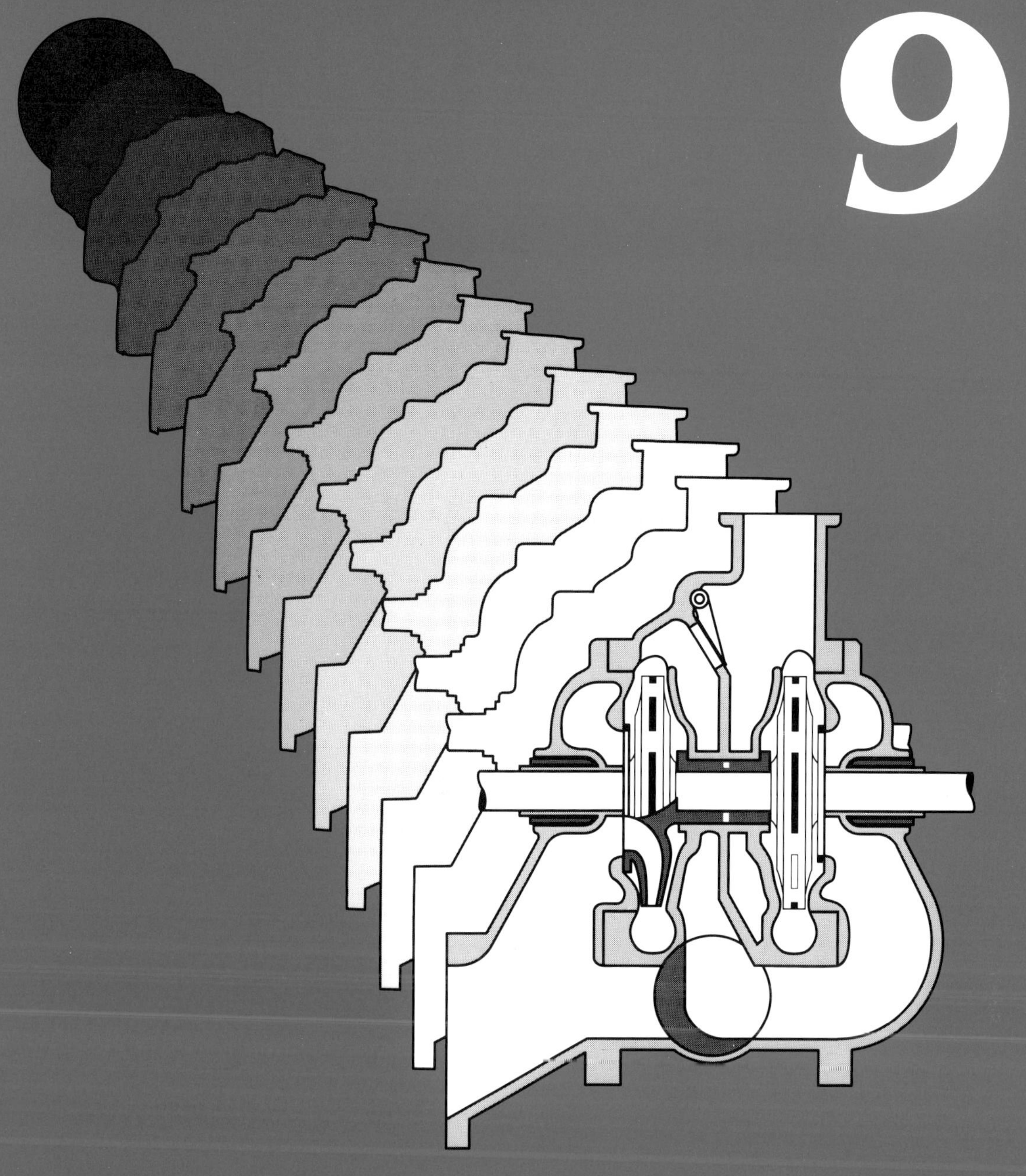

Apparatus Purchase and Specifications

LEARNING OBJECTIVES

After reading this chapter, the student should be able to accomplish the following learning objective:

- Outline the 10-step procurement process developed by the Federal Emergency Management Agency in their five-part *Guide For Preparing Fire Pumper Apparatus Specifications.*

Chapter 9

Apparatus Purchase and Specifications

When purchasing fire equipment, three factors must be considered: present equipment, required equipment, and future equipment needs. To obtain apparatus that achieves acceptable levels of performance, fire apparatus purchasers need to be familiar with NFPA 1901, *Standard on Automotive Fire Apparatus*. Specifications for equipment should be written incorporating these standards and individual department needs while soliciting the best price. Thousands of dollars are wasted each year because departments purchase equipment without using in-depth specifications.

Before writing specifications, fire apparatus purchasers must be familiar with the following definitions from NFPA 1901, *Standard on Automotive Fire Apparatus:**

1-5.1 Aerial Ladder Truck. A ladder truck equipped with a permanently mounted, power operated aerial ladder.

1-5.6 Elevating Platform Apparatus. A ladder truck equipped with permanently mounted, power operated booms of articulating construction or telescoping construction; or a combination of articulating and telescoping construction; and with a passenger carrying platform attached to the uppermost boom.

1-5.15 Light Attack Vehicle. A unit equipped with only a booster pump, with no more than three designated seating positions, carrying a limited amount of equipment.

1-5.18 Mobile Water Supply Apparatus. A piece of fire apparatus whose primary purpose is transporting water with a water tank 1000-gal capacity or larger. Truck should include a pump and have a limited hose body capacity.

1-5.21 Pumper. A piece of fire apparatus with a permanently mounted fire pump with a rated discharge capacity of 500 gpm or greater and complying with Chapter 3 of this standard; and may also include a water tank and hose body.

1-5.22 Pumper-Ladder Truck. A ladder truck plus the addition of a fire pump; and may also include a water tank and hose body.

1-5.23 Quadruple Ladder Truck. A service ladder truck with the addition of a fire pump, water tank, and hose body.

1-5.24 Quintuple Ladder Truck. A quadruple ladder truck with the addition of an aerial device.

1-5.25 Service Ladder Truck. A ladder truck carrying at least the standard complement of ground ladders, but neither an aerial ladder nor an elevating platform.

PREPARING APPARATUS SPECIFICATIONS

The Federal Emergency Management Agency (FEMA) has produced a 5-part *Guide for Preparing Fire Pumper Apparatus Specifications*. The guide advocates a 10-step procurement process for pumpers. The basic information presented in these guides can be used to help develop specifications

*Reprinted with permission from NFPA 1901, *Automotive Fire Apparatus*. Copyright © 1985, National Fire Protection Association, Quincy, MA 02269. This reprinted material is not the complete and official position of the NFPA on the referenced subject which is represented only by the standard in its entirety.

for any type of apparatus. The steps are listed below and each is discussed in greater detail in the sections that follow:

Step 1: Determine Performance Requirements

Step 2: Prepare the Bid Specification

Step 3: Advertise for Bidders

Step 4: Hold a Pre-Bid Conference

Step 5: Revise the Specifications

Step 6: Solicit Formal Bids

Step 7: Evaluate Bids and Award Contract

Step 8: Monitor Manufacture and Conduct Acceptance Tests

Step 9: Evaluate Bid Procedures

Step 10: Collect Data on Equipment Operation and Maintenance

Step 1: Determine Performance Requirements

An apparatus committee should be appointed to determine what the performance requirements of the new apparatus should be. The composition of the committee will vary from department to department, but it is recommended that members from every level of the fire department be represented. Firefighters and drivers are an important resource in helping the committee purchase an apparatus that will be accepted and fully utilized when delivered. Other helpful people are fire department mechanics, fire protection engineers, department or municipal finance officers, and anyone else who is knowledgeable about fire apparatus.

The committee should examine the current and future fire protection needs of the community and listen to the suggestions of firefighters who will use the new equipment (Figure 9.1). The committee should consider where the new apparatus will fit into the community's total fire protection plan. In some communities, fire apparatus is purchased on a scheduled replacement basis or when funds are available. Tight municipal budgets have forced fire departments in some areas to prove that a new piece of apparatus is *needed*, not just wanted. Help for formulating a community fire protection plan is available in a book distributed by the Federal Emergency Management Agency (FEMA), *A Basic Guide for Fire Prevention and Control Master Planning.*

As the performance requirements of the new apparatus are being determined, the committee should consider how the apparatus will be expected to perform in the following categories:

- Pump — capacity and pressure requirements.
- Aerial device — height, load capacity, horizontal reach, type (tower, aerial ladder, elevating platform).

Figure 9.1 Firefighters of all levels should be involved in meetings to develop apparatus specifications.

- Performance — This concerns acceleration, maximum speed, gradability (maximum speed up a specified grade), turning radius, angle of approach (the angle formed by level ground and a line from the point where the front tires touch the ground to the lowest projection at the front of the apparatus, at least 16 degrees), angle of departure (the angle formed by level ground and a line from the point where the rear tires touch the ground to the lowest projection at the rear of the apparatus, at least 8 degrees), breakover angle (the angle formed by level ground and a line from the point where the rear tires touch the ground to the bottom of the frame at the wheelbase midpoint, at least 10 degrees), and drivewheel horsepower (the power available at the wheels to move the vehicle) (Figure 9.2).
- Equipment to be carried — ground ladders, SCBA's, hose, fittings, and the weight and size of all equipment that needs to be carried on the apparatus.

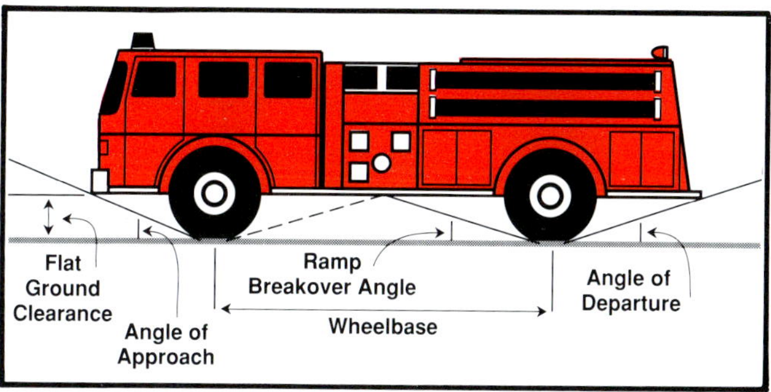

Figure 9.2 Performance is affected by many factors including the angle of approach, angle of departure, and breakover angle.

- Climate — deep snow, sandstorms, ice, bitter cold, extreme heat — all affect the operation of the apparatus and must be considered when specifying new apparatus (Figure 9.3).
- Tank capacity — based upon the fire flow needs and the availability of water in the area to be protected.

The work sheets provided in Figure 9.4 (on the following pages) are intended to help the apparatus purchase committee put together a complete list of desired features.

Figure 9.3 Apparatus may have to be specially designed for the climate in which it will serve. This pumper is designed for an area that receives heavy snowfalls, so it has all-wheel drive capabilities. *Courtesy of Joel Woods.*

SPECIFICATION WORK SHEET

Type of Apparatus: ____________________

Engine: ____________________

Electrical System:
- Size Alternator: ____________________
- Batteries Single/Dual: ____________________
- 110 Volt Power: ____________________

Warning Equipment:
- Lights: ____________________
- Siren: ____________________
- Horn: ____________________

Brakes: ____________________

Wheels/Tires: ____________________

Axles: ____________________

Transmission: ____________________

Compartments:
- Right Side: ____________________
- Left Side: ____________________
- Rear: ____________________
- Other: ____________________

Cab:
- Design: ____________________
- Seats: ____________________

Color: ____________________

Pump:
- Capacity: ____________________
- Type: ____________________
- Suction Inlets: ____________________
- Discharges: ____________________
- Pressure Control: ____________________

Figure 9.4 Detailed work sheets such as these will help the apparatus committee determine the exact specifications needed for new apparatus.

Pump Controls:

Drive: __________

Primer: __________

Gages: __________

Instruments: __________

Suction Hose: __________

Booster Reels:

Location: __________

Power Rewind: __________

Size/Capacity: __________

Tank Construction:

Flow Tank to Pump: __________

Fill Rate: __________

Vent: __________

Performance Requirements:

Top Speed: __________

Acceleration: __________

Stopping Distance/Speed: __________

Fuel 100% Pumping Time: __________

Turning Radius: __________

Maximum Weight: __________

Per Axle: __________

Maximum Height: __________

Maximum Length: __________

Angle of Departure: __________

Additional Items: __________

Figure 9.4 Continued.

Step 2: Prepare the Bid Specification

At this stage, performance requirements are translated into specific equipment specifications. The apparatus design should be a realistic balance between the amount of money available to build the apparatus and the department's performance requirements. Chiefs and members of other departments may relate problems and solutions to problems that they have encountered with their apparatus.

The committee should develop a very specific written set of apparatus specifications (Figure 9.5). Well-written specifications ensure that both the apparatus committee and the apparatus manufacturer understand how the apparatus is expected to perform. Apparatus manufacturers may be able to spot potential cost or maintenance problems at this stage and suggest design alternatives.

The draft set of specifications should provide a detailed, comprehensive picture of the new apparatus. Performance specifications allow the apparatus committee to clearly establish how they expect the apparatus to perform. Rather than specifying the type of engine and transmission the apparatus should have, the committee should include a requirement in the specification that the completed apparatus perform in a certain way. For example, a specification may require that fully loaded apparatus be able to come to a complete stop within 30 feet (9.1 m) from a speed of 20 mph (48.3 km/h) on even terrain.

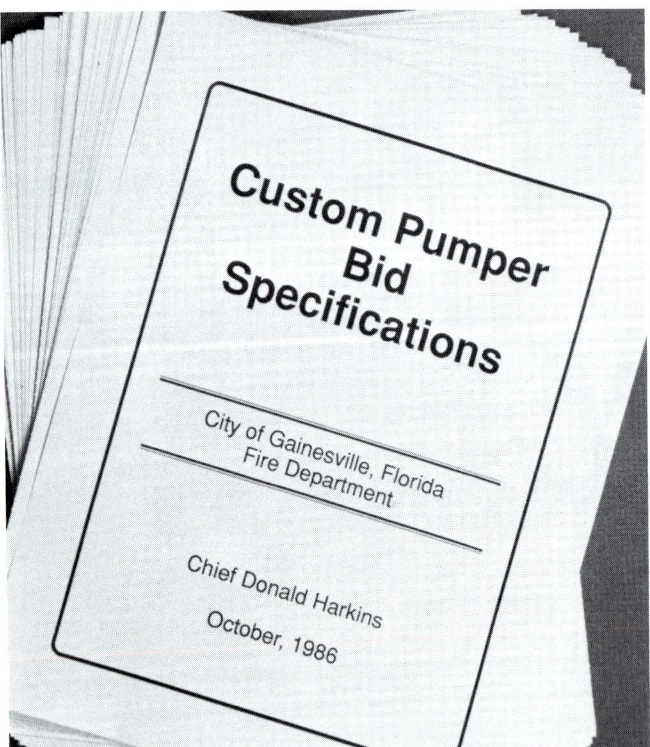

Figure 9.5 The specifications manual should be distributed to any manufacturer interested in submitting a bid on the apparatus. This manual should include all requirements for the work to be done. *Courtesy of Gainesville (Florida) Fire Department.*

An easy-to-use specification format will facilitate understanding between the apparatus committee and the manufacturer. The following is a possible format that can be used in writing the specifications. Consideration must be given to including each of the items discussed.

- A *statement of intent* that generally describes the apparatus sought.
- A general *statement of purchase* indicating it is an engineer, design, construct, and deliver type of specification. The statement of purchase should also indicate that it is not the intent of the purchaser to write out any vendor or manufacturer of similar or equal equipment to the types specified. Proposals or alternate bids for any equipment that will accomplish the same performance will be given careful consideration. The purchaser shall be the sole judge of equipment that is most advantageous and the decision of the purchaser shall be final.
- Minimum design *requirements* should be outlined indicating the apparatus shall be constructed with due consideration to the nature and distribution of the load to be sustained and to the general character of the service that the apparatus will be subjected to when placed in service. All parts should be strong enough to withstand the general service under full load. Design shall be such that various parts are readily accessible for lubrication, inspection, adjustment, and repair. Minor details or construction of materials, where not specified, are left to the discretion of the manufacturer who shall be solely responsible for the design and construction of all features. It is the responsibility of the manufacturer

to indicate if any unsafe or poorly designed criteria are contained in the specification and thoroughly explain them to the purchaser.

- *Definitions,* as necessary, should be stated, including one for "equal." "Equal" can be defined as the same level of quality, standard of performance, or design. Equal does not necessarily mean identical. A manufacturer of equal material should not be discouraged from bidding. The purchaser shall be the sole judge of equality and its decision shall be final.
- Adopt all applicable chapters of NFPA 1901 as the minimum specifications for items not specifically indicated.
- A statement of the purchaser's *right to reject* any or all bids or to accept any bid that meets or exceeds the specifications and is deemed to be in the purchaser's best interest, whether or not the accepted bid is the lowest bid.
- A *bid bond* of a stated amount of the total bid price may be desired to ensure the bidder will take the bid made if offered. Bond is returned when the contract is executed. The bond is forfeited to the purchaser for liquidated damages because of failure or default if a contract is not signed.
- The date and location of the *bid opening* are essential. Include the mailing address.
- A statement of the length of time the bid *prices* shall be *valid* from the bid opening.
- A *performance bond* to ensure the successful bidder will build the apparatus to specifications may be desired. This can be for varying dollar amounts and lengths of time. Generally, the bond would pay the difference that would be required to have another manufacturer build the apparatus specified.
- The *contractor* shall defend any and all suits and assume all *liability* for any officials or agents for the use of any patented process, device, or article forming a part of the apparatus or any appliance furnished under the contract.
- A *warranty* statement on materials and workmanship. The warranty should start at acceptance and cover a stated period of time, usually one or two years. Some components may be placed under warranty for longer periods, such as five years for the frame and engine. If component manufacturers have longer warranties, they should apply. Both parts and labor may be applied in early years. Extended warranties may be desired on some components. Extended warranties will cost appreciably more. However, they may be specified for some items; for example: the water tank shall be covered against any defects for a period of 10 years.
- *Manufacturer's capability* to produce may be specified by seeking a list of recently produced apparatus or evidence of financial stability.
- *Payment terms* should be outlined detailing that all acceptance tests must be passed before money is transferred. Prices should be for an FOB destination, delivered, accepted piece of equipment that includes the specified warranties.
- A *delivery date* can be specified by the purchaser or requested from the manufacturer. This date should be binding and a penalty clause can be included to enforce it. Another approach is that a stated amount may be deducted from the monies due the manufacturer for each calendar day any work remains incomplete, not as a penalty, but as liquidated damages. The manufacturer should not be held liable if performance failure arises out of causes beyond the control and without the fault or negligence of the manufacturer.
- The specifications should detail *factory visit(s)*. This should include the number of visits, the number of people, and the length of time that the manufacturer will be responsible for expenses.

- Any special *delivery* considerations should be indicated. This would include whether or not department drivers will drive the apparatus to its destination. It should be noted that the manufacturer has insurance liability (comprehensive, collision, and liability) until the unit is accepted. If a delivery engineer will be provided by the manufacturer and a length of time for training is required, this must be specified.

- Each bid should be accompanied by a detailed description of the apparatus and equipment, including a working *drawing* that shows body compartments with locations and sizes, ladder arrangements, and piping systems. A certified engine brake horsepower curve showing the maximum no-load governed speed should be supplied.

- Upon delivery of the apparatus, electrical wiring and piping *diagrams* and lubrication *charts* should be supplied. At least three sets of operator's *manuals*, two service manuals, and two parts manuals should be required.

- The specification should list and detail all *tests* required and to be performed before acceptance. These include Underwriter's Laboratory Certification, acceptance tests outlined in NFPA 1901, and any special local tests such as exterior noise and sound levels, performance and top speed, acceleration, gradability, fording, braking, load distribution, flow rates for various piping arrangements, and usable tank water. Alternative plans available in case of failure should be covered.

- All *equipment* that the apparatus manufacturer must supply should be listed, including quantity and manufacturer's model number when necessary.

- If the manufacturer is to supply prices for an *alternative* or *option item* as part of the bid, a separate proposal page should be included.

- A form for *changes* in construction after the contract has been signed should be provided (Figure 9.6).

- A fire department *Apparatus Evaluation and Bid Analysis Form* should be included. This form requires bidders to indicate major components and all exceptions to the specifications referenced by paragraph number so that a quick comparison and evaluation of all bids can be made (Figure 9.7 on the following pages).

- The major portion of the specifications will be the development of the *Special Provisions*. It is recommended that these be organized following NFPA 1901 with paragraphs numbered accordingly. It is in the Special Provisions that the specially desired options are spelled out in detail.

- If necessary to clarify an item to the manufacturer, the purchaser should supply *drawings* or *photos* of particular design configurations specified in the Special Provisions. Some examples would be a hose bed arrangement or special compartment.

- A *contact person* and mailing address should be provided by the purchaser for the use of all prospective manufacturers for clarification of the contents of the specifications.

- A *preconstruction conference* may be required by the purchaser with the selected manufacturer to finalize all construction details. If so, the conference should be outlined in the specifications as to location, expenses for transportation, per diem, number of people, and period of time.

Step 3: Advertise for Bidders

Local laws governing purchases by municipal organizations will dictate how and where advertisements for bidders should appear. The advertisement should contain basic information about the new apparatus and should name a contact per-

CLIFTON PARK FIRE DISTRICT #1
FIRE ROAD
CLIFTON PARK, NEW YORK

CHANGE ORDER

CHANGE ORDER NO. ____________ DATE ____________

ITEM NO. ____________ TITLE ____________

TO: FIRE APPARATUS, INC.

PLEASE MAKE THE FOLLOWING CHANGE ON OUR ORDER FOR TWO (2) 1,500 GPM (6 000 L/min) TRIPLE COMBINATION PUMPERS AS PER ABOVE REFERENCE:

THE AMOUNT OF CONTRACT WILL BE (INCREASED) (DECREASED) (NO CHANGE) IN THE SUM OF: ____________________

($____________) PLUS OR MINUS NEW YORK STATE SALES TAX.

____________________ ____________________

CLIFTON PARK FIRE DISTRICT #1 FIRE APPARATUS, INC.

PLEASE RETURN ONE (1) COPY OF THIS CHANGE ORDER TO FIRE DISTRICT #1.

NOTE THE ABOVE CHANGE APPLIES TO BOTH PUMPERS. THE AMOUNT OF INCREASE OR DECREASE IS FOR BOTH PUMPERS.

Figure 9.6 Any change in specifications or other details should be spelled out in writing, with all concerned receiving a copy of the details. *Courtesy of Clifton Park (New York) Fire Department.*

son whom apparatus salesmen can contact for more information. A sample ad is given below:

> The Town of Clifton Park, New York, seeks qualified bidders for the construction of one (1) 1,500 gpm (6 000 L/min) custom fire department pumper. Copies of the apparatus specifications and the date of the pre-bid conference are available from supervisor Jim Smith at the Clifton Park Town Hall, (518) 555-1200.

The apparatus committee has the right to set certain prequalifications of apparatus manufacturers in order to obtain a quality piece of apparatus. The committee may require that the manufacturers have a number of years' experience building the type of apparatus they are planning to purchase. They may also require a list of fire departments in their state or region that have purchased apparatus from that manufacturer.

The committee can set any requirement it feels necessary to facilitate the purchase of a reliable piece of apparatus. Requirements that limit

FIRE DEPARTMENT APPARATUS EVALUATION AND BID ANALYSIS FORM

The following proposal sets forth the details of pumping apparatus to be contracted for with the fire department, in accordance with their specifications.

Date ____________________

Bidder ____________________ Address ____________________

Manufacturer ____________________ Location ____________________

Special Provisions

a. All deviations and exceptions to these specifications are attached as page No. A-1, A-2, A-3, etc., a total of ________ pages.

b. All parts of the proposed apparatus manufactured and sold by the bidder are listed with required information on attached pages No. B-1, B-2, etc., a total of ________ pages.

Detail Description

Make ____________________ Model ____________________

Pumping capacity ________ Two sets of contractor's Specifications are attached, including:

a. Apparatus arrangement drawings
b. Four views, top—front end—rear end—left side

Overall length ____________________ Wheelbase ____________________

Extreme Height ____________________ Extreme Width ____________________

Front end to center line of axle ____________________

Track of front and rear wheels ____________________

Delivery Date ____________________ Date construction will start ____________________

Number of working days for completion ____________________

Carrying capacity: There will be ________ pounds on the front axle, which is ________% of the total weight. There will be ________ pounds on the rear axle, which is ________% of the total weight.

The following engine is being installed:

Make ____________________ Model ____________________

Type ____________________ No. Cylinders ________ H.P. ____________________

Cu. In. Displacement ____________________ Maximum Torque ____________________

Bore ____________________ Stroke ____________________ Compression Ratio ____________________

Number of main bearings ____________________ Projected bearing area ____________________

Is this engine the manufacturer's latest model? ____________________

Oil capacity ____________________ Make oil filter ____________________

The following described cooling system is being furnished:

Equipped with shutters of the following make ____________________

and the following type of shutter control ____________________

The make of thermostat is ________ with a setting to open at ________ degrees F.

Radiator make ____________________ Frontal area ____________________

Type of construction ____________________

Thickness ____________________ Flow rate ____________________

Fuel Tank, constructed of ____________________ Gauge Metal ____________________

Capacity ____________________ Location ____________________

Figure 9.7 The apparatus evaluation and bid analysis forms will enable the fire department to easily compare the submitted bids.

Braking System:

Air compressor, Make ____________ Model ____________ Capacity ____________

Air tanks, Description Dimensions

(1) ____________ ____________

(2) ____________ ____________

(3) ____________ ____________

(4) ____________ ____________

Brake Lining, Dimensions

	Front Wheels		Rear Wheels	
	Front Shoe	Reverse Shoe	Front Shoe	Reverse Shoe
Width				
Length				
Area				
Thickness				
Make of lining				

Parking Brake, Make ____________ Model ____________

Type ____________

Wheels, Make ____________ Type ____________ Model ____________

Tire, Make ____________ Size ____________ Ply ____________

Make of tube ____________ Maximum tire load ____________

Recommended tire load by the Tire and Rim Assoc. ____________

All wheels and tires are interchangeable? ____________

Adequate clearance for full tire chains on all wheels? ____________

Axles

	Front	Rear
Make		
Model		
Rated Capacity		
Clearance from road surface		

Track ____________ Gear Ratio ____________

Angle of Departure ____________

Power Steering, Make ____________ Model ____________

Drive Line, Clutch make ____________ Model ____________

Diameter ____________ Rated torque capacity ____________

No. of plates ____________

Transmission make ____________ Model ____________

Gear Ratios: First ____________ Second ____________ Third ____________

Fourth ____________ Fifth ____________ Top Speed ____________

Input torque capacity ____________

Propellor shaft, Make ____________ Model ____________

Universal joints, Make ____________ Model ____________

Figure 9.7 Continued.

Springs:		Front	Rear
	Pre-stressed	______	______
	Capacity	______	______
	Material	______	______
	No. of Leaves	______	______
	Width	______	______
	Length	______	______
	Type of shackle	______	______

Frame:

Material ______ Channel or "Z" type section, ______

Cross member bolted/riveted. Tensile strength ______ psi

Outside width of frame ______

Fire Pump:

Make ______ Model ______

Capacity ______ GPM at ______ psi. No. stages ______

Water Tank:

Steel gauge ______ type of steel ______

Describe construction details of cover ______

______

Describe configuration of bottom and describe how water is withdrawn including pipe size.

______

Figure 9.7 Continued.

the number of eligible manufacturers to one or two may be a source of trouble and possible lawsuits.

Step 4: Hold the Pre-Bid Conference

The apparatus committee should conduct a pre-bid conference to discuss the draft specification with potential bidders. As an alternative to pre-bid conferences, qualified bidders can be sent a set of draft qualifications for review and comment. After comments are returned and adjustments made to the specifications as necessary, the department can send out a formal request for bid. This method may be necessary for small communities where setting up a pre-bid conference may be difficult. In an age of rapid changes in prices and features, this continuous dialogue with several bidders helps the department to:

- Learn of new developments
- Check on technology limits
- Get accurate cost estimates
- Consider trade-offs among various performance options and/or design features
- Avoid misunderstandings and inconsistencies in the specifications
- Obtain several expert (if conflicting) opinions on the merits of options under consideration
- Minimize the number of "exceptions" or deviations from the specifications

As many pre-bid conferences as necessary should be held to develop a final bid specification. Normally, though, no more than two are needed. No matter how much effort is put into drafting the initial set of specifications, it may be necessary to revise them after each pre-bid conference.

PROPOSAL ON ALTERNATE BID

It may be necessary to ask for alternate bids in order to obtain information and prices for items that deviate from standard specifications. The information that follows is an example of such a proposal:

The bidder hereby proposes and agrees to furnish an apparatus in accordance with the attached specifications, except as listed below. It is understood that the District may, as its option, accept or reject any or all bids or alternate bids.

In lieu of closed canopy cab forward driver's compartment and custom chassis, bidder will provide Navistar Fleetstar 2050A or Ford Louisville L-9000 or Chevrolet Series 80 with custom cab on a commercial chassis of proper size.

Standard seats as available with commercial chassis will be acceptable.

Transverse hose beds to be in narrow 4-inch (100 mm) compartments ahead of the pump with swiveled outlets in the bottom of each compartment.

All other paragraphs of above specifications will apply to this chassis.

From bid for custom chassis with five-man cab for either:

	Add	Deduct
Navistar Fleetstar 2050A	$	$
Ford L-9000	$	$
Chevrolet Series 80	$	$

A working drawing and bid and evaluation analysis are not required of those bidders having submitted them for the custom chassis.

Step 5: Revise the Specifications

At this point, any changes the committee feels necessary are added to the specifications. The final set of specifications should be prepared and reviewed, line by line, by the apparatus committee. Now the specifications are ready for mailing.

Step 6: Solicit Formal Bids

The final set of apparatus specifications are sent to bidders who have been found qualified according to Step 3. The apparatus manufacturer will provide all the documents outlined in the apparatus specifications as well as a price for the specified apparatus.

Step 7: Evaluate Bids and Award Contract

The first step in bid evaluation is to make sure that all bids contain the documents, drawings, and bonds required in the apparatus specifications. If one of these elements is not present, the bid may be disqualified. The apparatus manufacturer may elect to provide options in the specifications or bid.

The committee should require the manufacturer to complete an apparatus evaluation and bid analysis form (Figure 9.7). This form outlines the major components of the apparatus chassis, engine, pump or ladder, and body. The committee can use these forms to carefully evaluate each bid according to the expected performance and compatibility of each manufacturer's apparatus. Performance can be defined as how well a bid agrees to meet or exceed the specified performance requirements outlined in Step 1. The new apparatus must also be compatible with existing equipment. The apparatus committee reviews all bids and bid evaluations and then awards the contract to the manufacturer they feel will provide the best apparatus.

After the contract has been awarded, the apparatus committee and the manufacturer's representative should meet and review the specifications and bid. Options may be decided upon at this point and a line of communication between the committee and the manufacturer can be established.

Step 8: Monitor Manufacture and Conduct Acceptance Tests

The apparatus committee should monitor the progress of construction. This can be accomplished by telephone or by visits to the factory at various points in the assembly process (Figure 9.8). The completed apparatus should be tested according to the requirements of NFPA 1901 as well as any requirements set forth in the performance specifications. A nationally recognized testing laboratory should be chosen to perform the pump or aerial device tests. Performance specification tests can be performed by the manufacturer, the fire department purchasing the apparatus, or a testing laboratory. Apparatus that cannot pass these tests should not be accepted unless corrected or an alternate agreement can be made by the manufacturer and the fire department.

Figure 9.8 When possible, fire department personnel should visit the factory to make sure work is proceeding according to specifications. *Courtesy of Gainesville (Florida) Fire Department.*

Step 9: Evaluate Bid Procedures

After the apparatus has been received and accepted, the apparatus committee should meet and evaluate the bid process. Problems and positive points in the bid process should be determined and recorded in a report. This report will be an asset to future apparatus committees. The report should also contain comments from the successful and unsuccessful apparatus bidders.

Step 10: Collect Data on Equipment Operation and Maintenance

This is really the first step in gathering the basic data to begin the next procurement cycle. Data will come chiefly from four sources: The "lessons learned" report described in Step 9, maintenance reports, budgeted versus actual operating and capital expenditures, and records of field performance by firefighters who have worked with the equipment. These four data sources make up the management information upon which future budgets, policies, and procurement can be based. Expected performance should be compared to actual operating experience.

The 10-step procurement program outlined provides a context in which performance specifications can work very effectively. Multistage procurement, while not the only system, provides a good mix of modern management and old-fashioned "eyeball-to-eyeball horse trading." The actual procurement program adopted should be tailored to suit a given department's needs and capabilities.

Chapter 9 Review

Answers on page 364

TRUE-FALSE: Mark each statement true or false. If false, explain why.

1. Federal laws governing purchases by municipal organizations dictate how and where advertisments for bidders should appear.

   ☐ T ☐ F ______________________________

   ____

2. Formal bid solicitation may begin immediately following the pre-bid conference.

   ☐ T ☐ F ______________________________

   ____

3. The first step in bid evaluation is to make sure that all bids contain the documents, drawings, and bonds required in the apparatus specification.

   ☐ T ☐ F ______________________________

   ____

4. Once the new apparatus has been accepted, the duties of the apparatus committee are complete.

   ☐ T ☐ F ______________________________

   ____

LISTING

5. What three levels should the purchase of fire apparatus be seen from?

   A. ______________________________

   B. ______________________________

   C. ______________________________

6. List at least three categories in which the committee should consider how the apparatus will be expected to perform (after performance requirements have been determined).

   ____

   ____

   ____

7. List at least three advantages of holding a pre-bid conference or sending manufacturers a set of draft qualifications for review and comment.

   ____

   ____

   ____

FILL IN THE BLANK: Fill in the blanks with the correct response.

8. A well-written set of specifications ensures that both the ____________ ____________ and the ____________ ____________ understand how the apparatus is expected to perform.

SHORT ANSWER: Answer each item briefly.

9. Name the NFPA document that purchasers of apparatus must be familiar with.

____

10. Name the agency that has prepared a five-part guide for preparing fire pumper apparatus specifications.

____

Appendix A

Description and Operating Practices for American Godiva Fire Pumps 5-89

(Note: Unless otherwise indicated, all photos in this Appendix are printed Courtesy of Bill Eckman.)

The operating practices for the Barton American Pump or the American Pump follow those of the Godiva Company. The American Fire Pump 5-89 Company has constructed many front mount pumps in capacities up to 1,500 gpm (6 000 L/min). Some were mounted by the American Fire Apparatus Company on apparatus they fabricated, but many were used by outside suppliers. The American Fire Pump Company also constructed single-stage and two-stage midship pumps up to 3,000 gpm (12 000 L/min) in capacity. In order to provide pressures up to 850 psi (5 860 kPa), they have added a small capacity two-stage pump to either the single-stage or two-stage midship mounted pump.

FRONT MOUNT PUMP

The front mount pump has been very popular on apparatus from tankers to rural pumpers and even some major city pumpers. Since it is installed by many different individuals, there is nothing really standard about the installation or instrumentation, but certain elements are typical.

Engaging the Pump

A typical front mount pump is shown in Figure A.1. It is driven by a shaft that is connected through two universal joints from the engine fan pulley to the pump gear box. The power is transmitted to the pump by a combination friction and positive drive clutch. The friction clutch is used to bring the pump up to the speed of the engine shaft smoothly and easily. When the pump is fully engaged, a positive locking arrangement takes over and eliminates any possibility that the clutch will slip or burn out.

The 500, 750, 1,000, and 1,250 gpm (2 000 L/min, 3 000 L/min, 4 000 L/min, and 5 000 L/min) pumps are all different in design. The 1,500 gpm (6 000 L/min) impeller is similar to, but slightly different from, the 1,000 and 1,250 gpm (4 000 L/min and 5 000 L/min) impellers. A number of different gear ratios are available for the gear box; these are used to match the operating characteristics of the engine to the desired capacity of the pump.

The maximum engine speed used while pumping is generally limited to the rpm required to reach about 35 mph (56 km/h) road speed. Engagement of the pump is usually made by using a manual lever mounted to the side of the pump. When engaging the pump, limit the engine speed to a low idle (less than 700 to 800 rpm). Slowly raise the clutch lever to the vertical position. As the friction clutch is engaged, the pump will begin to rotate. When the clutch lever is in the vertical position, that is, 90 degrees of rotation, the pump should be up to full speed. Pause a moment, then continue to move the clutch lever until it is fully engaged 185 degrees from the disengaged position. 5-89

If the lever offers resistance when trying to go past the vertical position, the clutch interlocking surfaces are in the wrong position. If this hap-

Figure A.1 A typical Barton American front mount pump.

pens, return the clutch lever to the horizontal position, allow the pump to stop, and engage it again. Never try to force the clutch lever to engage because damage to the clutch or pump can result.

Very few front mount pumps have been equipped with a power arrangement for transfer. Generally, an air cylinder or vacuum cylinder of some kind is used to accomplish the transfer. A toggle switch operates to the pump position until the light comes on, indicating that the pump is fully engaged (Figure A.2). To disengage it, move the toggle switch to the TOP (ROAD) position until the light goes out on the pump. It is then disengaged.

Figure A.2 Power is transferred to the pump by activating the toggle switch. The light indicates when the transfer is complete.

Pump Cooling and Heat Transfer

A warning light mounted in the cab that indicates pump engagement is critical on a front mount pump. If the pump is engaged as the vehicle is moving, the impeller will develop excess speed, with no water moving for cooling. The residual water will overheat and damage the shaft, the
5-89 packing glands, the impeller, and wear rings. Anytime the indicator light shows the pump is engaged, there should be water movement.

The American front mount pump is equipped with a frost-proof feature. Water at the engine temperature is circulated through the cored suction head of the pump from the cooling system. Since there are no valves in the lines to shut off the circulation of the radiator coolant through the pump housing, the pump will always feel warm to the touch when the engine has been operated. This warmth is not to be confused with a pump overheated from a lack of water movement when the pump is turning in a normal condition.

Most front mount pumps are equipped with an auxiliary cooling device to keep the engine cool during extended pumping operations. The exact installation method for these devices will vary with the apparatus fabricator that mounts the pump. Coolant is usually supplied to the cooler through a strainer with a removable cap. The cooler is mounted under the hood in the engine compartment and connected to the discharge side of the pump by ½-inch (13 mm) copper tubing through a strainer (Figure A.3). The line from the strainer goes through a quarter-turn or other type of valve to the heat exchanger. As the cooling water from the pump and the radiator coolant flow through the heat exchanger, the coolant temperature is lowered. The cooling water from the auxiliary cooler then is returned through another line to the intake side of a fire pump.

Figure A.3 On many Barton American pumps, cooling is provided by a marine-type cooler.

The heat exchanger is constructed in such a way that at no time does water from the fire pump come in contact with the radiator coolant inside the unit. Water will not contaminate the antifreeze solution. There are some cases where a second valve is supplied coming off the strainer or heat exchanger that allows some water from the fire pump directly into the radiator; this is called a radiator fill valve. Opening this valve will contaminate the coolant, and it should only be used when the fire pump is operating from a hydrant or pumping clean water. The strainer also provides a supply of water under pressure to the pressure governor if one is supplied. When the pump is pumping dirty water, the strainer tends to become clogged. In severe cases, the strainer can become so clogged that no water is able to reach either the pressure governor or the auxiliary cooler (Figure A.4).

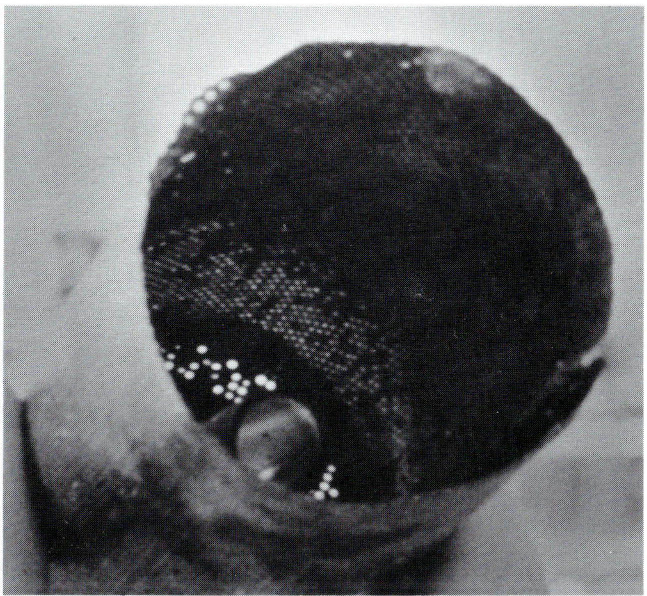

Figure A.4 A completely clogged strainer, like this one, can prevent the cooling system from functioning properly.

Discharges

The American front mounted pump is equipped with a bolt-on discharge head that goes on the top side of the pump to provide the discharge valves. Newer pumps have one quarter-turn valve with 2½-inch (65 mm) National Standard Threads for each 250 gpm (1 000 L/min) capacity of the pump, up to a maximum of four valves. The valves will lock in any position by turning the handle one quarter turn clockwise after it has been set.

The discharge head contains a gravity-operated valve (Figure A.5). This valve permits priming the pump without closing the discharge valve(s). As long as the pressure inside the pump is greater than the pressure in the discharge head, the valve will remain open. Anytime the pressure inside the pump is less than that in the discharge

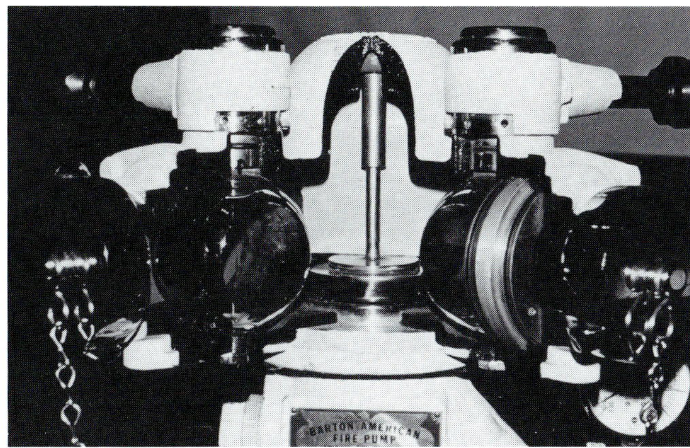

Figure A.5 This cutaway shows the gravity operated valve in the discharge manifold.

head, it will be closed. For this reason, if the pump has been operated under pressure supplying the hose reel, pressure may be trapped inside the discharge head. Each of the 2½-inch (65 mm) fittings should have a drain valve to bleed off this pressure. The drain valve is also used to drain the hoseline after the discharge valve has been closed. An individual line gauge may also be connected on the line side of each valve to provide pressure monitoring of each attack line.

An additional discharge connection is provided just below the main discharge head or on the discharge head itself to allow for connection to preconnected lines, tank fill connections, or hose reels located at the back of the fire truck body. At least one valve is usually provided for this line. In some cases, a number of different lines may be connected to this fitting, with individual valves that allow the pump operator to control hoselines at the rear individually from the front of the vehicle (Figure A.6).

Figure A.6 A 3-inch (77 mm) line coming off the additional fitting below the main discharge head.

Intakes

A strainer is supplied for a standard suction connection. The strainer is constructed with a fine mesh and is easily blocked with grass and weeds when operating from draft. If the pump goes into cavitation while operating from draft with a high vacuum rating, the first problem to suspect is that the strainer is blocked. The pump must be shut down long enough to inspect the strainer and, if necessary, to clean it out. If the proper strainer is used on the end of the suction hose, this problem will be minimized.

The water tank on the apparatus is connected to the suction manifold of the pump on one or both sides. A single 2½-inch (65 mm) line from the tank to the pump generally provides about 250 gpm (1 000 L/min). A single 3-inch (77 mm) line will provide approximately 350 gpm (1 325 L/min). If additional flow is required, dual lines will be needed. In addition to the tank-to-pump lines, most pumps will have a 2½-inch (65 mm) gated intake for use in relay or hydrant operation. Each tank-to-pump line will have a valve located near the pump. In some cases, there is also a valve in each line located at the tank itself. By closing both these valves and opening the drain at the low spot in the tank-to-pump line, freezing can be prevented when operating in cold weather. During this procedure, the drain must be closed and both valves opened in order for the pump to be able to operate from the water carried on the apparatus. Where two tank-to-pump lines are supplied, opening the valves in only one line will restrict the maximum flow that can be obtained from the pump.

Pump Packing

The American front mount pump has one stuffing box located at the rear of the pump between the impeller and the gear box for packing around the shaft. Some models provide for an adjustment in this packing by tightening two gland nuts, one on each side of the impeller shaft (Figure A.7). These nuts must be adjusted equally to keep an even amount of pressure on the packing all around the shaft. (**NOTE:** the pump must be turn-

Figure A.7 One of the packing nuts located on each side of the impeller shaft.

ing when adjusting this packing.) Proper adjustment of the packing can be determined when a small amount of water drips around the shaft, but not a steady stream. Some models use an injection type of packing that can be adjusted by turning the packing regulating screw clockwise one turn at a time. Pause between each turn for a few seconds to allow the added packing to form in the stuffing box. The regulating screw is a long cylinder with a hexagonal surface on it to allow use of a wrench for adjustment (Figure A.8). Figure A.9 shows a cutaway drawing of the injection-type packing around the shaft. Since at that point there will be no water supply around the shaft for cooling purposes, be careful not to overtighten the regulating screw to the point where the leakage stops completely. If the regulating screw is overtightened, the shaft will overheat and over a period of time become damaged. It will then become impossible to obtain a good seal around the packing.

Figure A.8 The regulating screw has a hexagonal surface to allow the use of a wrench for adjusting the injection packing.

SINGLE-STAGE MIDSHIP

The American Fire Pump Company also manufactures a single-stage pump design to be mounted midship on the pumper using the split drive shaft arrangement. These pumps use a double suction impeller capable of supplying as much as 2,000 gpm (6 000 L/min), depending on the impeller, piping, the gear ratio, and the power of the vehicle engine.

Transferring Power to the Pump

The pump transfer case is mounted between the transmission and the rear axle of the vehicle. It is usually controlled from the cab of the ap-

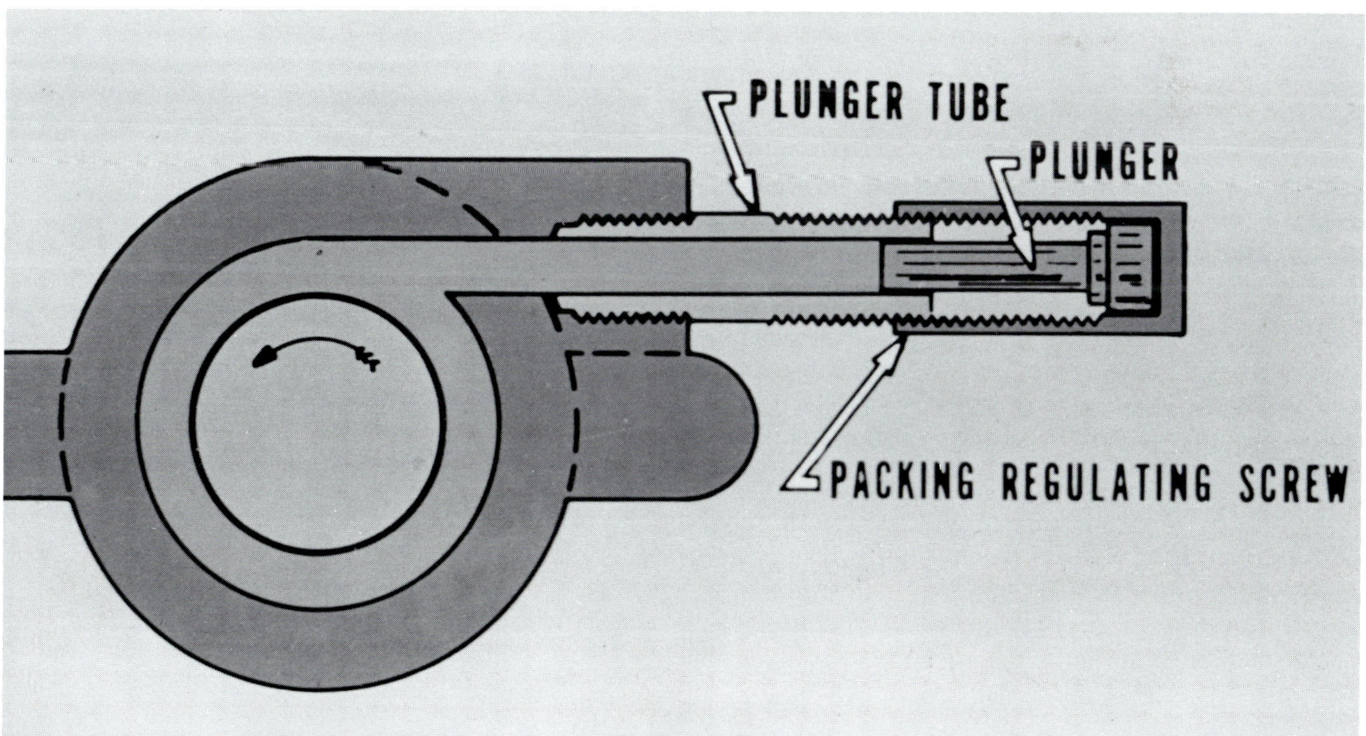

Figure A.9 A cutaway showing the injection-type packing around the impeller shaft.

paratus by some type of power shift arrangement (Figure A.10). If the apparatus is gasoline powered, the most common type of control uses the vacuum from the intake manifold of the engine. Diesel powered vehicles usually make use of air pressure from the braking system. In either case, to engage the pump, depress the clutch pedal, turn the shift control to the pump position, put the road transmission in proper gear for pumping, and release the clutch. The pump should now be turning.

Figure A.10 Power shift arrangement for controlling the pump transfer case on a midship pump.

For pumping operations, the road transmission on a manual shift vehicle should be in the direct drive. This is usually the highest gear in the transmission unless an overdrive transmission is used. There will be a designation plate on the dash specifying the proper gear for operation of the pump. As a check to make sure the transmission is in the correct gear, the operator can observe the speedometer before leaving the cab. The expected reading varies with different vehicles, but is usually between 10 and 16 mph (16 km/h and 24 km/h). Again, the discharge piping from the pump must have at least one 2½-inch (65 mm) gated discharge for each 250 gpm (1 000 L/min) pump capacity. At least one of these must discharge on the right side of the apparatus, but be controlled from the pump panel on the left side. Additional valves may be supplied for preconnected lines, master stream devices, lines to the rear of the apparatus, and so forth. This midship pump also uses a gravity operated check valve on the discharge similar to the front mount to permit priming the pump without closing the hoselines.

Pump Discharges

Older units use a lever controlled push/pull type valve with a locking mechanism as part of the discharge valve (Figure A.11). Once the valve has been set, a small locking lever should be pulled out. This will prevent the valve from closing or the setting changing during operation. Pulling out the locking lever is especially important when a line is being operated partially gated down to obtain a pressure differential. Individual line gauges will indicate the proper setting of each valve when it is desired to operate at different pressures on different hoselines. A drain valve should also be provided on the line side of each discharge valve. This type of valve enables pressure to be bled off individual lines that are shut down while the pump is still operating.

Hose reels are generally controlled by a screw-type valve with a knob on the pump panel. To supply water to the hose reels, turn the knob counterclockwise until it is fully opened. A similar type

Figure A.11 A lever controlled push/pull discharge valve with a locking mechanism. Locking lever prevents valve setting from changing.

of valve may be used to supply a tank fill line (Figure A.12). This line can be used to fill the tank when the pump is being supplied from another source. It is more commonly used to maintain a flow of water during an intermittent attack to prevent the pump from overheating. Another way to maintain a flow of water is by using a circulator valve (Figure A.13). The pump is equipped with a copper line from the discharge side of the pump to this circulator valve that directs the water either back into the tank of the apparatus or to the ground. The valve in either position will present an air leak that may prevent successful operation from draft since the spill line opens either into the tank or the open air.

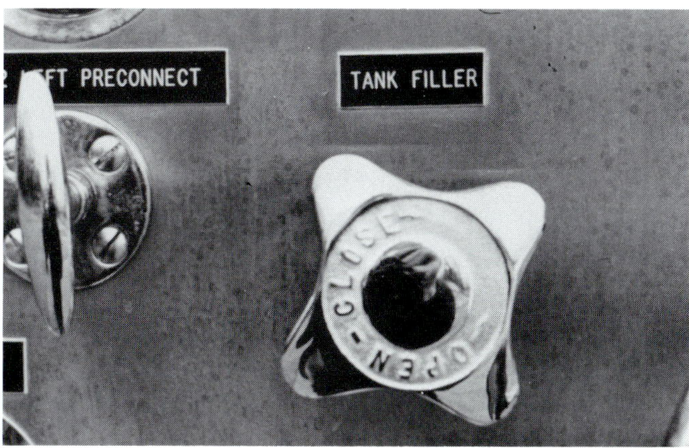

Figure A.12 A screw-type tank fill valve.

Figure A.13 A circulator valve.

While it does not specify on the valve, if the handle is placed in the vertical position halfway between the tank and the spill positions, the valve is closed and an air leak will be eliminated. Later apparatus models equipped with American Fire Pumps use valves supplied by Akron, Elkhart, or another manufacturer.

Other Valves

When operating from the tank on the apparatus, the tank outlet control must be turned to the OPEN position. Generally, it is not necessary to prime the pump when operating from the tank because the force of gravity will pull the water from the tank into the fire pump. Since the tank is mounted higher than the pump, operating the primer momentarily can speed the operation. This is especially true if the pump was drained when not in use and air was trapped inside the pump, delaying full priming. Again, there will be at least one gated 2½-inch (65 mm) intake fitting, usually equipped with some type of bleeder valve. This bleeder may be only a petcock on the body of the valve, or it may be a more elaborate drain valve mounted on the pump panel. In either case, the bleeder should be used to allow air to exhaust from the incoming supply line before the intake valve is opened. This procedure prevents pressure fluctuation in the attack lines that would occur if air trapped in the supply line enters the pump while it is operating. The master drain for the pump is located at a low point on a pump panel (Figure A.14). This drain is controlled by a disc valve and 5-89

Figure A.14 The master drain is a multi-part screw/disc valve. 5-89

the knob will be extended out to the panel. To drain the pump, turn the knob counterclockwise until it is fully open.

Many pumps have low spots in various lines that require additional drain valves. When draining the pump in freezing weather, be sure to *open* all the drains to prevent damage to the pump. When putting the pump in operation, be sure that all drains are *closed* to prevent air leaks that would prevent priming the pump. It is a common practice to provide a vacuum clutch control on the pump panel for those pumps mounted midship (Figure A.15). By operating this control when no water is being used, the pump operator can disengage the clutch to stop the pump without getting back into the cab of the vehicle. This prevents overheating and lessens wear and tear on the pump.

Figure A.15 A vacuum clutch control on the pump panel.

TWO-STAGE MIDSHIP

The two-stage midship mounted Duplex pump is very similar to the single-stage pump just described, with one major addition to the pump panel: the pump shift lever (Figure A.16). The single-stage pump turns the impeller faster to obtain higher pressures; the Duplex pump uses two separate impellers, one for capacity, and the other for pressure. Each of these pumps uses a double suction impeller and is controlled by an automatic pressure-operated valve, sometimes called a poppet valve (Figure A.17). The gear train is arranged so that by operating a shifted gear under control of the pump shift lever on the pump panel, either the capacity impeller turns, the pressure impeller turns, or both turn together (Figures A.18 and A.19). When only the capacity impeller is turning, the pressure valve associated with the capacity pump opens and water is allowed to enter into the

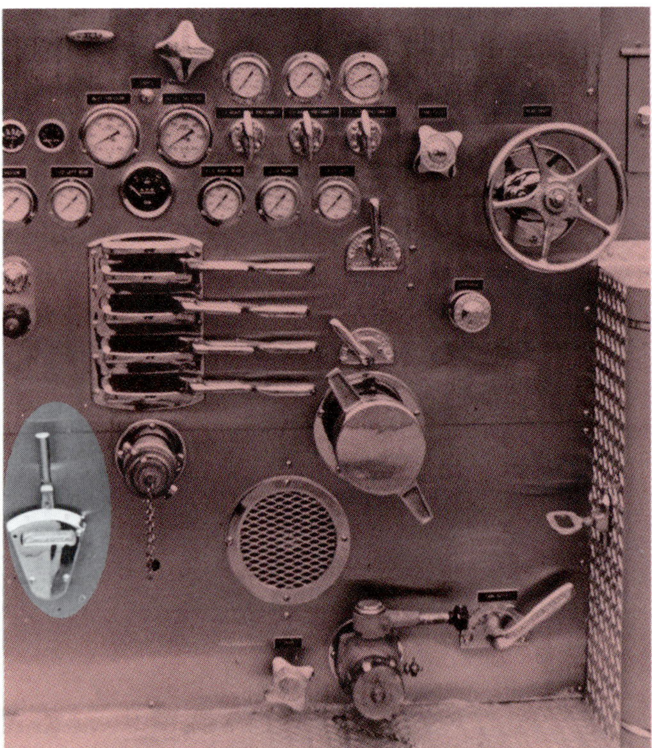

Figure A.16 The circle highlights the pump shift lever.

Figure A.17 An automatic pressure-operated valve (poppet valve) shown in a cutaway.

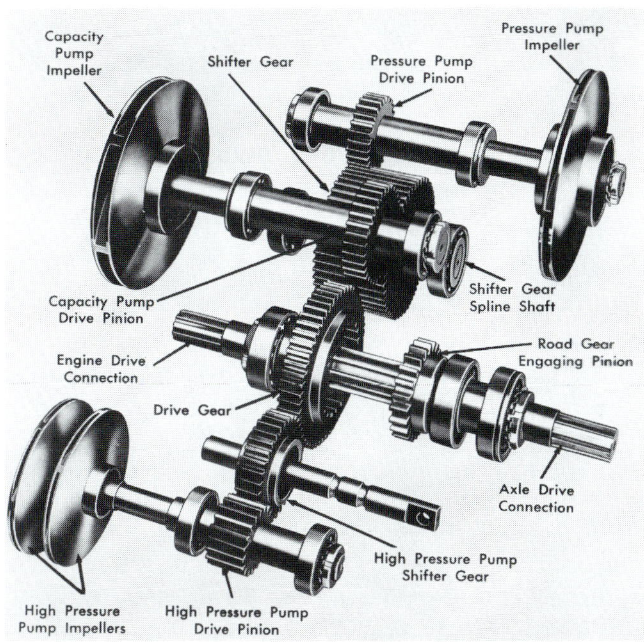

Figure A.18 The arrangement of a Duplex multistage gear train.

Figure A.19 A close-up of the pump shift lever on the pump panel.

discharge manifold (Figure A.20). The automatic valve on the pressure side is held closed by the pressure in the discharge manifold supplied by the capacity pump. When only the pressure impeller is turning, water comes into the pump through the intake manifold, goes through the connecting manifold on the top of the pump, and enters the pressure pump (Figure A.21). As the pressure is built up by the pressure impeller, the automatic valve opens, thus allowing the water to enter the discharge manifold. Once again, the pressure in the discharge manifold keeps the capacity valve closed. This is because the capacity impeller is not turning and no pressure is being developed in that pump.

In the SERIES mode of operation, both impellers are turning (Figure A.22). As the water enters the

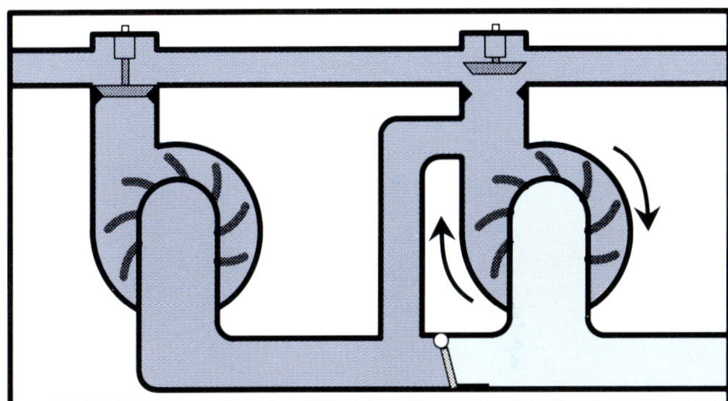

Figure A.20 Water entering the discharge manifold in the CAPACITY mode.

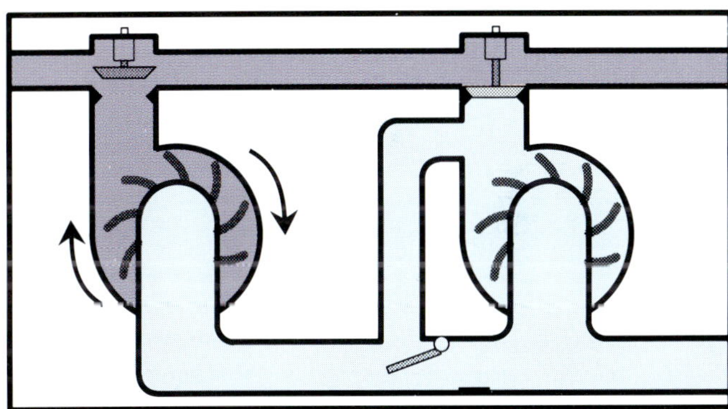

Figure A.21 Water entering the discharge manifold in the PRESSURE mode.

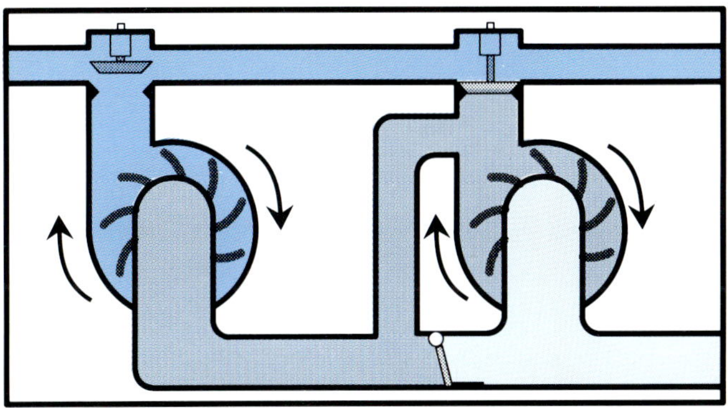

Figure A.22 SERIES mode of operation: both impellers are turning.

pump, the capacity impeller discharges directly into the connecting manifold. The higher pressure in the discharge manifold coming from the pressure pump keeps the capacity valve closed. When the water enters the connecting manifold, the pressure closes the swing-check-type valve to prevent the water from flowing back into the intake manifold and churning. The pressure impeller then adds to the incoming pressure that was produced by the capacity impeller and discharges through the pressure valve into the discharge manifold. Since the pressure in the discharge manifold is more than twice the pressure being developed by the capacity impeller, the capacity valve remains closed. A typical operation might provide for the capacity pump to build up a 100 psi (700 kPa) pressure. The pressure pump then adds 150 psi (1 050 kPa), so a total of 250 psi (1 750 kPa) is available in the discharge manifold.

Change from Pressure to Volume

It is common practice to operate with a shift lever in the SERIES position (vertical). (**NOTE:** refer back to Figure A.19). This permits the pump to develop the desired pressure with minimum engine rpm. At the same time, it does limit the maximum capacity of the pump to the gallonage (liters) that the pressure impeller can supply. Once the pump is in operation, it may be necessary to change the setting of this shift lever. To do this, operate the vacuum clutch control on the pump panel to disengage the clutch and allow the pump to stop turning. Move the lever to the desired position, to the rear for CAPACITY (the C position) to the front for PRESSURE (the P position), or in the middle for the SERIES position (vertical).

When operating in SERIES, there is a locking cam behind the panel that must be operated to keep the shift lever from slipping from the SERIES position into the CAPACITY position. The force of the gears would naturally tend to cause it to do that. When the shift lever is in the desired position, re-engage the clutch and adjust the throttle to obtain the desired pressure. Before disengaging the clutch to change operations, reduce the throttle to the IDLE position. *Making this change means that the fireground operation must be shut down.* For this reason it is important to have the pump in the proper pumping configuration when putting it into service and making the initial attack.

Since changing the setting of a pump involves interrupting the fire flow, it should be set initially to provide the maximum amount of water that can be called for by the hoselines rather than the amount actually being used. A good rule of thumb is to consider each 2½-inch (65 mm) line at 250 gpm (1 000 L/min) and each 3-inch (77 mm) line at 400 gpm (1 500 L/min). The pressure impeller in the Duplex pump is designed to supply as much as 50 percent of the rated capacity of the pump. Anytime more than 50 percent of the rated capacity may be required, the shift lever must be in the CAPACITY position rather than series or pressure. In general, anytime the pump is being used as a water supply pumper it should be in the CAPACITY position. This allows additional supply lines to be added as the fire progresses without any interruptions to the flow.

When functioning as an attack pumper, the Duplex pump is most commonly used in the SERIES position. Initial attack is usally made with 1½-inch (38 mm) or 1 ¾-inch (45 mm) lines. If the attack is made with 2½-inch (65 mm) lines, the pump may have to be operated in capacity. The pump operator must take this possibility into consideration in making the initial setup.

An interstage discharge pressure fitting may be installed in the connecting manifold between the discharge of the capacity impeller and the intake of the pressure impeller. When a pump is operating in the SERIES position, this interstage pressure will be approximately 40 percent of the discharge pressure of the pump. This permits the use of high pressure lines such as a hose reel from the discharge side of the pump while at the same time supplying larger lines at lower pressure from the discharge of the capacity impeller. When the pump is operating in capacity, the interstage pressure will be the same as the discharge pressure. When it is operating in the PRESSURE position, the interstage pressure will be zero. Any discharge lines connected at this point cannot be used.

If there is a problem encountered with the capacity impeller on the pump, the pump will work in an emergency at a reduced capacity by operating only the pressure impeller. The pump will function in the normal manner except for drafting. The construction of the pump and the connecting manifold is such that the pressure impeller is not capable of operating from draft independently of the capacity impeller.

PRIMER

A vacuum primer was the standard primer for Barton American fire pumps (Figure A.23). It uses the gasoline engine vacuum in the intake manifold. This primer connects the highest point of the pump suction manifold to the intake manifolds of the engine. This allows the vacuum developed there to evacuate the air from the pump. The primer actually serves two purposes. First, there is a check valve in the line that comes from the intake manifold to the primer. If there is a backfire or other types of explosive gases enter the intake manifold, the check valve operates to prevent those gases from entering the fire pump and causing damage. Second, as the pump becomes primed, the air is evacuated and water enters the primer (Figure A.24 on next page). There are two separate floats to prevent that water from entering the intake manifold of the engine. When the primer lever is pulled out, the valves are opened — one to the pump and one to the engine.

As the intake manifold vacuum evacuates the air from the primer and from the pump, the vacuum gauge beside the primer lever begins to register the vacuum reading present at the intake to the fire pump. As the pump is primed, water enters the wet chamber of the primer (Figure A.25 on next page). There, a metal float operates a ball valve when the chamber has filled with water; the valve seals off that water from the vacuum line. At the same time, an exhaust valve, which is a little ball valve, operates allowing air to be drawn into the primer from the outside. The sound of air passing through the exhaust valve provides an audible indication that the pump has been primed. If the float in the wet chamber becomes defective, the water will continue to rise inside the primer

Figure A.23 The vacuum primer was standard for Barton American fire pumps.

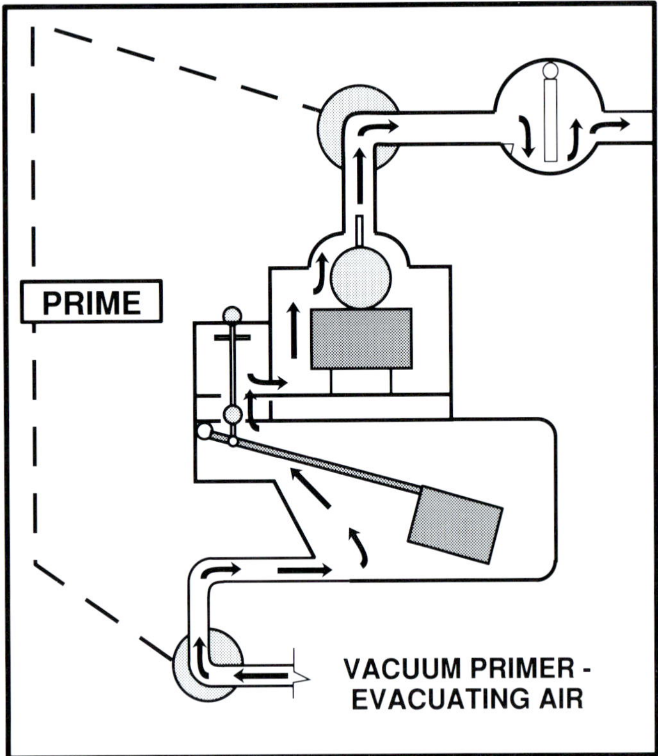

Figure A.24 As the pump becomes primed, air is evacuated and the pump fills with water.

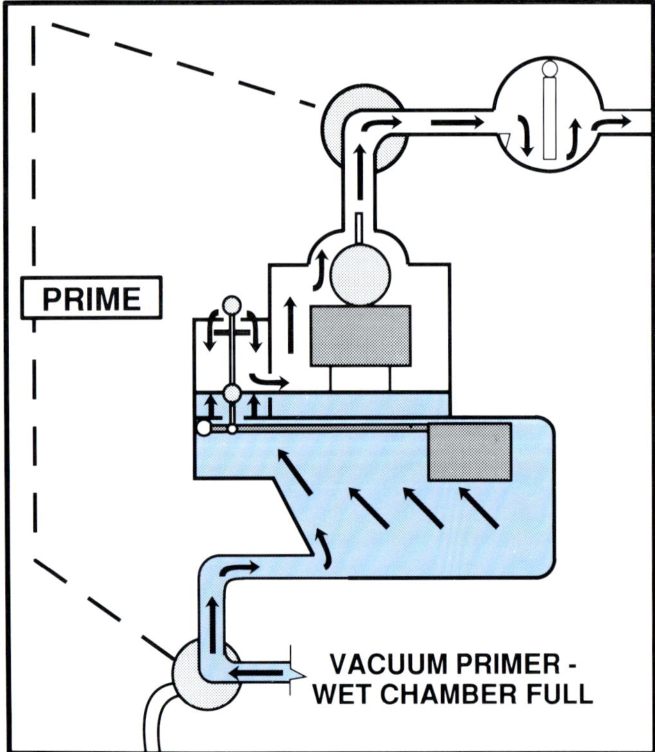

Figure A.25 Water enters the wet chamber of the primer.

into what is normally the dry chamber. There, a cork float rises with the water level, again by means of a ball-type valve, and closes off the water from entering the vacuum line to the engine (Figure A.26). Since 1983 all American Fire Pumps are equipped with positive-displacement primers.

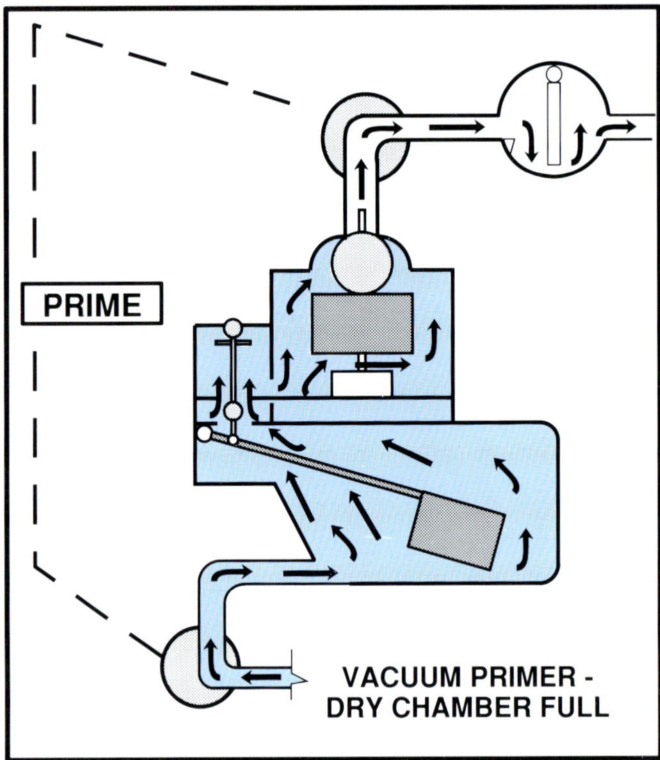

Figure A.26 A defective float allowing water to rise inside the primer into the dry chamber.

DRAFTING

To operate the pump from draft, it is necessary to make the pump airtight. Close all drain valves, tank fill valves, and the tank-to-pump line. Connect the suction hose to the steamer connection. Make all connections airtight. The pump should be engaged at 800 rpm or less. Then, set the engine throttle for a fast idling speed, normally somewhere around 1,000 rpm. After the pump has been engaged and is turning, pull out the primer handle. If the pump is equipped with a priming vacuum gauge, it should gradually increase toward 20 inches (508 mm) of vacuum.

When the primer handle is pulled out, the primer drain valve closes, the valve to the intake manifold opens, and the line to the pump suction connects through the primer to the intake manifold. As the pump fills with water and the discharge pressure gauge begins to show a steady pressure, increase the engine speed until the pressure indicates more than 50 psi (350 kPa). Open a discharge valve and allow some water to flow while releasing the priming handle. The valve to the vacuum source and the pump suction will close and the drain valve from the primer will open, allowing the water to drain out. Remember that the vacuum primer is not normally supplied with a diesel engine. It could be used with one, but it would require a separate vacuum pump. This is because the diesel engine operates under positive pressure into the cylinders rather than negative pressure. It does not seem practical to supply a separate vacuum pump, so in most cases a positive displacement primer is supplied. The American PR-51 has been commonly used, although any type of primer can be used for this purpose.

PRESSURE CONTROL

The pressure governor that has been developed for use with the American Fire Pump was also supplied as a standard on the older gasoline powered units. It consists of a butterfly valve mounted between the carburetor and the intake manifold, which is controlled by the pump pressure. By changing the setting of this butterfly valve, the pump pressure governor is able to control the pump pressure by changing the engine speed as necessary. This governor is a fixed spring reference-type governor (Figure A.27). The diaphragm is acted upon by water pressure from the pump against a reference spring. When the pressure from the pump becomes high enough to over-

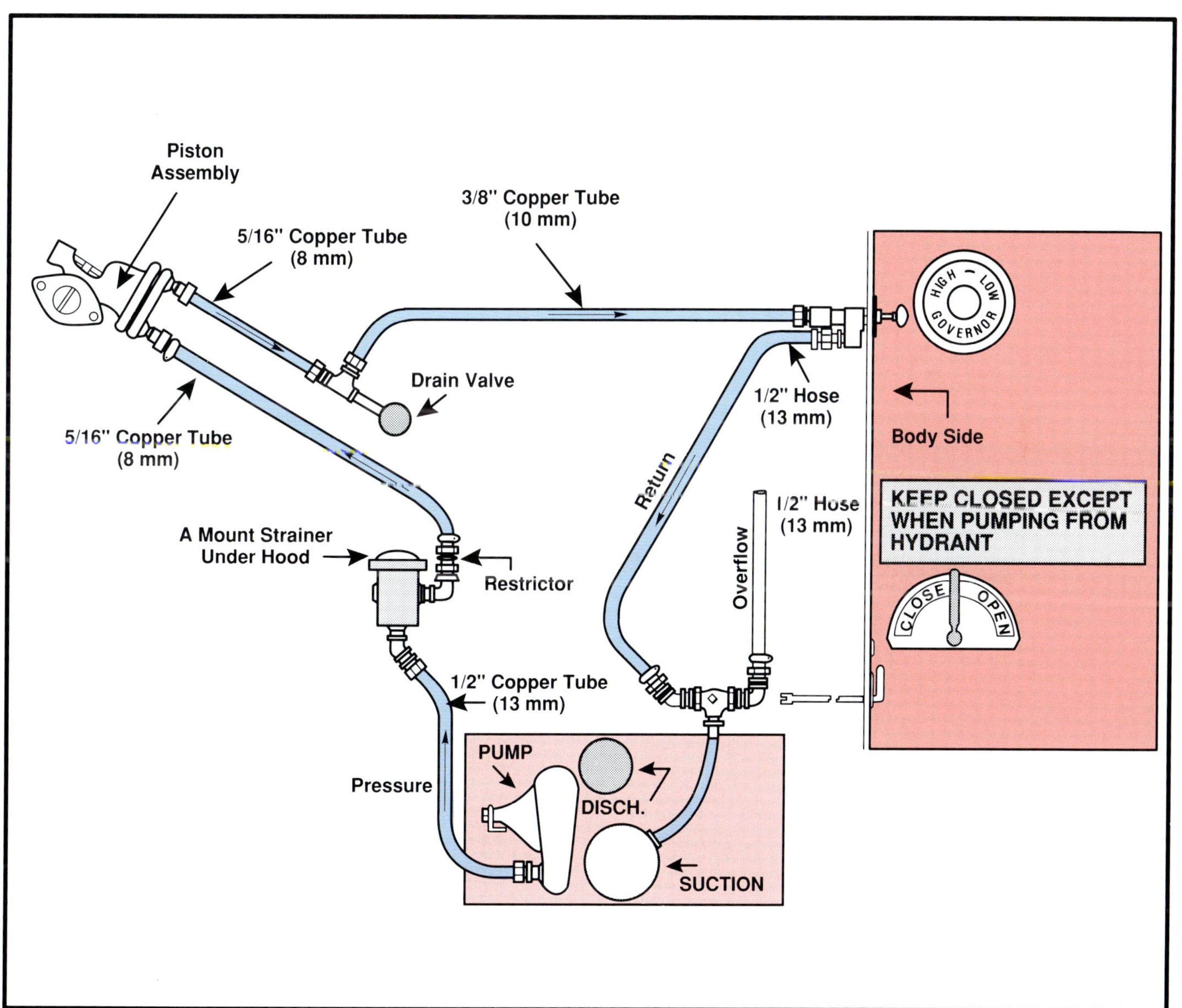

Figure A.27 The pressure governor is a fixed spring reference-type governor.

come the tension of the reference spring, a movement in the diaphragm causes a shaft to move. The shaft is coupled to a butterfly valve mounted between the carburetor and the intake manifold of the engine. This arrangement will reduce the engine speed until the pressure returns to the original value. To make the operating pressure variable, a governor control valve is located in the return line from the governor itself.

During operation, water enters the control chamber of the governor from the discharge side of the pump through the strainer that is normally mounted under the hood of the apparatus. The water flows through the governor, applying pressure to the diaphragm, and escapes from the control chamber of the governor through a return line. This return line is under the control of the governor control valve and the governor return valve.

As long as the control valve is fully open, the water entering the governor will be allowed to escape with no restriction. The pressure will not build up enough to make the governor shaft operate, since it normally requires somewhere between 85 and 110 psi (595 kPa and 770 kPa) to overcome the basic spring tension in the governor. By adjusting the governor control knob by partially closing the needle valve, a valuable friction loss is introduced into the return line. The operating point can be set to whatever pressure is needed.

The governor return valve provides the option of allowing the water from the governor to return to the intake side of the pump or to be discharged outside the pump, depending on the method of operation. This valve is labeled HYDRANT or DRAFT on front mount pumps or OPEN or CLOSE on the midship pump. The midship pump normally includes a direction plate saying, "Keep Closed Except When Pumping from Hydrants" (Figure A.28). This label is deceptive, however. Anytime the pump is operating with changing positive pressure on the intake, the valve should be in the OPEN position. This allows the needle valve to work against a constant pressure on the governor return line. It also allows the governor to provide a constant pressure on the discharge.

Figure A.28 The governor return valve.

While the manufacturer refers to the governor return valve as a way to provide for the difference between operating at hydrant or draft, relay-type operations are even more critical. When operating in a relay, the reduction in flow results in a decrease of friction loss in the supply line, a higher residual pressure, and an increase in discharge pressure. When the discharge pressure increases, the governor should operate to reduce the engine speed and bring the discharge back to the preset value. If the governor return valve is in the OPEN position, it will do that. If the governor return valve is in the DRAFT or CLOSED position, the increase in intake pressure as well as the discharge pressure will cause a decrease in flow through the control chamber. The throttle will increase until the pressure builds up to the operating point on the governor and the discharge pressure will continue to increase, rather than returning to the preset value. In other words, operating with the governor return valve in the CLOSED or DRAFTING position will maintain a constant net *pump* pressure, not a constant *discharge* pressure. Again, a standard American governor does not work well with diesel engines. Diesel engines are relatively slow to respond to changes in the throttle. There is no way to install a butterfly valve because there is no carburetor. Therefore, most American pumps equipped with diesel engines use a relief valve or some other type of governor. American Godiva builds a relief valve for their pumps. Also the Hale or Ross relief valve has been used extensively (Figure A.29). One governor that has been used with some success by American Godiva Fire Pump is the electronic governor built by Fire Research.

Figure A.29 A Ross relief valve connected between the discharge and the intake of a front mount.

Appendix B

Description and Operating Practices for American LaFrance Pumps

(Note: Unless otherwise indicated, all photos in this Appendix are printed courtesy of Bill Eckman.)

American LaFrance pumps are found primarily on the custom and commercial chassis fire apparatus manufactured by the American LaFrance Fire Apparatus Co. (Figure B.1). All American LaFrance fire pumps are two-stage pumps, and they are all mounted midship using a split drive shaft power transfer arrangement. The following section highlights the operation of these pumps.

ENGAGING THE PUMP

In most cases, pump engagement is done by a manual lever of some type. Usually, a T-type handle is mounted on the dash or in front of the seat off the floor. A quarter-turn locking arrangement is used on the lever. To engage the pump, turn the lever to the RELEASE position and pull out fully. The pump may not be completely engaged if the lever is not pulled out completely. The lock should also be operated while the pump is operating. In most cases, the pump's engagement is locked by turning the T-handle one-quarter turn clockwise.

After the pump is engaged, the road transmission is put into pumping gear with the transmission in direct drive. If a 5-speed transmission with overdrive is used (such as the one used in the 700 series), the pump will normally operate in fourth gear. A plate on the dash specifies the proper pump gear for operation.

One peculiarity of the road transmission of some American LaFrance trucks is that the pump gear is fourth gear, and it is located directly beside the reverse gear on the shift pattern. The pump will not operate if the road transmission is inadvertently put into the reverse gear. Therefore, a manual lock is provided to keep the transmission from going into reverse. This lock should be in place at all times except when the road transmission is in REVERSE position and the truck is actually moving backwards.

Once the road transmission is in the proper pumping gear, the transmission lock should be operated while the pump is in service. This also holds true for units equipped with automatic transmissions. Once the automatic transmission is put in the proper gear, the manual locks should be operated to keep it there.

Figure B.1 A typical American LaFrance custom pumper.

Another caution for automatic transmission operated units is to ensure that the transfer from the ROAD position to the PUMP position has actually taken place. The vehicle may start to move if the power has not been completely transferred to the pump. The automatic transmission torque converter may build up power in priming the apparatus if the lever has not been operated all the way. The power may be built up if the transfer did not complete when the throttle is increased to bring up the engine speed.

TRANSFER FROM PRESSURE TO CAPACITY

The primary difference between the older American LaFrance, the triple flow model, and the later twin flow unit is the means of transferring from the PRESSURE to the CAPACITY position. The older triple flow units were equipped with a manual transfer arrangement control by a handwheel (Figure B.2). As the handwheel was turned in the clockwise direction in toward the panel, the pump was in the PRESSURE position. As it was operated counterclockwise, and the handle moved away from the panel, the pump moved into the CAPACITY position.

Since the handwheel protrudes from the panel in the CAPACITY position, it is common practice to carry these pumps in the PRESSURE position when they are not in service. This can be a problem because the construction of the transfer valve is such that it is impossible to move the control with any appreciable amount of pressure being maintained by the fire pump. If the pump was initially put in service in the PRESSURE position, and there is a need for more than 50 percent capacity of the pump, the fireground operation will have to be interrupted. The pump must be shut down long enough to make the transfer.

To prevent this from happening, move the pump to the CAPACITY position before beginning to supply water in a relay from a drafting operation when multiple hoselines or master streams are going to be supplied. The twin flow pump uses a hydraulically operated transfer valve. The actual transfer takes place with the valve controlled by water pressure.

The three-way control valve mounted on the pump panel directs the water to the transfer valve as shown in the waterflow diagram (Figures B.3 and B.4). Water from the first stage of the pump comes in from the intake, goes through the impeller and the transfer valve in the CAPACITY position, and out through the discharge manifold. Water from the second stage also comes in through the intake, the check valve, the impeller, and to the discharge manifold.

The transfer valve is held in the CAPACITY position because the discharge pressure is being

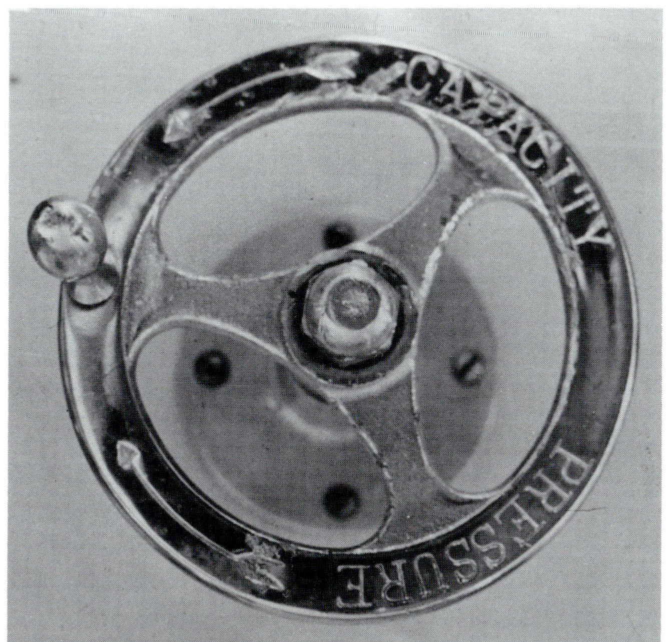

Figure B.2 Older American LaFrance triple flow pumpers were equipped with a manual handwheel pump shift.

Figure B.3 The hydraulic transfer control in the CAPACITY position.

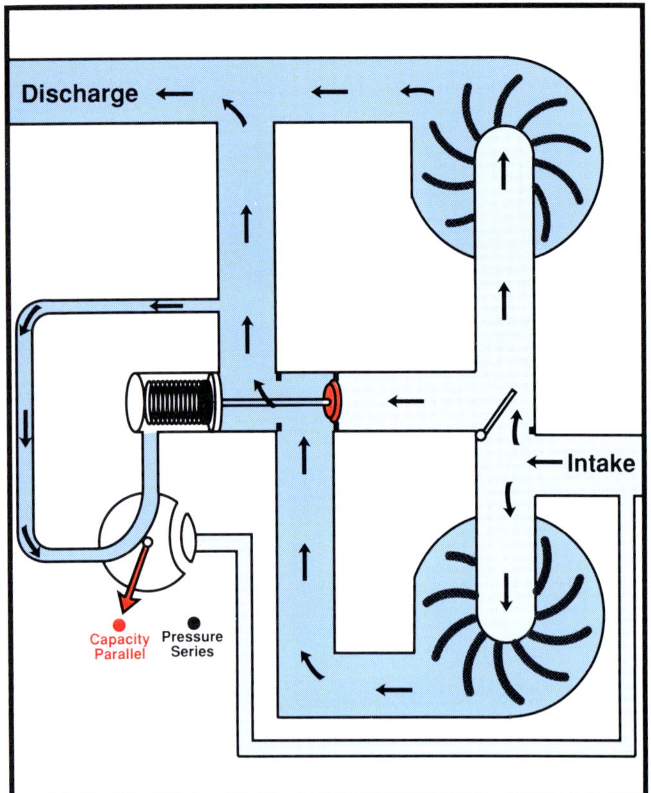

Figure B.4 Schematic of the American LaFrance hydraulic transfer valve in the CAPACITY mode.

applied to the pressure side of the valve. The pump discharge pressure pushes against intake pressure on the other side of the valve when it is in the CAPACITY position. To transfer into the PRESSURE position, move the three-way valve to the PRESSURE position (Figure B.5). This transfer directs the water from the intake side of the pump through the valve and over to the control side of the transfer valve (Figure B.6).

As water being discharged from the first stage of the pump acts against the control side of the transfer valve, it is opposed by the intake pressure, which is much lower. This causes the transfer valve to move into the PRESSURE position. At this point, water coming from the discharge side of the first stage acts against the face of the transfer valve. The transfer valve is opposed by second-stage pressure against the other side of the face. At the same time, the control end of the valve has second-stage pressure acting against the control surface being opposed by intake pressure.

The surfaces on the control end of the valve are greater than the surfaces of the valve face itself. The pressure differential on the control portion of the valve is enough to keep the valve in the PRESSURE position even though the pressures on the face would tend to close it back to CAPACITY. At this point, first-stage pressure is isolated from the discharge manifold by the transfer valve. First-stage pressure is isolated from the intake pressure by the operation of the check valve

Figure B.5 The hydraulic transfer valve in the PRESSURE (SERIES) position.

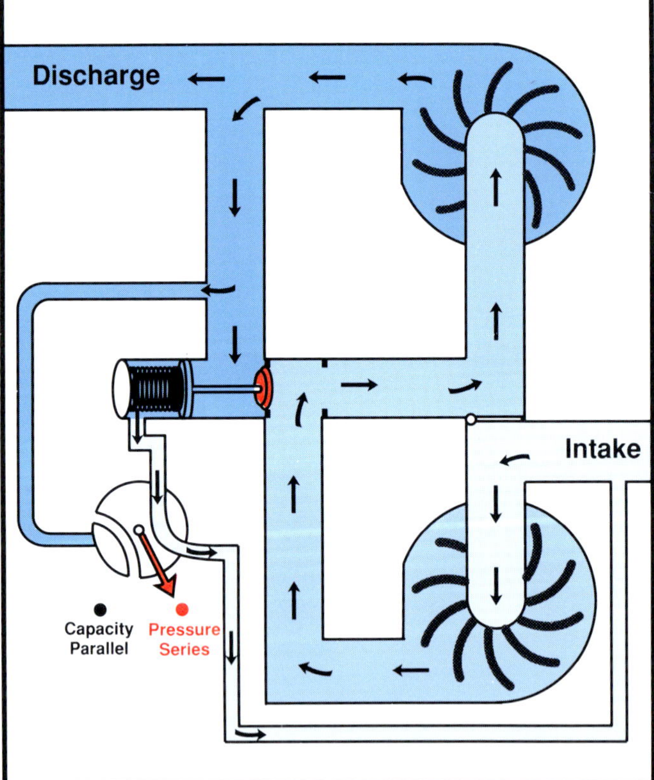

Figure B.6 Schematic of the hydraulic transfer valve in the PRESSURE position.

on the intake to the second stage. Water is now delivered from the discharge of the first stage to the intake of the second stage. Pressure builds up to the higher level and the pump is in the PRESSURE position instead of the CAPACITY position. Now the total capacity of the pump is limited to approximately 50 percent of the rated capacity, but the pressure is increased to nearly double that of the CAPACITY position.

When the pump is dry, there is not enough pressure to move the transfer valve to either position. It is held in the CAPACITY position by the force of the spring against the control surface of the valve. Therefore, when the pump is dry, it is always in the CAPACITY position, no matter what position the three-way valve is in. The pump will not transfer regardless of the setting until enough pressure builds up in the pump to overcome the force of the bypassing spring. This will normally happen around 50 psi (350 kPa).

To check the operation of the transfer valve, close observation of the discharge pressure gauge will indicate a sudden jump as pressure builds up. As the throttle is increased, pressure will go from 0 to around 50 psi (0 kPa to 350 kPa) where a sharp increase will take place. This indicates that the transfer valve has operated and the pump is in the PRESSURE position.

Since these valves operate in close tolerances in water, there is a tendency for corrosion to build up and for the valve to stick. If this happens, the pump may not transfer when the control is moved. If the operator is unaware that the transfer has not taken place, the pump may be operated in the PRESSURE position supplying hoselines with the throttle increased to make up for the difference in the pump operating mode. Later on, as the pump operates, the vibration of the pump and the pressure buildup may cause the transfer to take place. This will cause a sudden increase in pressure that may damage hoselines and equipment or injure the nozzlemen.

To ensure that the transfer has taken place when initially setting the pump, the pump should be changed from PRESSURE to CAPACITY to ascertain that the pressure does actually decrease. If the pump will not transfer to the PRESSURE position for fireground operation, it can be operated in capacity at higher engine rpm. If this is done, always leave the transfer control in the CAPACITY position to prevent a sudden change if the pressure were to suddenly break the valve loose.

Operating the pump in capacity is the only alternative since there is no manual override on this type of valve. If the transfer valve sticks, it will not transfer. Occasionally, increasing the pressure before putting the pump into operation causes the valve to break loose. From that point on it should operate properly.

Even though it is possible to transfer this pump at high pressure, it should never be done at more than 50 psi (350 kPa) net pump pressure to prevent damage to the hoselines and equipment or injury to the nozzlemen.

PRESSURE CONTROL DEVICES

Relief Valve

Over the years, American LaFrance has used different devices to provide automatic pressure control. Some of the custom apparatus, and most of the commercial, use an automatic pressure relief valve for pressure control. The dump-type relief valve works similar to the transfer valve. The dump-type relief valve has a double-ended valve with a control side and a valve side (Figure B.7).

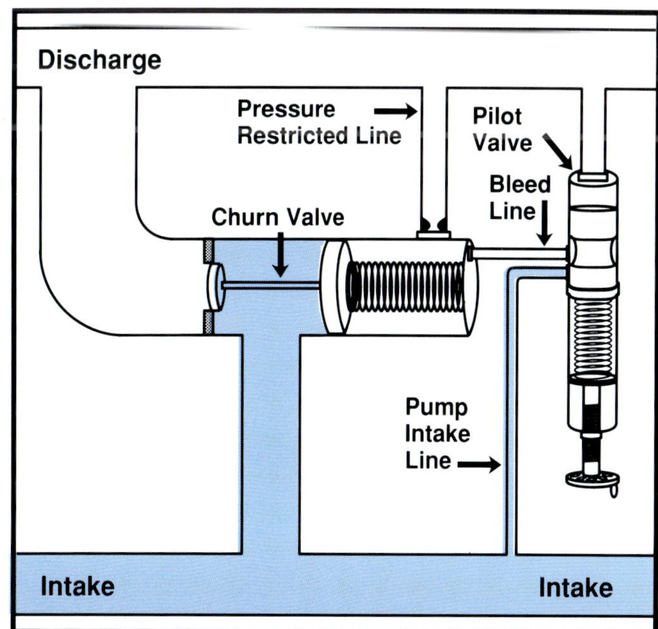

Figure B.7 Schematic of the relief valve.

Water from the discharge pressure of the pump acts to push against the face of the valve to open the relief valve. At the same time, water is connected to the control side of the relief valve operating against a larger area that keeps the relief valve closed. The operate point is set by adjusting the spring tension on the pilot valve by turning the handwheel. When the discharge pressure exceeds the setting of the pilot valve, the pilot valve will open. This allows water to dump from the control chamber of the relief valve through the pilot valve. The water goes back to the intake side of the pump faster than when entering the control chambers from the pressure side of the pump.

As the water escapes, the pressure on the control side of the valve decreases until the discharge pressure pushing against the face of the valve causes it to open. At this point, the discharge pressure will decrease until a balanced point is reached. Now pressure returns to the original operating point. As the pressure decreases, the pilot valve closes. Pressure builds back up in the control side of the relief valve and the relief valve closes.

To set the relief valve, take the following steps:

Step 1: Adjust the pump to the desired operating pressure with all handlines flowing.

Step 2: Turn the handwheel of the relief valve adjustment counterclockwise until the relief valve opens. (When the relief valve opens, a sharp decrease in discharge pressure of the pump will occur).

Step 3: Slowly turn the handwheel of the relief valve adjustment clockwise until the pump returns to its original pressure setting.

Step 4: The relief valve is now set. To keep the setting from changing during operation, turn the little knurled nut on the shaft of the adjustment control clockwise until it is tightened against the panel. This will prevent any change in the adjustment as the pump operates.

When the pump pressure overcomes the spring tension in the pilot valve, the relief valve will open (Figure B.7). The maximum pressure the pump can develop is then limited by the tension of the spring in the pilot valve. Another problem that can occur is if the pilot valve sticks open. If this happens, the relief valve will open and it will be impossible to build up pressure.

To overcome these problems, a relief valve shutoff control may be supplied. The relief valve control closes the return line. This prevents any water from escaping from the relief valve so that it will not open regardless of the pressure. The relief valve is held closed by the biasing spring when the pump is dry. As the pressure builds up, the same pressure is present on both ends of the relief valve. Because the control side is larger, pressure control is maintained.

Pressure Governor

American LaFrance also uses pressure governors. The earliest American LaFrance pumpers used a fixed-spring reference governor. These were especially common on the 700 series and were used on some commercial chassis as well. As the pressure changes, pushing against the bottom of the piston, the threaded rod moves back and forth through a linkage that goes to the throttle (Figure B.8). By locking the linkage on the threaded rod as the piston moves back and forth, engine speed can be changed.

The governor is normally prevented from operating by a governor shutoff valve. To put it in operation, it is necessary to disengage the pressure set control before opening the governor. This is done by pulling out on a T-type handle with a locking collar located on the pump panel. By pulling out on the handle, the governor is disengaged and can be locked in that position by turning the collar clockwise. Once the linkage has been released, the governor shutoff valve can be opened, allowing water to go into the governor. Before setting the governor, it is necessary to get all lines flowing that will be supplied by the pump in order to set the desired discharge pressure.

Once the governor is set, put it in service by releasing the governor set control and pushing the lever in all the way. Now the linkage is locked to the threaded rod on the governor. Anytime the

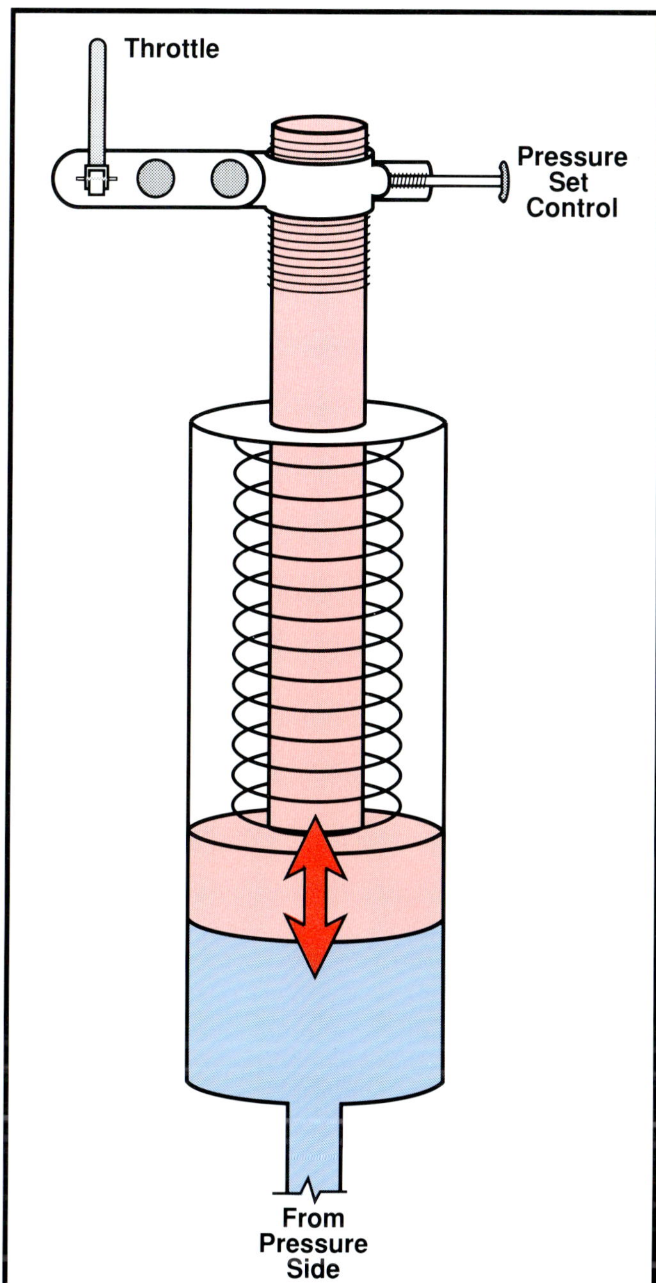

Figure B.8 The governor contains a piston that moves against a fixed spring.

pump discharge pressure increases above that point, the threaded rod will move. The linkage operates, and the throttle decreases and slows the engine to restore the pressure to the original setting. This governor, as normally installed, will not compensate for increased flow but will only decrease the engine rpm to prevent pressure surges.

Minor variations can be handled by increasing the normal throttle one or two turns after the governor is set. The discharge pressure should not change because, as the hand throttle is increased, the governor should decrease the pressure and remain constant. Minor variations will be compensated for by the governor. Major changes in flow will still involve resetting the hand throttle before the pressure can be brought back to the original setting.

Stored Pressure Reference Governor

Later American LaFrance pumpers use a more sophisticated governor. They operate as a stored pressure reference governor with a water governor control valve to control the operation of the governor. The governor shutoff valve takes it out of service, and the governor pressure gauge indicates the reference pressure that has been established in the pressure reservoir.

The governor control valve is shown in Figure B.9. This governor uses a balancing cylinder to control the operation of the throttle through a hydraulic clutch in the disengaged position (Figure B.10). It also uses a reference pressure cylinder

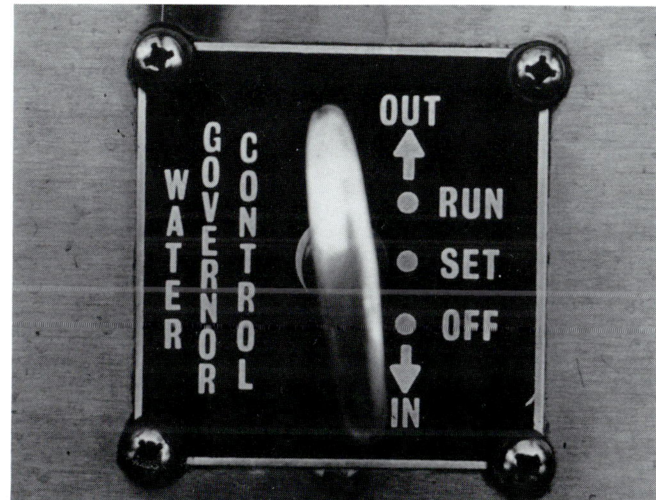

Figure B.9 Water governor control valve.

Figure B.10 The hydraulic clutch on a pressure governor system.

where the water pressure enters the cylinder under the discharge pressure of the pump and compresses the air inside the cylinder. This establishes a reference pressure that will be used to maintain the operating pressure by the governor.

In contrast to the other governor control, this governor needs to be set with minimum water flow. Operating pressure should be established by the hand throttle with the smallest line that will be used. An alternative way to do this is by flowing the booster line or even setting a tank fill valve to circulate a minimum amount of water. The governor will never reduce the engine throttle below the point at which it is originally set. It will only increase the engine speed.

To set the governor:

Step 1: Adjust the desired pressure with minimum flow.

Step 2: Pull the water governor control out to the SET position.

Step 3: Open the governor shutoff valve.

Step 4: Allow the system to stabilize until the governor gauge pressure reads the same as the discharge pressure. This will take some time, possibly as long as 60 seconds, for the stabilizing to take place.

Once the reference pressure has been established, move the water governor control valve to the RUN position, close the hand throttle, and the pressure should remain the same under the control of the governor. If the water supply is interrupted and the pump goes into cavitation when the pressure drops below 50 psi (350 kPa), the hydraulic clutch will disengage. The engine speed will drop back to an idle, preventing damage to the pump.

To change the operating pressure of this governor, open the hand throttle again until a resistance is felt or until the pressure on the discharge begins to increase. The water governor control valve should then be returned to the SET position and the throttle adjusted for the new pressure that is desired. Again, wait for the system to stabilize and the governor pressure gauge to become the same as the discharge pressure. Then operate the water governor control back to the RUN position. Close the throttle again.

To shut down the governor or to shut down operation of the pump with the governor in service:

Step 1: Turn the throttle out again until resistance is experienced or until the discharge pressure begins to increase.

Step 2: Set the water governor control valve to the OFF position.

Step 3: Reduce the speed of the engine slowly with the hand throttle back to an idle speed.

Step 4: Close the governor shutoff valve and open the governor drain valves to clear the system.

Effective use of any of these automatic pressure control systems depends on actual fireground experience or frequent practice during drills or routine inspections of the apparatus.

PRIMING DEVICES

Exhaust Primers

All American LaFrance pumps are equipped with some type of priming device. Older models may still be equipped with exhaust primers like the ones used on the 700 series and some commercial apparatus. These primers used the exhaust gases going through the exhaust system. The gases went through an ejector mechanism to create a vacuum that evacuated air from the fire pump and caused it to be primed (Figure B.11).

When the primary priming valve is opened, it opens a waterway from the suction side of the pump through the ejector mechanism that allows the air to be drawn from the pump. The vacuum is established by the venturi action of the exhaust gases passing by high velocity through the primer.

Two check valves are incorporated into the system. One prevents explosive gases from the exhaust system, such as a backfire, from being drawn into the pump. The other provides an escape of water under pressure if the primer is opened when the pump is receiving water from a relay or from a hydrant under pressure. This check valve allows the water to escape out onto the ground rather than be discharged into the exhaust system where damage might result.

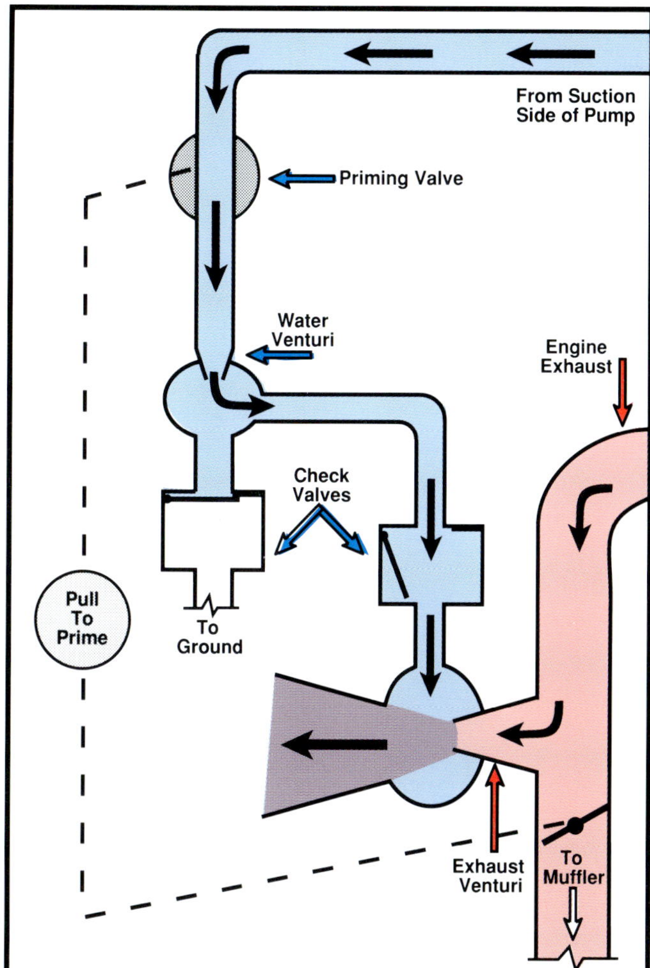

Figure B.11 Most older American LaFrance pumps utilize exhaust priming systems such as the one in this diagram.

To operate this primer, high engine rpm is required to get an adequate amount of velocity in the exhaust system. Usually the rpm needs to be 2,000 rpm or higher for the primer to operate effectively. Open the valve to deflect the exhaust gases through the primer instead of out of the tail pipe of the apparatus. This is frequently done by a separate lever, possibly with a black handle on it, mounted either on the pump panel or under the running board that controls this deflecting valve.

Since the exhaust gases go through the primer mechanism, the priming valve can then be opened. This valve usually has some kind of a T-handle mounted on the pump panel labeled PULL TO or some similar designation. When the priming valve is opened, the vacuum developed in the primer will bring the gases from the pump and discharge them out through the primer. When the pump is primed, as indicated by a buildup of pressure on the master discharge gauge, close the priming valve and open the exhaust deflector so the gases can be discharged normally through the exhaust system.

Positive Displacement Primers

Exhaust primers are very inefficient and will not tolerate any type of air leak in the system. They are also susceptible to deterioration in the exhaust system of the apparatus. Maintenance has been very high. Older American LaFrance pumps have been retrofitted with positive displacement primers like those on the later units.

These primers are electrically driven rotary vane positive displacement pumps (Figure B.12). Since the priming pump is electrically driven, the primer control serves two functions (Figure B.13). As the handle is pulled out, the priming valve is opened. This allows the priming pump to be con-

Figure B.12 These rotary vane priming pumps are much more reliable than the older exhaust primers.

Figure B.13 Primer control on the rotary vane priming system.

nected to the intake side of the fire pump. As the priming valve operates, the electrical switch also operates. This causes the electrical motor to start turning the priming pump and the prime to begin evacuating the air to the fire pump.

To use this primer, make sure the hard suction connections are airtight, set the pump for drafting, pull out on the priming control, and hold and operate it until the pump is primed. As the pump fills with water, pressure will begin to build on the master discharge gauge. At the same time, water is drawn through the primer and discharged on the ground underneath the apparatus. When this happens it is advisable to continue to operate the primer for a few seconds to evacuate any remaining pockets of air that might still be in the pump. Then release the primer, making sure it goes completely into the panel.

Positive displacement primers use oil both for lubrication and to provide a better seal in the positive displacement pump. Some are arranged so that the oil will enter the primer by gravity. Others require a primer pump operation (Figure B.14). In either event, it is necessary that the reservoir have an adequate supply of oil before operating from draft. The reservoir should always be refilled after each drafting operation.

Figure B.14 Operating the primer pump lubricator when beginning to prime the pump provides the lubrication necessary for proper operation of positive displacement primers.

DISCHARGE VALVES

The discharge valves used by American LaFrance are the locking type. The direct operating valve is locked by turning the handle clockwise (Figure B.15). The remote operating valve to control the discharge on the other side of the pumper is a push/pull type. It also contains a locking collar; it can be locked by turning clockwise and unlocked by turning counterclockwise.

Discharge valves should always be left unlocked when the pumper is not in service. Discharge valves should always be locked when being used to supply water to a hoseline. This becomes especially important if the discharge line is being operated with a valve partially closed and gated down to establish pressure differentials on various discharges. The movement of water through the valve will tend to cause the valve to change its setting if it is not locked adequately.

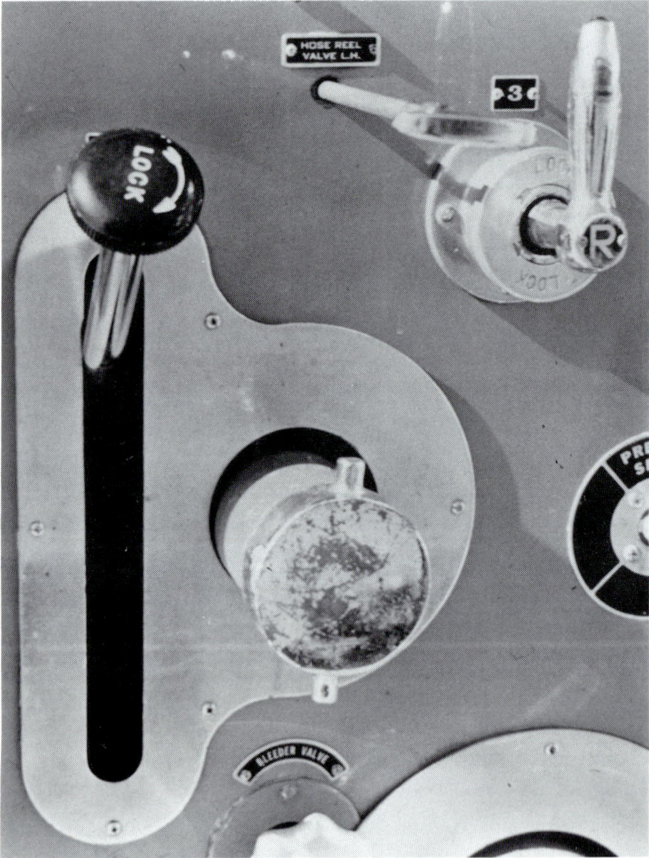

Figure B.15 Direct operating discharge valves use a quarter-turn locking mechanism. The remote hose reel valve in the top center of this picture is a push/pull type valve.

COOLING

Provisions are made to cool the pump and cool the engine. The pump is kept cool by maintaining some amount of flow even when all the lines are shut down. This may be labeled the booster line cooling valve or it may be labeled the cooling valve. Whatever its name, it maintains a constant amount of flow through the pump. The booster line cooling valve is a relatively small capacity line (Figure B.16). It will have a copper line of no more than ½- or ¾-inch (12 mm or 19 mm) diameter that flows not more than 5 gpm (19 L/min). This flow may be adequate, but if the pump is operated for extended periods of time with no water movement or at unusually high pressures, more flow may be required.

Additional water movement may be required to keep the pump from overheating. Pump overheating can be eliminated by opening the tank fill

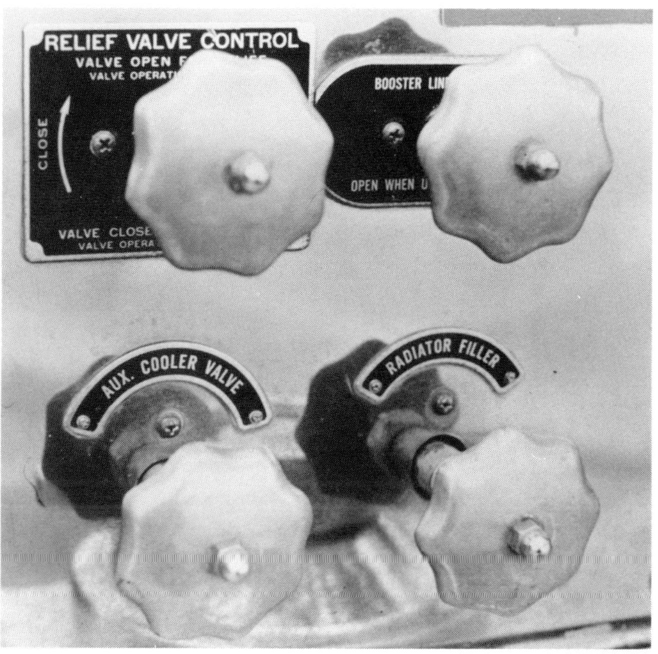

Figure B.16 The booster line cooling valve and auxiliary cooling valves are shown here.

valve (if the pump is so equipped). It can also be done by using a booster line to discharge water into the tank or in a safe place (if the pump is being supplied from a hydrant or from a relay).

Engine cooling can also be a problem if the pump is operated for extended periods of time under heavy loads. This is especially true during extreme weather conditions. An auxiliary cooler has been provided, generally a marine-type heat exchanger, and mounted somewhere on the apparatus. Some custom apparatus may have a heat exchanger actually installed inside the radiator. Whichever type has been supplied, an auxiliary cooler valve is provided to put the cooler in service. The auxiliary cooler valve should not just be left open; it is undesirable for the engine to run too cool. It only needs to be put in service when the engine begins to overheat. The auxiliary cooler valve will not contaminate the antifreeze solution in the radiator since the water from the pump never actually comes into contact with the coolant. It does its job through a heat exchanger that isolates the two types of water.

The radiator fill valve, which is shown right beside the auxiliary cooler valve, does contaminate the cooling system. When the radiator fill valve is opened, water will be discharged from the pump directly into the radiator. This valve should rarely be used as part of the fireground pumping operation. It should be used only when the fire pump is being supplied from a hydrant or some other source of clean water. It is provided for emergency use only and should not be needed during normal pumping operations.

Appendix C

Description and Operating Practices for Darley Fire Pumps

(Note: Unless otherwise indicated, all photos in this Appendix are printed courtesy of W.S. Darley & Co.)

The W. S. Darley Company manufactures a full line of fire pumps under the Champion name. They can be arranged for use as direct engine drive, front mount drive, midship mounted for a split shaft, or transmission PTO drive systems. They range from booster-type pumps with a capacity as low as 60 gpm (240 L/min) to high-capacity applications up to 2,000 gpm (8 000 L/min) for use in attack pumpers or water supply units.

Volume pumps are available as single-stage or two-stage pumps, midship or front mounted. There is an option available to add a completely independent third stage for discharge pressures up to 900 psi (6 300 kPa). A four-stage centrifugal pump is also offered that will provide discharge pressures up to 1,500 psi (10 500 kPa) for high-pressure fog fire fighting. One offering unique to the Darley pump line is a two-stage pump for front mount application. This allows the same versatility usually associated with midship pumps to be extended to front mount driven units.

A special gearing arrangement is available with two different systems to permit pump and roll operation at a reduced capacity, up to 250 gpm (1 000 L/min). Up to 1,000 gpm (4 000 L/min) is available for stationary pumping using the split shaft drive option with the same pump.

Darley pumps feature a wide range of options, which means that there is very little standardization in the way they are installed. Operating practices also vary with the method of installation. Along with the variety of drive systems and pump configurations, there are two different types of primers, two different relief valves, a pressure governor on some older units, and various power-operated accessories for pump engagement and two-stage transfer valves.

FRONT MOUNT PUMPS

The Darley Company offers a wide variety of front mount pumps (Figure C.1). The KDF series is a conventional single-stage front mount that can be set up for a pump capacity of 500, 750, or 1,000 gpm (2 000 L/min, 3 000 L/min, or 4 000 L/min). Others are constructed to get more than 250 psi (1 750 kPa) discharge pressure for high pressure fire fighting. The H series is designed to supply 100 gpm (400 L/min) at a discharge pressure of 500 psi (3 500 kPa). A variation of this basic pump, the HF series, will supply 500 gpm at 150 psi (2 000 L/min at 1 050 kPa) and develop 300 psi at 100 gpm (2 100 kPa at 400 L/min). The HF series is also designed for simplicity and minimum weight. These features are critical in such installations as a minipumper or a tanker (Figure C.2). For even more versatility, the RF, a two-stage front mount, will supply up to 750 gpm at 150 psi (3 000 L/min at 1 050 kPa) in PARALLEL and 60 gpm at 800 psi (240 L/min at 5 600 kPa) in SERIES.

The front mount pump is generally mounted behind the bumper, just in front of the radiator of the apparatus. The frame rails are extended ap-

Figure C.1 A Darley RF750 Model front mount pump.

Figure C.2 An example of a front mount pump installation.

proximately 2 feet (0.6 m), and the pump is mounted on a support between the rails. This permits easy access to the pump for operation and maintenance, and protects the pump from damage in case of a front-end collision. Power to drive the pump is provided by a power take-off unit fastened to the front of the crankshaft of the engine. The pump is independent of the drive train of the vehicle and its discharge pressure is controlled only by the engine speed, with no relation to road transmission or clutch. For this reason, the front mount pump can be used effectively for pump and roll operation.

Darley pumps use a floating type of drive system, getting power from the engine through a drive shaft with a universal joint on each end to allow for isolation between the pump and the engine. This makes the mounting of the pump much simpler. The pump is driven by a dry-type multiple plate clutch controlled by a lever on the side of the gear case.

Various gear ratios are available to match the required impeller shaft speed for the desired capacity to the torque and horsepower curve of the engine being used. The ratio may vary from 1.81:1 to 3.95:1. As a general rule, the gear ratio is chosen so that the rpm of the engine needed to supply the rated capacity of the pump will not exceed the equivalent of 35 mph (65 km/h) road speed in the direct drive position of the road transmission.

Three different pumps are used for front mount Class A pumpers. The RF model is a two-stage pump available as a 500 to 750 gpm (2 000 L/min to 3 000 L/min) pumper. The KDF model is a single-stage pump that will provide 500, 750, or 1,000 gpm (2 000 L/min, 3 000 L/min, or 4 000 L/min) capacity. The HF series described earlier also provides Class A capabilities.

To set the capacity, the gear ratio in the gear box is changed to turn the impeller faster when a higher rating is needed. The size of the inlet varies from 4 to 5 inches (100 mm to 125 mm), depending on rated capacity, and from two to four outlets will be provided on the discharge manifold. NFPA 1901 requires one 2½-inch (65 mm) outlet for every 250 gpm (1 000 L/min) of capacity, but this requirement can be met by the use of preconnected lines or rear discharges to supplement the outlets on the discharge manifold. In this way, a 1,000 gpm (4 000 L/min) pump may have only three outlets on the discharge manifold with a rear discharge providing the additional flow capability.

Piping, Valves, and Gauges

Older Darley pumps may have valves with quarter-turn nonlocking handles, but later models are equipped with self-locking ball-type valves (Figure C.3). To put in service, the valve can be turned to the desired position and will remain there. Each 2½-inch (65 mm) outlet should be equipped with a drain valve on the line side of the valve. In addition, an individual line pressure gauge may be

Figure C.3 This cutaway shows the Darley self-locking ball valves.

connected to the same location (Figure C.4). The gauge is important when two or more attack lines are in service simultaneously with different pressures required for each one. In this situation, the discharge pressure of the pump should be adjusted to the highest pressure that will be needed. Other lines can be adjusted to lower pressure by partially closing the valve until the individual line gauge registers the desired setting.

Figure C.4 Each discharge is equipped with a drain located on the underside of the discharge piping.

Darley pumps are equipped with an automatic check valve in the discharge manifold. The check valve will be closed anytime the pressure in the manifold exceeds the pressure being developed by the pump. This permits the pump to be primed with valves opened and discharges uncapped when drafting. Priming is possible because the vacuum in the pump causes the atmospheric pressure from the outside to keep the check valve closed and prevent air leaks.

The check valve can also cause pressure to be present in the discharge manifold when the pump is not in operation. If this pressure is not relieved when the pump is shut down (either by opening a discharge valve or individual drain valve), water under pressure can be retained within the manifold indefinitely. To prevent injury when removing a cap from a discharge outlet, the individual line drain valve should always be opened first.

Each pump is equipped with at least two gauges: a master pressure gauge and a compound gauge connected to the inlet of the pump. These gauges may be mounted directly on the pump housing itself, but this is not a good practice. The vibration that is always present in the pump will be communicated directly to the gauges. This vibration will make it difficult, if not impossible, to keep the gauges calibrated accurately. A better arrangement is to mount the gauges somewhere on the truck body, connected with flexible lines to the fire pump to isolate them from vibration. A remotely located gauge panel also provides the flexibility to install individual line gauges.

Wherever gauges are mounted, it is important to keep them properly calibrated. This is especially so with the compound gauge. The physical size of the vacuum scale is such that small errors in zero calibration can make the gauge useless.

A master drain valve is available but is not commonly used on front mount pumps. Individual fabricators use different types of valves and systems to drain the pump. Whatever the type, it is important that they be fully closed when the pump is not in use. Since they may be installed to discharge below the bumper, under the truck, it is possible for them to be open or leaking without the pump operator being aware of it. The pump operator should always be aware of any unexplained leakage under the vehicle and take the necessary steps to eliminate it as soon as possible. Drains that are not completely closed will make it difficult, if not impossible, to prime the pump from draft due to the resulting air leaks.

All of the Champion front mount pumps are equipped with a cored heating jacket that uses engine coolant to prevent freezing in cold weather. Because of this, the pump housing will always feel warm to the touch when the engine is running. This warmth should not be confused with the overheating that may occur if the pump is used for extended periods of time without adequate water movement.

Operating the Front Mount Pump

Immobilize the vehicle before putting the pump in service. Put the road transmission in the NEUTRAL position, and carefully operate the lock that is provided to ensure that the transmission does not inadvertently shift into a drive gear while the pump is operating. This is especially important if the vehicle is equipped with an automatic transmission. If the transmission is engaged or left in the DRIVE position on arrival, the parking brake will initially keep the vehicle from moving. When the throttle is advanced preparatory to putting the pump in operation, the torque developed may build to the point that the parking brake will no longer hold the vehicle and it will begin to move. The operator of a front mount is especially vulnerable because he or she is standing directly in front of the vehicle, where the pump controls are. If the vehicle should move forward, injury is likely. Set the parking brake and put blocks under the wheels as an additional safety factor.

With all valves closed and all caps in place, connect the water supply to the inlet side of the pump. Because of the automatic check valve, it is not necessary to have the main discharge valves closed. However, it is a good policy to keep them closed to prevent damage to the pump or injury to personnel due to a sudden discharge when pressure is first developed. An exception to this rule is that the priming action can be speeded somewhat by leaving one discharge open slightly. This allows a minimum amount of water to flow and the air that has been trapped in the pump to be exhausted from the housing.

If the pump will be operating from the tank on the apparatus, the tank-to-pump valves must be opened. The manner of installation of this piping will vary with the fabricator who has installed the pump. In general, the tank-to-pump line will be connected to one of the 2½- or 3-inch (65 mm or 77 mm) openings on the bypass manifold at the side of the pump.

If the pump will be expected to supply more than 400 gpm (1 600 L/min) from the tank, there will probably be more than one tank-to-pump line, each with its own valve installed where the line enters the pump. If both tank-to-pump valves are not opened before putting the pump in operation, the maximum flow that can be attained will be reduced. To prevent freezing in cold weather, the valve may be located at the outlet of the tank and controlled remotely from the pump operating position at the front of the vehicle (Figure C.5).

Another way to prevent freezing is to install a second valve at the outlet of the tank so that drains can be installed at the low point of the line. This will eliminate the danger of freezing while the apparatus is traveling over the highway. When the pump is taken out of service, both valves should be closed, the line drained, and the drain valve closed again. These actions will prevent problems when the pump is next placed in service. There is a potential danger with a second valve because both valves must be fully open to ensure an adequate supply of water to the pump when it is next used. This can be easily overlooked.

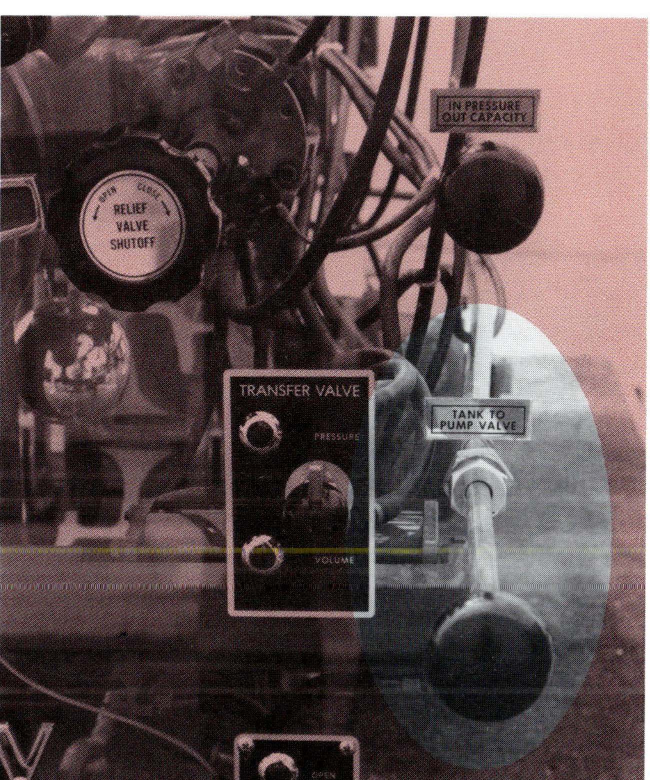

Figure C.5 This picture shows the location of the tank-to-pump valve on a midship pump installation.

Another opening on the bypass manifold will be used for a 2½-inch (65 mm) gated inlet to allow for water supply from a hydrant, from a relay, or from a tanker. This inlet will have some type of full flow valve and should also be equipped with a bleeder or drain valve on the line side of the main valve. This valve provides a means for bleeding off the air that has been trapped in the supply line as it fills with water before opening the valve into the pump. If the air is not eliminated, the pump will probably lose prime, resulting in pressure fluctuations when the transition is made from the tank on the apparatus to receiving water from an external supply.

After the pump has filled with water, either from the tank, from a hydrant, a relay, or priming from draft, the clutch must be engaged to allow the impeller to turn and build up pressure. With the engine operating at idling speed, slowly move the clutch lever, located on the right side of the gear case toward the radiator until it is fully engaged and the impeller is turning (Figure C.6). Maintain a firm grasp on the lever while moving it to prevent a snap engagement and sudden strain on the gears.

Disengagement should also be done with the motor at idling speed to minimize stress on the drive train and clutch. After the impeller is turning, increasing the engine rpm by turning the throttle control in a counterclockwise direction should cause the pressure gauge to register pressure on the discharge. If a discharge valve has been opened slightly, water will begin to flow. If this does not happen within a few seconds, air may be trapped within the pump housing and a momentary operation of the primer may remove it and speed the priming of the pump.

The throttle should be adjusted to build the pressure above 50 psi (345 kPa) before the discharge valves are opened to begin flowing water. Open the discharge valves slowly while observing the pressure gauge closely. If pressure drops greatly, pause a moment, and in severe cases, reoperate the primer to reestablish discharge pressure. When the desired discharge lines have been placed in service, the throttle can be adjusted to set the desired operating pressure.

With the pump in operation, there is always the danger of overheating when the demand decreases to the point where there is not an adequate flow of water through the pump to keep it cool. This is especially true in the early stage of the attack when it is common for lines to be operated intermittently. If the pump is equipped with a circulator system, it should always be used, either in the TANK or SPILL position, depending upon the mode of operation being used.

In some cases, the circulator system will not be enough to keep the pump cool. A circulator valve generally allows approximately 10 gpm (40

Figure C.6a The circle highlights the location of the pump clutch engagement lever.

Figure C.6b A close-up of the pump clutch engagement lever.

L/min) to flow. This will not be adequate if the pump is operating at high pressure for extended periods with no water moving. If the pump is equipped with a tank fill line, it can be used for cooling by opening the line partially to sustain enough water movement to prevent overheating. Front mount pumps often do not have tank fill lines due to space limitations. Space is limited when piping is installed from the front of the vehicle, where the pump is, to the rear of the apparatus, where the tank is located.

A good alternative is to use the hose reel for cooling. The smaller flow from this 1-inch (25 mm) line will not seriously detract from the pump capacity, but will provide sufficient flow to keep the pump cool. When drafting or operating from a hydrant, it is a simple matter to direct the stream from the hose reel back into the source or under the vehicle to discharge in a safe location. When operating from the tank, where the water supply becomes critical, the hose stream can be directed into the tank fill opening. Whatever cooling method is chosen, it is essential that an adequate flow be maintained to prevent damage to the pump from overheating.

Darley front mount pumps are equipped with one stuffing box for packing where the drive shaft leaves the gear box and enters the pump housing. When the packing is adjusted properly, water will drip slowly from the packing gland when the pump is operating, but not run in a steady stream. If water is leaking steadily, the packing is too loose and air will leak into the pump during drafting. This air leak may make it impossible to prime the pump from draft. On the other hand, if there is no water leaking, no cooling water is being supplied to the packing. The pump shaft will overheat, causing permanent damage if the pump is operated that way for any appreciable length of time.

Some older Darley pumps may be equipped with graphite packing in the stuffing box. Two nuts, one on each side of the packing gland, must be adjusted evenly to get a good seal on the packing. A good practice is to tighten each nut one-sixth of a turn, then allow the pump to run for some time to allow the packing to adjust itself around the shaft. Check the leakage rate again and feel the gland to check for overheating. If the leakage is still too great, tighten the nuts another one-sixth turn. If the gland feels too warm, loosen the nuts a fraction of a turn and allow the packing to adjust itself. Continue to adjust the packing until the proper amount of leakage is attained.

Most of the Darley pumps in service use a special plastallic packing that is injected into the stuffing box by a screw-operated plunger (Figure C.7). This type of packing also requires a slow dripping from around the shaft to maintain cooling and lubrication of the shaft during operation. If the water is leaking in a steady stream, the packing should be adjusted by a light tightening of the knurled plunger screw. This should only be done when the pump is primed and running at low speed. Be careful not to overtighten the packing. With the pump primed and operating at 150 psi (1 050 kPa) discharge pressure, water should be dripping from around the shaft, probably from 15 to 60 drops a minute. Additional packing can be added by removing the threaded cylinder and injection plunger. It should not be necessary to remove the old packing.

Figure C.7 This cutaway shows the location of the injection packing gland.

A few front mounts may be equipped with a master drain, but even these may have additional drain valves at low points on the piping. Most of them have individual drain valves located at low points, both on the pump itself and its associated piping. It is extremely important that front mount pumps be completely drained in freezing weather due to their vulnerable location on the front of the vehicle where they are exposed to wind and weather. It is a good idea to make sure all drains are closed again when the vehicle is in the station to prevent difficulties in priming when the pump is next used.

Various accessories are normally supplied with front mount pumps. These are installed in accordance with desired methods of operation. Since Darley pumps are installed by various fabricators, accessories from other types of fire pumps may be installed. Instructions for these can be found in the section devoted to that manufacturer's equipment.

RF series of front mounts are two-stage pumps and have a series-parallel control valve located at about the middle of the pump on the left side. When this valve is in the VOLUME position, the pump is arranged for a maximum volume of water at a discharge pressure up to 150 psi (1 050 kPa). In the PRESSURE position, the two stages are in series, and a higher pressure can be obtained at a reduced capacity. At 250 psi (1 750 kPa), 50 percent of the rated capacity is available, and the pump has the ability to go as high as 800 psi (5 600 kPa), but at a capacity of only 60 gpm (240 L/min). Positioning of this control is critical. If it is in the wrong position, the pump may not be able to supply the needed attack lines. As a general rule, anytime more than one 2½-inch (65 mm) outlet is in use, the pump should be in the VOLUME position. Initial setup is important since the control lever should not be moved with a net pump pressure higher than 50 psi (350 kPa) for safety reasons.

An additional feature available with the two-stage pump is a separate outlet intended for 1½-inch (38 mm) attack lines. This outlet always operates from the first stage, regardless of the setting of the transfer control. When the pump is operating in series to provide maximum pressure for the hose reel, a 1½-inch (38 mm) line connected to this interstage outlet would be operating at approximately one-half the discharge pressure. This eliminates the need to gate down this attack line to a more workable pressure.

MIDSHIP MOUNTED PUMPS

Midship mounted pumps offer much more flexibility in capacity and design since they are not limited to the small space that the front mount pumps have available for installation of piping and accessories. Midship mounted pumps can also provide higher rated flow since the full power of the engine is available to drive the pump. The maximum capacity of the front mounted unit is limited by the crankshaft bearings and the allowable torque capabilities of the PTO unit. Midship pumps are manufactured by Darley from 60 to 2,000 gpm (240 L/min to 8 000 L/min) capacity (Figure C.8).

Midship mounted pumps can be powered by a transmission mounted power take-off unit or a split shaft arrangement with a pump drive transmission located in the drive train between the truck transmission and the rear axle.

Figure C.8 A full view of a Darley midship pump.

Transmission PTO

A midship mounted pump can be driven by a transmission mounted PTO with power being transmitted to the pump through a drive shaft equipped with Universal joints. This provides a great deal of flexibility as to where the pump can be mounted on the chassis. The better the drive shaft on the pump lines up with the PTO, the less wear that can be expected. Since the PTO is driven from an idler gear in the transmission, it is not affected by the position of the other gears and can be used for pumping while in motion.

The PTO is under the control of the vehicle clutch, and the clutch must be disengaged when the PTO control is moved. The power rating of the PTO limits the size of the pump that can be used. In the past, PTO pumps were limited to about 250 gpm (1 000 L/min), but with the advent of more powerful engines and larger transmissions, higher capacity is possible. The Darley HM can be installed with a PTO to provide as much as 500 gpm (2 000 L/min). This makes it ideal for tankers and minipumpers.

Split Shaft Transmission

The split shaft transmission consists of a gear box inserted in the drive train between the road transmission and the rear axle. A sliding clutch gear can be moved forward to direct the power to the impeller shaft to operate the pump, or to the rear wheels for over-the-road travel. The shift can be controlled by a manual shift lever with a direct mechanical linkage to the control rod or by a vacuum or air-powered cylinder controlled by a selector switch.

If a power transfer arrangement is furnished, it will include a green indicator light that comes on when the shift is completed and power has been disconnected from the rear wheels and transferred to the pump. This is especially important on vehicles equipped with an automatic transmission. The parking brake will keep the vehicle from moving while the motor is operating at idle speed. When the automatic transmission is put into the DRIVE position, even after the pump has been engaged, the parking brake will initially keep the vehicle from moving. When the throttle is opened to build up pressure, torque increases, and the power is transmitted to the rear wheels if the shift has not completed. At approximately 1,500 to 2,000 rpm, sufficient torque is developed to overcome the drag of the parking brake. When this happens, the vehicle may move away from the operator. This can happen even if the wheels have been blocked before operating the pump. The vernier throttle has a red quick-release button in the center. The operator should push this button all the way in immediately if the truck begins to move.

Pump and Roll Operation

Darley uses two different methods of providing pump and roll capability with a midship split shaft pump. The SP model provides up to 1,000 gpm (4 000 L/min) operating in a stationary mode using the split shaft transmission. It also has a second drive shaft that can be connected to a PTO unit for pump and roll operation. When using the PTO for pump and roll, the capacity is limited to 250 gpm (1 000 L/min). This arrangement requires two different drive systems, one driven from the drive shaft and the other from a transmission PTO unit.

The SPR model also provides for pump and roll capability, but does it with only one drive system. The split shaft transmission has an additional sliding clutch gear that allows the pump to be driven through a different gear train for pump and roll. It also provides for the rear propeller shaft to be disconnected for stationary pumping and connected for pump and roll. While this system uses only one drive line, it has a much more complex gearing arrangement. Pump and roll mode of operation limits the maximum pressure to 250 psi (1 750 kPa) at a safe maximum of 200 gpm (800 L/min).

Putting the Pump in Operation

The vehicle should always be brought to a complete stop before the pump is put into operation. Even though the engine may not be driving the pump transmission because the clutch is disengaged or the automatic transmission is in neutral, movement of the rear wheels will cause the gears in the transfer case to turn. This will cause the gears to clash if the sliding gear is moved to transfer the drive from road to pump at that time. After stopping, immobilize the vehicle by setting the parking brake and putting blocks under the wheels at the earliest opportunity. The road transmission must be in the NEUTRAL position while the shift is made. If the vehicle is equipped with a manual transmission, the clutch must be disengaged while shifting either into the ROAD or PUMP position. With an automatic transmission, the gears will not be turning while it is in neutral and the transfer can take place.

If the vehicle is equipped with a manual transfer control, the shift lever must be moved all the way to the PUMP position and locked in place before putting the pump in service. If it is difficult to complete the shift, it may be because of pressure on the gears that will not allow the sliding clutch gear to move. To relieve the pressure, move the pump shift lever to the NEUTRAL position, halfway between the PUMP and ROAD position, and engage the clutch or put the automatic transmission in drive momentarily. Disengage the clutch or automatic transmission and try again.

With a power transfer, the green light on the dash and on the pump panel next to the throttle control will come on indicating it is safe to increase the throttle speed to put the pump in operation. If this light does not come on, it indicates that the shift has not completed or that there is a problem with the light circuit. Always verify that the shift has actually completed before increasing the throttle.

Put the road transmission in the recommended position. With a manual transmission, this will be in direct drive, the highest gear, unless the vehicle has an overdrive transmission. An automatic transmission will usually pump in DRIVE or 2-5 position. Once the transmission is in gear, lock it in place with the safety latch that is provided. Engage the clutch on a manual transmission. At this point, the speedometer should be registering a road speed of somewhere between 10 and 30 mph (16 km/h and 48 km/h), depending on the rpm of the motor at idle and the gear ratio being used.

If the truck is equipped with an automatic transmission and does not register any reading on the speedometer, depress the accelerator and verify that the speedometer registers. This reading on the speedometer with the vehicle stationary is an indication that power has been transferred from the rear wheels to the pump drive gears. There may be an audible indication that the shift has taken place. Listen for the sound of operation of the pump shift and of the gears turning in the pump transmission after the transfer has taken place as another indication that the pump is turning and ready for operation.

If the pump is equipped for pump and roll operation using a separate PTO unit, the mode of operation is determined by which pump engagement control is operated. If it uses the gearing within the pump transmission, additional operations are necessary. In this configuration, the mode of operation is controlled by two air pressure switches. For stationary pumping, use the same procedures as above, but make the shift by pulling out the stationary pump switch. The stationary pump light will come on, indicating that the gears have shifted fully. For pump and roll, leave the stationary pump switch in the IN position. Disengage the clutch, shift the truck transmission to its lowest gear. With the clutch disengaged, pull the pump and roll switch out. The pump and roll light should come on. Engage the clutch slowly. The vehicle will begin to move as the pump begins to turn. Closely observe the pressure gauge and limit truck speed so the pump discharge pressure does not exceed 250 psi (1 750 kPa).

If the pump will be operating from the water tank on the apparatus, it can be engaged and primed immediately upon arrival. If the water supply will be from draft, hydrant, or a relay, the pump should not be engaged until all connections have been made and the water is readily available. If the pump turns for any appreciable length of time before water is available to cool the packing, damage to the impeller shaft may result. Water will flow into the pump by gravity when operating from the tank, but in some cases, air will be trapped in the pump housing and cause difficulty in building up pressure. If this happens, momentary operation of the primer may speed up the operation.

Darley midship pumps use an injection packing system similar to the front mount, but some of them have two stuffing boxes instead of one

(Figure C.9). Adjustment criteria is the same, that is, water should be dripping from the shaft where it goes through each stuffing box, but not running a steady stream. Since midship pumps are usually enclosed by the body, it is more difficult to tell whether the packing is adjusted properly. Water may be leaking from a drain valve or other location and be confused with leakage from packing. Careful observation will be required to prevent damage from incorrectly adjusted packing.

Figure C.9 Some Darley midship pumps have two packing stuffing boxes.

Valves, Piping, and Gauges

Each of the Darley midship pumps can be supplied in different capacities. While the same impellers and other essential features in the construction of the pump are used, three variables are employed in changing the rating. Different gear ratios are available to match the engine torque curves to the required impeller speed, which will vary for different flow requirements. The suction inlet can be furnished with 4- to 6-inch (100 mm to 150 mm) threads for the suction hose that is needed to handle different flow rates. Finally, 2½-inch (65 mm) discharges are added to handle the additional hoselines for the increased water movement.

NFPA 1901 requires that one 2½-inch (65 mm) outlet or larger be provided for every 250 gpm (1 000 L/min) of capacity. This requirement can be met by the use of rear discharges or preconnected lines, but they must be controlled from the pump operator's panel by use of remote control linkages where required. Older Darley pumps may have valves with quarter-turn nonlocking handles, but later models are equipped with self-locking ball-type valves. These valves can be adjusted to any position and will remain there. Each valve should be equipped with a drain valve on the line side of the valve to relieve the pressure if the valve is closed and the hoseline shut down while the pump is still operating. An individual line gauge can be connected at the same point. These gauges may be installed on the pump panel at a location remote from the valve or outlet, associating them with the proper outlet.

Darley pumps are equipped with an automatic check valve in the discharge manifold (Figure C.10). The check valve will be closed anytime the pressure in the manifold exceeds the pressure inside the pump. The closed check valve permits priming the pump with the discharge outlets un-

Figure C.10 A cutaway of the automatic check valve located in the discharge manifold.

capped. This is because the vacuum in the pump creates a differential with the atmospheric pressure outside that will keep the check valve closed. The check valve can also cause pressure to be present in the discharge manifold when the pump is not in operation. If the pressure within the manifold is not relieved when the pump is shut down by opening a discharge valve or individual drain valve, water under pressure can be retained within the manifold indefinitely. To prevent injury when removing a cap from a discharge outlet, the individual line drain should always be opened first.

Centrifugal pumps are prone to heating when operating with little or no flow through them. Large midship pumps present a problem, especially in the early stages of a fire attack when hose streams are applied intermittently, because of the amount of water moved by the impellers as they churn. Operating at high pressure also causes a problem. As a general rule, the higher the pressure, and the larger the pump, the more water that has to move to prevent overheating.

If the pump has been equipped with some type of circulator valve to prevent overheating, it should always be left open except when priming the pump to operate from draft. It should be recognized, however, that a circulator valve will probably not provide enough water movement to prevent overheating when high pressure is required for extended periods of operation with little or no water flowing. The tank fill line, where provided, or a small hoseline, can make it possible to move enough water to operate indefinitely without damage to the pump.

When operating from the tank on the apparatus, opening the tank fill line too far can subtract from the flow available from the tank to the point that there will not be sufficient capacity when nozzles are opened. When setting the tank fill valve for cooling, open it far enough to get an appreciable water flow, but not enough to have a significant impact on the discharge pressure.

In addition to the large inlet to the pump, each midship pump is also equipped with at least one gated 2½-inch (65 mm) fitting. These are connected to companion fittings on the inlet fitting and are independently controlled. The 2½-inch (65 mm) inlets greatly restrict the capacity of the pump and are used mostly for relay operations. These operations are limited in flow rate anyway by the size of the supply hoselines. These gated inlets are equipped with a drain or bleeder valve on the line side of the gate valve. This allows the air that was trapped in the hoseline as it fills with water to bleed off harmlessly before it enters the fire pump and causes fluctuation in the discharge pressure.

The large inlet can also be gated. A butterfly valve is available that mounts behind the panel. This can be used when setting up to operate from draft while supplying attack lines from the tank. Closing the valve allows the cap to be removed from the large inlet and suction hose connected without interrupting the flow from the tank. If a front or rear suction has been provided, it too will be gated. It may not be possible to operate at full capacity from a front or rear suction. Due to the difficulty in routing the line from the pump to the front or rear of the vehicle, the size of the line may be reduced to the point that it restricts the flow. If this is the case, the operator needs to be aware of these limitations when using this inlet for drafting.

The pump panel will always be equipped with at least two gauges. One is a master pressure gauge connected directly to the discharge manifold of the pump. The other is a compound gauge connected directly to the inlet manifold. Both of these gauges, as well as the individual line gauges, can be compound gauges since a centrifugal pump has a vacuum throughout the system while priming, but only the inlet gauge must be compound. Some installations have a lower maximum pressure on the intake gauge than on the discharge, but in others they may be identical. A tachometer must also be installed on the pump panel, along with various engine indicators such as oil pressure, water temperature, and an ammeter.

Most midship pumps are equipped with a master drain that should drain all critical points of the pump. This is a drain valve and not a pressure relief valve. It should never be opened when the pump is pressurized, since damage may result to the drain valve by doing so. When there is a complex system of piping with associated low spots in the system, additional drain valves may be installed. These are usually located under the running board for convenience of installation and operation. During freezing weather, it is important that all these valves be opened when draining the pump; otherwise, water trapped within the pump or piping may freeze and cause physical damage. It is a good practice to close all these drain valves after the pump has been drained so that the pump is ready for service. Air leaks that will prevent a pump from being primed are often caused by open drain valves that were not closed after the pump was last drained.

Darley midship pumps also use a cored heating jacket to prevent freezing. Some use the radiator coolant for heating in a manner similar to the front mount. Others use the exhaust gases for this purpose. With either system, the pump casing will always be warm when the engine is running.

TWO-STAGE PUMPS

Darley two-stage pumps are conventional in design. They use two identical impellers, each designed to handle half the rated capacity of the pump. They can be arranged in series for maximum pressure with reduced volume, or in parallel for maximum volume at reduced pressure. The transfer is controlled by a lever marked PRESSURE and CAPACITY. When the control is moved, a bronze valve rotates within a bronze sleeve to direct the discharge of the first stage, either to the intake of the second stage for maximum pressure or to the discharge manifold for maximum capacity (Figure C.11). In the SERIES position, there is a suction check valve which operates to isolate the intake of the first stage from the intake of the second stage to allow for increased pressure in the second stage.

Due to mechanical limitations of the shift mechanism, it is difficult to make the transfer with pressure in the pump. For reasons of safety to personnel and equipment, this transfer should never take place when the net pump pressure exceeds 50 psi (345 kPa). A vacuum or air pressure powered transfer control is also available. If this is the case, the transfer is made the same way, by putting the

Figure C.11 The transfer valve.

control in the PRESSURE position for series operation, or in the CAPACITY position for parallel operation.

The setting of the transfer control is very important. The best guide for use of this control is the number of hoselines being supplied. Since the pump is designed to be used in the PARALLEL position when more than 50 percent of the capacity is needed, it should be set to the CAPACITY position anytime more than one-half the required outlets are in use. This means that anytime a 500 to 750 gpm (2 000 L/min or 3 000 L/min) pump is supplying more than one line it should be in CAPACITY. A 1,000 or 1,250 gpm (4 000 L/min or 5 000 L/min) could supply two lines and a 1,500 gpm (6 000 L/min) or larger could supply three hoselines before changing to CAPACITY.

HIGH-PRESSURE OPERATION

A third stage can be added to the SE two-stage pump for high-pressure operation. This additional stage will provide up to 100 gpm (400 L/min) at 900 psi (6 300 kPa) pressure or 200 gpm (800 L/min) at 600 psi (4 200 kPa). The sliding gear clutch that drives the third stage is operated by a shift lever marked *fog* for pumping position and *out* for disengaged. The shift lever must be latched in one of these positions before the main pump is engaged, and should never be moved while the volume pump is running.

To put the third stage in operation, make sure the third stage discharge line is closed tightly and the clutch is latched in the PUMP position. Prime the pump in the normal manner, making sure the pump is fully primed before engaging the main pump. Adjust the pressure to the set point by using the hand throttle. Any changes must be made slowly due to the high pressure being developed. The third stage has a small bypass line going to the water tank on the apparatus. Always open this bypass valve when operating the third stage.

SINGLE-STAGE PUMPS

Single-stage pumps have traditionally been associated with low-capacity front mount or PTO pumps. Darley single-stage pumps are available for higher capacity than the two-stage pumps they offer, as high as 2,000 gpm (8 000 L/min). To get the large flow capability, double suction impellers are used. These provide a better hydraulic balance as well as increased capacity.

Operation of the single-stage pump is much simpler than the two-stage pump since there is no possibility of being in the wrong position at a critical stage in the pumping operation. Maintenance is also easier, since there is no transfer valve, suction check valve, or transfer mechanism.

A single-stage pump is more efficient than a two-stage when it is operating at or near 100 percent of its rated capacity, but fire pumps rarely are called upon to operate in this range. At low flow rates, such as would be encountered when supplying one or two handlines, the two-stage pump is more efficient. This is not a significant concern in fire apparatus today. The diesel engines in common use for fire apparatus have sufficient reserve power that inefficiency is not important. In fact, the single-stage pump is very well suited for use with a diesel engine. Many diesel engines are difficult to control at low rpm under light load. The higher rpm required to develop the necessary pressure when coupled to a single-stage pump instead of a two-stage pump provides a smoother and better controlled operation.

PRIMING DEVICES

Darley pumps use vacuum from the intake manifold of the engine for priming as a standard installation. They also have an electrically driven primer that is optional for auxiliary use with the vacuum primer, as the only primer on diesel engines, or when so specified by the customer.

Vacuum Primers

Older pumps use a vacuum primer with a single float. A line from the intake manifold of the engine is connected through the primer to the intake side of the pump. When the pump is primed and full of water, water is forced into the primer by the pressure differential created by the engine vacuum. When the primer fills with water, the float valve rises to close off the line to the intake manifold, thus preventing water from getting into the engine.

Later models of the vacuum primer have a second float added (Figure C.12). This provides an added safeguard to prevent water from entering the engine after the pump is primed. If the first float fails to properly seal off the vacuum line, water enters the second chamber and the second float valve will do the same job. Under normal operation, the second float never operates since the primary one closes off the line before any water enters the secondary chamber.

Figure C.12 This picture shows a later model primer that utilizes two floats.

The pump should be primed before the impeller begins to turn. To prime the pump, all booster line valves, tank fill valves, drains, and other openings into the pump must be closed. Main discharge gates can be left open since the automatic check valve will prevent air from getting into the pump. Front mount pumps generally use a rotary-type priming valve, while midship models use a panel-mounted push/pull control. All newer models use push/pull type valves.

To prime the pump, leave one discharge gate slightly open and uncapped, and open the primer shutoff valve with the pump shift set to the PUMP position. Leave with the road transmission in neutral. Since vacuum is highest in a gasoline engine at low speeds, set the engine for a fast idle, approximately 500 to 600 rpm. Open the primer shutoff valve and wait about 15 seconds for the pump to fill with water. Close observation of the suction hose during this period should show that the hose begins to sag from the weight of the water. When the primer valve is opened, the engine will probably run rough due to the drop in vacuum in the intake manifold as air is drawn from the pump through the primer.

After the pump has been filled with water, put the road transmission in direct drive and start the pump turning. Within 15 seconds, a steady stream of water should be discharged from the pump. Close the primer shutoff valve, increase the engine speed until the discharge pressure climbs above 50 psi (350 kPa), and open the discharge gate slowly until water is flowing. If pressure is lost while opening the discharge gates, reprime the pump by opening the primer shutoff valve until the pressure comes back up.

If for some reason the primary float valve in the primer does not close when the pump is primed and the secondary float operates, it will not be possible to reprime the pump until the vacuum is broken by venting the line. With the rotary valve, this can be done by closing and then opening the shutoff valve again. With the panel mounted valve, venting is accomplished by pushing the lever in past the OFF position against the spring.

Rotary Vane Priming Pump

The electric primer uses a rotary vane positive displacement pump driven by an electric motor to evacuate the air from the pump. The primer shutoff valve not only opens the line from the priming pump into the fire pump, but includes an electric switch on the valve to start the electric motor running when the valve is opened. This primer uses an oil reservoir to provide a better seal within the rotary vane pump for more effective priming. The oil reservoir also lubricates the parts of the pump to ensure a more reliable operation. The oil reservoir holds 6 quarts (5.7 L) and should be kept full of SAE 30 motor oil. The line from the reservoir to the priming pump has a #62 drill vent hole at a high point to stop the siphon action when the priming pump is not running. This hole must be kept open at all times.

The priming procedure for the fire pump is the same with the electric primer as it is for the vacuum primer. When a steady stream of water is discharged from the priming pump, close the primer shutoff valve. Do not run the primer any longer than necessary.

If the vacuum primer and an electric primer are connected in parallel, opening the primer shutoff valve will connect the vacuum to the pump. The electric primer may then be started and used to prime the pump faster or enable it to operate from a higher lift. A separate control will be supplied for this purpose.

AUTOMATIC PRESSURE CONTROL DEVICE

Some of the oldest Darley pumps may use a pressure governor for automatic pressure control. Later models use an automatic relief valve. Two versions of this valve are presently available.

Pressure Governor

The pressure governor limits pressure surges by controlling the engine speed of the vehicle. Spring tension is adjusted by means of a handwheel to limit the maximum pressure that will be developed by the pump. When the pressure of the pump is great enough to overcome the tension of the adjustable spring, a piston within a cylinder will move. The piston is linked to the carburetor so that movement of the piston will change the speed of the engine. In some installations, the governor will actually be mounted on or near the carburetor, and it may be necessary to open the hood to make the adjustment.

To use the pressure governor, all lines that will be supplied must be flowing and the pressure set to the desired level by use of the hand throttle. Then the governor shutoff valve is opened to allow water to flow to the governor and put it into operation. Turn the adjustment handwheel counterclockwise until engine speed decreases. Then slowly turn it clockwise until engine speed comes back to the established rpm and pressure returns to the desired level. The governor now has control. If one of the lines is shut down, pressure will rise, causing the piston to move. The piston in the governor will move far enough to reduce the engine speed to bring the pressure back down to where it was set. If the line is opened again and the pressure drops, the governor will return to its original setting and the pressure will rise. If the pressure drops when the governor shutoff valve is opened, it indicates that the pressure adjustment handwheel is set below the operating pressure. Turn the valve off, turn the handwheel clockwise several turns, then try again.

Pressure Relief Valve

Front mount pumps have a pressure relief valve mounted directly on the pump itself (Figure C.13). The pilot valve is located within the same housing as the main relief valve. It is mounted between a port on the discharge side of the pump and the bypass suction manifold on the side of the pump. There is a strainer in the line from the discharge of the pump to the pilot valve that has to be kept clean. This can be done by opening the flush valve on the strainer while operating the pump from a clean water supply. This relief valve is controlled entirely by the adjustment handwheel and no other shutoffs have been provided. It is a dump-type relief valve, and reacts to pressure surges by opening the waterway from discharge to intake, increasing the flow, and bringing the pressure back to the set point (Figure C.14).

Figure C.13 A pressure relief valve mounted directly on a front mount pump.

Figure C.14 A cutaway of the pressure relief control valve.

To set the relief valve, first turn the handwheel clockwise to raise the setting higher than the operating pressure to be used. One turn of the handwheel raises the operating point approximately 50 psi (350 kPa). Put all hoselines in operation flowing the desired amount, then set the pressure to the desired operating point with the hand throttle. Turn the handwheel counterclockwise until the pressure drops 5 to 10 psi (35 kPa to 70 kPa) below the set point. Then slowly turn the handwheel clockwise, allowing time for the pressure to stabilize as it changes, until the gauge returns to the set point. Any pressure changes due to decreases in flow will be compensated for by the relief valve and should not affect the operating pressure. If additional lines are put in service and the pressure drops, the hand throttle will have to be increased to bring the pressure back to the set point. If the required operating pressure changes, repeat this procedure.

Remote Control Pressure Relief Valve

The remote control relief valve is a dump-type relief valve and operates on the same general principles as the one-piece valve (Figure C.15). This device has the main relief valve mounted on the pump between the discharge manifold and the inlet manifold, but the pilot valve is mounted on the pump panel. Since the pilot valve and the relief valve are connected with flexible tubing, this allows for maximum flexibility in mounting.

Figure C.15 The remote control pressure relief valve operates on the same principle as the one-piece relief valve.

The pilot valve gets its supply of water from the pump through a strainer from the discharge side of the pump (Figure C.16). On some of the older Darley relief valves, the strainer is accessible only from behind the panel. The current relief valve strainer is easily accessible from the operator's panel for cleaning or replacement. There is a relief valve flush valve on the operating panel that enables the strainer to be flushed during operation. This should be done often when the pump is operating from clean water.

A relief valve shutoff is also part of the pilot valve assembly. This is a four-way valve that supplies water directly from the discharge of the pump to the main relief valve when it is in the OFF position. This keeps the relief valve closed, regardless of the setting of the pilot valve. When this four-way valve is in the ON position, water goes through the pilot valve to the relief valve. If the discharge pressure exceeds the setting of the pilot valve, it will open and allow water to be dumped back into the inlet side of the pump. When enough water has been allowed to escape through the drain on the pilot valve, the relief valve will open. This allows water to bypass from the discharge side of the pump to the intake to bring the pressure back to the set point.

In addition to the shutoff valve, there are two mechanical locks to make the relief valve inoperative. When the adjustment handwheel is turned all the way clockwise, the pilot valve is made inoperative and will not open regardless of the pressure. There is also a mechanical lock for the relief valve itself. Turning the relief valve shutoff control all the way clockwise holds the relief valve closed. This lock is provided in case the strainer becomes clogged, a line breaks, or the water supply from the pump to the relief valve is interrupted. If this is the case, the relief valve would stay open and it would be impossible to build up pressure. Under ordinary operating conditions, this shutoff will seldom, if ever, be used. It must be kept fully open to allow sufficient travel of the piston to handle the capacity of the pump.

The remote control relief valve may be equipped with pilot lights to indicate whether the valve

Figure C.16 A cutaway showing the pilot portion of the relief valve.

is open or closed. This can be a help in setting the valve since the light may operate before a change in pressure on the discharge gauge is noted. It will also provide a quick indication to the operator as to whether the valve is open or closed at any given time.

To set the remote control relief valve, adjust the pump to the desired pressure with all hoselines flowing and the four-way valve in the OFF position. Turn the pressure adjust handwheel several turns clockwise. Turn the four-way valve on. If the pressure drops, it indicates that the relief valve is set below the operating pressure. The handwheel will have to be turned clockwise until the pressure returns to the set point. If the pressure does not change when the four-way valve is turned on, turn the handwheel counterclockwise until the pressure drops 5 to 10 psi (35 kPa to 70 kPa). Slowly turn the handwheel clockwise, allowing time for the pressure to stabilize as it changes, until the pressure returns to the set point. The relief valve is now set and will prevent the discharge pressure from rising above that for which it has been set. If a different operating pressure is desired, repeat the setting procedure. If additional lines are placed in service and the pressure drops, readjust the hand throttle until the pressure returns to the set point.

If the pump is equipped for high-pressure operation, above 250 psi (1 750 kPa), the pressure adjustment handwheel will have to be turned to the maximum clockwise position to keep the valve from opening when operating in this range.

AUXILIARY COOLING

There is no standard arrangement for auxiliary cooling with Darley pumps. However, all UL-rated pumps have a special screened outlet to ensure against any debris entering the auxiliary heat exchanger. Most of them use a marine-type heat exchanger mounted in the engine compartment or some other convenient location. In this type of cooler, coolant from the radiator passes through copper tubing that is immersed in water being supplied from the fire pump. Water inside the housing cools the coolant in the tubing by conduction.

The water from the pump is controlled by a shutoff valve on the pump panel that can be used to keep the operating temperature of the engine within the desired range. The auxiliary cooler will not contaminate the coolant within the system and can be used anytime it is needed without concern for damage to the system. The auxiliary cooler may be equipped with a separate drain valve, often in some location remote from the other drains. If this is the case, special attention will be needed to ensure it is completely drained in freezing weather to prevent damage.

There may be a radiator fill valve furnished. If there is, it will allow water from the fire pump to be supplied directly into the radiator. This will dilute the antifreeze solution. If the pump is supplying contaminated water, serious damage to the cooling system may result. The radiator fill valve is an emergency control and should only be used in situations where it is required to protect firefighters or when human life is endangered.

Appendix D

Description and Operating Practices for FMC Pumps

(Note: Unless otherwise indicated, all photos in this Appendix are printed courtesy of Bill Eckman.)

FMC fire pumps were first manufactured under the "John Bean" label. They originally came to the fire service as high pressure fog fire pumps. A volume pump was added to the high-pressure pump to provide more versatility. Pumpers were available in the "HP" version equipped with a high-pressure pump only, in the "V" version, equipped with a volume pump only, or as the "HPV" model, with both a high-pressure and a volume pump. The high-pressure pump is a positive displacement pump, constructed with three cylinders, designed to operate at 850 psi (5 860 kPa). The John Bean volume pumps are two-stage pumps, available in capacities from 500 to 1,000 gpm (2 000 L/min to 4 000 L/min). FMC does not manufacture a front mount pump, but will install a Hale front mount pump when specified. Other FMC pumpers have been equipped with Hale midship pumps when a higher capacity was required, a single-stage pump was specified, or when requested by the customer. Operation of pumpers equipped with Hale fire pumps is covered in the section on Hale pumps (Appendix E). FMC now offers a single-stage pump known as an FMC Ram fire pump that will provide a Class A rating from 750 to 1,500 gpm (3 000 L/min to 6 000 L/min) without changing the basic components.

John Bean and FMC volume pumps are driven by a midship transfer arrangement and cannot be operated while in motion. The high-pressure pump is driven by a transmission mounted PTO and can be used for pump and roll operation.

FMC pumps are normally installed only on apparatus constructed by the fire apparatus division of the FMC Corporation. They are installed on both commercial and custom chassis and have a number of different, but standardized, installations. They use conventional relief valves for automatic pressure control and a positive displacement pump for priming.

HIGH-PRESSURE PUMPS

The FMC high-pressure pump is operated from a transmission-mounted PTO unit. It can be supplied from the tank on the apparatus, under pressure from a hydrant or relay, or from draft. It has a preset relief valve that limits the discharge pressure to no more than 850 psi (5 860 kPa). Since it is a piston-type positive displacement pump, it is self-priming and needs no external primer. When used in conjunction with a volume pump, it can be used to prime the volume pump in case of failure of the priming system.

Since the high-pressure pump has a maximum capacity of approximately 60 gpm (227 L/min), it can only be used to supply two hose reels containing ¾-inch (19 mm) high-pressure hose. If more than two reels are needed, an additional pump will have to be installed (Figure D.1). If a second pump is installed, it will require its own set of gauges and controls.

The high-pressure pump is designed so that the engine operates at a constant rpm. While in operation, the engine speed must be kept within 50 rpm of the absolute speed indicated on the nameplate mounted on the pump panel near the tachometer (Figure D.2). The pressure then will

Figure D.1 Units with more than two high-pressure booster lines will have to be equipped with additional sets of high-pressure gauges and controls.

Figure D.2 The pump panel should be equipped with a plate that indicates the maximum engine speed for high-pressure operation.

be maintained at or near 850 psi (5 860 kPa) by the high-pressure relief valve. A manual pressure release valve is furnished to allow a gradual buildup when the pump is initially put in service (Figure D.3). This valve bypasses the relief valve and allows water from the high-pressure pump to escape into the water tank on the apparatus to limit the pressure buildup.

Water enters the high-pressure pump from the tank through a 3-inch (77 mm) tank shutoff valve. The normal position of the tank shutoff valve is *in* to the panel with the valve open and water

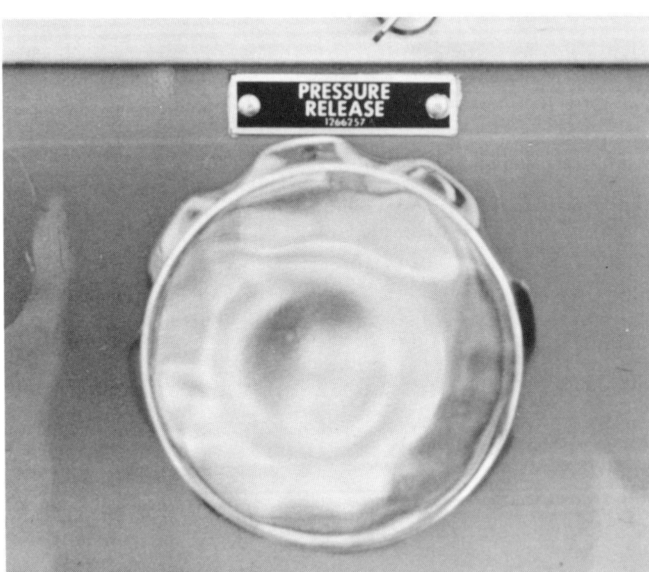

Figure D.3 The pressure release valve allows water from the high-pressure pump to escape into the water tank on the apparatus to limit the pressure buildup.

being supplied to the intake of the high-pressure pump. If it is desired to run with the pump dry during freezing weather, the tank shutoff control must be pulled *out,* which closes the valve, and opens all drains. Figure D.4 shows the new style tank valves.

Most high-pressure pumps will have an optional gated 2½-inch (65 mm) intake to connect another water source without shutting down the pump. This arrangement will allow the pump to operate from a relay or a hydrant, as well as from the water tank on the apparatus. By use of hard suction hose connected to this intake, it is also possible to draft from a portable reservoir or static source.

Engagement of the transmission PTO can be by a manual or a power shift control, operated by air pressure derived from the air brake system on the vehicle, or by engine vacuum. Late models of

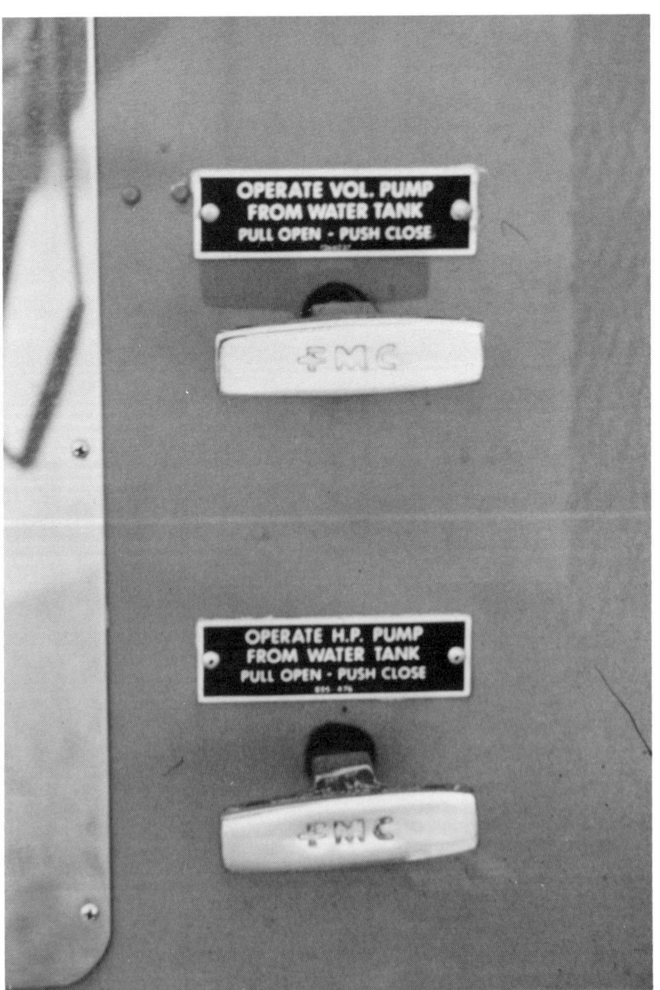

Figure D.4 The high-pressure pump and volume pump may have separate controls for the tank-to-pump water supply lines.

either type control have an indicator light to let the operator know that the PTO is fully engaged. Earlier models do not have an indicator light; therefore, it is essential that the operator closely observe other indications to make sure the PTO has been properly engaged (Figure D.5).

Only one pressure gauge is provided for the high-pressure pump. It has a scale reading from 0 to 1,000 psi (0 kPa to 6 985 kPa) and is connected to the discharge of the pump. A tachometer is also furnished and a tank level gauge to provide a measure of the water remaining in the tank that can be used.

Figure D.5 An example of an older power transfer control without an indicator light.

Operating the High-Pressure Pump

To put the pump in service, immobilize the vehicle. With the apparatus in position, set the parking brake and position the wheel chocks. Engage the PTO unit to start the pump turning.

If the vehicle is equipped with a manual transmission, put the road transmission in neutral. With the clutch depressed, engage the PTO by operating the shift control. With a manual control, turn the handle counterclockwise to unlock it. Pull the handle out as far as possible, then lock it in position by turning it clockwise. If the vehicle is equipped with a power shift, lift the locking gate and operate the switch to the right. Lock the switch in the OPERATE position by use of the locking gate. Slowly release the clutch. The PTO engaged light located near the handle should come on to show that the PTO is fully engaged. If it does not, depress the clutch and try again.

With an automatic transmission, set the engine speed to idle and put the transmission in neutral. Operate the PTO pump shift control while shifting the automatic transmission into reverse or drive at the same time with the other hand. Shift the transmission back into neutral and the pump should be engaged. If the PTO engaged indicator is not lit, repeat the process.

With the tank shutoff valve in the OPEN position, increase the engine speed slowly with the throttle control until the tachometer indicates the required rpm for high pressure operation. Slowly close the pressure release control. As the pressure builds up when the pressure release valve is closed, it may be necessary to readjust the throttle to maintain the desired engine rpm. When the valve is completely closed, the discharge pressure should be 850 psi (5 860 kPa). If the pressure exceeds 850 psi (5 860 kPa), reduce the engine rpm to prevent damage to the apparatus and the high-pressure relief valve should be checked as soon as possible. If the pressure is less than 850 psi (5 860 kPa) with the correct engine rpm, the relief valve may be stuck open, and should be inspected.

Open the hose reel valve on the line or lines that will be used and unlock the hose reel brake. Pull out the needed hoseline and relock the hose reel brake to prevent additional unwinding of the reel. The pump is now ready for service.

To shut down the high-pressure pump, rewind the hose reel, while still under pressure, and slowly close the hose reel valve. Turn the throttle clockwise to bring the engine to an idle while slowly opening the pressure release valve at the same time. Disengage the PTO shift control and lock in the RELEASED position.

To operate the high-pressure pump from a relay or a hydrant, use the same procedure as pumping from the tank except for the intake controls. To change over from tank to pressurized source, make the connection to the gated 2½-inch (65 mm) intake fitting. When connections are made and the source of water is in place, open the 2½-inch (65 mm) intake valve. Open the volume pump tank-to-pump valve, then slowly close the tank shutoff valve. (**NOTE:** If the operation is from a pressurized source from the beginning, all of these changes should be made before putting the pump in operation. All other procedures are the same.)

The high-pressure pump can be operated to draft from a folding tank, a stream, or other static source. First, position the vehicle in a safe location as close to the water source as possible. The lift should be no more than 20 feet (6 m). Set the parking brake and position the wheel chocks with the transmission in neutral. Attach the hard suction hose to the 2½-inch (65 mm) gated intake to the pump. A strainer, preferably the float dock type, should be connected to the end of the hard suction and all connections must be airtight. If a barrel strainer is being used, it should be submerged at least 2 feet (0.6 m) below the water and 2 feet (0.6 m) off the bottom. Pull out the tank shutoff control to completely close the valve. Open the 2½-inch (65 mm) intake valve and the volume pump tank-to-pump valve. Engage the high-pressure pump. Increase the engine speed to 1,200 to 1,500 rpm. Within 15 to 20 seconds, the pressure gauge should register a slight pressure. Slowly increase the throttle until the tachometer indicates the required rpm for high pressure operation. Slowly close the pressure release valve. If there is a pulsation or hammering noise in the pump, there are probably pockets of air trapped inside the pump. Rapidly open and close the pressure release valve one-half turn 8 to 10 times in succession. If the pulsation persists, check for air leaks.

For pump and roll operation of the high-pressure pump, close the pressure release valve fully and open the hose reel valve. Engage the high-pressure pump, put the transmission in the lowest gear, and drive. Maintain a steady speed, as close to the high-pressure pumping engine rpm as possible. Never exceed the absolute engine speed specified.

MIDSHIP MOUNTED PUMPS

John Bean or FMC volume pumps are mounted midship and can be supplied with capacities from 500 to 1,500 gpm (2 000 L/min to 6 000 L/min) (Figure D.6). They use a split-shaft arrangement with a pump drive transmission located in the drive train between the truck transmission and the rear axle.

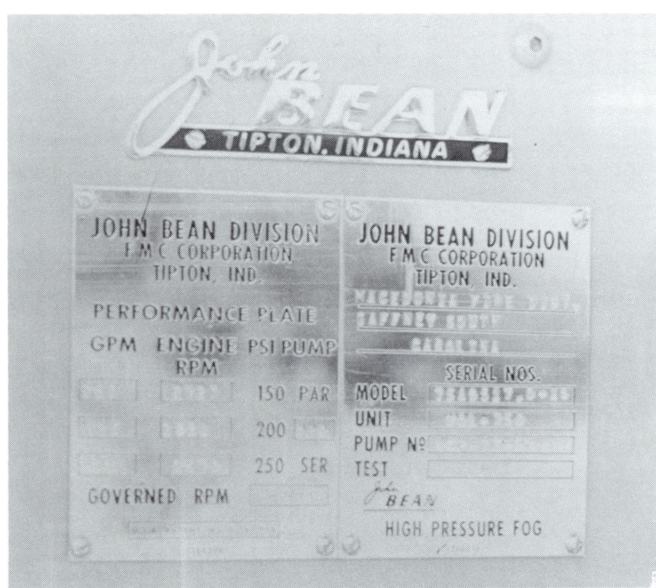

Figure D.6 The pump panel performance plate for a Bean volume pump.

Split Shaft Transmission

The split shaft transmission consists of a gear box inserted in the drive train between the road transmission and the rear axle. A sliding clutch gear can be moved forward to direct the power to the impeller shaft to operate the pump, or to the rear to provide power to the wheels for over-the-road travel. The shift can be controlled by a manual shift lever with a direct mechanical linkage to the control rod or by a vacuum-powered or air-powered cylinder controlled by a selector switch. Other apparatus may be equipped with an electric shift control activated by a toggle switch located inside the cab. If a power transfer arrangement is furnished, it will include a green indicator light that comes on when the shift is completed. The light indicates that power has been disconnected from the rear wheels and transferred to the pump. This is especially important on vehicles equipped with an automatic transmission.

The parking brake will keep the vehicle from moving while the motor is operating at idle speed. When the automatic transmission is put into the DRIVE position, even after the pump has been engaged, the parking brake will still keep the vehicle from moving. When the throttle is opened to build up pressure, torque will be increased, and the power transmitted to the rear wheels if the shift has not completed. At approximately 1,500 to 2,000 rpm, sufficient torque will be developed to overcome the drag of the parking brake, and the vehicle may move away from the operator. This can happen even if the wheels have been blocked before operating the pump. The vernier throttle has a red quick-release button in the center. The operator should immediately push this button all the way in if the truck begins to move.

Putting the Pump in Operation

The vehicle should always be brought to a complete stop before the pump is put into operation. Even though the engine may not be driving the pump transmission because the clutch is disengaged or the automatic transmission is in neutral, movement of the rear wheels will cause the gears in the transfer case to turn. This will cause the gears to clash if the sliding gear is moved to transfer the drive from road to pump at that time. After stopping, the vehicle should be immobilized by setting the parking brake and putting blocks under the wheels at the earliest opportunity. The road transmission must be in the NEUTRAL position while the shift is made. If the vehicle is equipped with a manual transmission, the clutch must be disengaged while shifting either into the ROAD or PUMP position. With an automatic transmission, the gears will not be turning while it is in neutral and the transfer can take place. If there is an automatic transmission brake, it should also be engaged to stop the gears from turning while the transfer is made.

If the vehicle is equipped with a manual shift, engage the pump and disengage the rear axle by turning the handle counterclockwise. Pull out on the control to the end of its travel, then lock it in position by turning the handle clockwise. If the manual control will not complete its travel, there may be pressure on the gears that will not allow the sliding clutch gear to move. To relieve the pressure, move the pump shift control to the NEUTRAL position, halfway between the PUMP and ROAD position, and momentarily engage the clutch or put the automatic transmission in drive. Disengage the clutch or automatic transmission and try again. With a power transfer, the green light on the dash should glow to indicate that the shift has completed. If this light does not come on, either the shift has not completed or there is a problem with the light circuit.

Put the transmission in the proper gear for pumping as designated on the face of the shift control box. This will normally be in the direct drive gear with a manual transmission or in the DRIVE or 3-6 position of an automatic transmission. Lock the transmission in position with the shifter lock. If the transmission is in the proper gear, the shifter lock should drop in place easily. Do not force it. If it does not move in place, the transmission may be in the wrong gear.

Always verify that the shift has actually completed before increasing the throttle. This can be done by observing the speedometer while releasing the clutch on a manual transmission. The speedometer should read 5 to 15 mph (8 km/h to 24 km/h) on the speedometer. If it does not, the shift may not have completed, or the road transmission could be in the wrong gear. When the vehicle is equipped with an automatic transmission, the speedometer should read some slight speed indication. If it does not, momentarily use the accelerator to increase the engine speed. The speedometer should then register a reading in proportion to the speed of the engine. The pump panel will also have an indicator light to ensure that the shift was complete (Figure D.7).

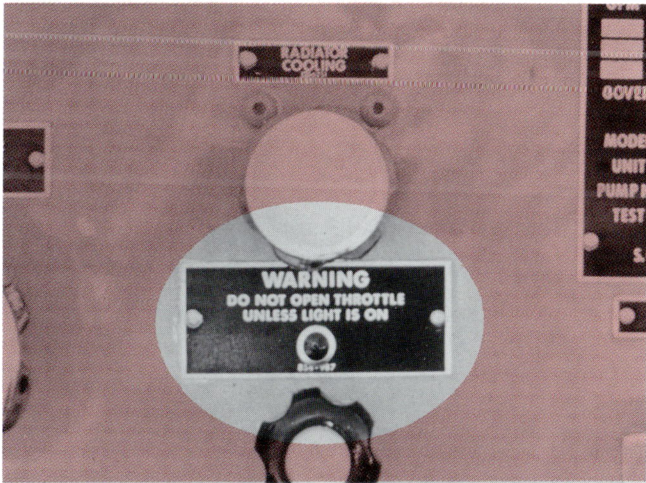

Figure D.7 The circle highlights the power transfer indicator light. Do not increase the throttle if this light is not on.

If the pump will be operating from the water tank on the apparatus, it can be engaged and primed immediately upon arrival. When the water supply is from draft, hydrant, or relay, the pump should not be engaged until all connections have been made and the water is readily available. If the pump turns for more than 2 minutes at 800 rpm or higher or for more than 5 minutes at idle before water is available to cool the packing, damage to the impeller shaft may result. Water will flow into the pump by gravity when operating from the tank, but in some cases, air will be trapped in the pump housing and cause difficulty in building up pressure. If this happens, momentarily operating the primer may speed up the operation.

FMC pumps use a single stuffing box for packing the pump. The packing consists of graphite impregnated packing rings, a lantern ring to apply water for cooling and lubrication, and a packing gland to adjust the packing to the proper pressure. When the packing is properly adjusted, water should be dropping from the shaft where it goes through the stuffing box, but not running a steady stream. Since midship pumps are usually enclosed by the body, it is difficult to tell whether the packing is adjusted properly. Water may be leaking from a drain valve or other location and be confused with leakage from the packing. Careful observation will be required to prevent damage from incorrectly adjusted packing.

To adjust the packing, set the pump to operate with a discharge pressure of 150 psi (1 034 kPa). In order to prevent overheating, the tank fill valve can be partially opened to assure a constant flow of water through the pump while adjusting the packing. Observe the leakage rate from around the pump impeller drive shaft. Water should be dropping from around the shaft, approximately 12 to 60 drops a minute, but not running a steady stream. If the leakage is excessive, tighten the package by adjusting the two elastic stop nuts holding the packing gland in place. It is important that adjustment of the two nuts be made evenly to ensure a good seal by the packing. It is a good idea to adjust each of the nuts one-sixth of a turn at a time. Then wait for the packing to seat before making further adjustments. This procedure can be repeated until the leakage is at the proper rate. Be careful not to overtighten the packing or overheating and damage to the shaft may result.

Valves, Piping, and Gauges

Each of the FMC midship pumps can be supplied in different capacities. While the same impellers and other essential features in the construction of the pump are used, three variables are employed in changing the rating. Different gear ratios are available to match the engine torque curves to the required impeller speed, which will change for different flow requirements. The suction inlet can be furnished with 4- to 6-inch (100 mm to 150 mm) threads to accept the suction hose needed for different flow rates. Finally, 2½-inch (65 mm) discharges are added to handle the additional hoselines for the increased water movement.

NFPA 1901, *Standard on Automotive Fire Apparatus*, requires that one 2½-inch (65 mm) outlet or larger be provided for every 250 gpm (1 000 L/min) of capacity. This requirement can be met by using rear discharges or preconnected lines, but they must be controlled from the pump operator's panel. Remote control linkages can be used where required. Older FMC pumps will have valves with quarter-turn locking handles (Figure D.8). Turn the handle clockwise to lock the valve in position or counterclockwise to unlock it.

It is a good idea to leave the valves in the UNLOCKED position when they are closed, but

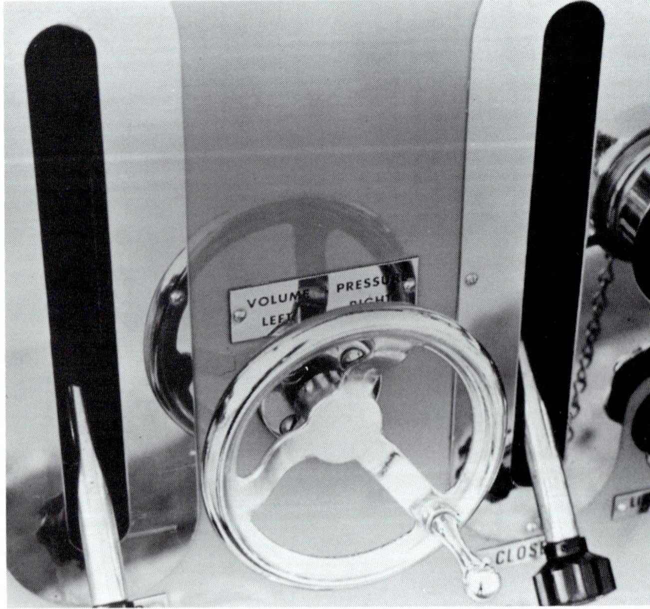

Figure D.8 Older Bean pumps are equipped with quarter-turn, locking discharge valve controls. Also shown here is the handwheel-type volume-pressure transfer control.

always lock them in position when they are opened and in use. This is particularly important when operating with a valve partially closed to reduce the pressure on an individual line. Locking the valve in place is important since movement of the water through the valve will tend to open it further and increase the pressure to the nozzle. Later models are equipped with self-locking ball-type valves (Figure D.9). These valves can be adjusted to any position and will remain there as long as pressure is being developed by the pump. Each valve should be equipped with a drain valve on the line side of the valve to relieve the pressure if the valve is closed and the hoseline shut down while the pump is still operating. An individual line gauge can be connected at the same point. These gauges may be installed on the pump panel away from the valve or outlet, but they will be marked with an identifying number to associate them with the proper outlet (Figure D.10).

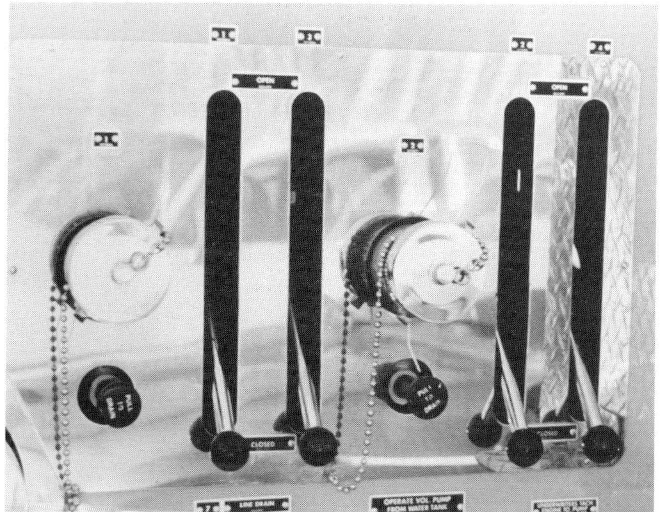
Figure D.9 An example of the newer style self-locking valves.

Centrifugal pumps are prone to heating when operating with little or no flow through them. Large midship pumps present a greater problem, especially in the early stages of a fire attack when hose streams are applied intermittently, because of the amount of water moved by the impellers as they churn. Operating at high pressure also causes a problem. As a general rule, the higher the pressure, and the larger the pump, the more water that has to move in order to prevent overheating. If the pump is equipped with some type of circulator valve to prevent overheating, it should always be left open except when priming the pump to operate from draft. Be aware, however, that a circulator valve will probably not provide enough water movement to prevent overheating when high pressure is required for extended periods of operation with little or no water flowing. The tank fill line, where provided, or a small hoseline, can make it possible to move enough water to operate indefinitely without damage to the pump. When operating from the tank on the apparatus, opening the tank fill line too far can lessen the flow available from the tank to the point that there is insufficient capacity when nozzles are opened. When

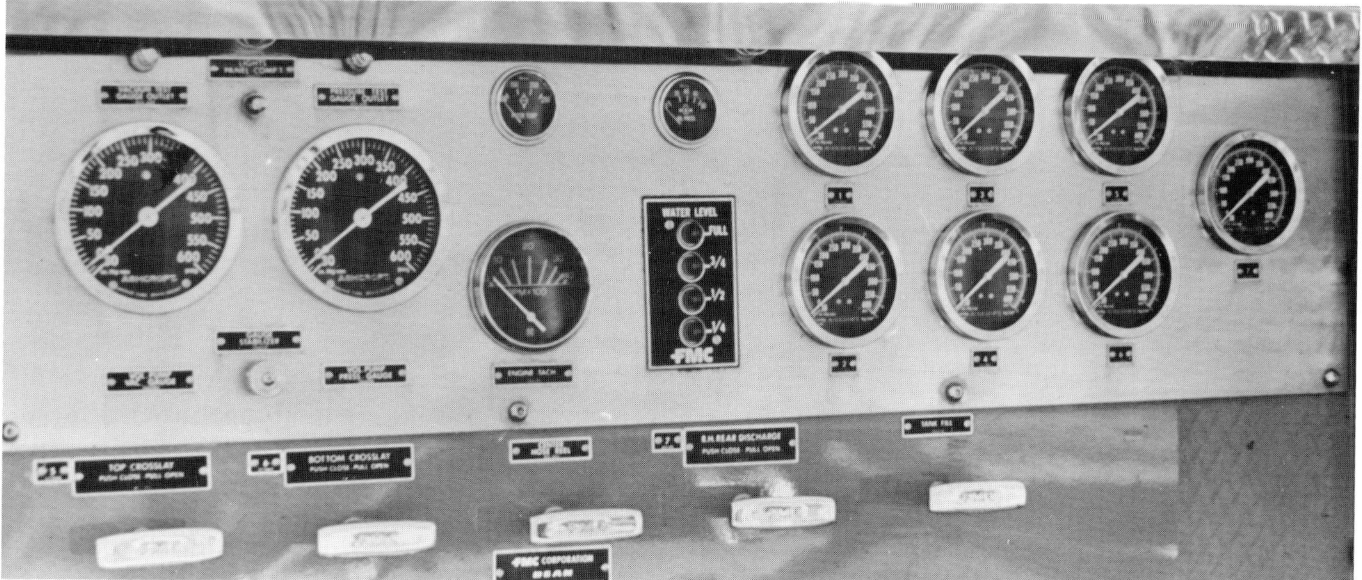

Figure D.10 A view of the gauge arrangement on the FMC pump panel.

setting the tank fill valve for cooling, open it far enough to get an appreciable water flow, but not enough to have a significant impact on the discharge pressure.

In addition to the large inlet to the pump, each midship pump is also equipped with at least one gated 2½-inch (65 mm) fitting. These are connected to companion fittings on the inlet fitting and are independently controlled. The 2½-inch (65 mm) inlets greatly restrict the flow of the pump and are used mostly for relay operations that are limited in flow rate by the size of the supply hoselines. These gated inlets are equipped with a drain or bleeder valve on the line side of the gate valve. This allows the air trapped in the hoseline to bleed off harmlessly as the hose fills with water. Bleeding off air is necessary before the water enters the fire pump and causes fluctuations in the discharge pressure.

If a front or rear suction has been provided, it will probably be gated. It can be used when setting up to operate from draft while supplying attack lines from the tank. Closing the valve allows the cap to be removed from the large inlet and the suction hose connected without interrupting the flow from the tank. It may not be possible to operate at full capacity from a front or rear suction. Due to the difficulty in routing the line from the pump to the front or rear of the vehicle, the size of the line may be reduced to the point that it restricts the flow. If this happens, the operator needs to be aware of the limitations when using this inlet for drafting. An external butterfly valve can be used on the intake manifold if no other gated intake has been provided to make this transition.

The pump panel is always equipped with at least two gauges. One is a master pressure gauge connected directly to the discharge manifolds of the pump, and the other a compound gauge connected directly to the inlet manifold (Figure D.11). Both of these gauges, as well as the individual line gauges, may be compound gauges since a centrifugal pump has a vacuum throughout the system while priming. Only the inlet gauge may be compound. Some installations have a lower maximum pressure on the intake gauge than on

Figure D.11 The master pressure and compound gauges.

the discharge, but in others the pressures may be the same. A tachometer must also be installed on the pump panel, along with various engine indicators such as oil pressure, water temperature, and an ammeter.

A pressure gauge stabilizer control may be supplied on the pump panel. This valve can be used to reduce fluctuations of the pressure gauge needle during pumping operations. Partially close this control as needed to stabilize the needle. If the control is completely closed, there will be no reading on the pressure gauge of the pump. Additional stabilizer valves for individual line gauges may be found behind the gauge panel. To adjust these, the panel needs to be opened and the stabilizer valves adjusted to minimize fluctuations in these gauges.

Most midship pumps are equipped with a master drain that should drain all critical points of the pump. This is a drain valve and not a pressure relief valve. It should never be opened when the pump is pressurized, since damage may result to the drain valve. Where there is a complex system of piping with associated low spots in the system, additional drain valves may be installed. These are usually located under the running board for convenient installation and operation. During freezing weather, it is important that all these valves be opened when draining the pump or water trapped within the pump or piping may freeze and cause physical damage. It is a good practice to close all drain valves after the pump has been drained so that the pump is ready for service. Air leaks that will prevent a pump from being primed are often caused by open drain valves that were not closed after the pump was last drained.

TWO-STAGE PUMPS

FMC two-stage pumps are conventional in design except that they do not require a suction check valve. They use two identical impellers, each designed to handle half the rated capacity of the pump. They can be arranged in series for maximum pressure with reduced volume, or in parallel for maximum volume at reduced pressure. The transfer is controlled by a handwheel marked PRESSURE and VOLUME. When the control is moved, a bronze rotating manifold valve turns. This connects the intake of the second stage either to the discharge of the first stage for maximum pressure or to the intake manifold for maximum capacity.

Due to mechanical limitations of the shift mechanism, it is difficult to make the transfer with pressure in the pump. For reasons of safety to personnel and equipment, this transfer should never take place when the net pump pressure exceeds 50 psi (345 kPa). An electric-powered transfer control is also available. If this type of control has been supplied, a toggle switch and two indicator lights will be on the pump operations panel. Push the switch down and hold until the pressure light comes on. In some cases, the indicator lights may also be supplied with a manual transfer control. This provides an added measure of certainty that the control has been operated fully. If the electric changeover mechanism should malfunction, a manual override is provided. To use the manual override, open the gauge panel door and remove the safety pin from the rod connected to the manual override handwheel inside the pump compartment. Turn the handwheel fully to the desired position and continue the operation (See Figure D.8).

The setting of the transfer control is very important. The best guideline for its use is the number of hoselines being supplied. Since the pump is designed to be used in the PARALLEL position when more than 50 percent of its capacity is needed, it should be set to the VOLUME position anytime more than half the required outlets are in use. This means that anytime a 500 or 750 gpm (2 000 L/min or 3 000 L/min) pump is supplying more than one line it should be in volume. A 1,000 or 1,250 gpm (4 000 L/min or 5 000 L/min) could supply two lines and a 1,500 gpm (6 000 L/min) or larger could supply three hoselines before changing to volume.

SINGLE-STAGE PUMPS

Single-stage pumps have traditionally been associated with low-capacity front mount or PTO pumps. FMC Ram single-stage pumps are available at the same capacity as the two-stage pumps they offer, as high as 1,500 gpm (6 000 L/min). To achieve the large flow capacity, FMC uses double suction impellers. They also use two different stripping edges to take water from the discharge of the impellers 180 degrees apart in the housing. These two features provide a good hydraulic balance in the pump as well as high capacity.

Operation of the single-stage pump is much simpler than the two-stage pump since there is no possibility of being in the wrong position at a critical stage in the pumping operation. Maintenance is also easier, since there is no transfer valve or shift mechanism.

A single-stage pump is more efficient than a two-stage when it is operating at or near 100 percent of its rated capacity, but fire pumps rarely are called upon to operate in this range. At low flow rates, such as would be encountered when supplying one or two handlines, the two-stage pump is more efficient. The diesel engines in common use for fire apparatus have sufficient reserve power that inefficiency is not important. In fact, the single-stage pump is very well suited for use with a diesel engine. Many diesel engines are difficult to control at low rpm under light load, and the higher rpm required to develop the necessary pressure when coupled to a single-stage pump provides a smoother and better controlled operation.

PRIMING DEVICES

FMC fire pumps have a positive displacement pump for priming. Two different types have been used. Older John Bean volume pumps may use a rotary gear-type primer driven from the pump transmission. Other pumps use a rotary vane primer driven by an electric motor. Both primers use a single control that opens the priming valve and engages the primer drive system to make the primer start turning (Figure D.12).

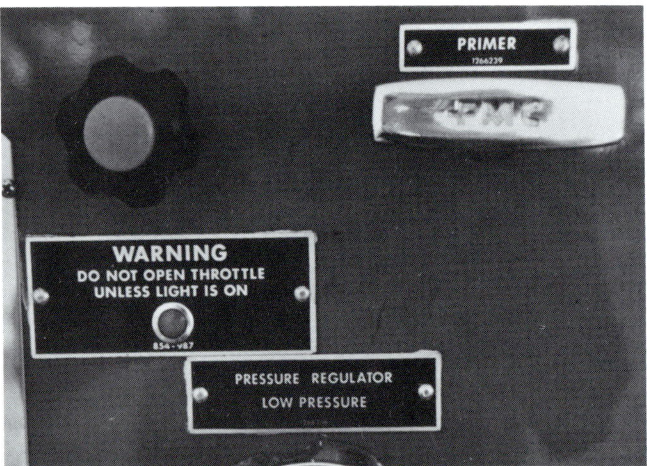

Figure D.12 The primer control valve is located near the throttle control.

Rotary Gear Priming Pump

This primer is belt driven from a sheave mounted on the volume pump impeller shaft extension. The rotary gear pump uses 20W motor oil to provide a better seal as well as to lubricate the pump during priming while it is running dry. The priming oil reservoir holds 1 gallon (4 L) of oil and must be closely checked to see that it is kept full. Since the oil reservoir is generally mounted higher than the priming pump, oil would continue to flow into the pump by siphon action after priming is completed. The copper line from the reservoir to the priming pump has a small vent hole at the high point of the line to break the siphon and prevent this from happening.

The priming valve handle opens the priming valve and at the same time closes the clutch face against the belt sides. This mechanism must operate freely to prevent the priming valve from sticking open and causing the volume pump to lose its prime when the priming pump stops turning. To prime the pump, it is only necessary to pull out on the priming valve handle until the priming pump begins to turn. This can be verified by the sound of the air being ejected from the priming pump under pressure.

Rotary Vane Priming Pump

The electric primer uses a rotary vane positive displacement pump driven by an electric motor to evacuate the air from the pump. The primer shutoff valve not only opens the line from the priming pump into the fire pump, but includes an electric switch on the valve to start the electric motor running when the valve is opened. This primer uses an oil reservoir to provide a better seal within the rotary vane pump for more effective priming and to lubricate the parts of the pump to ensure a more reliable operation. The oil reservoir holds 1 gallon (4 L) and should be kept full of SAE 20 motor oil. The line from the reservoir to the priming pump has a small vent hole at a high point to stop the siphon action when the priming pump is not running. This hole must be kept open at all times.

Priming the Pump

The procedure for priming the pump is the same for either type of primer. Position the vehicle as close to the water source as possible, set the parking brake, and position the wheel chocks. Be sure that all pump drains and other openings to the pump are closed. Attach the hard suction hose to the suction intake fitting. Attach a strainer to the end of the suction hose. Some type of floating strainer is preferable. If none is available, the barrel type strainer must be submerged under at least 2 feet (0.6 m) of water and should be suspended 2 feet (0.6 m) from the bottom. Verify that all connections on the suction hose are airtight.

Engage the pump, set the transmission to the proper gear, and set the transfer valve to the VOLUME position if the volume pump is a two-stage pump. Increase the engine speed with the hand throttle to approximately 1,500 rpm. Pull out on the prime control and hold it out until the pump is primed. When the pump is primed, pressure is indicated on the gauge and water is discharged from the primer under the vehicle. This should happen in about 15 or 20 seconds. If the pump is not successfully primed within 30 seconds, stop priming and check for air leaks. Operating the primer for more than 30 seconds can result in damage to the primer or the pump.

When the pump has filled with water, increase the engine speed with the throttle control until the pressure gauge registers at least 50 psi (345 kPa). Slowly open the desired discharge gate while increasing the throttle to attain the required operating pressure. If the pump loses prime while opening the discharge gates, wait a short time for it to stabilize and regain the pressure. Momentary operation of the primer when the pressure drops may evacuate the air pockets and enable the pump to continue to operate.

AUTOMATIC PRESSURE CONTROL DEVICES

Pressure surges are limited on FMC volume pumps by use of a pressure regulator control. The pressure regulator is actually an automatic relief valve. Two different types are in use. One is a single unit mounted on the pump with the pilot valve and relief valve built into the same housing. The other is a remote control valve with a pilot valve mounted on the pump panel and the relief valve mounted on the pump.

PRESSURE RELIEF VALVE

Some of the older John Bean volume pumps have a pressure relief valve mounted directly on the pump itself. The pilot valve is mounted within the same housing as the main relief valve. It is mounted between a port on the discharge side of the pump and the suction manifold of the pump. This relief valve is controlled entirely by the adjustment handwheel and no other shutoffs have been provided. It is a dump-type relief valve, and reacts to pressure surges by opening the waterway from discharge to intake. This increases the flow and brings the pressure back to the set point.

The discharge pressure of the pump acts against the face of the relief valve, attempting to open it. Water from the discharge side of the pump is supplied through the pilot valve to the opposite side of the relief valve, holding it closed. When the discharge pressure becomes greater than the force of the spring in the pilot valve, the pilot valve opens. This allows enough water to bleed off from the opposite side of the relief valve to let the discharge pressure force it to open. The increased flow that is now being bypassed from discharge to intake reduces the net pump pressure and brings it back to the set point.

To set the relief valve, first turn the pressure regulator handwheel clockwise to raise the setting higher than the operating pressure to be used. Put all hoselines in operation flowing the desired amount, then set the pressure to the established operating point with the hand throttle. Turn the handwheel counterclockwise until the pressure drops 5 to 10 psi (35 kPa to 70 kPa) below the set point. Then slowly turn the handwheel clockwise, allowing time for the pressure to stabilize as it changes, until the gauge returns to the set point. Any pressure changes due to decreases in flow will be compensated for by the relief valve and should not affect the operating pressure. If additional lines are put in service and the pressure drops, the hand throttle will have to be increased to bring the pressure back to the set point. If the required operating pressure changes, repeat this procedure.

REMOTE CONTROL PRESSURE RELIEF VALVE

The remote control relief valve is a pressure operated type relief valve with the pilot valve mounted on the pump panel. This device has the main relief valve mounted on the pump between the discharge manifold and the inlet manifold, and water will be transferred back to the intake when the valve opens. Since the pilot valve and the relief valve are connected with flexible tubing, there is maximum flexibility in the position where it can be mounted.

When the pump is dry, the main relief valve is held closed by a biasing spring mounted outside the valve itself. As pressure builds up in the pump, the pressure acting against the face of the relief valve keeps it closed because of the way in which it is constructed. Water from the discharge side of the pump is supplied to a pilot valve that is held closed by an adjustable spring. Spring tension is set by a pressure regulator handwheel on the pump panel. When the discharge pressure of the pump increases enough to overcome the spring tension in the pilot valve, it will open, allowing water to pass through to the opposite side of the main relief valve. This will cause the relief valve to open and allow water to bypass from the discharge of the pump back into the intake, reducing the pressure. The relief valve will open far enough to allow water to bypass until the discharge pressure is brought back to the set point.

Either of these relief valves may be equipped with pilot lights to indicate whether the valve is open or closed. This can be a help in setting the valve since the light may operate before a change in pressure on the discharge gauge is noted. It will also quickly indicate to the operator whether the valve is open or closed at any given time.

To set the remote control relief valve, adjust the pump to the desired pressure with all hoselines flowing. Turn the pressure regulator handwheel several turns clockwise. Turn the handwheel counterclockwise until the pressure drops 5 to 10 psi (35 kPa to 70 kPa). Slowly turn the handwheel clockwise, allowing time for the pressure to stabilize as it changes, until the pressure returns to the set point. The relief valve is NOW set and will prevent the discharge pressure from rising above the pressure for which it has been set. If a different operating pressure is desired, repeat the setting procedure. If additional lines are placed in service and the pressure drops, readjust the hand throttle until the pressure returns to the set point.

AUXILIARY COOLING

There is no standard arrangement for auxiliary cooling with FMC pumps, but most of them use a marine-type heat exchanger mounted in the engine compartment or in some other convenient location. In this type of cooler, coolant from the radiator passes through copper tubing that is immersed in water being supplied from the fire pump. The coolant within the tubing is cooled by conduction to the water inside the housing. The water from the pump is controlled by a shutoff valve on the pump panel. This valve can be used to keep the operating temperature of the engine within the desired range. This valve may be labeled radiator cooling not to be confused with the radiator fill valve. The auxiliary cooler will not contaminate the coolant within the system, and can be used anytime it is needed without concern for damage to the system. The auxiliary cooler may be equipped with a separate drain valve, often at a location remote from the other drains. If this is the case, special attention will be needed to ensure it is completely drained in freezing weather to prevent damage.

There may be a radiator fill valve furnished. If there is, it will allow water from the fire pump to be supplied directly into the radiator. This will dilute the antifreeze solution. If the pump is supplying contaminated water, serious damage to the cooling system may result. The radiator fill valve is an emergency control and should only be used in situations where it is required to protect firefighters or when human life is endangered.

A gear case cooler is also provided to prevent overheating of the lubricant in the gear case under prolonged operations. This heat exchanger is connected directly between the discharge side and the intake of the volume pump, and cannot be shut off.

AUTOMATED CENTRAL COMMAND CONSOLE

An optional feature of FMC pumps is the console that moves control of the pump to a rear-facing midship location just behind the cab. All valves, except drain valves, are controlled from this location by toggle switches mounted on the console. Indicator lights are provided to verify whether the valve is in the OPEN, CLOSED, or PARTIALLY OPEN position. In the PARTIALLY OPEN position, both the open and closed lights will be on.

Operation with the console is the same as with a standard panel except that electrically operated controls are operated by toggle switches instead of manual operation. A master power switch is provided that controls power to all of the individual toggle switches. The switch must be in the ON position to activate any of the controls.

Appendix E

Description and Operating Practices for Hale Fire Pumps

(Note: Unless otherwise indicated, all photos in this Appendix are printed courtesy of Bill Eckman.)

The Hale Pump Company manufactures a full line of fire pumps. These include portable pumps with separate engine drive, small booster-type pumps, front mount pumps, and both single and multistage pumps for midship mounting driven through a split shaft arrangement. They range in capacity from 100 gpm (400 L/min) to as much as 2,000 gpm (8 000 L/min) for high-capacity attack pumpers or water supply units.

For some time, the two-stage pump was generally accepted as the standard in the fire service. As higher horsepower engines became more readily available, the Hale Fire Pump Company was instrumental in the movement toward high-capacity single-stage pumps on many modern pumpers.

FRONT MOUNT PUMPS

The Hale front mount pump is generally mounted behind the bumper and in front of the apparatus radiator (Figure E.1). The frame rails are extended approximately 2 feet (0.6 m), and the pump is mounted on a bar between the rails. This provides some measure of protection to the pump in case of a collision. Power to drive the pump is provided by a front mount PTO unit fastened to the front of the crankshaft. For this reason, the pump is independent of the drive train of the veh-

Figure E.1 Typical installation of a Hale front mount pump.

icle and is only controlled by the engine speed. This provides for pump and roll operation while the vehicle is in motion.

Power is transmitted from the PTO unit to the pump by means of a drive shaft equipped with universal joints and synchromesh clutch. A gear box using the required gear ratio to match the horsepower curve of the engine being used provides the desired rpm of the impeller needed to obtain the specified capacity of the pump. In most cases, the maximum rpm of the engine is no more than the equivalent of 35 mph (56 km/h) road speed at rated 100 percent pump capacity.

Hale front mount pumps come in 500, 750, or 1,000 gpm (2 000 L/min, 3 000 L/min, or 4 000 L/min) capacity. The same basic pump is used for all three sizes, but the gear ratio is changed to turn the impeller at a different speed. The intake manifold varies from 4½ to 6 inches (115 mm to 150 mm) depending on specifications. The discharge manifold varies from two to four 2½-inch (65 mm) outlets. NFPA 1901 requires one 2½-inch (65 mm) outlet for each 250 gpm (1 000 L/min) capacity. This requirement can be met by the use of preconnected lines or rear outlets, so a 1,000 gpm (4 000 L/min) pump may only have two or three outlets on the discharge manifold.

Discharge valves are of the ball-type with quarter-turn locking handles (Figure E.2). These valves can be locked in position by turning the handle clockwise. The valves should always be locked when the outlet is in use and in the UNLOCKED position when they are not in use. Each valve is equipped with a drain valve on the line side of the valve. Valves may also be equipped with an individual line pressure gauge mounted on the pump panel.

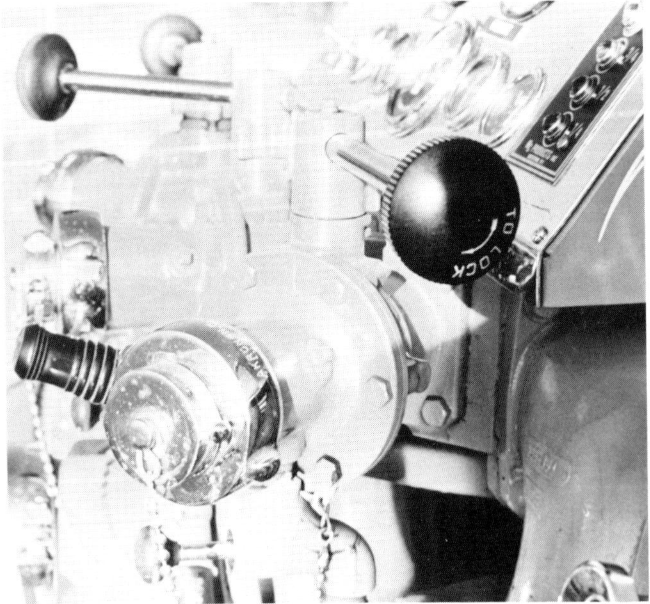

Figure E.2 Ball-type, discharge valves with quarter-turn locking handles.

When different types of attack lines are used simultaneously, the pump pressure must be set for the highest required pressure. Other lines can be established at a lower pressure by partially closing the valve until the individual line gauge registers the desired amount. When this is done, it is important to lock the handle to prevent the setting from changing as the water flows through the valve.

The individual drain valves operate with a push/pull operation. They are equipped with an O-ring mounted on a plunger to prevent leakage. If the O-ring needs to be replaced, it is only necessary to remove a snap ring mounted inside the valve housing and pull out the entire valve assembly. Since the drain valves may be piped below the bumper, it is possible for them to be leaking without being noticed. The operator should always be aware of any unexplained leakage underneath the pumper and take steps to eliminate it.

Water at engine temperature constantly circulates through the head of the pump to prevent freezing when exposed to very cold temperatures. For this reason, the pump casing will always feel warm to the touch when the pump is operating. This warmth should not be confused with the overheating caused by a lack of water circulating through the pump. The drive unit is cooled by water from the fire pump circulating through a heat exchanger inside the drive unit. This prevents dangerous overheating of the gears during prolonged periods of pump operation. In neither case does the radiator coolant come in contact with the water inside the pump, nor does the cooling water from the pump contaminate the lubrication in the drive unit.

To operate the pump from the tank on the apparatus, it is first necessary to open the tank-to-pump valve (Figure E.3). The tank-to-pump line

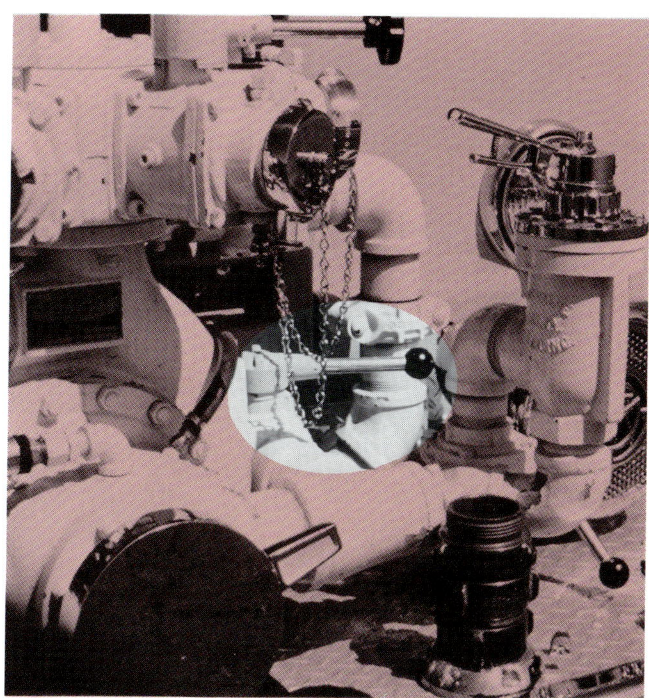

Figure E.3 Tank-to-pump valve.

Figure E.4 An example of a discharge outlet drain.

is generally connected to either the side or the bottom of the intake manifold to the pump. If the pump will be required to supply more than 400 gpm (1 600 L/min) from the tank, there will probably be more than one line from the tank to pump, each with its own valve.

If all valves are not opened before putting the pump in operation, the maximum flow available will be reduced. To provide more flexibility in operation, tank-to-pump lines may be combined with 2½-inch (65 mm) gated intakes, relief valve returns, or intake relief valves (See Figure E.3). If this is the case, the tank valves must be in the proper position for the associated accessory to work.

To allow for draining the tank-to-pump lines at the lower points and eliminate danger of freezing, the valve may be located at the opening of the tank and remotely controlled from the pump operating position (Figure E.4). In other cases, there may be a second valve located where the piping leaves the tank as well as the intake to the pump. In freezing weather, both valves should be closed and the line allowed to drain completely. When the line is dry, the drain valve should be closed to prevent possible problems when the pump is next put into service.

After the pump is full of water, engage the synchromesh clutch to make the impeller turn. The shift lever is located on the right side of the pump and is shifted to the PUMP position by pushing down. Since the clutch is of the syncromesh type, very little resistance should be encountered in making the change. To minimize stress during this change, the motor should be at minimum rpm when the lever is moved either to the PUMP or ROAD position (Figure E.5).

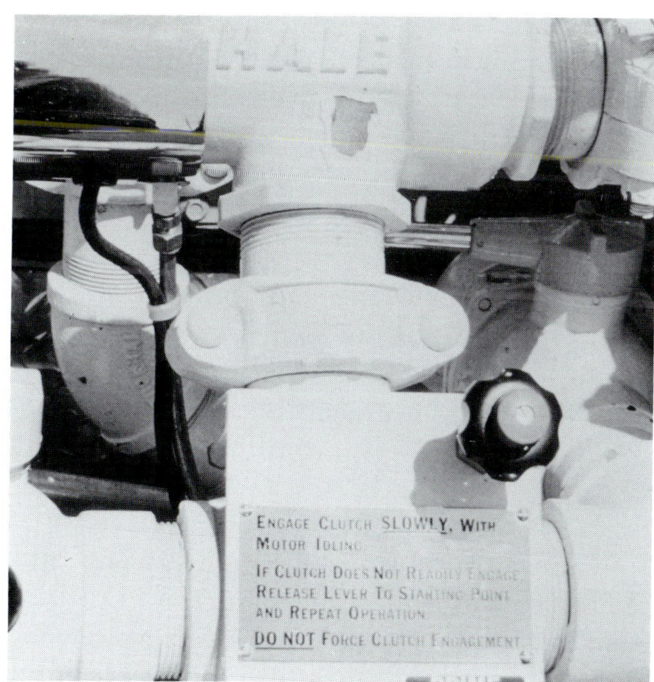

Figure E.5 A message may be put directly on the pump to remind the operator that the engine should be running at a minimal speed when the transfer of power is made.

Once the pump is engaged, adjusting the engine throttle in a counterclockwise direction should cause the pressure gauge to indicate an increasing pressure. If this does not happen immediately, there may be an air lock in the pump or lines, especially if the tank-to-pump line has been drained. Operating the primer momentarily may speed the operation.

The Hale front mount pump is equipped with one stuffing box for the packing where the drive shaft leaves the gear box and enters the pump housing. If the packing is adjusted properly, approximately 8-10 drops a minute will leak out when the pump is operating between 60 and 100 psi (420 kPa and 700 kPa). The exact leakage rate is not important, but when the pump is in operation, water should be dripping from the stuffing box, not leaking in a steady stream. If it is leaking steadily, the packing is too loose and would probably cause an air leak which could make the pump impossible to prime. On the other hand, if there is no leakage, no cooling water is being provided to the packing, and the pump shaft will very likely overheat. Overheating will permanently damage the shaft if it is operated for any length of time.

Adjust the packing by turning two nuts, one on each side of the packing gland. These nuts must be adjusted evenly to get a good seal with the packing. Adjust each nut a fraction of a turn, then allow the packing to seat before making further adjustment. This method of adjustment exerts an even pressure on the packing material and provides a good seal without overheating the impeller shaft.

Hale front mount pumps use standard Hale accessories such as the SMV primer, the Engine Master Governor, or the relief valve mounted between the discharge and intake piping of the pump.

MIDSHIP MOUNTED PUMPS

Midship mounted pumps are much more flexible since they are not limited to the small space that front mount pumps are. They can also provide a higher rated flow since the full power of the engine is available to drive the pump. The maximum capacity of the front mount is limited by the crankshaft bearings and the allowable torque for the PTO unit. Midship pumps are available from 500 to 2,000 gpm (2 000 L/min to 8 000 L/min) capacity.

Most midship mounted pumps are powered by splitting the drive shaft between the road transmission and the rear axle and inserting a gear case in the drive train. A shifting mechanism then directs the power either to the pump impeller shaft for pump operation, or to the coupling shaft to the rear wheels for operation over the road. With this arrangement, power is only fed to the pump or the drive train. The vehicle cannot be driven while the pump is operating, and it cannot be used for pump and roll operation.

The shift between the PUMP and ROAD positions of the gearbox can be accomplished in many ways, depending primarily upon which fabricator mounts the pump on the apparatus. Older models are usually manually operated by using a lever or cable mechanism to accomplish the shift. Many later models use a power shift arrangement, controlled either by engine vacuum or compressed air from the air brake system. Along with the power shift mechanism, two green lights should be supplied, one mounted on the dash of the apparatus and the other on the pump panel in the vicinity of the hand throttle. The lights should come on when the shift has been completed, indicating that it is safe to increase the throttle.

These indicator lights are very important when the apparatus is equipped with an automatic transmission. With manual transmissions, the clutch is engaged after the road transmission is placed in the proper gear and the pump transfer has been engaged. Then the pump will begin to turn. If the shift has not been completed and power is still being supplied to the rear wheels, the engine will stall. With an automatic transmission, very little torque is actually developed at an idle speed of the engine. Therefore, there might not be any indication that the transfer was not completed until the hand throttle is increased at the pump panel to prime the pump or build up the pressure.

As the engine rpm increases, the torque converter in the road transmission begins to transmit more power to the rear wheels, and the parking brake or wheel chocks may not be sufficient to prevent the vehicle from running away. For this reason, the hand throttle should never be used to build up engine speed until the green safety light is on, indicating that it is safe to do so.

Later models may use a full torque power take-off, utilizing the flywheel of the engine or another means of power transfer to drive the pump. These vehicles can operate while in motion. In addition, they do not require the road transmission to be in the DRIVE position to operate the pump. The speed of the pump is controlled only by the engine rpm and not by any setting of the road transmission. This is a common arrangement on apparatus equipped with heavy-duty automatic transmissions and large diesel engines.

To put the split shaft pump in service, operate the pump engagement control. If a manual transfer control has been supplied, be sure it is operated completely. Place the road transmission in the proper gear for pump operation as marked on the dash of the vehicle. This is usually in the DIRECT DRIVE position.

When the clutch is engaged, the speedometer will begin to indicate a road speed, probably somewhere between 10 and 30 mph (16 km/h and 48 km/h), depending on the idle rpm of the engine and the gear ratio of the vehicle. The pump operator should be aware of the correct speedometer reading. Observation of the speedometer enables the operator to check that the transmission is in the proper gear and that the shift has been completed before leaving the driver's seat.

If the pump will be operated from the tank on the apparatus, the pump can be engaged immediately upon arrival. If water supply will be from draft, a hydrant, or a relay, the pump should not be engaged until all connections have been made and the water is readily available. If the pump begins to turn before water is available to cool the packing, damage to the shaft may result. After the pump has been engaged, the shift mechanism should be locked in position if a manual lock has been provided, and the manual or automatic transmission of the vehicle locked into the proper gear. This is especially important when the pumper is equipped with an automatic transmission and a PTO pump that requires leaving the transmission in neutral. If the selector is inadvertently moved into a DRIVE position, the pumper may run away from the operator.

Discharge valves and individual line gauges are similar to those used on the front mount pump. For the most part, they are equipped with quarter-turn locking ball valves, drain valves, and connections for line gauges connected to the line side of the valve. Some of the newer Hale pumps use a full flow valve with a floating seat that is self-locking when pressure is applied. No further locking action is required with this type of valve (Figure E.6). On the opposite side of the vehicle, 2½-inch (65 mm) outlets are controlled remotely by a handle on the pump panel.

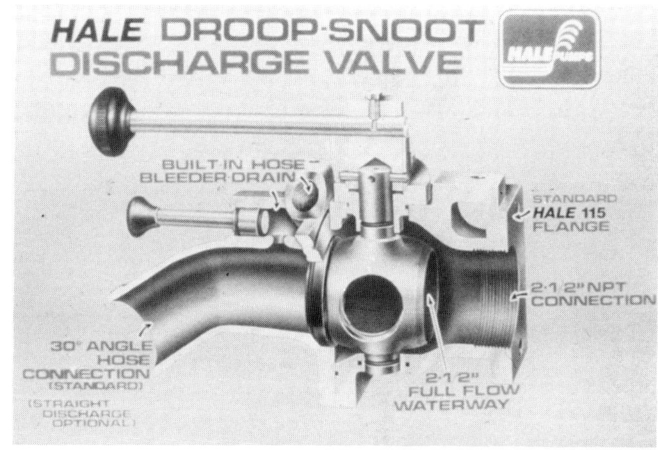

Figure E.6 Newer Hale pumps are equipped with full flow, self-locking valves.

Pressure gauges—master as well as individual line gauges—may be equipped with dampener valves to minimize fluctuation of the needle. These valves can be mounted on the pump panel or behind it with access through some type of door. The gauge will not indicate any pressure if the valves have been closed. The proper way to adjust them is to close them by turning clockwise; then, open a part of a turn until the gauge follows pressure changes readily.

Hale midship pumps also use only one stuffing box for packing, but the construction is not the same as the front mount unit. Adjustment is made by use of an adjustment collar with slots in it, which eliminates the need for special tools (Figure E.7). By releasing the packing locking lever and inserting a screwdriver blade in the slot, the adjustable packing gland can be turned to exert the proper pressure on the packing rings. This is done to provide a good seal with adequate lubrication and cooling. Once adjustment is complete, reinsert the locking pin to keep the gland in the proper position. The same amount of leakage should be present in this pump as in the front mount—8 to 10 drops a minute with 60 to 100 psi (420 kPa to 700 kPa) discharge pressure.

Various drain valves are standard equipment. Most of the critical points on the pump are connected to a master drain valve. This valve enables the pump to be drained in one operation by pulling out on the handle. This is a drain valve, not a pressure relief valve, and should never be operated with pressure on the pump. If it is opened with the pump pressurized, damage to the valve may result.

Other drains are required since the master drain valve only takes care of the pump itself and the normal accessories. Additional drain valves must be supplied at any low point in the piping or other location where water may be trapped. Drains are not very critical in warm weather, but it is good operating practice to use them regularly to prevent overlooking them during freezing weather where damage to the pump and piping may result. It is also good operating practice to close all drains when the apparatus is ready for another call to prevent air leaks from an open drain valve while attempting to operate from draft.

TWO-STAGE PUMPS

The typical Hale two-stage pump is conventional in design. It uses two identical impellers, each designed to handle half the rated capacity of the pump. The impellers can be put in series for pressure operation or in parallel for full rated vol-

Figure E.7 This cutaway shows the adjustment collar on the packing stuffing box.

ume of the pump. Making the change manually requires 2½ turns of a handwheel to operate the transfer valve 90 degrees and change the flow through the pump (Figure E.8). While the transfer valve is moving, a pointer on the pump panel indicates its position. The pointer is geared to the handwheel through a small cable that can easily become loose. If the pointer does not move, it is still possible that the transfer has been completed.

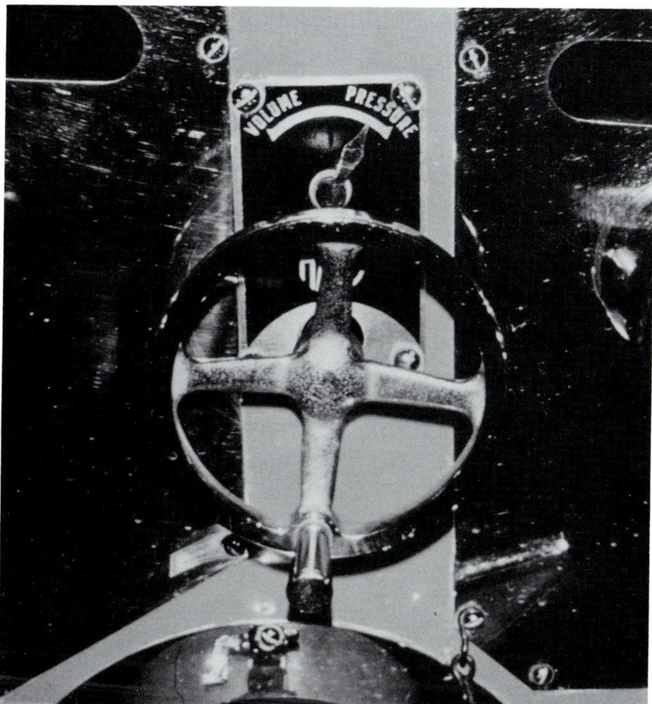

Figure E.8 This handwheel is used to manually transfer the two-stage pump between the volume and pressure modes.

To find out if the transfer has taken place, put the pump in operation and turn the handwheel counterclockwise to the end of its range. At that point, the pump should be in the VOLUME position. Adjust the pressure to a reference point on the master pressure gauge and observe the engine rpm on the tachometer. Reduce the pressure below 50 psi (350 kPa), then turn the handwheel clockwise to the end of its range. Adjust the throttle to the same rpm that was established in the VOLUME position and read the pressure gauge again. The new reading should be approximately twice as high as the original reading. Due to the mechanical limitations of the manual transfer, it is difficult to make with pressure on the pump and probably impossible to accomplish with more than 50 psi (350 kPa) net pump pressure.

The earliest form of power transfer used by the Hale Pump Company makes use of engine vacuum to operate the transfer valve (Figure E.9). When the "volume button" is pushed, engine vacuum for the intake manifold is connected to one end of the power transfer cylinder and atmospheric pressure on the other end causes the transfer valve to be rotated 90 degrees. At the same time, the indicator on the pump is in the VOLUME position (Figure E.10).

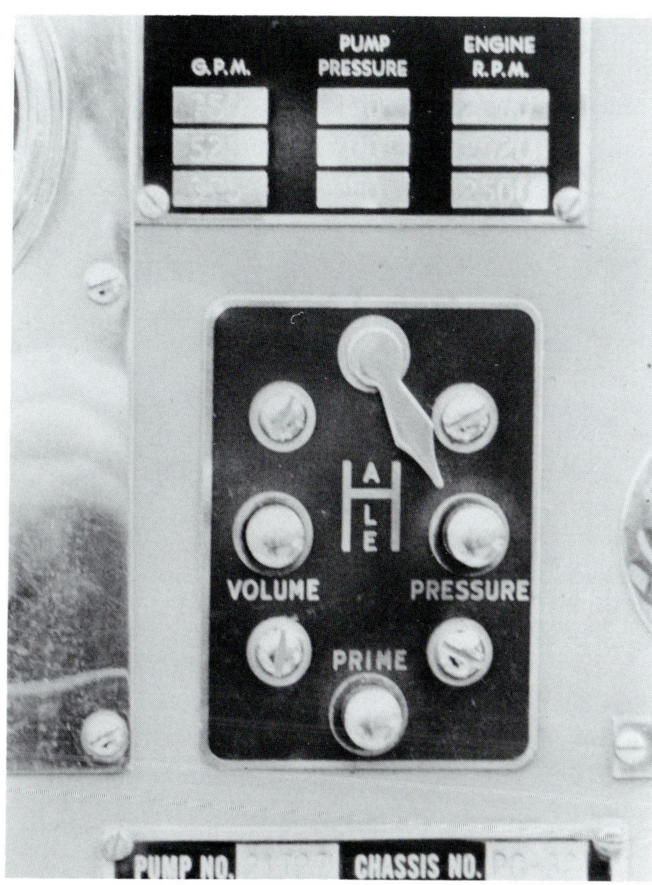

Figure E.9 Early type of transfer control that utilized engine vacuum.

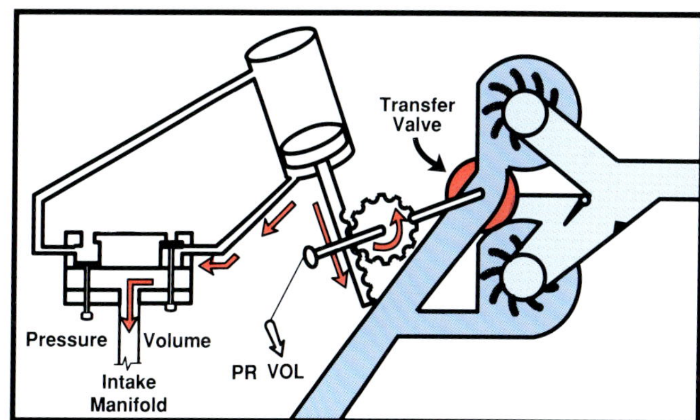

Figure E.10 Schematic of the Hale Vacuum Power Transfer in the VOLUME position.

To change the mode of operation, the "pressure button" is pushed, air is evacuated from the other end of the cylinder. Atmospheric pressure causes the piston to move in the opposite direction and rotate the transfer valve to the PRESSURE position (Figure E.11). Again, the indicator on the pump panel shows that the transfer has been completed. Due to the limited amount of force available with engine vacuum, the lower the pressure, the better the transfer will work. Later variations of this system involve an air pressure cylinder instead of a vacuum cylinder using the pressure present in the air brake system to accomplish the transfer. This type of operation involves more available pressure and a more positive action.

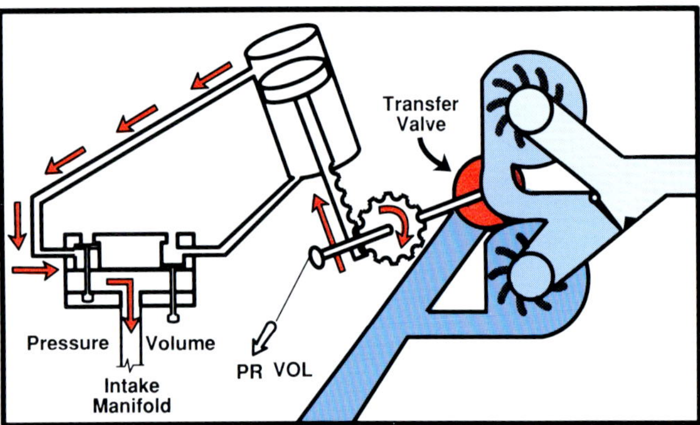

Figure E.11 Schematic of the Hale Vacuum Power Transfer in the PRESSURE position.

The hydraulic transfer is another form of power operation. The system is controlled by a four-way valve labeled VOLUME and PRESSURE on the pump panel (Figure E.12). In the VOLUME position, the four-way valve connects the low pressure side of the pump to one end of the transfer cylinder and the discharge pressure to the other, causing the transfer valve to be in the VOLUME position (Figure E.13). When the control valve is moved to the PRESSURE position, the connections are reversed (Figure E.14). The piston moves to the other end of the cylinder, changing the transfer valve from the VOLUME to the PRESSURE position while doing so. The drive shaft that the power transfer system uses to move the transfer valve has a pointer on the end. This pointer provides a positive indication that the transfer has been completed (Figure E.15). It also provides a manual override of the system if the power transfer does not work. There is a hole through the shaft, and a pin can be inserted to provide sufficient leverage to turn it manually. If no pin is available, the shaft is hexagonal in shape and a standard wrench can be used to provide control of the system.

Figure E.12 Control switch for the Hale Hydraulic Transfer operation in the VOLUME position.

This hydraulic transfer system varies from the vacuum power transfer and the manual transfer in that the actual movement of the valve is done

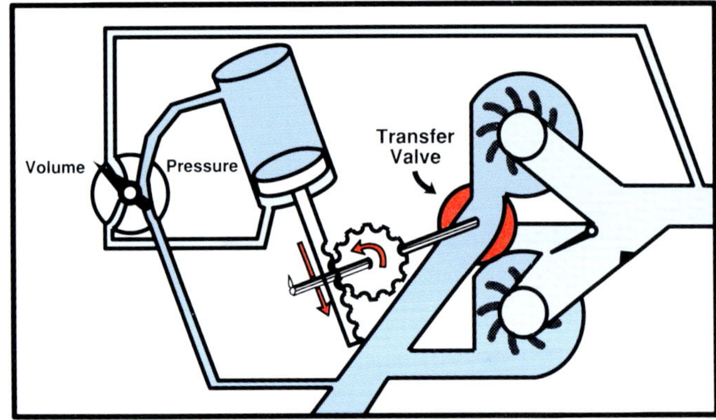

Figure E.13 Schematic of the Hale Hydraulic Power Transfer in the VOLUME mode.

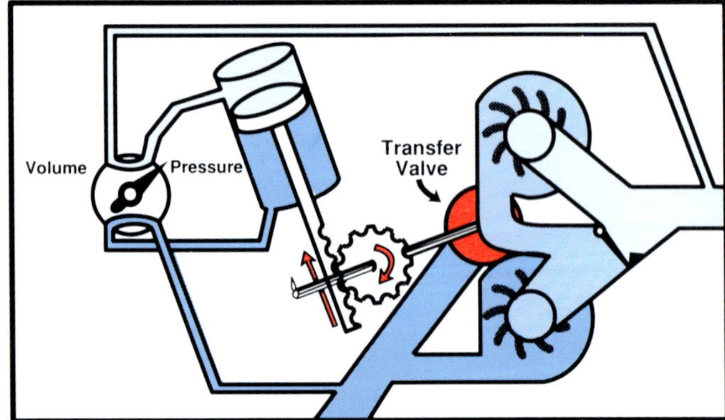

Figure E.14 Schematic of the Hale Hydraulic Power Transfer in the PRESSURE mode.

Figure E.15 The pointer attached to the end of the drive shaft indicates the position of the power transfer system. The hex nut acts as the manual override.

by the water pressure developed by the pump. The transfer will not usually take place until there is at least 35 to 50 psi (245 kPa to 350 kPa) pressure differential between the intake and the discharge of the pump. It may be possible to operate this transfer device at relatively high pressure due to the higher pressure differential; however, operating it at more than 50 psi (350 kPa) net pump pressure may cause damage to the pump. More importantly, it may cause excessive pressure surges on any lines being supplied.

SINGLE-STAGE PUMPS

While at first glance it seems that a single-stage pump would have less capacity than a corresponding two-stage pump, this is not necessarily so. There are some important differences in construction between the two types of pumps. The Hale single-stage pump uses a dual suction impeller. This impeller takes water into both sides of the impeller, rather than on one side as found in a typical two-stage pump. Not only does this provide additional capacity, it also balances the hydraulic force of the water as it enters the impeller from opposing sides. The result is that the forces effectively cancel each other. In addition, there are two stripping edges in the volute and water comes off two points on the impeller, 180 degrees opposed to each other. This provides a better balance and puts less strain on the drive train and critical parts of the pump (Figure E.16).

Operation of the single-stage pump is simpler than a two-stage pump since there is no possibility of being in the wrong position at critical times. There are other advantages with a single-stage pump. One thing that has been emphasized is that a single-stage pump is more efficient at 100 per-

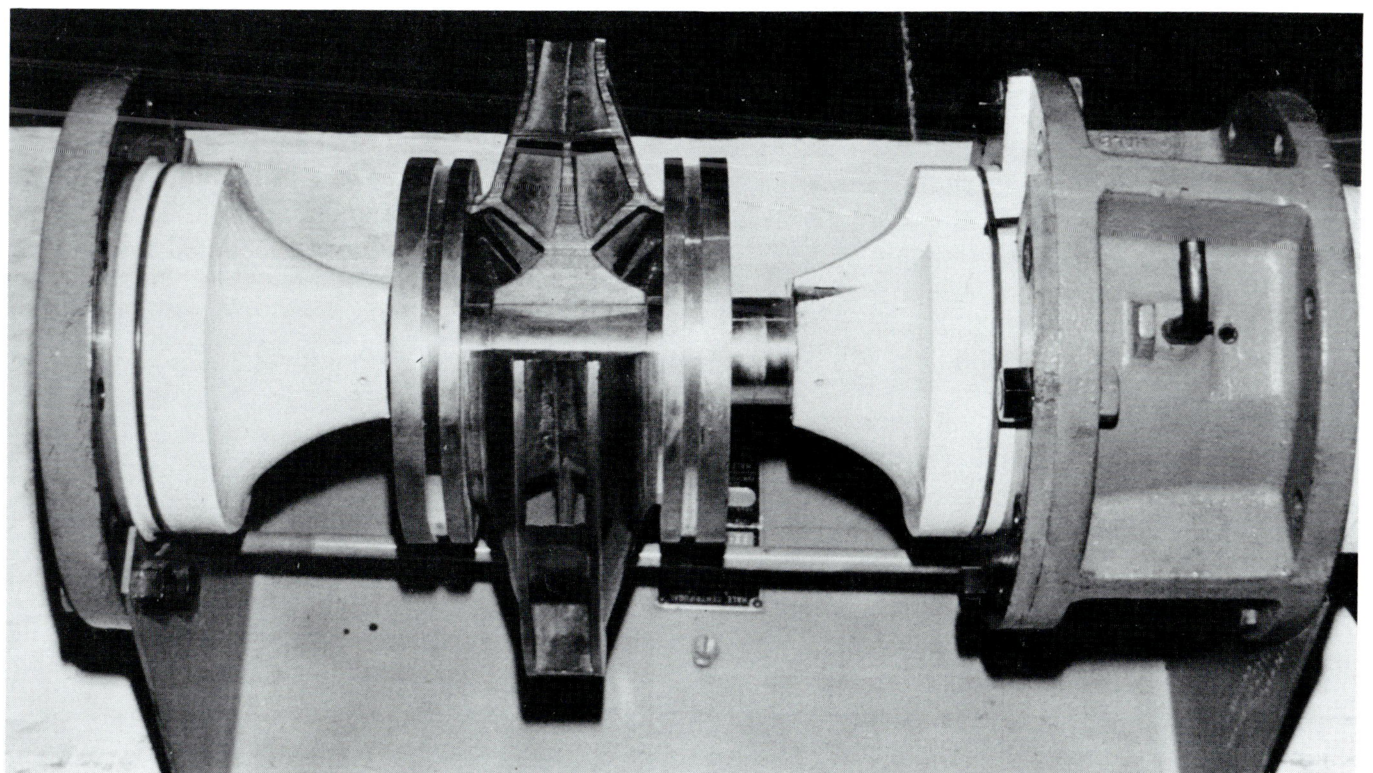

Figure E.16 Cutaway of a single-stage impeller.

cent of its capacity. This is not really significant, since a fire pump rarely operates at or near 100 percent capacity. At low flow rates, supplying one or two handlines, the two-stage pump is more efficient. The most important advantage is that the single-stage pump is very well suited to modern diesel engines, especially those that are designed to operate at a relatively high rpm. Many diesel engines are difficult to control at low rpm with a light load, and the relatively higher rpm required to develop adequate pressure with a single-stage pump makes for a smoother and better controlled operation.

PUMP ACCESSORIES

The Hale Pump Company does not build fire apparatus, but rather supplies pumps and equipment to various fabricators. Therefore, accessories are frequently designed for flexibility in order to make them adaptable to a maximum amount of different applications. They are convenient for use with other manufacturer's pumps and are a popular choice when retrofitting apparatus to modernize its operation.

Priming Devices

In the interest of weight, cost, and simplicity of operation, Hale portable pumps are generally equipped with an exhaust primer. By operation of a lever, the exhaust gases are diverted from the muffler and directed through a venturi mechanism where enough vacuum is developed to evacuate the air from the pump. All other Hale fire pumps use a positive displacement pump for priming, either rotary gear or rotary vane type.

ROTARY GEAR PRIMER

The rotary gear primer is standard on most of the older Hale midship mounted pumps and is used on large capacity pumps. The priming pump is a rotary gear pump that has the capacity to remove large quantities of air from the pump (Figure E.17). It is a very effective primer and sufficient vacuum can be developed in spite of appreciable air leaks. This can create a problem in operation. It is not uncommon to obtain water from draft while the primer is operating, then lose it when water starts to move. To prevent this, special care is necessary to eliminate all air leaks before priming the pump.

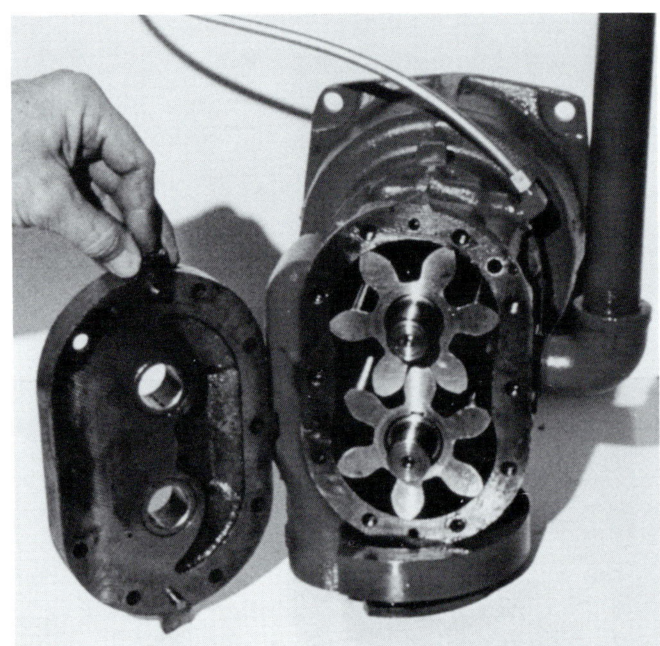

Figure E.17 Cutaway of a positive displacement rotary gear priming pump. This is a type of positive displacement pump. *Courtesy of Hale Fire Pump Co.*

The rotary gear priming pump is driven from the pump transmission. The internal gears are turning all the time the main fire pump is in operation, but the priming pump does not operate until the diaphragm-type clutch is engaged from the pump panel (Figure E.18). To begin priming, the control push button is operated (Figure E.19). When the line from the intake manifold is connected to the priming valve and the priming clutch, enough air will be evacuated to create a

Figure E.18 One type of primer control is the push/pull, T-handle control.

Figure E.19 Another primer control is the push button.

pressure differential sufficient to cause air pushing against a diaphragm to operate both of these devices. The clutch causes the priming pump to turn and the priming valve opens to allow the air to be evacuated from the intake of the pump.

When enough air has been removed and the pump is full of water, the prime button can be released. The air pressure equalizes on both sides of the diaphragm, and both the clutch and priming valve will release. Due to the close tolerances in the construction of the push-button type vacuum priming control, it may not completely close when released. If this happens, the priming pump will not stop operating, and the priming valve will remain partially open. This creates an air leak that will cause a loss of prime. This can be prevented by frequent lubrication with some type of graphite material.

When this primer is used with a diesel engine, no source of vacuum is readily available. Air pressure from the braking system can be used in place of a vacuum. Where this has been done, it can be identified by the physical appearance of the control.

As with any power-operated accessory, it is possible for a broken vacuum or air line to make the primer inoperative while the basic system is intact. Manual override controls make it possible to operate in an emergency. These controls are of two types. In some cases, both the priming valve and the clutch have been connected to a single manual control. In other cases, the priming valve and clutch have been connected to the pump panel separately. Where individual controls have been provided, the priming valve should be closed before the clutch is disengaged. If this is not done, the prime will be lost when the air feeds back into the pump through the priming pump after it stops turning.

To prime the pump, shift to the PUMP position, put the road transmission in the proper gear, engage the clutch, and set the hand throttle to 1,000—1,200 rpm. Push the prime button and hold it until the pump is completely primed.

ROTARY VANE PRIMER

The other type of positive displacement primer that Hale uses is the SMV-type rotary vane pump (Figure E.20). This primer consists of a rotary vane type pump with four fiber vanes mounted on a rotor within the housing. As the pump turns, the vanes are forced against the housing by centrifugal force, forming an airtight seal. Centrifugal force is constantly forcing the vanes against the housing while the pump is turning, and they are self-compensating for wear in the pump.

The SMV primer is driven by an electric motor. It can be mounted nearly anywhere on the apparatus and connected to the fire pump through

Figure E.20 The Hale SMV primer is a rotary vane pump.

vacuum lines. To minimize the possibility of air pockets in the system and to permit discharging water from the pump directly onto the ground when the pump is successfully primed, it will usually be mounted low on the chassis.

The SMV primer uses a single control for its operation. Pulling on a priming handle opens a valve from the primer into the intake manifold of the fire pump. At the same time, an electrical switch operates to begin turning the electric motor. Engine rpm is not critical to the operation of the primer, and the pump can be filled with water without turning. By setting the throttle for 1,000–1,200 rpm before operating the primer, the alternator will provide enough current to keep the battery from being discharged by the heavy electrical requirements of the motor. Maintaining this rpm will also ensure that the fire pump is turning fast enough to build up pressure and prevent loss of prime when the primer is shut off.

Priming Oil System

Both positive displacement primers use lubricating oil in the system. This oil serves two purposes. First, it forms a film of oil on the housing and the pump rotors or gears to provide a more airtight seal and prime the pump more quickly. As the pump wears and the tolerances become greater, the sealing action becomes more critical. Also, the oil lubricates the moving parts while they are running dry before the pump is primed and acts as a preservative.

It is a good practice to operate the primer dry for a few seconds after the pump has been drained after a pumping operation. This will form a thin coat of oil on the metal parts and inhibit deterioration due to rust or corrosion. A supply of SAE 30 motor oil is carried in a reservoir on the apparatus. Two vents are associated with the system. The cap on the reservoir has a vent in it that allows the oil inside the tank to be replaced with air as it is removed. If this vent becomes blocked, a vacuum will form inside the reservoir as the oil is used, and the flow will be interrupted.

The vent in the high point of the line to the primer serves a different purpose (Figure E.21). In most cases, the discharge of the priming pump

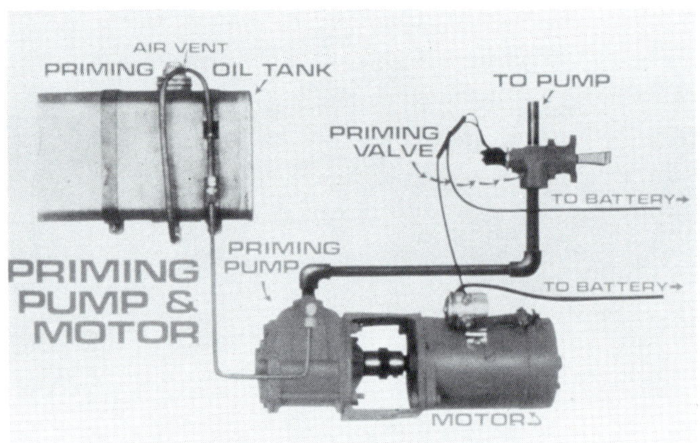

Figure E.21 Vent holes are placed in the oil reservoir cap and the line to the primer to assure proper primer operation. *Courtesy of Hale Fire Pump Co.*

is lower than the oil reservoir. Because of this, the priming oil will continue to be siphoned out of the tank when the primer stops turning if the system is airtight. A small hole, approximately a #60 diameter, is located at a high spot in the tubing. This hole has to be small enough that the vacuum from the primer will be able to overcome the air leak and draft the oil from the reservoir. Since the hole is so small, it can be easily clogged with paint, dirt, or grease, so should always be kept clean. One indication that the hole is clogged is priming oil leaking from the primer when it is not operating.

AUTOMATIC PRESSURE CONTROL DEVICES

The Hale Pump Company uses three different devices for automatic pressure control. Two relief valves are used, one integral to the midship pump, the other an "in-line" type for mounting in any application, and a pressure governor. The device used depends upon the type of pump, how it is mounted, the type of engine in the apparatus, and the preference of the user.

Engine Master Governor

The engine master governor is a stored pressure reference governor. This means that it has the ability to increase engine rpm to maintain the desired pressure when additional lines are added. It can also prevent pressure surges by reducing engine rpm when the flow is decreased. Construction and installation of this governor has been simplified by combining the hand throttle, the balancing cylinder, and the operating controls in one housing mounted on the pump operator's panel (Figure E.22). The only additional equipment required is a pressure reservoir and in-line strainer. These two items help to protect the governor and the entire governor system by maintaining the same pressure at all points in the system. This allows the throttle to move freely within the balancing cylinder since the water can flow from one side of the control piston with little or no resistance to movement. Air trapped within the pressure reservoirs will be compressed to the same pressure as the water pressure being applied by the pump.

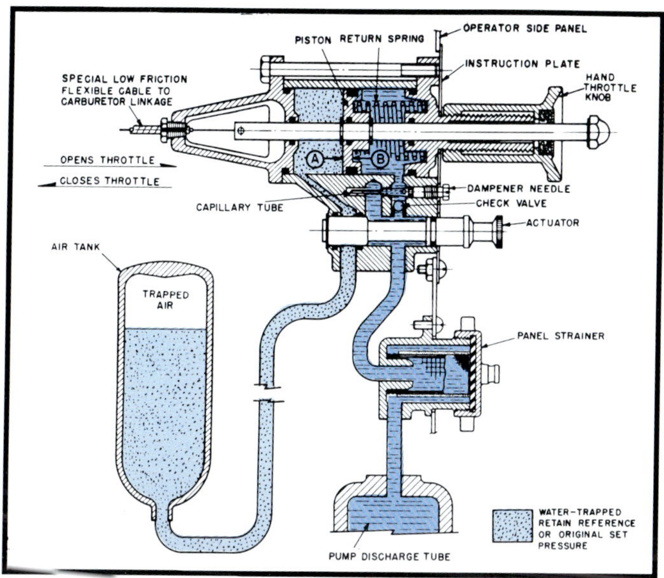

Figure E.22 A schematic of the Hale Engine Master Governor.

To put the governor in service, turn the hand throttle counterclockwise until the discharge gauge reads approximately 10 psi (70 kPa) above desired operating pressure. Pull out the actuation knob (Figure E.23). This isolates the pressure on the reference side of the system from the discharge pressure from the fire pump on the operational side of the system. The throttle can then be turned clockwise to the end of its travel. The engine speed will not decrease significantly since the water pressure on both sides of the balancing cylinder will maintain the throttle linkage in the same position.

When the pressure being exerted by the throttle knob is released, the throttle return spring mounted within the balancing cylinder causes the pressure to decrease slightly. This decrease is the

Figure E.23 Governor actuation knob.

reason for initially setting the pressure slightly above the desired operating pressure. How much this increase needs to be depends upon the construction of the particular apparatus. Experience will enable the operator to determine how much to allow for this change.

Once the governor has been set, any pressure differential between the reference and the operating portion of the system will cause the piston within the balancing cylinder to move. The engine rpm will change and the pressure on the discharge side of the pump will be maintained at the reference value. For example, if the discharge pressure in the pump drops 15 psi (105 kPa) from the set value, the compressed air pocket in the reservoir expands, forcing the piston in the direction of the lower pressure in the balancing cylinder. When that happens, the throttle linkage causes engine rpm to increase until the pressure on both sides of the balancing cylinder are equal.

If the discharge pressure from the pump increases, the increased force against the operating side of the piston within the balancing cylinder will cause it to move and force additional water into the reservoir. This additional water compresses the air in the tank until the pressure is again equalized within the system. With the movement of the throttle, engine speed will decrease until the pressure has been reestablished at the original setting.

There are two inherent problems in a device of this type. One limitation is the speed of response to changes of throttle setting in an internal combustion engine. Since the governor is a hydraulic device, there is very little delay in responding to changes in operating pressure. Since changes in engine speed cannot take place as rapidly, the tendency is for the governor to overcorrect for pressure changes. Compensating for the initial overcorrection may again cause the governor to overshoot the mark. Constantly trying to reach the correct pressure is known as a hunting action. The practical effect of this phenomenon is a fluctuation in pressure anytime the governor acts to compensate for pressure changes.

One solution to the problem of pressure fluctuation is to restrict the flow of water by using some type of needle valve for dampening. Limiting the movement of water in and out can slow the operation of the governor to match the speed of response of the engine being used. The problem with this approach is that the speed of response may be slowed to the point that it is not possible to meet the NFPA requirement of no more than 30 psi (210 kPa) increase when all lines are closed. Hale's approach to this problem has been to slow down the increase of speed while maintaining the fastest possible shutdown. Since the primary purpose of a pressure control device is to protect the personnel operating attack lines from pressure surges, this approach is very effective.

In order to slow the operation of the governor, water flows through a capillary tube to the operating side of the balancing cylinder, which greatly restricts the rate of flow. A dampening adjustment is used with this waterway to further match the operating time of the governor to the speed of response of the engine. To allow for a more rapid shutdown in case of pressure surges, a ball-type check valve is used to bypass the capillary tube.

If the operating pressure exceeds that within the operating side of the balancing cylinder, the check valve will open. This allows a rapid inflow of water, which creates a pressure differential and a rapid decrease in engine rpm. With a lower pressure in the operating side of the governor, the ball valve will remain closed and the throttle can only increase as rapidly as the water can be forced out of the balancing cylinder through the restricted waterway of the capillary tube. It is not unusual for this governor to require as much as 15 seconds to bring the pressure back to its original setting when flow is resumed.

A more serious problem is the lack of any safeguard to prevent damage to the pump or apparatus if the water supply to the pump is interrupted. If the pump runs dry while the governor is set, pressure from the reference side of the system causes the throttle to increase. This increase in throttle is an attempt to bring the system without resistance from the pump back to its original setting which will cause it to move until it reaches the end of its travel. Since the pump is dry, no pressure buildup occurs. If the engine governor is not operating properly, the engine may increase its speed until damage results. Only close observation by the pump operator can prevent this from happening.

While the engine master governor can be taken out of operation by simply pushing in on the actuator control, a much smoother operation will result if the hand throttle is turned out until it comes in contact with the acorn nut on the governor shaft. The actuator can then be pushed in, and the engine slowed under the control of the hand throttle smoothly and gently.

To keep this governor operating reliably, two operations must be performed regularly. Most Hale governors have a strainer mounted somewhere on the pump panel. This strainer should be removed after every pumping operation involving dirty water. The strainer should also be cleaned on a regular basis (Figure E.24).

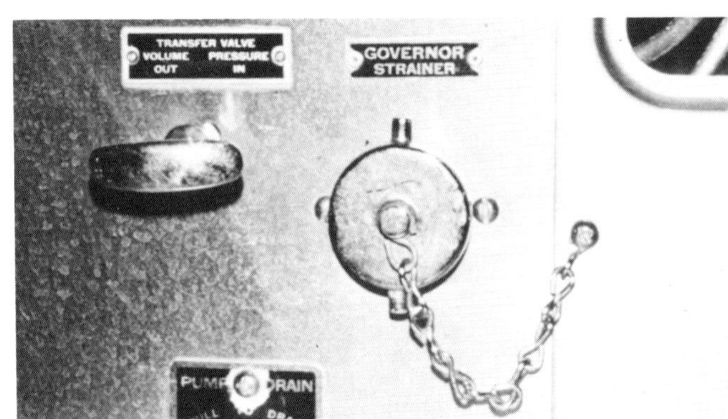

Figure E.24 The governof strainer prevents foreign objects from damaging the governor system. It should be cleaned every time dirty water is pumped.

Another concern is the pressure reference reservoir. For the governor to work well, there must be an adequate amount of air trapped in the tank to act as an air spring and establish a good solid reference. It is possible for this tank to become waterlogged after frequent operations. To keep this from happening, the reservoir should be drained frequently. In some cases, this drain may be connected to a master drain valve, although it is usually brought out to a separate location and requires special attention.

HALE RELIEF VALVES

Hale uses a pressure-type relief valve. Each of their relief valves consists of two parts. There is a main relief valve inserted in a waterway between the discharge and intake side of the pump. There is also a pilot valve with an adjustment handwheel mounted on the pump panel that controls the operation of the relief valve.

When the pump is dry, the relief valve is held in the CLOSED position by a biasing spring mounted outside the valve (Figure E.25). The main relief valve contains a double-ended valve installed in such a way that water pressure against the valve face acts to keep the valve closed. When the discharge pressure is great enough to overcome the spring tension set by the pressure adjustment handwheel on the pilot valve, the control valve opens (Figure E.26). The discharge pressure is applied to the control chamber of the main relief

Figure E.25 This biasing spring holds the relief valve closed when the pump is dry.

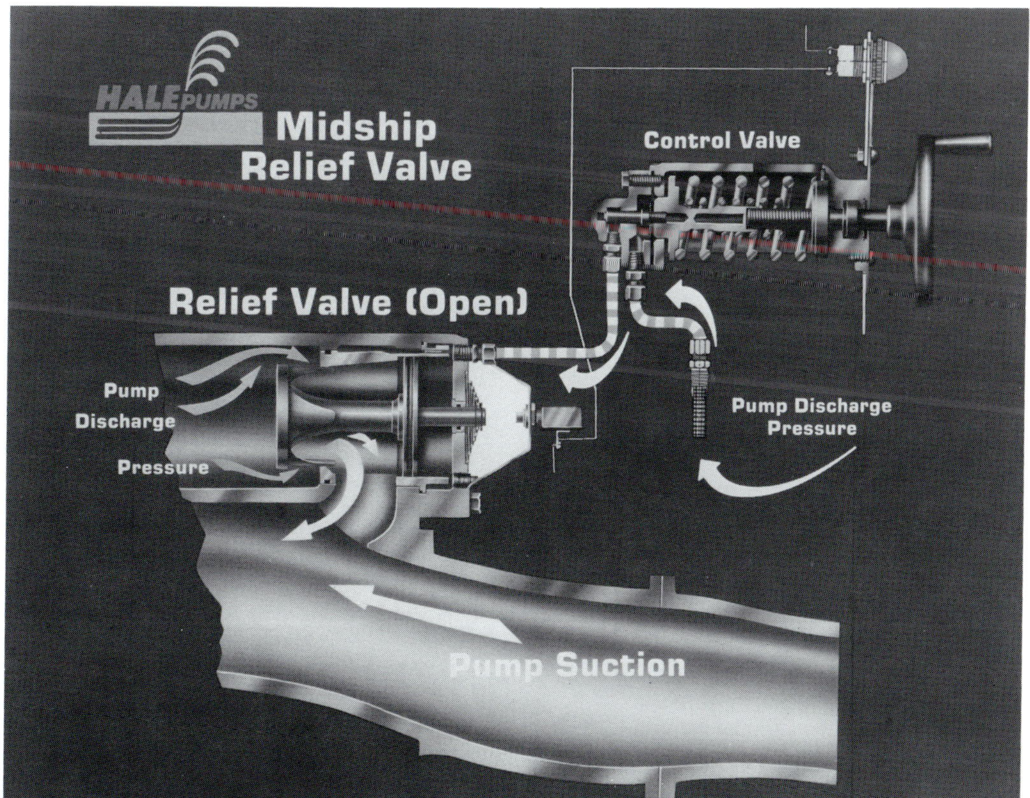

Figure E.26 The relief valve assembly includes the valve and biasing spring mounted on the pump and the pilot valve with pressure adjusting handwheel mounted on the pump panel. *Courtesy of Hale Fire Pump Co.*

valve where it acts against the piston (large end of the relief valve). The same amount of pressure in psi (kPa) is applied to the valve end of the main valve and the piston end.

Since the piston end is larger than the valve end, more total force will be generated in that direction. The valve will open, allowing water to bypass the pump, decreasing the net pump pressure. The valve will open far enough to allow just enough water to flow through the restricted opening. This maintains a friction loss pressure differential equal to the desired net pump pressure with the water actually flowing at the time.

To set the relief valve, all lines must be flowing at the desired rate of flow. Turn relief valve control handwheel counterclockwise until the valve opens (Figure E.27). The most reliable indication is the amber light on the pump panel. An electrical switch is mounted on the relief valve where it can be operated by the main relief valve shaft when the valve opens.

The construction of the valve is such that it is easy for the switch to get out of adjustment and provide a false indication. After the valve has been opened, it is a good practice to turn the handwheel clockwise until the valve just closes. This final adjustment should be made slowly and carefully, allowing time for the valve to stabilize as the wheel is turned.

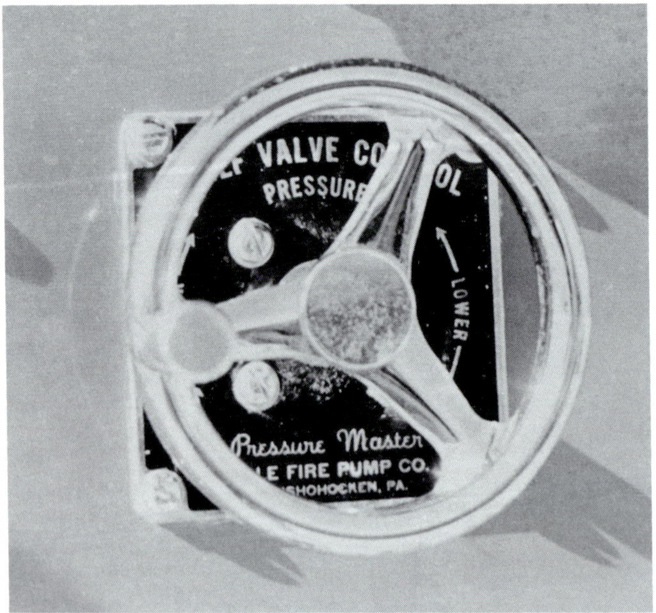

Figure E.27 Relief valve control handwheel.

When the pressure drops enough for the pilot valve to close, pressure trapped in the control chamber will bleed off in one of two ways. It bleeds off either through the bleed hole in the face of the main valve or through the relief port in the housing. Then the main relief valve will close. If the valve does not close when the pilot valve closes, this hole may be clogged and will have to be cleaned. This is a maintenance problem and the pump will have to be taken out of service to take care of it.

Unlike the pressure governor, the relief valve is not capable of compensating for a decrease in pressure. If additional lines are to be added, the throttle must be adjusted. No change will have to be made to the setting of the pilot valve unless the discharge pressure is to be changed.

Both Hale relief valves work in the same manner and use the same control valve and adjustment handwheel. The difference between the two is that the in-line relief valve can be mounted away from the fire pump itself and provides more flexibility in construction of the apparatus (Figure E.28).

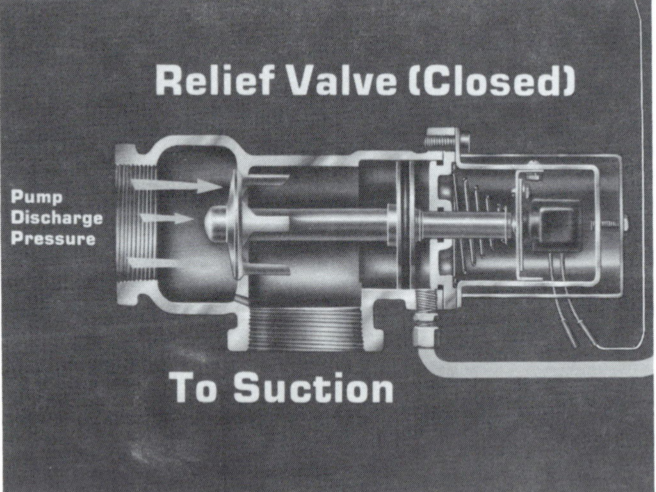

Figure E.28 The in-line relief valve can be installed in piping remote from the pump casing. *Courtesy of Hale Fire Pump Co.*

AUXILIARY COOLING

There is no standard arrangement for auxiliary cooling with the Hale pumps. Most manufacturers use a marine-type auxiliary cooler, (Figure E.29) mounted externally in the most convenient location. The auxiliary cooler is controlled by a valve on the pump panel. It should be operated to keep the operating temperature of the engine within the most desirable range. The radiator fill valve, where supplied, is an emergency control and should not be used in normal pump operation. This is because it will contaminate the antifreeze solution in the engine cooling system.

Figure E.29 Tho Hale "K" cooler circulates hot water from the radiator around a coil with cooling water from the pump. *Courtesy of Hale Fire Pump Co.*

Appendix F

Description and Operating Practices for Seagrave Fire Pumps

(Note: Unless otherwise indicated, all photos in this Appendix are printed courtesy of Bill Eckman.)

Seagrave fire pumps are installed only on custom and commercial chassis apparatus constructed by Seagrave prior to about 1975 (Figure F.1). Seagrave no longer manufactures their own pump; most newer Seagrave apparatus are equipped with Waterous pumps with either Seagrave or Waterous accessories. The standard installation used one of two basic pumps. Both two-stage midship mounted pumps were driven by a split shaft arrangement.

Figure F.1 A typical Seagrave custom pumper.

OPERATION OF THE TWO-STAGE PUMP

Putting the Pump in Gear

The midship mounted pump used by Seagrave contains the transmission and drive gears as an integral part of the pump. The midship mounted pump is controlled by a shifting lever located on or near the pump panel of the apparatus (Figure F.2). The pump is driven from the tail shaft of the road transmission. By movement of the shifting clutch inside the gear case the tail shaft either floats in neutral or drives the impellers in the PUMP position. Alternately, the transfer case connects the power from the tail shaft to a drive shaft in the ROAD position.

Pumping operations should take place with the road transmission in the direct drive gear as

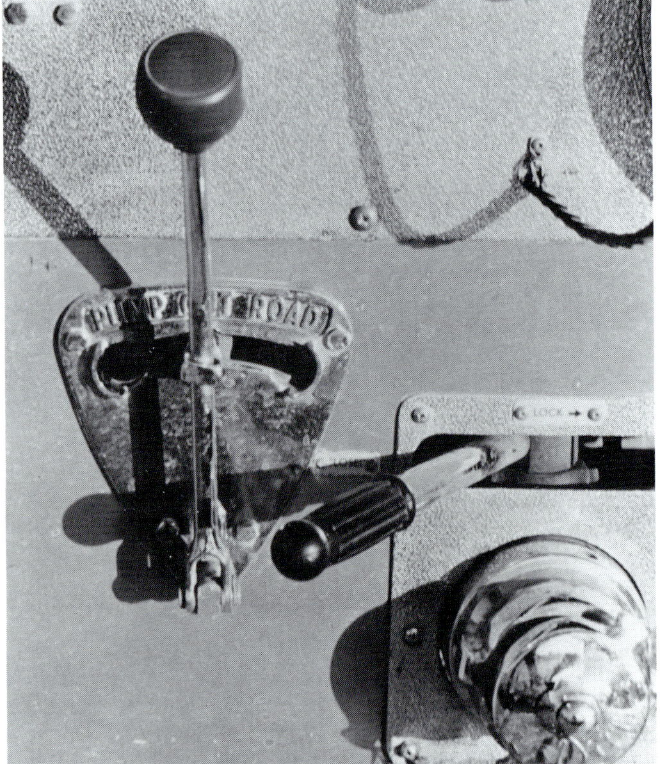

Figure F.2 The Seagrave pump shift lever is usually located in a lower corner of the pump panel.

specified on the dash of the vehicle (Figure F.3). This will normally be in fourth with fifth gear being an overdrive.

To put the pump in gear, pull out on the knob associated with the pump shift lever. Pulling out on this knob activates a vacuum cylinder which will disengage the clutch. After the gears have stopped turning, move the lever to the center position. Since the speedometer is driven from the rear axle on most Seagrave pumpers, it will not register when the shift lever is in the PUMP position. For this reason the speedometer cannot be used as an indication that the transmission is in the proper gear.

The tachometer is driven from the transmission and will register anytime the clutch is engaged. To prevent clashing the gears while shifting, it is best to wait until the tachometer indicates

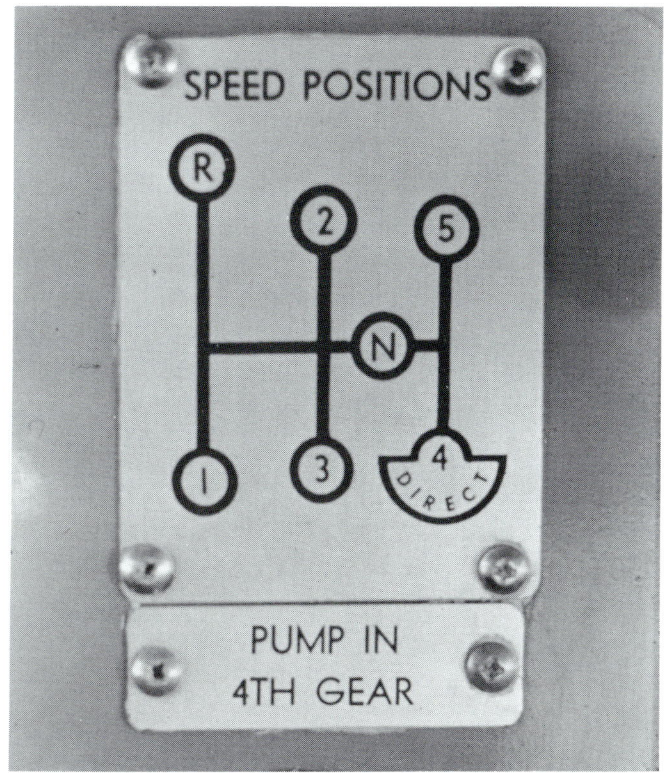

Figure F.3 A plate on the dashboard usually shows the shifting pattern and pumping position for manual transmissions.

that the gears have stopped turning before moving the shift lever. The shift lever is located on the 400 and 500 series in a position that permits the operator to move it to the OUT position while still seated in the cab. By reaching down, pulling out on the knob, and shifting to neutral, it is possible to put the road transmission in gear before leaving the cab.

Later model units with the shift lever located on the pump panel do not permit this. These models have a clutch control located in the cab on the dash (Figure F.4). By operating the clutch control

Figure F.4 Later models have a clutch control toggle switch mounted on the dashboard.

to the PUMP position, control of the clutch is with the pump shift lever. After operating the clutch switch to the PUMP position, leave the transmission in neutral. Move the shift lever to the OUT position, then put the road transmission in fourth gear. The pump will start to turn when the shift lever is moved to the PUMP position.

Discharge Valves

Discharge valves are of the twist-lock type with one 2½-inch (65 mm) discharge outlet for every 250 gpm (1 000 L/min) of capacity (Figure F.5). These valves can be locked in position by turning the rubber handle clockwise and released by turning counterclockwise. The locking mechanism has a tendency to get out of adjustment and not to lock securely. This can be a serious problem since it will not be possible to maintain a stable pressure on a hoseline operating with the valve partially closed to reduce the nozzle pressure. When it is difficult to lock the valve in position, it should be referred to the maintenance department for prompt attention.

Figure F.5 Seagrave valves are of the twist-lock type. Often, the operating handles for different valves will all be located together.

Discharge outlets located on the other side or at the rear of the pumper are remotely controlled from the pump operator's position with a similar type of locking arrangement. Each discharge outlet is equipped with a hoseline bleeder valve located on the discharge side of the valve. This permits draining a hoseline that has been taken out of service without disturbing the operation of other lines that may still be operating.

Individual line gauges may also be connected to the line side of the valves. These are essential where standard operating practices involve using various size attack lines with a different operating pressure requirement. The pump should be set for the highest pressure needed. Then the gate valve (associated with each line requiring a lower pressure) should be partially closed until the individual line gauge associated with that outlet indicates the desired pressure. It must then be locked in that position. Any changes in flow through the nozzle connected to one of these lines will change the friction loss through the gated valve, and the discharge pressure.

Intake Valves

The pump will also be equipped with one or more gated 2½-inch (65 mm) intake fittings. Use of these inlets to the pump provides flexibility since the associated gate valve permits making and breaking connections without disturbing the operation of the pump. At the same time, it is good to keep in mind the flow limitations imposed by the associated piping. These intakes generally use 2½-inch (65 mm) pipe, 90-degree bends, and a T-type fitting into the pump. As a general rule, these components limit the maximum flow to less than 500 gpm (2 000 L/min). A suction clappered siamese or a gated 2½ x 2½ x 6-inch (65 mm x 65 mm x 150 mm) wye on the large intake is more effective where large flows are frequently encountered.

Drains

A master drain valve is supplied with the pump. The majority of the drain points on the pump are connected to one master drain valve, and the pump can be drained by pulling out on the handle. This is *not* a relief valve and should *not* be opened under pressure. Damage to the valve may result if this is done. There may also be individual drain valves at low points on the piping or accessories where it would not be possible to drain them completely by connecting to the master drain (Figure F.6).

The pump should be drained after each operation, especially when operating in freezing weather. All drain valves should be closed before making the apparatus ready for service. If this is not done, an open drain can be overlooked when attempting to operate from draft, making it impossible to prime the pump.

Figure F.6 The master drain control valve is usually located on the lower portion of the pump panel.

Series-Parallel Valve

The biggest difference between the two basic Seagrave pumps is in the method of transferring from PARALLEL to SERIES operation. The older barrel-type pump, usually installed on the 400 and 500 series of conventional chassis custom apparatus, uses a piston-type valve to accomplish the changeover by hydraulic action (Figure F.7). When the pump is dry, the piston is held in the VOLUME position by the force of gravity. When the pump is put in operation, a four-way valve connects one end of the piston-type valve to the low pressure side of the pump and the other to the high pressure side.

Discharge pressure pushing against the face of the piston keeps it in the VOLUME position. To transfer to pressure, the four-way valve is changed and the connections are reversed. This

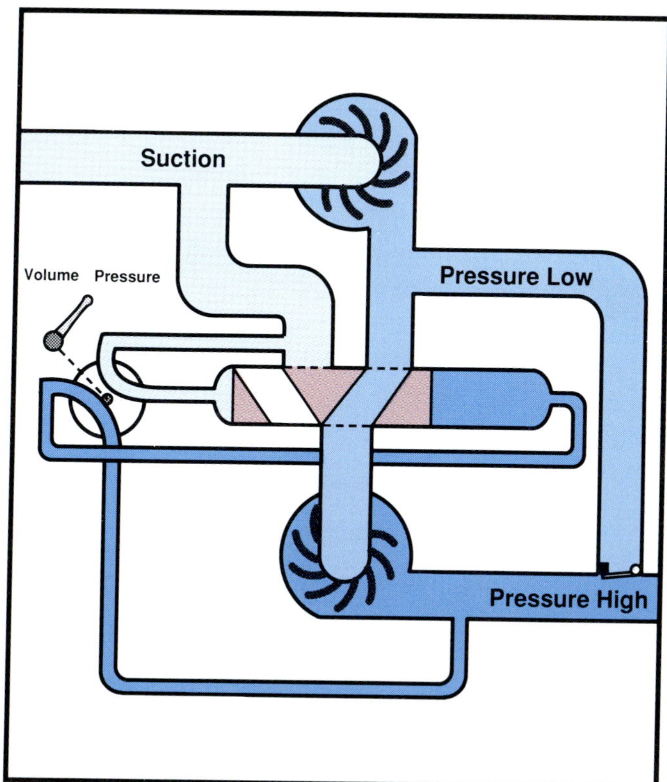

Figure F.7 Older type hydraulic changeover system.

discharge pressure pushing against the piston now forces it to move to the other end of the cylinder. This changes the direction of water movement through the pump and puts the two impellers in SERIES instead of PARALLEL. The discharge pressure of the pump will increase shortly when the change has been completed. A check valve is provided to isolate the intake side of the two stages during SERIES operation.

To transfer the pump, it is only necessary to pull out on the volume-pressure control. This will change the position of the four-way valve and cause the pump to transfer to the PRESSURE position. The increased pressure from the discharge of the first stage pushing against the check valve will keep it closed. Using water pressure from the pump allows the transfer to take place at any pressure.

While there will be more resistance to the movement of the piston when the pump is operating at high pressure, there will be enough force available to make it move. Seagrave specifies that the transfer can take place at any pressure. Good operating practice, however, dictates that the net pump pressure should be less than 50 psi (345 kPa) when it is done. This will prevent pressure surges in the system and reduce stress and strain on the pump and its accessories. Most important, it will protect hoseline personnel from injury. The discharge pressure will remain relatively constant during the transfer if the governor is properly set and in good operating condition. Limiting the pressure while making the changeover provides a greater margin of safety, however.

The transfer valve, the four-way valve, and the connecting lines are all internal to the pump. This means that there is no way to manually override the transfer if it does not work. The only way to ensure that it will work when needed is to exercise it frequently. As with any device that involves moving metal parts within a metal housing, the possibility of rust and corrosion is always present. Frequent use prevents a buildup that will inhibit or prevent operation altogether.

If the valve seems sluggish, building the pressure up to a somewhat higher level and operating the transfer a number of times may clean the surfaces and allow for a quicker operation. If the valve is stuck, it may not transfer immediately but may do so suddenly when the pressure builds up. Only close attention on the part of the pump operator can prevent a sudden pressure change when this happens.

Many Seagrave pumpers have a red arrow on the master discharge pressure gauge indicating the point at which the pump should be changed from VOLUME to PRESSURE. The UL test plate will also specify which position the pump was in for the initial tests. This changeover point is usually somewhere between 180 and 200 psi (1 260 kPa and 1 400 kPa). From the standpoint of effective fireground operations, a better indication of how the transfer valve should be set is the amount of outlets in use. The pump is designed to supply up to 50 percent of its capacity with one impeller. As a rule of thumb, anytime more than 50 percent of the required outlets are in use, the pump should be operating in VOLUME. If less than 50 percent of the outlets are used, the pump should be in the PRESSURE position.

The pump used on the cab-forward custom apparatus or commercial chassis uses an external cylinder to transfer from SERIES to PARALLEL (Figure F.8). It also uses water pressure to move a piston within the power transfer cylinder. By connecting discharge pressure to one end of the cylinder and intake pressure to the other, the piston is forced to one end of the cylinder. By reversing the connection through the four-way valve, the piston moves to the other end of the cylinder.

Figure F.8 The external cylinder pictured in the top of the photo is used to make the changeover on custom, cab-forward Seagrave apparatus.

The power cylinder is connected to a transfer valve that turns 90 degrees with each movement of the piston. This directs the output of the first stage impeller either into the discharge of the pump for PARALLEL operation or to the intake of the second stage impeller for SERIES operation. This transfer can also take place at a relatively high working pressure since the pressure of the pump is being used to cause it to move.

The same safety considerations apply here as with the other Seagrave pump. Since the power transfer cylinder and the four-way valve are mounted external to the pump, it is possible to assist the transfer manually if difficulty is experienced. However, it is far more effective to exercise it frequently and keep it operating properly.

Later models of this pump used compressed air from the braking system to accomplish the transfer instead of water pressure. Using compressed air eliminates the problem of metal parts moving in water and provides a more maintenance-free and reliable system. To use this transfer, just pull on the volume-pressure control and leave it in that position as long as the pump will be operating in series. Pushing it in will cause the pump to return to volume. A similar arrangement of some compressed air transfer systems uses a knob with a pointer that can be turned to the VOLUME or PRESSURE position (Figure F.9). Other air pressure transfers use a spring-loaded key that must be held in the proper position until the transfer has completed, then released (Figure F.10).

Figure F.9 The pointer on the knob indicates which mode the pump is in.

Figure F.10 This spring-loaded key is pushed in the direction of the desired mode. The key is held in that position until the changeover is completed.

PUMP ACCESSORIES

Since the Seagrave pump is only installed on apparatus constructed or contracted by Seagrave, accessories are relatively standardized in construction and operation.

Priming Devices

All Seagrave pumps use a rotary vane vacuum pump for priming (Figure F.11). It has four vanes mounted in a rotor positioned off center within the housing. As the rotor turns, the vanes are forced out against the housing by centrifugal force forming an airtight seal. The area where the air enters the pump has a large space between the housing and the rotor. This space becomes smaller as the vanes turn, compressing the air and building up pressure, until it gets to the discharge of the priming pump where it is expelled.

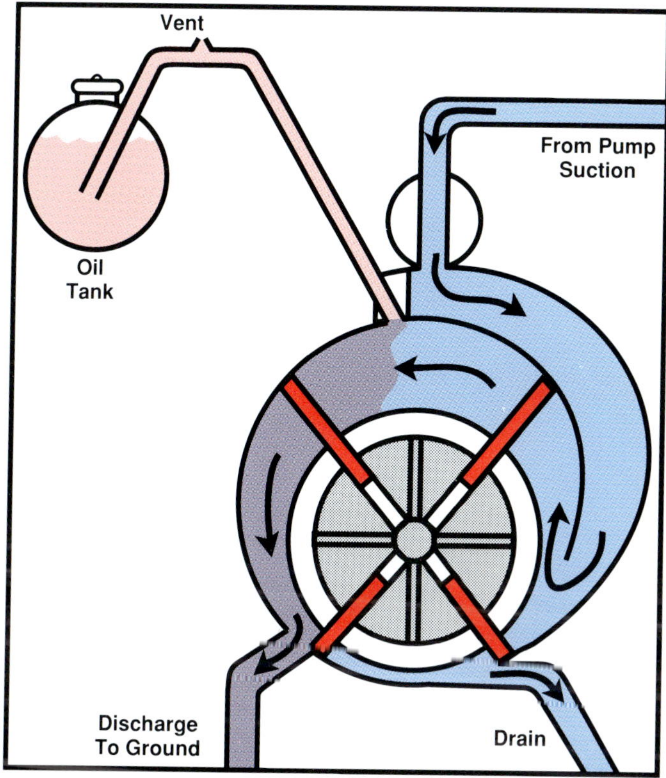

Figure F.11 All Seagrave pumps utilize the rotary vane vacuum pump for priming.

The priming pump is driven by a cone-type clutch from the gear box in the pump. The clutch is engaged by pulling on a "prime" control on the pump panel and energizing a vacuum cylinder. This control engages the clutch to make the priming pump turn and at the same time opens a valve connecting the vacuum pump to the intake side of the fire pump. This allows the air to be evacuated from the pump and water to be forced in by atmospheric pressure. As the pump fills with water, the priming pump will also fill. Water will be discharged to the ground under the pumper.

Effective priming requires an airtight seal within the priming pump. The vanes on the rotor are self-compensating for wear since centrifugal force keeps them against the housing. To compensate for irregularities in the housing, each of the vanes is equipped with a sealing strip. The sealing strip is easily replaceable where it comes in contact with the pump housing.

The Seagrave primer also uses priming oil both as a lubricant and to provide a better seal while priming. The priming pump is connected through copper tubing to an oil reservoir filled with SAE 90 oil. In extremely cold weather, a lighter grade oil such as SAE 50 may be used for free flowing and better sealing. Two vents are required in the system. As the oil is drawn from the tank, air has to replace it to prevent a buildup of vacuum that will restrict the flow of priming oil to the pump. A vent in the cap of the reservoir takes care of this.

Since the priming pump is generally mounted physically lower on the apparatus than the reservoir, there is a tendency for the oil to continue to flow by a siphon action after the priming pump has stopped turning. A small vent at a high point in the line is used to break the siphon. If this vent becomes clogged, oil will continue to leak out of the primer while the apparatus is not in service. It is a good practice to operate the primer dry for a few seconds after the pump has been drained at the end of a pumping operation. This leaves a protective coating of lubricating oil on the moving parts of the vacuum pump.

To prime the pump, after an airtight waterway has been established to the water source, put the pump in the PUMP position, the transmission in the proper gear, and set the engine rpm to 700-800 rpm. Pull out on the prime control and hold it until the pump is primed. Water discharging from the primer and a buildup of pressure on the discharge gauge are two indications that the pump has been successfully primed. The prime control can then be released and the primer will stop turning.

Automatic Pressure Control

All Seagrave pumpers use a pressure governor for automatic pressure control. The same basic device is used for controlling gasoline or diesel engines with only minor changes in the governor cylinder itself.

The Seagrave governor is of the adjustable spring reference type with discharge pressure pressing against a piston connected to a throttle linkage. In earlier gasoline engines, control of the engine speed was done by a separate butterfly valve mounted between the carburetor and the intake manifold. Later gasoline-powered units and all diesel engines use a governor control cylinder with a linkage from the piston to the throttle mechanism (Figure F.12). A rubber diaphragm transmits the pressure to the control piston without allowing it to come in contact with water.

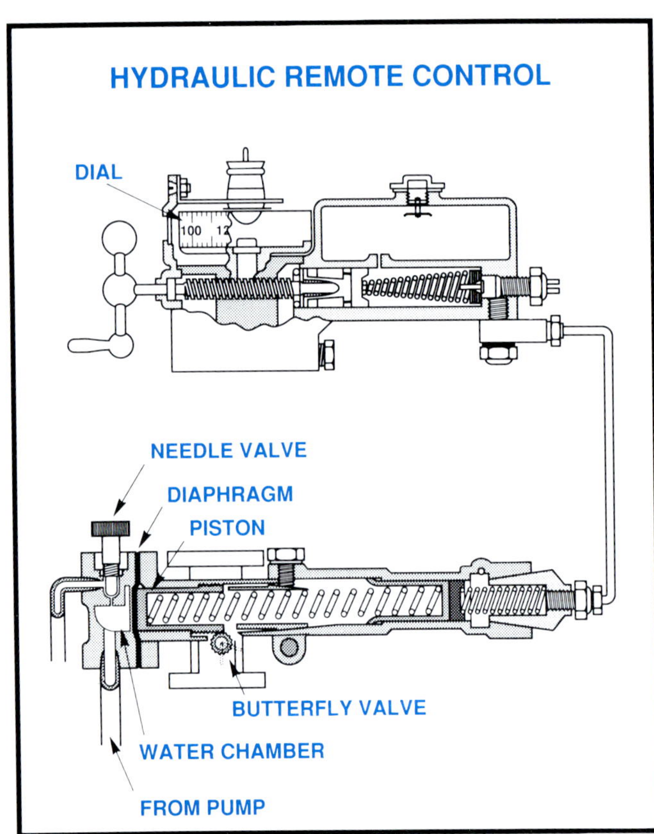

Figure F.12 Diagram of the Seagrave pressure governor.

The Seagrave governor is unique in that the spring tension adjustment is done remotely. The adjustment control is connected through a pressure line filled with hydraulic brake fluid to a slave cylinder within the governor itself. When the hand crank is adjusted, the change in position is transmitted through the hydraulic system to the slave cylinder, which adjusts the spring tension in the governor (Figure F.13). The position of the operating governor cylinder is determined by the setting of the hand crank on the control mechanism.

Mounting the governor in the immediate proximity of the carburetor or fuel system eliminates lengthy and complex mechanical linkages and makes the governor extremely accurate and fast acting. The hand crank has a dial connected to it that provides an approximation of the pressure setting. It also has a dial light that comes on anytime the hand crank is set above 0 psi (0 kPa).

Figure F.13 A hand crank is used to adjust the governor.

Setting this governor is much like setting a relief valve. It does not have the capability of increasing engine rpm above the setting of the hand throttle but will only act to limit pressure increases. To set the governor, all lines must be flowing at the maximum flow rate. The hand crank should be turned to a point approximately 50 psi (350 kPa) above the desired setting. Set the discharge pressure approximately 10 psi (70 kPa) above the desired operating pressure with the hand throttle. Open the governor shutoff valve. If any change in pressure is observed when this is done, it is an indication that the hand crank on the governor has been set below the actual operating pressure of the pump. The handle should be turned clockwise until the pressure comes back to the original setting.

Once the governor control has been set higher than the discharge pressure of the pump, it should be turned counterclockwise slowly until the pump pressure decreases to the desired setting. The pump is then operating under the control of the pressure governor. Increasing the hand throttle setting will not have any effect on the engine rpm or the discharge pressure. To change the discharge pressure, the governor control should be changed to the new setting and the hand throttle increased to attain the new operating point. If the pressure is to be decreased, the hand throttle should be turned in until the pump is operating 10 psi (70 kPa) above the new pressure. Then the crank on the governor control can be readjusted as before.

A measure of flexibility can be attained by increasing the hand throttle several turns once the pump is operating under the control of the governor. This will allow the engine to speed up enough to compensate for increased flow or existing lines or the addition of other hoselines. By increasing the hand throttle to its maximum setting, the full range of engine power would be available for changes in flow with no additional adjustment, but there is a danger in doing this. With the throttle setting at maximum and the engine rpm limited only by the water pressure in the governor, a loss of water supply could have serious consequences.

If the pump runs dry, the pressure drops, and the governor will increase engine rpm to try to recover the established pressure setting. In doing so, the engine may run away with itself and damage to the engine or the pump may result. It is a good operating practice to limit any increase of the hand throttle to one or two turns to provide for minor changes in flow automatically, while limiting the amount of change that may take place to a safe operating range.

Like all pressure governors, the Seagrave governor is susceptible to hunting when the speed of response of the governor is more rapid than the response of the engine to changes in setting the throttle. To take care of this problem, the Seagrave governor has a needle valve in the control chamber. This valve has a flattened point so that a certain amount of water will always flow through the system. The wider the valve is opened, the more water can escape and the longer it will take for the pressure to build up in the governor and cause it to take control. At the same time, the more it is opened, the slower the response will be to changes in pressure.

To match the governor to a particular engine, the needle valve should be closed completely. If the governor tends to overshoot the mark in making pressure changes, the valves should be opened slowly. This should be done until a sudden change does not cause the engine to fluctuate, hunting for the proper setting. Once the needle valve is set for a particular engine, it should require very little attention. However, it should be flushed out by opening the needle valve fully after pumping dirty water, or there is an indication that the governor is becoming clogged with dirt.

Extra Cooling

Seagrave pumps will be equipped with an auxiliary cooling device. This will be a heat exchanger where pump water flowing through copper tubing comes in contact with engine coolant and decreases the temperature of the coolant. Custom apparatus may have the heat exchanger built into the radiator. Coils of copper tubing are submerged in the radiator coolant with pump water circulating through the tubing. The degree of cooling depends on the amount of water flow through the cooler as controlled by the auxiliary cooler control on the pump panel (Figure F.14).

Figure F.14 The amount of pump cooling can be adjusted by manipulating the auxiliary cooler control on the pump panel.

Other models use some type of external heat exchanger, many of the marine type. Whatever the construction, the auxiliary cooler provides a means of extra cooling of the engine during prolonged pumping operations. The radiator coolant is not contaminated because the water is circulated through the fire pump. The auxiliary cooler is constructed so that the pump water does not contact the coolant in the radiator and will not dilute the antifreeze solution. The auxiliary cooler should be set to maintain the operating temperature of the engine within the recommended range as specified in the operator's instruction manual for the engine.

The radiator fill valve will contaminate the antifreeze solution if it is opened (Figure F.15). It is an emergency control and should only be opened when the engine is overheating. Overheating occurs when the water level in the radiator is low. Opening the radiator fill valve allows water from the pump to fill the radiator.

Figure F.15 The radiator fill valve should only be used in extreme emergencies because it will contaminate antifreeze solution.

The circulator valve that has been installed on some apparatus is used to prevent the pump from overheating (Figure F.16). It is opened when intermittent operation is experienced, such as occurs during initial attack of a fire with small handlines. The circulator valve will maintain a flow of water through the pump when all hoselines are shut down. It is limited in its ability to keep the pump from overheating by the size of the cooling line. Most of these valves use 1/4 or 3/8-inch (6 mm or 10 mm) copper tubing and will only flow approximately 10 gpm (40 L/min). This will not be sufficient for prolonged operation at high pressure and additional cooling such as flowing a booster line should be considered.

Figure F.16 The circulating valve, when open, allows a constant flow of water through the pump when all attack lines are shut down.

An intake relief valve may be provided to protect the pump during relay operations. This valve is connected to the intake manifold of the pump. It is set so that when the incoming pressure exceeds a predetermined amount, the valve opens, allowing water to escape from the intake. This increases the friction loss in the supply line, thereby limiting the residual pressure increase at the pump. These valves are commonly set to 150 psi (1 050 kPa) and left there for the protection of the pump.

A more effective way to operate is to set the valve to the operating residual pressure during each pumping operation. In this way a relatively constant pressure will be maintained on the intake allowing for a more stable operation. The intake relief valve should be equipped with 2½-inch (65 mm) NST threads and be capped when not in use. A shutoff valve is also useful and should be provided with the apparatus.

Appendix G

Description and Operating Practices for Waterous Fire Pumps

(Note: Unless otherwise indicated, all photos in this Appendix are printed courtesy of Bill Eckman.)

The Waterous Company manufactures a full line of fire pumps that can be arranged for use as direct engine drive, front mount drive, or midship mounted for split shaft, flywheel PTO, or transmision PTO drive systems. They range from booster-type pumps with capacity as low as 60 gpm (240 L/min) to high capacity applications up to 2,000 gpm (8 000 L/min) for use in attack pumpers or water supply units. High-volume pumps are available as single-stage or two-stage pumps, midship or front mounted. A third stage may be added for high-pressure operation with either a single-stage or two-stage volume pump. A four-stage centrifugal pump is also available that will provide ultra-high pressure for specialized applications.

Since 1955, many Waterous pumps have used an electrically powered pump shift mechanism and electric transfer valves for two-stage operation. Some applications even include electrically controlled discharge valves. A special gearing arrangement is available to permit pump and roll operation with a split shaft midship pump at a reduced capacity up to 250 gpm (1 000 L/min). Full capacity is available for stationary pumping.

Since the Waterous Company does not construct fire apparatus, there is very little standardization in the installation of their pumps. Pumps and accessories are installed by each fabricator to meet the needs of their particular design and construction practices.

FRONT MOUNT PUMPS

The front mount pump is generally mounted behind the bumper, just in front of the radiator of the apparatus. The frame rails are extended approximately 2 feet (0.6 m), and the pump is mounted on a support between the rails. This provides for easy access to the pump for operation and maintenance, and protects the pump from damage in case of a front-end collision. Power to drive the pump is provided by a power take-off unit fastened to the front of the crankshaft of the engine. The pump is independent of the drive train of the vehicle and its discharge pressure is controlled only by the engine speed; there is no relationship to the position of the road transmission or clutch. For this reason, the front mount pump can be used for pump and roll operation. Power is transmitted from the PTO unit, through a drive shaft and a dry clutch, to the gear box. The gear box provides the rpm needed to develop the desired pressure flow.

Two basic gear cases are available: medium duty and heavy duty. The medium-duty case has an 8-inch (203 mm) diameter clutch, and the heavy-duty gear case has a 9-inch (229 mm) clutch for pumps 500 gpm (2 000 L/min) and higher. Each of the gear cases can be supplied with several different gear ratios to better match the required impeller shaft speed requirements to the torque and horsepower curve of the engine being used. In most cases, the rpm of the engine required to supply the rated capacity of the pump will not exceed the equivalent of 35 mph (56 km/h) road speed in the direct drive gear of the road transmission.

Two basic pumps are used for front mount Class A pumpers. The CG or CL series is normally used to provide 500 or 750 gpm (2 000 L/min or 3 000 L/min), while the CX will supply either 1,000 or 1,250 gpm (4 000 L/min or 5 000 L/min) (Figure G.1 on next page). In establishing the rating of the pump, the gear ratio is changed to turn the impeller faster for higher capacity. The intake manifold is sized from 4½ to 6 inches (115 mm to 150 mm), and the discharge manifold varies from two to four 2½-inch (65 mm) outlets. NFPA 1901 requires one 2½-inch (65 mm) outlet for every 250 gpm (1 000 L/min) of capacity, but this require-

Figure G.1 Front mount pumps are available with up to 1,250 gpm (5 000 L/min) capacity. *Courtesy of Waterous Co.*

ment can be met by using preconnected lines or rear discharges to supplement the outlets on the discharge manifold. In this way, a 1,000 gpm (4 000 L/min) pump may have only three outlets on the discharge manifold with a rear discharge making up the difference.

Piping, Valves, and Gauges

Discharge valves are usually quarter-turn, quick-opening, locking-type ball valves (Figure G.2). To open or close the valve, the handle lifts up, releasing the lock, and the valve turns as much as a quarter-turn. The valve can be locked in any position by pushing down on the lever, thus applying pressure to the locking plate. These valves have stops that prevent them from turning past the two extreme positions. They should always be locked in position when the outlet is in use and unlocked when it is not. Each valve should be equipped with a drain valve on the line side of the valve to allow the pressure to be relieved when an individual line is shut down while the pump is still operating.

Each discharge valve may also have an individual line pressure gauge connected to the same location. This gauge is important when different types of attack lines are put in service simultaneously with varying pressure requirements. When this happens, the discharge pressure of the pump must be set to the highest pressure that will be required. Other lines can be adjusted to a lower pressure by partially closing the valve until the individual line gauge registers the desired setting.

Figure G.2 Quarter-turn locking ball valves are used on the front mount pump. *Courtesy of Waterous Co.*

When operating this way, it is important that the handle of the valve be locked in place to prevent the setting from changing as the water flows through the valve.

Drain valves can be of any type that the apparatus fabricator chooses to use, but it is important that they be fully closed when the pump is shut down. Since they may discharge below the pumper, it is possible for drain valves to leak without the operator knowing it. The pump operator should be aware of any unexplained leakage and take steps to eliminate it as soon as possible.

Each pump is equipped with at least two gauges: a master pressure gauge and a compound gauge connected to the inlet of the pump. These gauges may be mounted directly on the pump housing itself, but this is not a good practice. The vibration that is always present in the pump will be communicated directly to the gauges, and it will be difficult, if not impossible, to keep them calibrated accurately. A better arrangement is to mount the gauges somewhere on the truck body, connected with flexible lines to the fire pump to isolate them from vibration. A remotely located gauge panel also provides the flexibility to install individual line gauges as well. Wherever they are mounted, it is important to keep them calibrated. This is especially so with the compound gauge. The physical size of the vacuum scale is such that small errors in zero calibration can make the gauge useless.

Waterous front mount pumps are equipped with a suction fitting that is cored to permit the engine coolant to circulate within the housing. This prevents freezing in cold weather. The pump housing will always feel warm to the touch when the engine is running. This warmth should not be confused with the overheating that may occur if the pump is used for extended periods without adequate water movement.

Operating the Front Mount Pump

Before putting the pump in service, the vehicle must be immobilized. Put the road transmission in the NEUTRAL position, and carefully operate the shift lock. The shift lock is provided to ensure that the transmission does not inadvertently shift into a drive gear while the pump is operating. This is especially important if the vehicle is equipped with an automatic transmission. If the transmission is engaged or left in the DRIVE position on arrival, the parking brake will initially keep the vehicle from moving. When the throttle is advanced preparatory to putting the pump in operation, the torque developed may build to the point that the parking brake no longer holds the vehicle. The operator is especially vulnerable when using a front mounted pump since he or she is standing directly in front of the vehicle where the pump controls are located and is likely to be injured if the vehicle moves forward. Set the parking brake and put blocks under the wheels as an additional safety factor.

With all valves closed and all caps in place, connect the water supply to the inlet side of the pump. If the pump will be operating from the tank on the apparatus, the tank-to-pump valves must be opened. The manner of installation of this piping varies with the fabricator who has installed the pump. In general, however, the tank-to-pump line will be connected to one of the 2½- or 3-inch (65 mm or 77 mm) openings on the side of the suction fitting through some type of ball valve.

If the pump will be expected to supply more than 400 gpm (1 600 L/min) from the tank, there will probably be more than one tank-to-pump line, each with its own valve installed where the line enters the pump. At least one of these lines will have a T-fitting to enable a 2½-inch (65 mm) gated suction intake to be installed between the tank valve and the pump. This will allow for relay operation or receiving supply lines from other apparatus. This inlet will have some type of full flow valve and should also be equipped with a bleeder or drain valve on the line side of the main valve. The bleeder valve provides a means for bleeding off the air that has been trapped in the supply line as it fills with water before opening the valve into the pump. If the air is not eliminated, the pump will probably lose prime. In addition, pressure fluctuations will be experienced when the transition is made from operating from the tank on the apparatus to receiving water from an external supply.

If both of the tank-to-pump valves are not opened before putting the pump in operation, the maximum flow that can be attained will be reduced. To provide more flexibility in construction and operation, these tank-to-pump lines may be combined with relief valve returns, intake relief valves, or other fittings. If this is the case, the tank valves will have to be in the proper position for the associated accessory to operate properly.

In order to prevent freezing in cold weather, the valve may be located at the outlet of the tank, and controlled remotely from the pump operating position at the front of the vehicle. Another way of handling this is to install a second valve at the outlet of the tank. In this way, drains can be installed at the low point of the line and eliminate the danger of freezing while the apparatus is traveling over the highway. With this arrangement, when the pump is taken out of service, both valves should be closed, the line drained, and the drain valve closed again to prevent problems when the pump is next placed in service. There is a danger with a second valve in that both valves must be fully open to ensure an adequate supply of water to the pump when it is next used. This can be easily overlooked.

After the pump has filled with water, either from the tank, from a hydrant, a relay, or priming from draft, engage the clutch to allow the impeller to turn and build up pressure. With the engine operating at idling speed, slowly move the clutch lever, located on the left side of the pump, firmly toward the radiator and away from the operator. Push the lever firmly, and make sure it moves "over center" so that the clutch locks in the ENGAGED position. Disengagement should also be done with the motor at idling speed to minimize stress on the drive train and clutch.

After the impeller is turning, increasing the engine rpm by turning the throttle control in a counterclockwise direction should cause the pressure gauge to register pressure on the discharge. If a discharge valve has been opened slightly, water will begin to flow. If this does not happen within a few seconds, air may be trapped within the pump housing. A momentary operation of the primer may remove it and speed the priming of the pump. The throttle should be adjusted to build the pressure above 50 psi (350 kPa) before the discharge valves are opened to begin flowing water. Open the discharge valves slowly while observing the pressure gauge. If pressure drops greatly, pause a moment, and in severe cases, operate the primer to reestablish discharge pressure. When the desired discharge lines have been placed in service, the throttle can be adjusted to set the desired operating pressure.

With the pump in operation, there is always the danger of overheating when the demand decreases to the point where there is not an adequate flow of water through the pump to keep it cool. This is especially true in the early stage of the attack when it is common for lines to be operated intermittently. If the pump is equipped with a circulator system, it should always be used, either in the TANK or SPILL position, depending upon which mode of operation is being used. In some cases this will not be enough. A circulator valve generally only allows approximately 10 gpm (40 L/min) flow. This will not be adequate if operating at high pressure for extended periods of time with little or no water flowing.

If the pump is equipped with a tank fill line, this can be used for cooling by opening the line partially to sustain enough water movement to prevent overheating. Front mount pumps often do not have tank fill lines. This is due to space limitations imposed by installing piping from the front of the vehicle, where the pump is, to the rear of the apparatus, where the tank is located. A good alternative is to use the hose reel for cooling. The smaller flow from this 1-inch (25 mm) line will not seriously detract from the pump capacity, but will provide sufficient flow to keep the pump from overheating. When drafting or operating from a hydrant, it is simple to direct the stream from the hose reel back into the source or under the vehicle to discharge in a safe location. When operating from the tank, where the water supply is limited, the stream from the hose can be directed into the tank fill opening. Whatever method is chosen, it is essential that an adequate flow be maintained to prevent damage to the pump from overheating.

Waterous front mount pumps are equipped with one stuffing box for packing where the drive shaft leaves the gear box and enters the pump housing. When the packing is adjusted properly, water will drop (10 to 60 drops a minute) from the packing gland when the pump is operating under pressure, but not run in a steady stream. If water is leaking steadily, the packing is too loose and air will leak into the pump during drafting. This air leak may make it impossible to prime the pump from draft. On the other hand, if there is no water leaking, no cooling water is being supplied to the packing, and the pump shaft will overheat. This will cause permanent damage if the pump is operated that way for any appreciable length of time.

Adjustment of the packing is done by adjustment of two nuts, one on each side of the packing gland. These nuts must be adjusted evenly to get a good seal with the packing. A good practice is to tighten each nut one-sixth of a turn, then let the pump run for about 15 minutes to allow the packing to adjust itself around the shaft. Check the leakage rate again and feel the gland to check for overheating. If the leakage is still too much, tighten the nuts another one-sixth turn. If the gland feels too warm, loosen the nuts a fraction of a turn and allow the packing to adjust itself. Continue to adjust the packing until the proper amount of leakage is attained.

Some Waterous pumps are equipped with a mechanical seal instead of packing. No adjustment can be made to the mechanical seal. If the mechanical seal leaks at an excessive rate (more than one drop every five seconds), it must be replaced. This requires disassembly of the pump, which must be done by qualified maintenance personnel.

A few front mount pumps may be equipped with a master drain, but even these may have additional drain valves at low points in the piping. Most front mount installations have individual drain valves located at low points, both on the pump itself and its associated piping. It is extremely important that front mount pumps be completely drained in freezing weather due to their vulnerable location on the front of the vehicle where they are exposed to wind and weather. It is a good idea to make sure all drains are closed again when the vehicle is in the station to prevent difficulties in priming when the pump is next used.

Various Waterous accessories are normally supplied with front mount pumps, installed in accordance with desired methods of operation.

MIDSHIP MOUNTED PUMPS

Midship mounted pumps offer much more flexibility in capacity and design. Unlike front mount pumps, they are not limited to a small space for installation of piping and accessories. Midship pumps can also provide a higher rated flow since the full power of the engine is available to drive the pump. The maximum capacity of the front mounted unit is limited by the crankshaft bearings and the allowable torque capabilities of the PTO unit. Midship pumps are manufactured at Waterous from 150 to 2,000 gpm (600 L/min to 8 000 L/min) capacity (Figure G.3).

Midship mounted pumps can be powered by a transmission mounted power take-off unit, a fly wheel PTO unit, or a split shaft arrangement with a pump drive transmission in the drive train between the truck transmission and the rear axle.

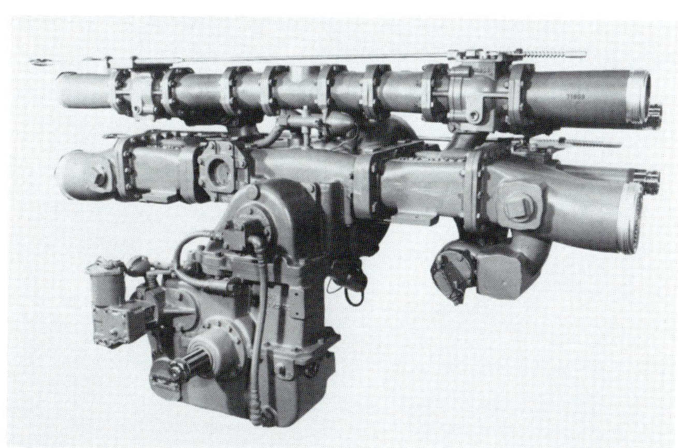

Figure G.3 A Waterous CM series midship pump. *Courtesy of Waterous Co.*

Transmission PTO

A midship mounted pump can be driven by a transmission mounted PTO with power being transmitted to the pump through a drive shaft equipped with Universal joints. This provides a great deal of flexibility as to where the pump can be mounted on the chassis, but the better the drive shaft on the pump lines up with the PTO, the less wear that can be expected. Since the PTO is driven from an idler gear in the transmission, it is not affected by the position of the other gears and can be used for pumping while in motion. The PTO is under the control of the vehicle clutch, and the clutch must be disengaged while the PTO control is moved. The power rating of the PTO limits the size of the pump that can be used. In the past, PTO pumps were limited to about 250 gpm (1 000 L/min), but with the advent of more powerful engines and larger transmissions, higher capacity is possible. Due to the lower cost and lighter weight, PTO pumps are often used on tankers and minipumpers.

Flywheel PTO

The flywheel PTO arrangement allows the pump to take power directly from the engine, with power to the pump through a drive shaft equipped with U-joints. This provides the same flexibility for mounting as the transmission PTO unit. It allows for pump and roll operation with speed ranges from 2 to 40 mph (3 km/h to 64 km/h) possible while the pump is operating. It has two primary advantages over the transmission PTO. Since it is driven directly from the engine, the inherent power losses in the road transmission are avoided, and the full horsepower of the engine is available to drive the pump. Since the drive system bypasses the road transmission and the vehicle clutch, it can be engaged or disengaged while the vehicle is in motion. Further, it is not affected by the need to disengage the vehicle clutch to change gears while pumping in motion. This type of drive is especially well adapted to high power diesel engines and automatic transmissions as installed in many later model pumpers.

Split Shaft Transmission

The split shaft transmission consists of a gear box inserted in the drive train between the road transmission and the rear axle. A sliding clutch gear can be moved forward to direct the power to the impeller shaft to operate the pump, or to the rear wheels for over-the-road travel. The shift takes place under the control of an electrically powered automotive type actuator that transfers power from the road to the pump and back. A two-position switch, mounted in the cab, actuates the shift unit. There is a spring-loaded guard over the switch to prevent accidental operation. A green indicator light shows that the transfer has been fully completed to the PUMP position and a red warning light shows if it has not. Pumps manufactured before 1955 may have some type of manual shift control, but all midship transfer pumps since that time feature some variation of this electric shift mechanism. Earlier models did not incorporate the indicator lights but the shifting action was the same.

The green pump engaged light may be extended to the pump panel in the vicinity of the hand throttle, with the warning notice "Do not open throttle unless light is on" (Figure G.4). This is particularly important if the vehicle is equipped with an automatic transmission.

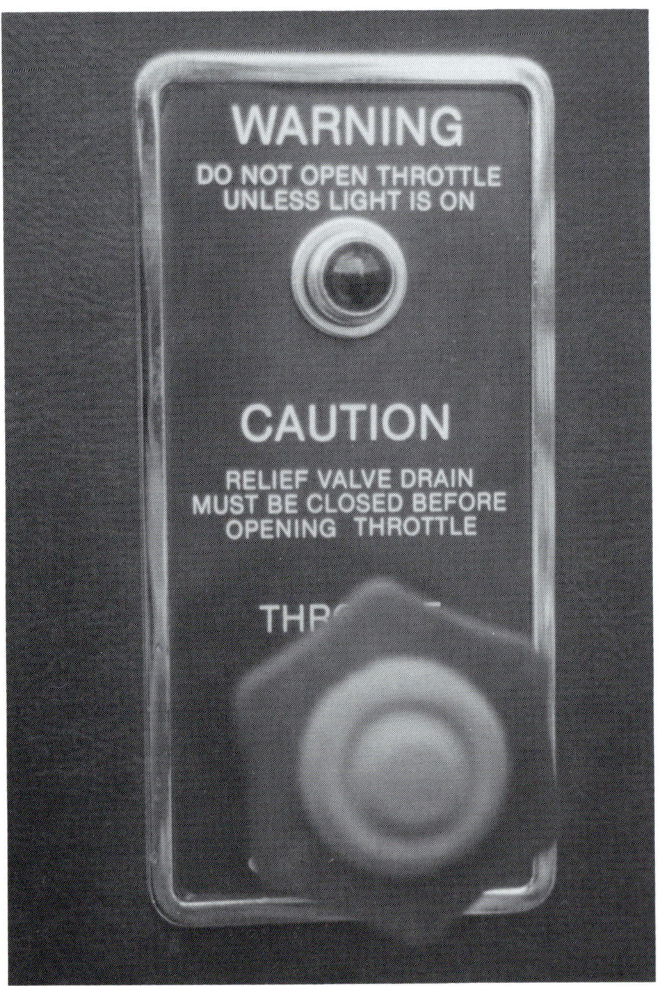

Figure G.4 The pump engaged pilot light indicates engine power has been transferred to the pump.

The parking brake will keep the vehicle from moving while the motor is operating at idle speed. When the automatic transmission is put into the DRIVE position, after the pump has been engaged, the parking brake will still keep the vehicle from moving initially. When the throttle is opened to build up pressure, torque increases, and the power will be transmitted to the rear wheels if the shift has not completed. At approximately 1,500 to 2,000 rpm, sufficient torque will be developed to overcome the drag of the parking brake, and the vehicle may move away from the operator. This can happen even if the wheels have been blocked before operating the pump. The vernier throttle has a red quick-release button in the center. The operator should immediately push this button all the way in if the truck begins to move.

The electric shift includes a provision for manually overriding the shift mechanism in case of electrical failure. The override may or may not be extended to the control panel, depending on the fabricator that installed the pump on the chassis and the specifications developed by the purchaser. The manual override involves two separate controls (Figure G.5). One of these disengages the electrical shifting mechanism, and the other controls the movement of the sliding collar inside the gear case, which actually transfers the power from road to pump. The mechanism is changed, either from road to pump or from pump to road. As a precaution, it is a good idea to have the transfer switch in the desired position while the pump is operating, even if the switch does not work and the manual controls are used. If the problem is an intermittent condition, sporadic operation could cause damage to the shift mechanism or the gear boxes. In some cases, sporadic operation can even cause the vehicle to move inadvertently.

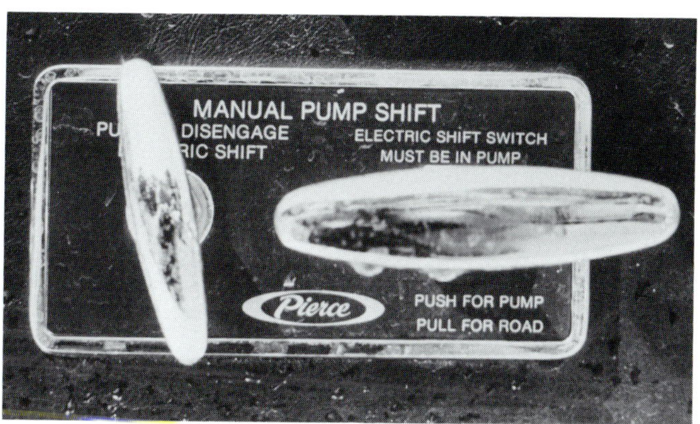

Figure G.5 Manual override pump shift controls.

Putting the Pump in Operation

The vehicle should always be brought to a complete stop before the pump is put into operation. Even though the engine may not be driving the pump transmission because the clutch is disengaged or the automatic transmission is in neutral, movement of the rear wheels will cause the gears in the transfer case to turn. This will cause the gears to clash if the sliding gear is moved to transfer the drive from road to pump at that time. After stopping, immobilize the vehicle by setting the parking brake and by putting blocks under the wheels at the earliest opportunity.

The road transmission must be in the NEUTRAL position while the shift is made. If the vehicle is equipped with a manual transmission, the clutch must be disengaged while shifting either into the ROAD or PUMP position. With an automatic transmission, the gears will not be turning while it is in neutral and the transfer can take place. If the automatic transmission is equipped with a transmission brake, operate the brake before shifting the pump transfer case.

Lift the guard and operate the toggle switch on the dash from the ROAD to PUMP position. The green light should come on, indicating that the pump is engaged. If the red light flashes, it indicates that there is pressure on the shift collar that will not let the shift complete. If this happens, shift the automatic transmission from neutral to reverse, back to drive, then back to neutral. If the vehicle is equipped with a manual transmission, engage the clutch momentarily, then disengage it. At that time, the shift should complete. If it still does not complete, it may be necessary to use the manual override controls. This can be done by first putting the transmission in neutral. With the clutch pedal depressed, disengage the electrical shifting mechanism and move the shift collar manually to the PUMP position. If this is done, be sure that the two-position switch remains in the PUMP position while the pump is operating. Always verify that the shift has actually completed before turning the throttle to increase the engine speed.

Put the road transmission in the recommended position. With a manual transmission, this will be in direct drive, the highest gear, unless the vehicle has an overdrive transmission. An automatic transmission will usually pump in the DRIVE or 3-6 position. Once the transmission is in gear, lock it in place with the safety latch that is provided. One indication that the transmission shift lever is in the proper position is that the manual shift lock falls into position easily to keep the transmission from slipping out of gear.

If it is difficult to get the lock to engage, check the transmission for the proper gear. It should match the plate on the dash that specifies the proper pump gear for the transmission. If the vehicle is equipped with a manual transmission, engage the clutch. At this point, the speedometer should be registering a road speed between 10 and

30 mph (16 km/h and 48 km/h), depending on the rpm of the motor at idle and the gear ratio being used. Only familiarity with the pumper will provide a reliable benchmark for this reading. If the apparatus is not equipped with shift indicator lights, the reading on the speedometer provides a verification that the shift has actually taken place. If the truck is equipped with an automatic transmission and does not register any reading on the speedometer, depress the accelerator and verify that the speedometer registers. This reading on the speedometer, while the vehicle is stationary, is an indication that power has been transferred from the rear wheels to the pump drive gears. There may be an audible indication that the shift has taken place. Listen for the sound of operation of the pump shaft and of the gears turning in the pump transmission after the transfer has taken place. These are other indications that the pump is turning and ready for operation.

If the pump is equipped for pump and roll operation, two controls are provided, a front shift and a rear shift. Operation of these controls are identified by four lights. In the ROAD position for normal driving, the toggle switch for the front shift will be in the DOWN position, while the toggle switch for the rear shift will be in the UP position. The two road lights will be on. To put the pump in service for stationary pumping, lift up the guard and operate the toggle switch for the front shift to the UP position. The two lights designated for stationary pumping will come on and the pump will be operating in a conventional manner for a midship transfer.

To shift from the ROAD position to pump and roll, lift up the guard and operate the toggle switch for the rear shift to the DOWN position. The lights designated pump and roll should not be lit. Put the road transmission in low gear, engage the clutch if furnished, and the vehicle will start to move. With an automatic transmission, the 1-2 position should be used for pump and roll. As the engine speed increases, pressure in the pump will build up. The operator will have to limit the engine speed to maintain the proper pressure rather than the desired road speed. To do this, a pressure gauge in the cab is a very desirable accessory. Table G.1 gives the proper switch positions and light indications for this system.

TABLE G.1
SWITCH AND LIGHT POSITIONS FOR POWER TRANSFER INDICATION

TRANSFER ***POSITION***	**FRONT** ***SHIFT***	**REAR** ***SHIFT***	***LIGHTS ON***
Road	Down	Up	Upper Right, Lower Left
Stationary	Up	Up	Upper Right, Upper Left
Pump & Roll	Down	Down	Lower Right, Lower Left

If electrical controls are furnished with this gear case, manual overrides will also be furnished that operate in the same manner as the conventional gear case. In this instance, there will be two sets of manual controls. The front shift will have the manual controls on the right side of the vehicle for stationary pumping, and the rear shift controls will be on the left side of the vehicle for pump and roll operation.

If the pump will be operating from the water tank on the apparatus, it can be engaged and primed immediately upon arrival. If the water supply will be from draft, hydrant, or a relay, the pump should not be engaged until all connections have been made and the water is readily available. If the pump turns for any appreciable length of time before water is available to cool the packing, damage to the impeller shaft may result. Water will flow into the pump by gravity when operating from the tank, but in some cases, air will be trapped in the pump housing and cause difficulty in building up pressure. If this happens, momentary operation of the primer may speed the operation.

Recent models of Waterous midship transfer pumps use a chain drive system from the drive shaft to the impeller shaft instead of gears (Figure G.6). While the pump is in operation, the oil pressure for the chain drive should indicate between 15 and 30 psi (105 kPa and 210 kPa) (Figure G.7). If this is not the case, damage to the drive system may result.

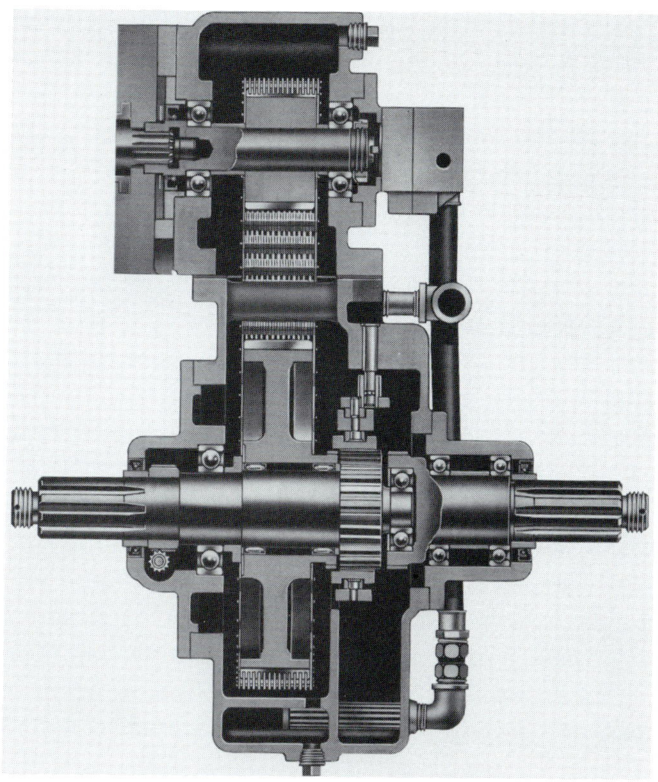

Figure G.6 Cutaway of Waterous chain-driven midship pump transmission. *Courtesy of Waterous Co.*

Figure G.7 The operator's panel pump transmission oil pressure gauge.

Waterous midship pumps use a similar packing arrangement to the front mounts. The primary difference is that most of them have two stuffing boxes, one on each end of the impeller shaft. Each of them has six graphite-impregnated packing rings and a lantern ring to provide water under pressure for cooling from the discharge side of the pump. The packing is adjusted by tightening two unbalanced nuts, one on each side of the packing gland, at the same rate to keep the pressure balanced on the packing rings (Figure G.8). When the packing is adjusted properly, there will be from 10 to 60 drops a minute coming from the packing gland. The procedure for making this adjustment is the same as with the front mount pump.

Figure G.8 The two adjustment nuts can be seen on the left of the pump packing.

To prevent water from entering the gear case through the ball bearing that is used to keep the impeller shaft in position, a flinger ring is mounted on the shaft. The flinger ring causes the water coming from the packing gland to be dispersed around the shaft (Figure G.9). This may make it difficult to gauge the amount of leakage accurately. Some water must be coming out, but should not be running in a steady steam. The exact amount is not critical.

Figure G.9 The flinger ring causes water coming from the packing gland to disperse from the shaft.

Valves, Piping, and Gauges

Each of the Waterous midship pumps can be supplied in different capacities. While the same impellers and other essential features in the construction of the pump are used, three variables are employed in changing the rating. First, different gear ratios are available to match the engine torque curves to the required impeller speed, which varies for different flow requirements. Second, the suction inlet can be furnished with 4½- to 6-inch (115 mm to 150 mm) threads for the size suction hose required for different flow rates. Third, 2½-inch (65 mm) discharges are added to handle the additional hoselines for the increased water movement.

NFPA 1901 requires that one 2½-inch (65 mm) or larger outlet be provided for every 250 gpm (1 000 L/min) of capacity. This requirement can be met by the use of rear discharges or preconnected lines. If necessary, these must be controlled from the pump operator's panel by remote control linkages. Some Waterous midship pumps may have valves with quarter-turn, quick-opening, ball valves with locking handles, but the standard installation is a rack and sector valve. These valves are opened and closed by pulling out or pushing in on a T-type handle. (**NOTE:** Refer back to Figure G.4). They can be set accurately in any position and locked in place by turning the T-handle to the vertical position (Figure G.10).

One variation on the standard rack and sector valve is the electric actuator valve. These valves are controlled by an electrical toggle switch that can be mounted in any convenient location. The valve can be set to any position by moving the spring-loaded toggle switch toward the OPEN or CLOSED position and releasing it when the valve is in the position needed to establish the required pressure. Manual override provisions are made for each valve. There are some cases where standard quarter-turn valves furnished by other manufacturers have been specified. Some are standard ball-type valves and can be locked in position by turning the handle in a clockwise direction. Others have been equipped with automatic locking valves. These can be placed in any position and will be held there as long as pressure is being supplied from the pump. Whichever type valve has been supplied, it should be locked in position when in use and left unlocked when the pump is not in service.

All discharge and intake valves on midship pumps should also be equipped with drain valves, sometimes referred to as bleeder valves, on the line side of the valve (Figure G.11). Individual pressure gauges are also connected to this point on the pump and used the same as on a front mount pump.

Centrifugal pumps are prone to heating when operating with little or no flow through them.

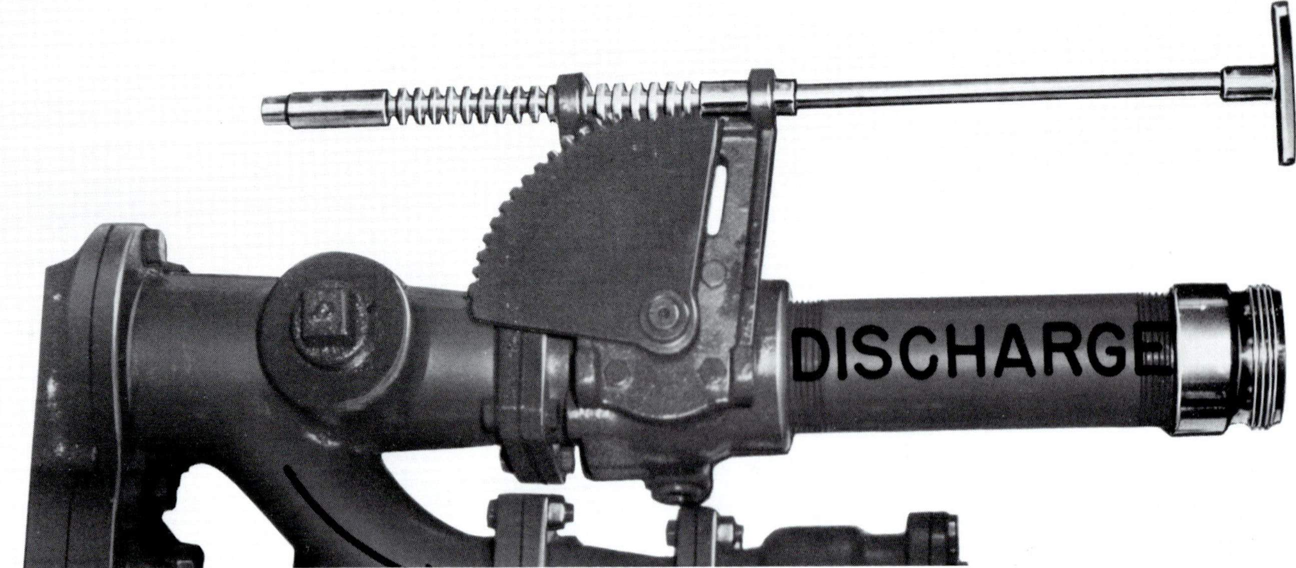

Figure G.10 Rack and sector control for discharge ball valve. *Courtesy of Waterous Co.*

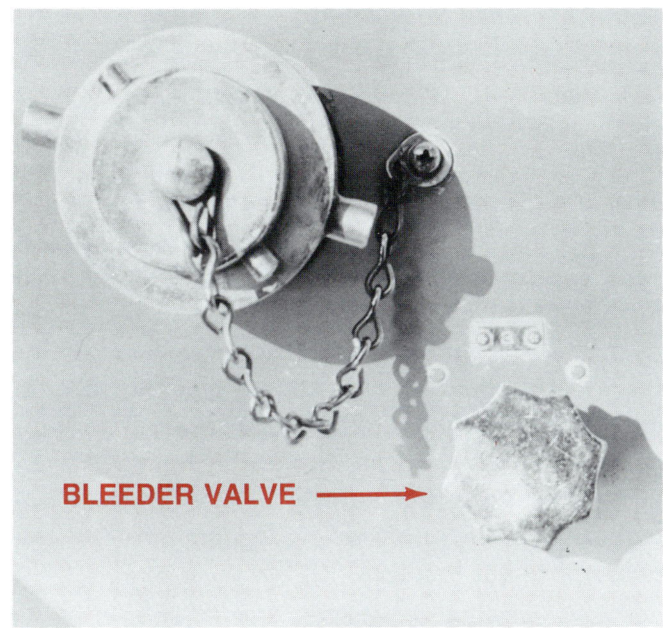

Figure G.11 Each discharge is equipped with a drain valve.

Large midship pumps present a greater problem, especially during the early stages of a fire attack when hose streams are applied intermittently. Overheating occurs because not enough water is being moved by the impellers as they churn. Operating at high pressure can also cause problems. As a general rule, the higher the pressure, and the larger the pump, the more water that has to move in order to prevent overheating. Most midship pumps are equipped with a tank fill valve. This valve is also used as a circulator line for keeping the pump cool during periods of low flow (Figure G.12). On a standard installation, this tank fill line is usually a 1- or 1½-inch (25 mm or 38 mm) line and may detract seriously from the capacity of the pump if fully opened. When opening from the tank, it is a good idea to open the valve partially, but not enough to cause the discharge pressure to drop appreciably.

There is a potential problem in using the tank fill line for cooling. Some midship pumps connect this line to the discharge of the first stage. This action limits the maximum pressure that can be applied to the tank and may prevent damage to the tank. However, it also means that when the pump is operated in the SERIES or PRESSURE position, all the water being circulated comes from the first stage. The second stage may be churning without any water movement, even if the tank fill valve is open.

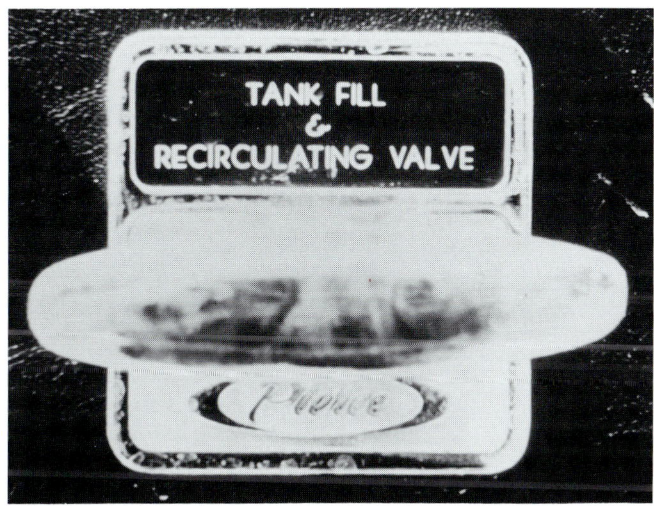

Figure G.12 The tank fill line can also be used as a circulator for pump cooling.

The pump panel is equipped with at least two gauges. One is a master pressure gauge connected directly to the discharge manifold of the pump, and the other a compound gauge connected directly to the inlet manifold (Figure G.13 on next page). Both these gauges and the individual line gauges can be compound gauges since a centrifugal pump has a vacuum throughout the system while priming. Only the inlet gauge must be

Figure G.13 A pump panel with larger compound gauges for the master discharge and intake gauges.

a compound gauge. Some installations have a lower maximum pressure on the intake gauge than on the discharge, but in others, they may be identical. A tachometer must also be installed on the pump panel, along with various engine indicators such as oil pressure, water temperature, and an ammeter.

Most midship pumps are equipped with a master drain that should drain all critical points of the pump (Figure G.14). This is a drain valve and not a pressure relief valve. It should never be opened when the pump is pressurized, since damage may result to the drain valve by doing so. If the O-ring in the master drain valve is damaged, there will not be a good seal when the valve is closed. This may make it difficult, if not impossible, to prime the pump to operate from draft. When there is a complex system of piping with associated low spots in the system, additional drain valves may be installed. These are usually located under the running board for convenience of installation and operation. During freezing weather, it is important that all these valves be opened when draining the pump of water; otherwise, water trapped within the pump or piping may freeze and cause physical damage.

It is a good practice to close all drain valves after the pump has been drained so that the pump is ready for service. Air leaks that will prevent a pump from being primed are often caused by open drain valves that were not closed after the pump was last drained.

The standard installation of the tank-to-pump line on the Waterous midship pump is through an

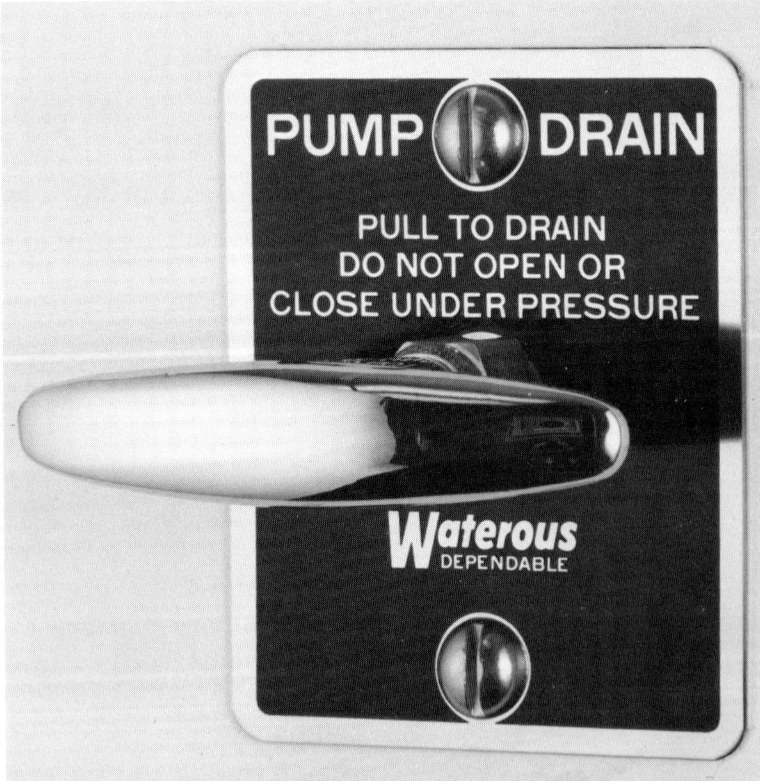

Figure G.14 The pump drain should not be operated under pressure conditions.

inlet on the rear of the intake manifold. This opening is equipped with a check valve built into the pump body to keep from causing damage to the tank if the tank-to-pump valve is opened while the pump is receiving water under pressure from another pump. This check valve makes it impossible to fill the tank through this line when water is available again after the tank has been emptied. The tank fill line can be used to refill the tank.

If the pump is expected to supply more than 600 gpm (2 400 L/min) from the tank, there will probably be more than one tank-to-pump line, each with its own valve installed where the line enters the pump (Figure G.15). If both the tank-to-pump valves are not opened before putting the pump in operation, the maximum flow that can be attained will be reduced.

Figure G.15 When large flows are required between the tank and pump, dual lines and controls are provided.

The second tank-to-pump line will be connected to another opening on the intake line. It will probably not be equipped with a check valve and can be used to fill the tank from a gated intake by opening the tank-to-pump valve while water is being supplied under pressure. Depending on the size of the two tank-to-pump lines and the method of installation, one of them may have a greater capacity than the other. This can affect the flow rate of the pump if only the line is being used.

In addition to the large inlet to the pump, each midship pump is also equipped with at least one gated 2½-inch (65 mm) fitting. These are connected to companion fittings on the inlet fitting and are independently controlled. The 2½-inch (65 mm) inlets greatly restrict the capacity of the pump. They are used mostly for relay operations, which are already limited in flow rate by the size of the supply hoselines. These gated inlets are also equipped with a drain or bleeder valve on the line side of the gate valve. This allows the air that is trapped in the hoseline as it fills with water to bleed off harmlessly before it enters the fire pump and causes fluctuations in the discharge pressure.

The large inlet can also be gated. A butterfly valve is available that mounts on the suction fitting, external to the pump. This can be used when setting up to operate from draft while supplying attack lines from the tank. Closing the valve allows the cap to be removed from the large inlet and suction hose connected without interrupting the flow from the tank. If a front or rear suction has been provided, it may be gated, thus eliminating the need to install an external valve. It may not be possible, however, to operate at full capacity from a front or rear suction. Due to the difficulty in routing the line from the pump to the front or rear of the vehicle, the size of the line may be reduced to the point that it restricts the flow. If this is the case, the operator needs to be aware of the limitations when using this inlet for drafting.

MULTISTAGE PUMPS

Two-Stage Pumps

Waterous two-stage pumps are conventional in design. They use two identical impellers, each designed to handle half of the rated capacity of the pump. They can be arranged in SERIES for maximum pressure with reduced volume, or in PARALLEL for maximum volume at reduced pressure. The transfer is controlled by a lever marked PRESSURE and VOLUME. When the control is moved, a ball transfer valve rotates within a cylinder to direct the discharge of the first stage either to the intake of the second stage for maximum pressure or to the discharge manifold for maximum capacity. In the SERIES position, there is a swing check valve, gravity operated, on each side of the pump. This valve isolates the intake of the first stage from the intake of the second stage to allow for increased pressure in the second stage.

Due to mechanical limitations of the shift mechanism, it is difficult to make the transfer with pressure in the pump. For reasons of safety to personnel and equipment, this transfer should never take place when the net pump pressure exceeds 50 psi (350 kPa). If the pump is equipped with only a manual transfer control, the only indication as to the mode of operation of the pump is the position of the control levers. It is normally arranged so that the handle is pulled out from the panel when the pump is in the VOLUME position, and pushed in for PRESSURE (Figure G.16). This mechanical linkage can be reversed for special applications. When it is a standard operating practice for the pump to be left in the VOLUME position between calls, the linkage can be arranged so that the transfer handle is pushed into the panel for this setting of the pump.

Figure G.16 A manual transfer valve.

If the operation of the transfer valve is suspect, it can be checked by setting the discharge to 100 psi (700 kPa) with the pump operating in the VOLUME position. Observe the engine rpm indicated on the tachometer. Reduce the pressure below 50 psi (350 kPa), move the transfer control from the VOLUME to the PRESSURE position, and adjust the throttle to the same engine rpm that was previously set. The discharge pressure should be appreciably higher than before, at least 1.5 times as much in PRESSURE as it was in VOLUME.

Electrically powered Waterous pumps may use a transfer valve actuator with a panel-mounted toggle switch to control the transfer from PARALLEL to SERIES (Figure G.17). Colored indicator lights are associated with the toggle switch to provide an indication of the position of the transfer valve. When the pressure light is on, the pump is arranged in SERIES with the discharge of the

Figure G.17 An electric transfer valve arrangement. **NOTE:** Pump is in PRESSURE.

first stage connected to the intake of the second (Figure G.18). To transfer the pump to the PARALLEL mode of operation, press the spring-loaded toggle switch toward the VOLUME position. As the transfer is made, the pressure light will be extinguished. As the transfer completes, the volume light will come on (Figure G.19).

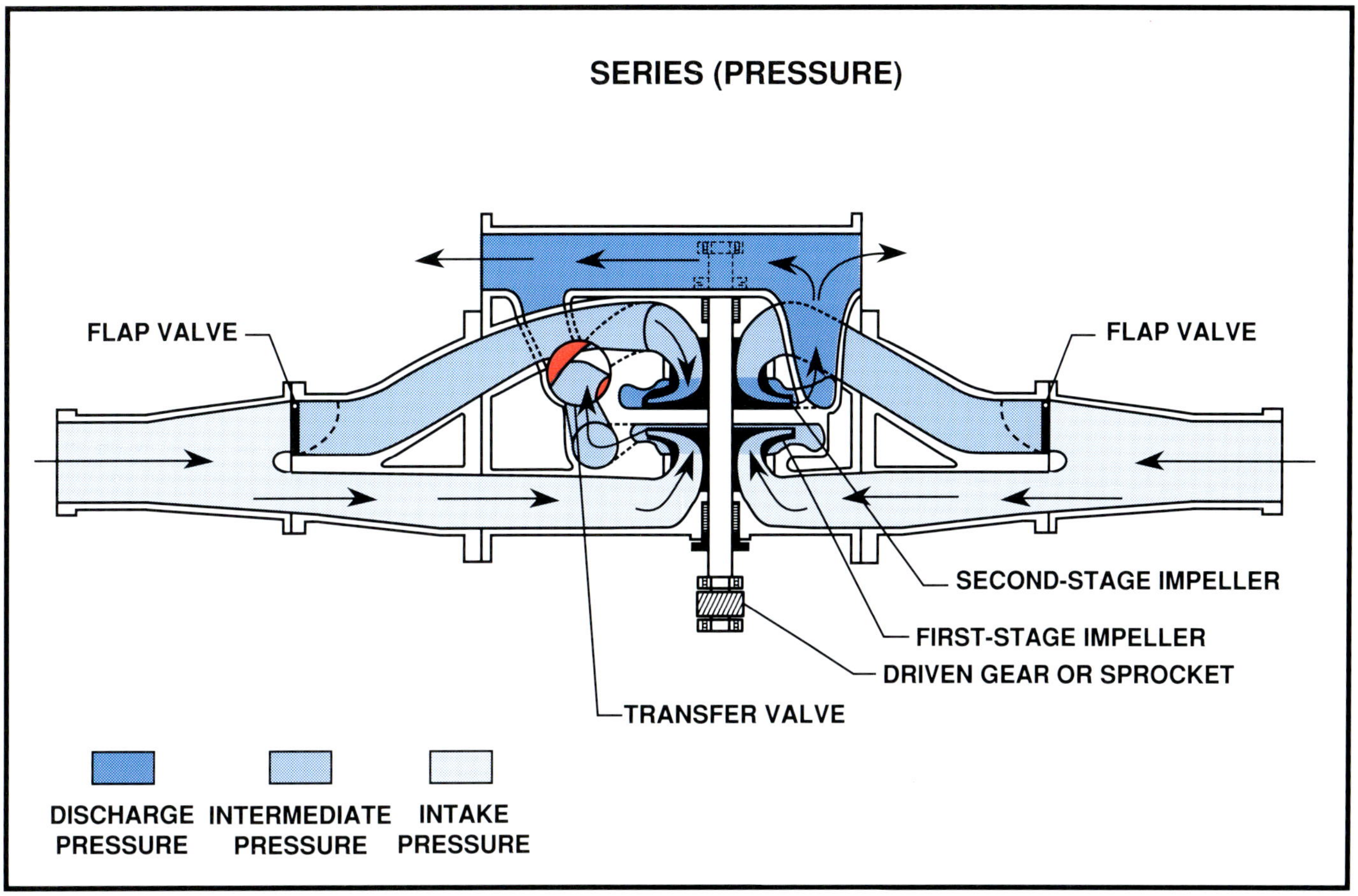

SERIES (PRESSURE)
FLAP VALVE
FLAP VALVE
SECOND-STAGE IMPELLER
FIRST-STAGE IMPELLER
DRIVEN GEAR OR SPROCKET
TRANSFER VALVE
DISCHARGE PRESSURE
INTERMEDIATE PRESSURE
INTAKE PRESSURE

Figure G.18 The diagram shows the valve arrangement when the pump is in the PRESSURE (SERIES) mode.

PARALLEL (VOLUME)
SECOND-STAGE IMPELLER
TRANSFER VALVE
FLAP VALVE
FLAP VALVE
FIRST-STAGE IMPELLER
DRIVEN GEAR
OR SPROCKET
DISCHARGE PRESSURE
INTAKE PRESSURE

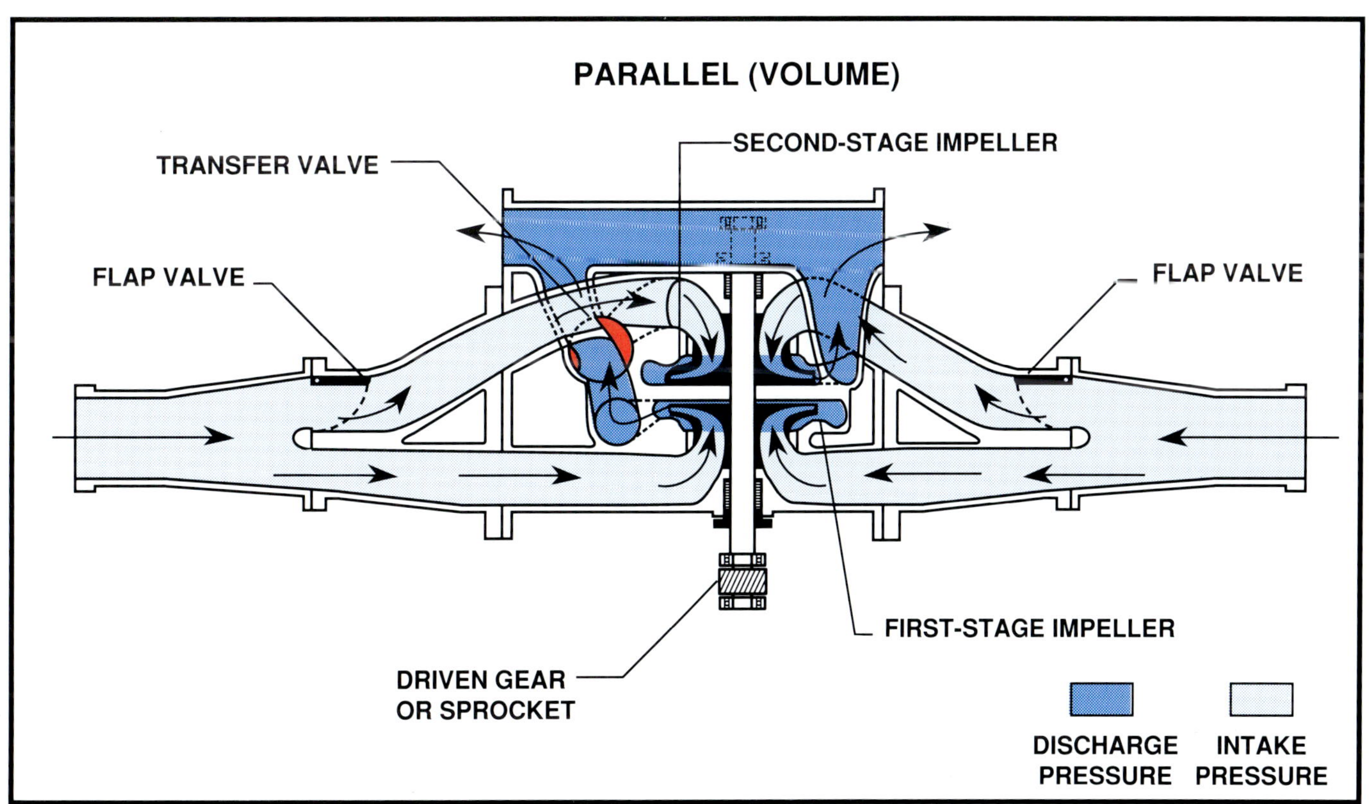

Figure G.19 This diagram shows the valve arrangement when the pump is in the VOLUME (PARALLEL) mode.

If the electric transfer does not work, the pump can still be changed by applying the hand crank that is supplied. In an emergency, apply a wrench to this hexagonal shaft and turn it until the transfer is completed (Figure G.20). Pressure indications can be used to verify that the transfer has been made. If the indicator lights do not come on as expected, the most likely cause is electrical trouble, possibly a burned-out bulb. Observe the manual override shaft and pressure gauges to verify the operation of the transfer mechanism.

The electric actuator applies considerably more force to the transfer valve and will operate up to 250 psi (1 750 kPa). It also accomplishes the transfer more slowly and with fewer pressure surges. In an emergency, the electric transfer can be made at pressures above 50 psi (350 kPa), but good operating practices still call for reducing the pressures below 50 psi (350 kPa) before making any change.

The setting of the transfer control is very important. The best guide for use of the transfer control is the number of hoselines being supplied. Since the pump is designed to be used in the PARALLEL position when more than 50 percent of the capacity is needed, it should be set to the VOLUME position anytime more than half the required outlets are in use. This means that anytime a 500 or 750 gpm (2 000 L/min or 3 000 L/min) pump is supplying more than one line it should be in the VOLUME position. A 1,000 or 1,250 gpm (4 000 L/min or 5 000 L/min) or larger pump could supply three hoselines before changing to volume.

Figure G.20 The transfer can be made manually in an emergency by turning the hexagonal shaft with a wrench.

Three-Stage Pumps

If a third stage has been provided, its impeller will be mounted on the same drive shaft as the other two impellers, but will only pump water when the control valve is opened. When the third stage control valve is closed, a drain cock on the bottom of the volute and a vent cock on the top automatically open. This keeps the third stage dry and prevents pressure buildup to dangerous levels when the first two stages are operating alone.

When the third stage is not pumping, water from the first stage is used for cooling the packing on the third stage. When the third stage is in operation, it provides its own water for the stuffing box. A check valve prevents water from being forced back into the first stage. The third stage can deliver water anytime the volume pump is operating, but maximum pressure will only be obtained when the transfer valve is in the PRESSURE position.

To put the third stage in operation and operate the pump for high pressure, turn the relief valve or pressure regulator to the OFF position. Set the transfer valve to the PRESSURE position and set the third stage control valve to the OPEN position. Open the valve to the high-pressure line and engage the pump drive system. Accelerate the engine to obtain the desired discharge pressure. To shut down, slow the engine to idling speed, close the third stage control valve, and disengage the pump. Always be sure to close the third stage control valve after operating the pump. This is necessary to cause the drain cock to open and allow the pump to drain.

SINGLE-STAGE PUMPS

Single-stage pumps have traditionally been associated with low-capacity front mount or PTO pumps. Waterous single-stage midship mounted pumps come in the same capacity range as two-stage pumps. To get a large flow with a single impeller, double suction impellers are used, receiving water from both sides. In addition to the increased capacity, this also provides a better hydraulic balance in the pump. The discharge side is also balanced, with water being taken off the impeller by two stripping edges located 180 degrees apart.

Operation of the single-stage pump is much simpler than the two-stage pump since there is no possibility of being in the wrong position at a critical stage in the pumping operation. Maintenance is also simpler since there is no transfer valve, suction check valve, or transfer mechanism.

Much has been made of the fact that a single-stage pump is more efficient than a two-stage pump when it is operating at or near 100 percent of its rated capacity, but fire pumps are rarely called upon to operate in this range. At low flow rates, such as would be encountered when supplying one or two handlines, the two-stage pump is more efficient. Efficiency is not a significant concern in fire apparatus today. The diesel engines in common use for fire apparatus have sufficient reserve power that inefficiency is not important. In fact, this extra horsepower is one thing that makes the single-stage pump well suited for use with a diesel engine. Many diesel engines are difficult to control at low rpm under light load. The higher rpm required to develop the necessary pressure when coupled to a single-stage pump provides a smoother and better controlled operation.

WATEROUS PUMP ACCESSORIES

The Waterous Pump Company does not build fire apparatus; rather, it supplies pumps and equipment to various manufacturers. Waterous accessories are intended to be used with Waterous pumps and are designed for flexibility to make them adaptable to a maximum number of applications. Waterous accessories can also be used with other manufacturers' pumps and are often installed when retrofitting older apparatus to modernize their operation.

Priming Devices

In the interest of weight, cost, and simplicity of operation, a portable pump, or some of the older front mounts, may use an exhaust primer. By operation of a lever, the exhaust gases are diverted from the muffler and directed through a venturi mechanism where enough vacuum is developed to evacuate the air from the pump. All other Waterous fire pumps use a positive displacement pump for priming, either a rotary gear or rotary vane type.

The rotary gear priming pumps (Figure G.21), and some of the rotary vane pumps, mount directly on the split shaft transmission or the pump itself. They can be driven by an automotive-type electric motor or by the pump drive train. Either power driver may employ internal gears or a drive belt if the pump uses a chain drive transmission. There is an option available that allows the pump to be driven by an electric motor under normal circumstances, and with a mechanically driven emergency system in case of electrical system failure.

Figure G.21 The rotary gear primer can be driven electrically, as shown, or mechanically. *Courtesy of Waterous Co.*

Waterous pumps use two different priming valves, either manually operated or fully automatic. Both valves are designed to allow the priming pump to evacuate air from several regions of the pump simultaneously for faster priming and elimination of air pockets that may be trapped inside the pump. At the same time the valve is operated to open the airway from the fire pump to the priming pump, an electrical switch, integral to the valve, closes, causing the priming pump to operate.

The manual electric priming valve accomplishes this by a single push or pull control operated from the pump operator's panel (Figure G.22).

The fully automatic priming valve uses a solenoid to put the priming system in operation; it is controlled by a single panel-mounted switch (Figure G.23). The priming valve is also provided with a lever to permit manual operation in case of an electrical failure. To prime the pump with an electrically driven primer, close all discharge valves, establish an airtight connection to the hard suction hose, and operate the prime control. Exact engine speed during priming is not critical, but the motor should be turning fast enough to keep the alternator charging. This is necessary because the priming motor requires a large amount of current to operate.

It is easier to prime a Waterous pump before the pump drive mechanism is engaged since the priming is done from the discharge of the pump. If the impellers are turning while priming, a false indication of priming can result when water is thrown into the discharge while there are still air pockets within the pump. Continue to operate the

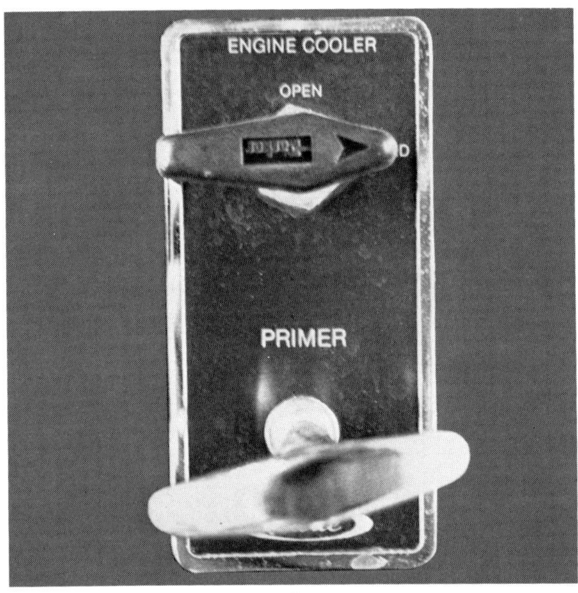

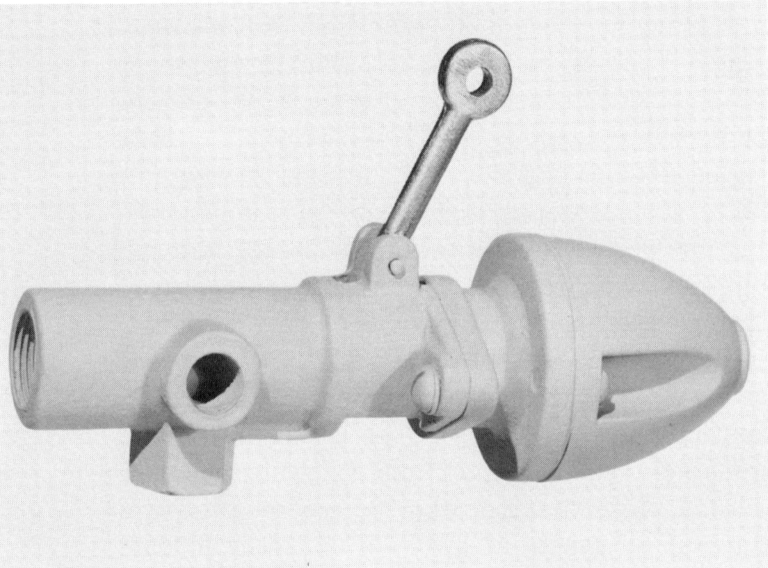

Figure G.22 The manual electric priming valve is operated by a pull handle control. *Courtesy of Waterous Co.*

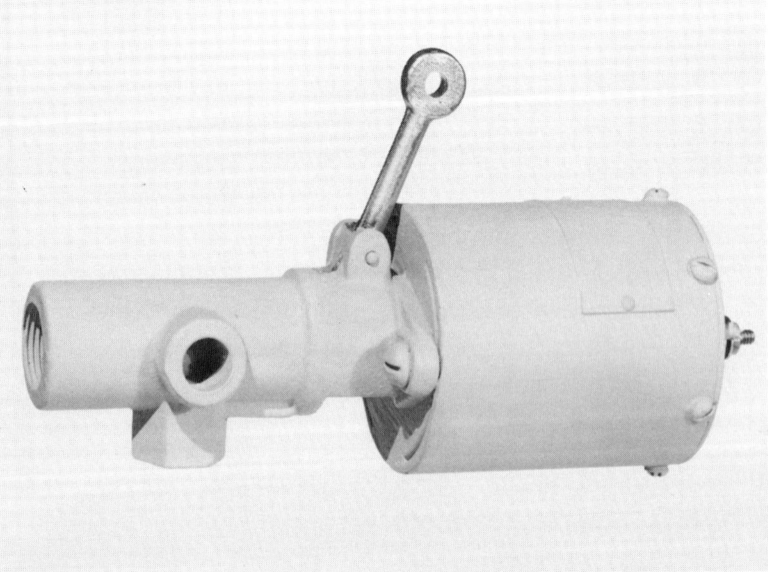

Figure G.23 The fully automatic priming valve is push-button operated with a manual override. *Courtesy of Waterous Co.*

primer until all the air has been evacuated and the pump is completely filled with water. The best indication that this has happened is a steady stream of water being discharged from the priming pump, usually on the ground under the pumper. Engage the pump. As the throttle is increased, pressure will build up as indicated by the master discharge pressure gauge. Build the pressure to at least 50 psi (345 kPa). Then open the discharge valve slowly so as not to lose the prime by an excessive demand for water when initially filling the hoseline. If the pressure should be lost as the discharge valve is opened, it is possible to operate the primer momentarily while flowing water. This will evacuate any small pockets of air that may have been trapped in the pump when the priming operation was completed.

If the pump has been equipped with the mechanical priming option, it should be possible to operate from draft in spite of a failure in the electrical system. The automatic priming valve has two manual override controls at the pump panel while the manual valve has only one. If there are two manual controls, one will open the priming valve while the other engages the priming pump with the drive system. If the pump is equipped with a manual priming valve, the normal push/pull control will have to be used to operate the priming valve, with the separate manual prime control operated to engage the pump with the gear drive system. In either case, the prime control should be released before the manual drive lever is released. This operation will be much smoother if the pump is not turning while the manual prime lever is moved. Either engage the priming pump before the pump transmission is shifted from the ROAD position to the PUMP position, or disengage the vehicle clutch or automatic transmission while the pump drive system is engaged. While priming with the mechanical drive system, the engine rpm should be maintained at approximately 1,000-2,000 rpm.

One of the rotary vane models is designed for separate mounting instead of being installed on the pump transmission. This priming pump has been designed this way to allow for greater flexibility as well as for use with front mount, PTO drive, or direct engine mounting. Either of the standard priming valves can be used with the separate mounted primer. It operates the same way as the transmission mounted unit, but there is no provision for a mechanical drive override.

It is always important to consistently maintain the electrical system in a pumper, but with an electrically driven primer, maintenance is critical. Allowing corrosion to build up on battery terminals or connections may decrease the voltage available to drive the primer to the point that it will not operate.

Priming Oil System

All Waterous primers use priming oil in the system. The oil serves two purposes. As the priming pump turns, oil is drawn into the pump and forms a film on the inside of the housing and on the moving elements. This film provides a better seal and allows the pump to prime faster. As the pump wears and the tolerance between the gears and the housing becomes greater, the sealing action becomes more critical.

The priming oil also lubricates the pump while it is running dry before the pump is primed. The oil forms a protection coating on all parts of the pump; this coating acts as a preservative while the apparatus is out of service.

It is a good practice to operate the primer for a few seconds after the pump has been drained at the end of a pumping operation. This will form a thin coat of oil on the metal parts and inhibit deterioration due to rust or corrosion. A supply of SAE 30 motor oil is carried on the apparatus in a reservoir that holds 5 quarts (5 L) and is equipped with a dipstick to measure the level.

The level of the priming oil should be checked after every pumping operation to ensure that an adequate supply will be available to prime the pump when it is used the next time. It is also a good practice to carry a quart (liter) of oil on the apparatus for use if the supply runs low while operating from draft.

There are two vents in the priming oil system (Figure G.24 on next page). The cap on the storage tank has a small hole in it to allow air to enter

Figure G.24 The priming oil reservoir has two vents to assure proper operation. *Courtesy of Waterous Co.*

the tank as the oil is withdrawn. If this vent becomes blocked, a vacuum may build up inside the tank. The vacuum will prevent a free flow of oil to the priming pump while it is turning. There is also a vent in the fitting where the oil supply line enters the top of the tank.

Since the tank may be mounted in any position on the apparatus with respect to the priming pump, the supply line enters the reservoir at the top and extends to the bottom of the tank. The oil is actually drafted from the reservoir by the vacuum created when the priming pump turns. If the discharge opening of the priming pump is located at a point lower than the oil reservoir, the oil would continue to flow from the opening by a siphon action after the pump stops turning. To prevent this, there is a very small hole, which is approximately #60 AWG (American Wire Gauge) in diameter in the fitting where the line enters the tank to break the siphon action. This hole must be small enough to allow the priming pump to overcome the air leak and continue to draft oil from the tank while it is turning. At the same time, either of these vents can be blocked by grease, dirt, or paint, and should always be kept clean and regularly inspected.

One indication that the vent in the cap has become blocked is the inability of the priming pump to get any oil when it is operating. Blockage of the vent in the supply line will cause oil to continue to leak out of the priming pump discharge when the pump is not operating. Any unexplained oil leakage should be promptly investigated.

AUTOMATIC PRESSURE CONTROL DEVICES

The Waterous Pump Company has used three different devices for automatic pressure control: a relief valve and two different governors. Which device is used depends on the age of the pumper, the type of pump, the mounting arrangement, the type of engine in the vehicle, and the preference of the user.

Pressure Relief Valve

The Waterous relief valve is a dump-type relief valve. It consists of two parts: a relief valve and a pilot valve. A relief valve is mounted on the pump or inserted in a waterway between the discharge and the intake sides of the pump (Figure G.25). A pilot valve is mounted on the pump panel (Figure G.26 on next page). It contains a control to set the operating pressure and a switch to put the relief valve in operation or to take it out of the system. The pilot valve may be associated with a set of indicator lights to let the operator know whether the relief valve is closed or open.

When the pump is dry, the relief valve is held closed by a biasing spring mounted inside the

Figure G.25 The relief valve is mounted between the discharge and intake sides of the pump.

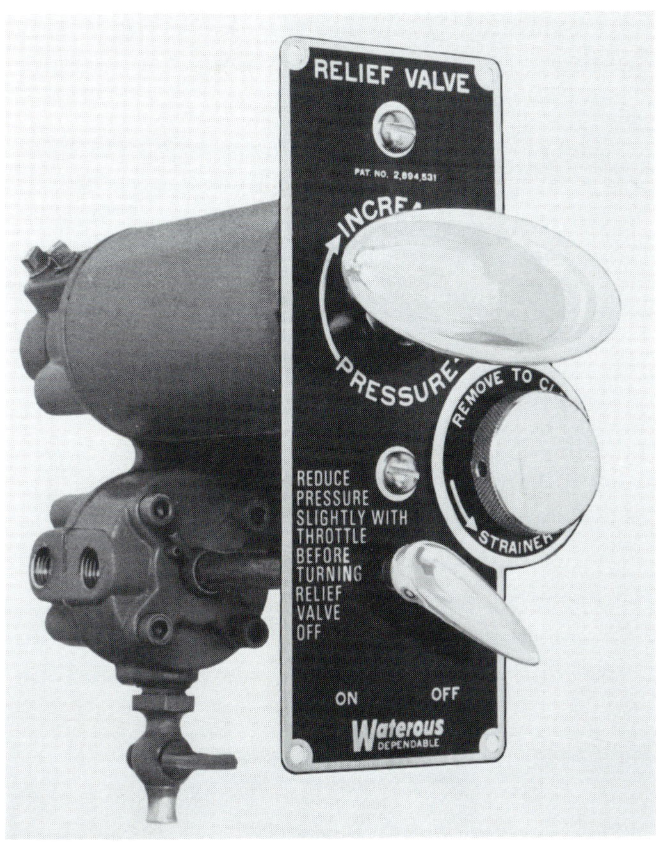

Figure G.26 The pilot valve control is mounted on the pump panel. **NOTE:** The four-way valve is off. *Courtesy of Waterous Co.*

valve housing (Figure G.27). When pressure is developed in the discharge manifold, it acts against the operating end of the double-ended relief valve, attempting to push it open. This will allow water to bypass from the discharge to the intake. When the relief valve control is in the OFF position, water under discharge pressure is piped through the four-way ON/OFF valve to the control side of the relief valve. Since the control end of the relief valve has a larger surface than the operating end, more force is exerted against the control side than against the operating surface. The valve will stay closed. The electrical switch that is coupled mechanically to the relief valve operates, indicating that the valve is closed.

To put the relief valve in service, the four-way valve is operated to the ON position (Figure G.28). This inserts the pilot valve into the supply line from the discharge of the pump to the control end of the relief valve. When the discharge pressure from the pump exerts more force against the diaphragm and the control valve in the pilot valve

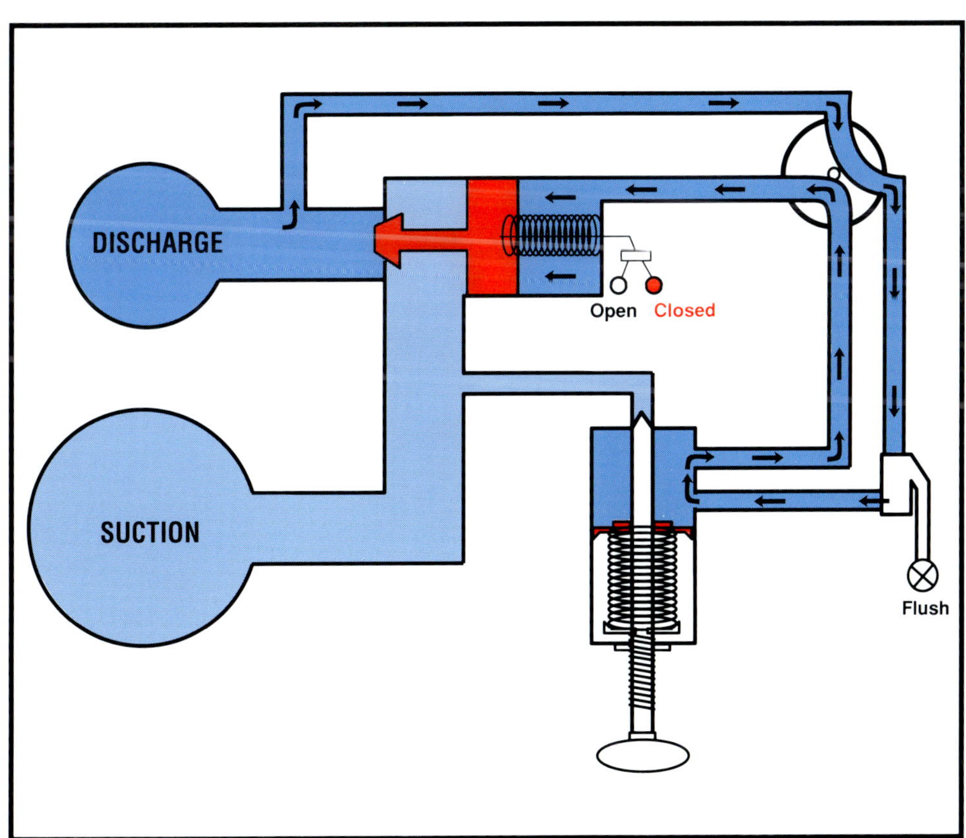

Figure G.27 This schematic shows the relief valve system in the CLOSED position with the pressure set.

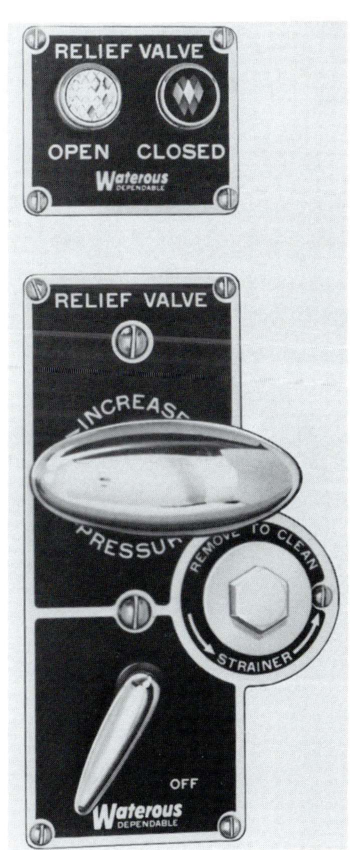

Figure G.28 The pilot light indicates the relief valve is OPEN. *Courtesy of Waterous Co.*

than the pressure established by the adjustable spring under the control of the pressure adjustment handle, the pilot valve will open (Figure G.29). This allows water to flow out of the pilot valve and the control side of the relief valve into the intake manifold of the pump. When this happens, the pressure per square inch (square millimeter) in this portion of the system will be reduced because the pilot valve is constructed with a restricted waterway into its housing. This waterway will not allow water to enter the valve from the pump as rapidly as it is being dumped back into the intake.

When the pressure on the control side of the relief valve is reduced enough to allow the force of the discharge pressure acting against the operating side of the valve to overcome it, the valve opens and allows water to bypass from the discharge manifold of the pump back into the intake. This increased flow reduces the net pump pressure in the fire pump.

The relief valve will stabilize when the total force against each end of the valve is equal. It will remain in that position until changes in flow or engine rpm cause a change in pressure. At that point, the balancing action of the pilot valve and relief valve will adjust the flow to bring the pump pressure back to its original setting. The relief valve will allow sufficient water to flow through to maintain a friction loss pressure differential equal to the desired net pump pressure that has been previously set.

When additional lines are put in service and the discharge pressure drops below the set point, the pilot valve will close. The pressure builds up in the control chamber of the relief valve and the

10-91

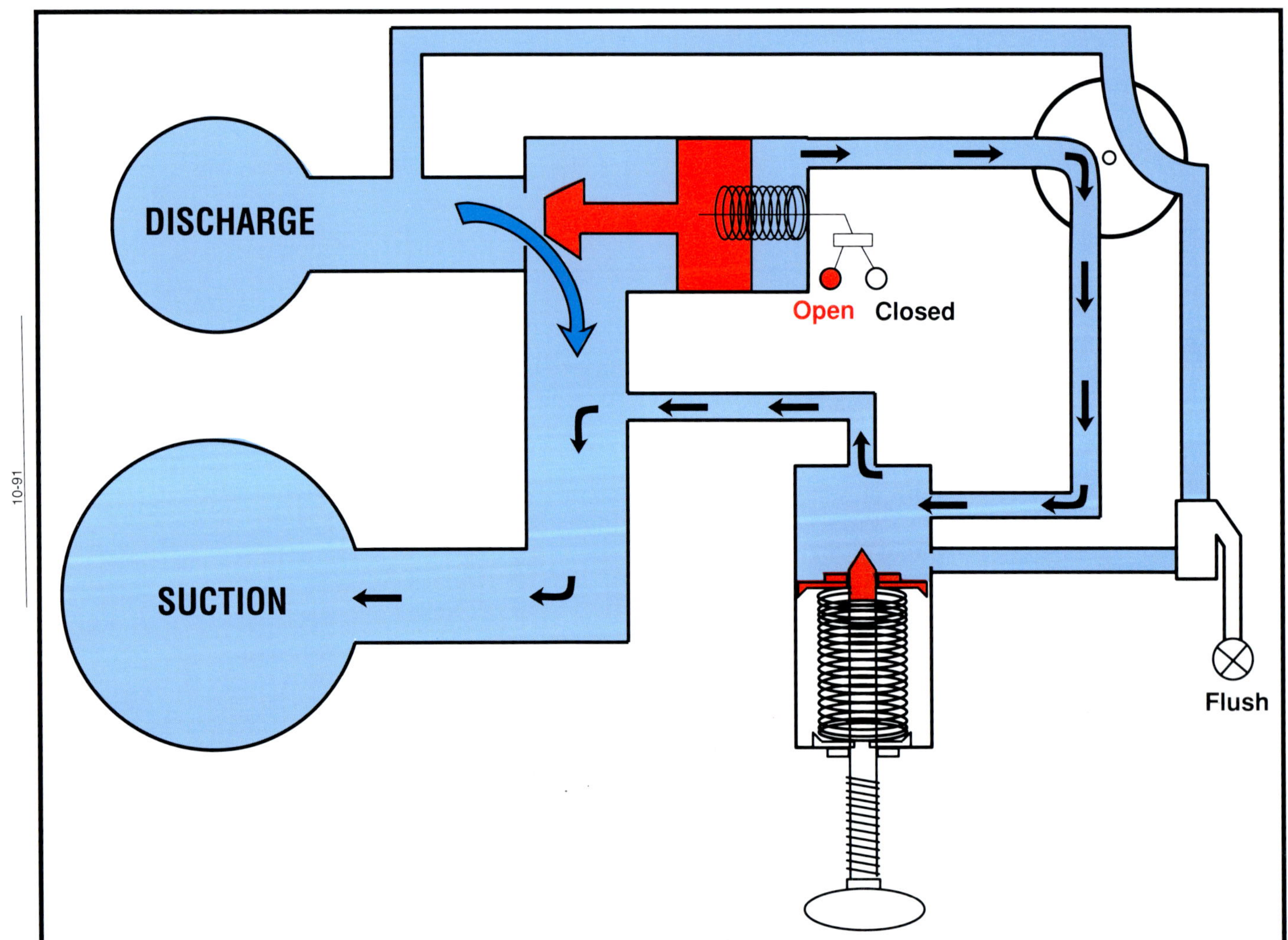

Figure G.29 This schematic shows the relief valve system in the OPEN position, holding discharge pressure as set by the pilot valve spring.

relief valve closes, thus restoring the pressure to its original setting. Since this relief valve prevents pressure surges by maintaining the flow through the pump, it cannot increase the pressure above the initial setting. It can only limit pressure increases.

SETTING THE RELIEF VALVE

Before setting the relief valve, all lines must be in operation at the maximum flow rate. The throttle can be adjusted to attain the desired operating pressure. To set the relief valve, first turn the pressure adjust control clockwise to raise the setting higher than the operating pressure. Put the four-way handle in the ON position. If the pilot valve is set to a higher pressure than the operating pressure, no change will take place when the relief valve is turned ON. The pressure will stay the same, and the green light indicating the valve is closed should be lit. Turn the pressure adjust knob counterclockwise until the relief valve opens. When this happens, the pressure drops 5 to 10 psi (35 to 70 kPa) below the set point and the amber light will come on, indicating that the valve is now open. Slowly turn the pressure adjust control clockwise, allowing time for the pressure to stabilize as it changes, until the gauge returns to the set point. The amber light should go out and the green light should come on, indicating that the valve is closed. The pump is now operating with the relief valve closed and no water being bypassed from the discharge to the intake of the pump.

Any pressure changes due to decreases in flow will be compensated for by the relief valve and should not affect the operating pressure. If additional lines are put in service and the pressure drops, the hand throttle will have to be increased to bring the pressure back to the set point. If the required operating pressure changes, repeat this procedure.

If it is difficult to get the relief valve to work it may be due to a dirty strainer. All Waterous relief valves are equipped with a strainer in the line between the four-way valve and the pilot valve (Figure G.30). Older models have the strainer mounted in back of the pump panel, inaccessible to the operator. Newer models have the strainer removable from the panel. Older models have a relief valve flush valve on the pump panel. To clean the strainer, turn the four-way to the ON position, then open the relief valve flush control. Water will back flush the strainer and clean out any sand and debris, but this can only be done when the pump is operating with clean water.

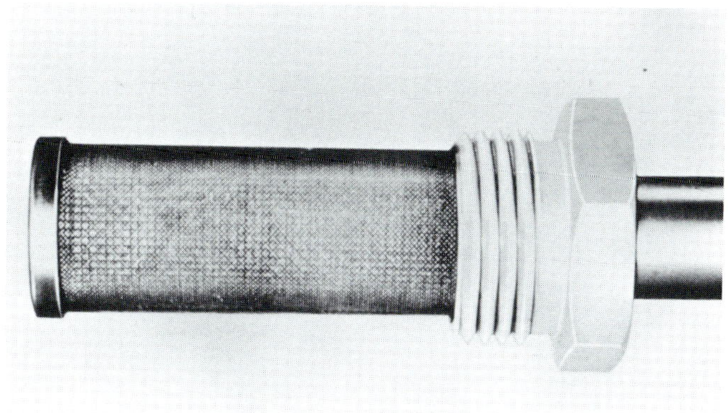

Figure G.30 A strainer is placed between the four-way valve and the pilot valve to keep dirt from the pilot valve.

Newer models provide for cleaning the strainer by removal from the panel. The strainer can be removed by turning it counterclockwise using an adjustable wrench or straight wrench. It can then be cleaned by washing it in clean water and reinstalling it.

Pressure Governor

The pressure governor limits pressure surges by controlling the speed of the vehicle. Earlier models of Waterous pumps use a variable spring reference governor. Newer models are much more sophisticated, using an inert gas for a reference and providing for an emergency shutdown in case of cavitation.

Older Waterous governors are simply a variation on the standard relief valve. They use a pilot valve similar to the one on the relief valve to establish an operating pressure by varying the pressure on a spring (Figure G.31 on next page). Instead of a relief valve, the pilot valve is connected to a balancing cylinder that is used to control the throttle mechanism of the truck. Like the relief valve, this governor only limits pressure increases and will not compensate for decreases in pressure when additional lines are put in service.

Figure G.31 The earlier governor was similar to a relief valve in operation.

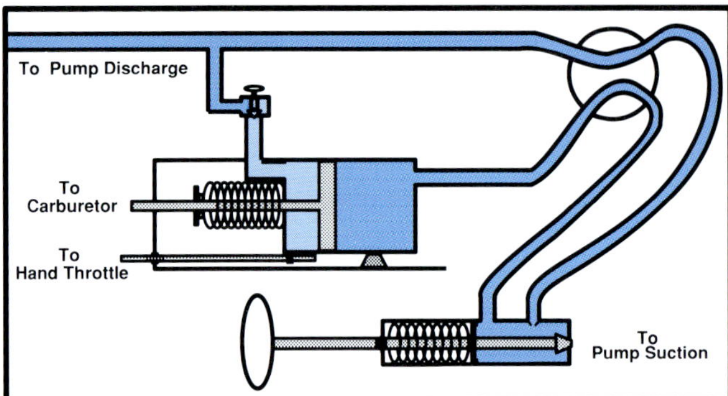

Figure G.32 A schematic of the governor system.

Before setting this governor, all lines need to be flowing at the maximum rate. When the four-way valve is in the OFF position, the piston in the balancing cylinder is held at the end of its travel in the maximum throttle position. When the four-way valve is turned to the ON position, nothing will happen until the discharge pressure causes the pilot valve to open. This allows the pressure to bleed off from the control end of the balancing cylinder. When this happens, the piston within the cylinder moves to reduce the setting of the throttle. This allows the engine speed to decrease and the pressure to return to the set point. The piston will stabilize in this position until the pressure changes again (Figure G.32).

The effectiveness of this governor is limited by the speed of response of the engine to changes in the throttle setting. Since pressure surges due to changes in flow take place very rapidly, while changes in engine speed take place much slower, the engine will often overshoot the mark. By the time the pressure returns to the correct value, the throttle is trying to reduce the speed still further. As the governor then corrects for this overshooting of the mark, it will again increase too much. As this fluctuation (known as "hunting") takes place, it will be very difficult to maintain effective fire streams. To minimize this effect, a needle valve in the supply line from the pump to the operating end of the balancing cylinder slows the movement of water in and out by restricting the waterway. This has the result of slowing the movement of the governor to match the engine on the apparatus. As the governor's response speed is reduced, its effectiveness in eliminating sudden pressure surges, such as occur when a nozzle is shut down suddenly, suffers.

The needle valve is normally installed on the pump operator's panel and should be kept open except when hunting occurs. At that time, it should be closed only as much as necessary to stop the hunting while allowing for optimum response to changes in pressure.

There is no danger of the governor causing the pump to "run away from the water" when the supply is interrupted, since it is only designed to reduce the speed of the engine, not increase it. It has the same limitations as the relief valve in that it will not compensate for decreases in pressure if additional lines are put in service. It can only minimize pressure surges or other increases in pressure.

Use the same procedures for putting the governor in service that apply to the relief valve, except that the pressure control will have to be adjusted more slowly. Changes in pressure can only take place as rapidly as the engine will respond to changes in the throttle setting. Therefore, it is necessary to wait after making any adjustments until the engine has had a chance to stabilize at its new setting. When operating under the control of the governor, the operator must be alert to fluctuations in engine speed or pressure due to hunting action and adjust the needle valve to compensate. **CAUTION:** If the needle valve is completely closed, there will be no changes in engine speed due to governor action. The governor will not respond at all to changes in pressure and will be inoperative, even though it may be turned on.

Waterous Pump Pressure Controller

The Waterous pressure controller is a stored pressure reference governor and has the ability to either increase or decrease the engine speed to compensate for changes in pressure (Figure G.33). It is designed to operate when the pressure reaches 75 psi (525 kPa) and will disengage anytime the discharge pressure drops below 30 psi (210 kPa). This provides a margin of safety in that the engine will not be driven to maximum rpm if the water supply to the pump is interrupted and the governor attempts to compensate for the reduced pressure by continuing to increase the engine speed.

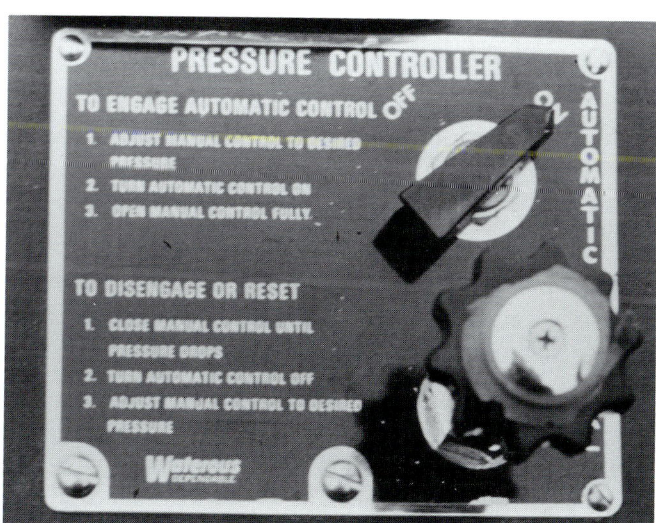

Figure G.33 The pressure controller can compensate for higher or lower pressure changes.

The pressure controller consists of three parts: a directional flow valve to control the setting and operating of the governor, an accumulator to establish a pressure reference for the system (Figure G.34), and a cylinder assembly to control the throttle setting of the motor (Figure G.35). The ac-

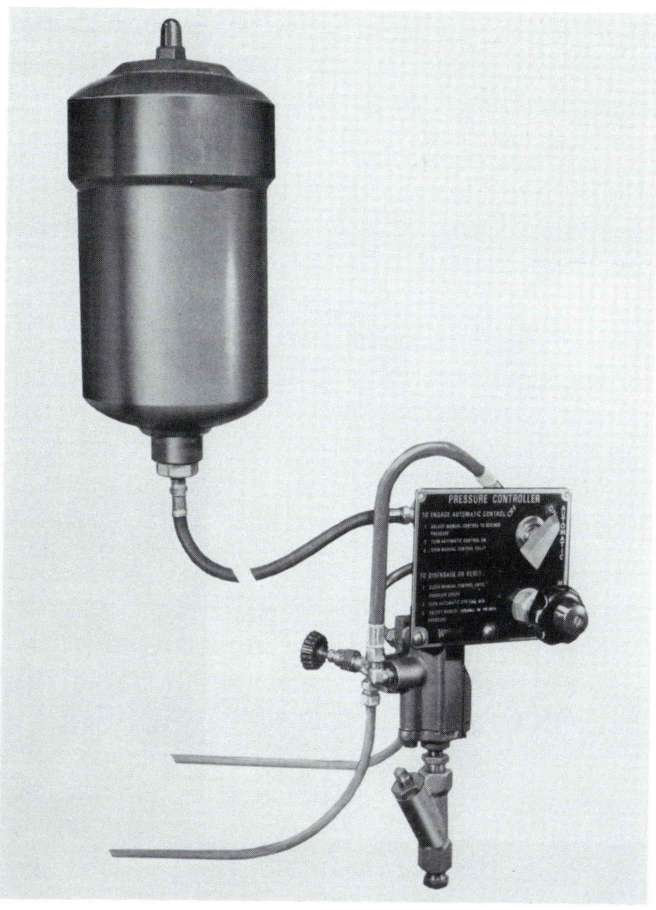

Figure G.34 A pressure reference test is set in the accumulator of the system. *Courtesy of Waterous Co.*

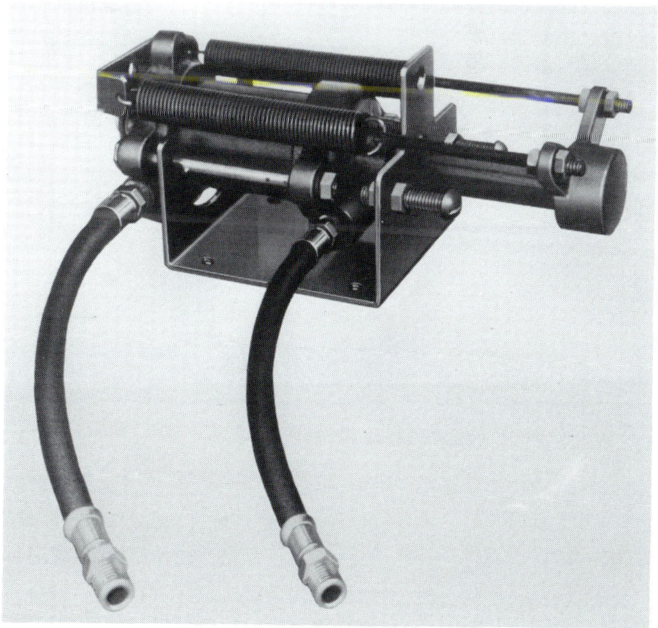

Figure G.35 This cylinder assembly controls the throttle for the pressure control system. *Courtesy of Waterous Co.*

cumulator contains a bladder to separate the air pressure from the water, thus ensuring that the system will not become waterlogged. This bladder is maintained at a pressure of 75 psi (525 kPa) when the pump is dry. This establishes the minimum operating pressure for the controller.

When the pump is dry, a piston located within the cylinder is held in the position for maximum throttle setting by a biasing spring. When the pump is charged, water under pressure from the pump enters the entire system, including the accumulator, and compresses the gas within the bladder to establish a reference pressure. Since the same pressure is present throughout the system on both sides of the piston in the cylinder, the biasing spring will be controlling the throttle. As the hand throttle is increased to set the operating pressure, the entire cylinder moves with the throttle.

When the desired pressure has been reached and the system has had enough time to stabilize, the ON/OFF switch controlling the directional flow valve can be turned on. The directional flow valve then isolates the two systems: the reference portion of the governor with the accumulator, and the control side of the piston from the active side of the piston where the operating pressure of the pump is present (Figure G.36).

Figure G.36 A diagram of the pressure control system in the SET position.

Once the governor has been set, the hand throttle is opened fully. As the hand throttle is opened and the cylinder moves, the increased pressure on the active side of the piston acting against the reference pressure on the other side causes it to move within the cylinder. This movement reduces the throttle setting to bring the pressure back to the set point. There may be slight pressure surges when the system adjusts itself while the throttle is being set, but these fluctuations should not exceed 5 psi (35 kPa) and will be momentary in nature. Any changes in discharge pressure will be compensated for by a corresponding change in engine speed, thus maintaining a constant pressure.

When set in this manner, the governor will compensate for added flow when additional hoselines are put in service as well as for decreased flow when lines are shut down. If the pump goes into cavitation for any reason, and the pressure fluctuates or is reduced below 30 psi (210 kPa), the directional flow valve will release. Releasing the flow valve deactivates the system and reduces engine speed to an idle. If this happens, close the hand throttle completely and turn the control valve to the OFF position before resuming operation. The pressure controller must then be reset in the normal manner.

To change the operating pressure while the governor is operating, reduce the setting of the hand throttle until the pressure decreases below the point for which it was set. Turn the control valve to the OFF position. No change in pressure should take place. Adjust the hand throttle to the new pressure setting on the discharge. Allow a few seconds for the governor system to stabilize, then turn the control valve to the ON position. Open the hand throttle to maximum. The pressure controller is again in control of the operation.

This governor also has a needle valve to match the speed of response of the governor to the type of engine in the vehicle, but it is not accessible from the pump operation panel. It is usually mounted near the directional control valve or the accumulator tank and is set when the governor is initially installed. The governor should not require any further attention unless mechanical problems are encountered or changes are made to the system. Excessive fluctuation when changes in flow are encountered, or a failure to respond quickly enough to limit pressure surges to less than 30 psi (210 kPa) are indications that the needle valve needs adjustment or other problems are being experienced by the pressure controller.

AUXILIARY COOLING

There is no standard arrangement for auxiliary cooling with Waterous pumps, but most of them use a marine-type heat exchanger mounted in the engine compartment or in some other convenient location. In this type of cooler, coolant from the radiator passes through copper tubing which is immersed in water being supplied from the fire pump. The coolant within the tubing is cooled by conduction to the water inside the housing. The water from the pump is controlled by a shutoff valve on the pump panel. This valve can be used to keep the operating temperature of the engine within the desired range.

The auxiliary cooler will not contaminate the coolant within the system and can be used anytime it is needed without concern for damage to the system. The auxiliary cooler may be equipped with a separate drain valve, often at some location remote from the other drains. If this is the case, special attention will be needed to ensure it is completely drained in freezing weather to prevent damage.

A radiator fill valve may be furnished. If there is, it will allow water from the fire pump to be supplied directly into the radiator. This will dilute the antifreeze solution. If the pump is supplying contaminated water, serious damage to the cooling system may result. The radiator fill valve is an emergency control and should only be used in situations where it is required to protect firefighters or when human life is endangered.

Review Answers

CHAPTER 1

1. The following are correct: completing fire reports or maintenance requests, reviewing manufacturer's operating instructions, studying pre-fire operations plans, reading maps.

2. The following are correct: depth perception, peripheral vision, identification of traffic signs and signals, glare vision and glare recovery time, night vision, reaction time, hand-eye coordination.

3. The fire apparatus driver/operator is responsible for the safe transportation of firefighters, apparatus, and equipment to and from the scene of an emergency or other call for service. Once on the scene, the driver/operator must be capable of operating the apparatus properly and *swiftly*. The driver/operator must ensure that the apparatus and the equipment carried on the apparatus are maintained in a ready condition at all times.

4. The driver/operator is often in command of the apparatus in an acting capacity while the officer is absent.

5. Driver/operators are most often promoted from the rank of firefighter. Promotion is often based upon knowledge or performance tests or a combination of these tests. Some departments are using assessment centers to determine promotional choices. Whatever the method used, promotions should be based upon skill and ability rather than seniority or position.

CHAPTER 2

1. False. The driver/operator and/or the fire department may be held liable for negligent acts.
2. False. The driver/operator is responsible for the safety of personnel riding on the apparatus.
3. C

5-89

4. B
5. D
6. B
7. A
8. A. Aim high in steering: find a safe path well ahead
   B. Get the big picture: stay back and see it all
   C. Keep your eyes moving: scan - do not stare
   D. Leave yourself an "out": be prepared - expect the unexpected
   E. Make sure others see you: do not gamble - use lights, horn, and signals
9. The following are correct: state and local driving regulations for emergency and nonemergency situations, departmental regulations, hydraulic calculations, specific operational questions with regard to pumping.
10. Speed
11. left
12. 200 to 300 rpm
13. As local policy dictates and as long as the driver/operator does not endanger life or property.
14. Emergency vehicles are not exempt from laws requiring all vehicles to stop for school buses flashing signal lights to indicate that children are boarding or disembarking. Fire apparatus should proceed only after a proper signal is given by the bus driver or police officer.
15. Fire apparatus may travel with caution against a red traffic signal or stop sign if state regulations and department policy permit. However, it is strongly recommended that even if local policies permit travel against a red light or stop sign, that the apparatus be brought to a *complete* stop before proceeding through the intersection.
16. When all lanes of traffic in the same direction as the responding apparatus are blocked, the apparatus driver/operator should move the apparatus into the opposing lane of traffic and proceed through the intersection at an extremely reduced speed.

17. To combat a skid, release the brakes, allowing the wheels to rotate freely. Turn the steering wheel so the front wheels face in the direction of the skid and gradually let up on the accelerator. If using a standard transmission, do not release the clutch until the vehicle is under control and just before stopping the vehicle. Once the skid is controllable, gradually apply power to the wheels to further control the vehicle by giving traction.
18. During adverse weather conditions, brakes should be pumped, not jammed, for a gradual slowdown or stop. Suddenly applying the brakes can lock the wheels and throw the apparatus into a dangerous skid.
19. Remember that it takes 3 to 15 times more distance to come to a complete stop on snow and ice than it does to come to a complete stop on dry concrete.
20. Low oil level.
21. Shut the vehicle down immediately and investigate.

CHAPTER 3

1. False. The smallest recognized capability for a fire department pumper is 500 gpm (1 892 L/min).
2. B
3. A. Terrain
   B. Bridge weight limits
   C. Monetary constraints
   D. Size of other tankers in the area
4. The following are correct: adequate but reasonable water tank capacity, adequate filling rate, adequate dump time, adequate suspension and steering, properly sized chassis, properly sized engine for tank size and terrain, sufficient braking ability, proper tank mounting, proper and safe tank baffling.
5. A. Water rescue
   B. Fire fighting
   C. Supplying water to land-based apparatus
6. A. Water tank
   B. Hose bed
   C. Fire pump
7. nonstandard
8. When apparatus conform to the specifications set forth in NFPA 1901, *Standard for Automotive Fire Apparatus,* they are classified as standard apparatus.
9. The main difference between a midipumper and a minipumper is their size, and therefore the amount of equipment they carry and the capacity of the pump.
10. To transport water to areas beyond a water system or to areas where water supply is inadequate and needs to be supplemented.

CHAPTER 4

1. False. It is best to use the hydrant because the distance from the hydrant to the curb may vary.
2. False. When using front or rear intake connections, the vehicle should be aimed or angled in the direction of the hydrant. This angle should be 45 degrees or less.
3. A
4. The following are correct: initial fire size-up and potential spread, rescue situations, water supply, apparatus exposure (radiant heat, building collapse, occupancy hazards, overhead power lines, backdraft), method of attack, fire exposures, wind direction, terrain, relocation potential.
5. A. Available hard suction hose and lift
   B. Wet ground approaches (entrapment)
   C. Stability and support for pumper
   D. Access for additional apparatus
   E. Tidal movement
6. The following are correct: amount of water needed at the fire, distance between water source and fire, hose size, pumper capacity, terrain.
7. The following are correct: location of water supply to be used, distance from street to fire department connection, positioning requirements of other apparatus, pump pressure readings that indicate malfunctions and damaged systems.

8. The following are correct: availability and amount of water needed at a given fire, the immediate need for effective fire streams, available hose and pumping capacity.
9. Pulling past the front door allows incoming aerial apparatus better position.
10. To hold the strainer off the bottom, tie a rope to the apparatus or to a nearby object. Floats, such as a spare tire or plastic bucket, can be used with rope to hold the strainer at an appropriate depth.
11. In dual pumping, the pumpers are connected intake-to-intake. This is used when a pressurized water source has more water available than a single pumper can use. In tandem pumping, the pumpers are connected discharge-to-intake. This is used when there is a need to boost the pressure of a water supply between the supply source and its destination.
12. Level I staging is used on all responses. It means that only the first-arriving engine, truck company, and other support units should go directly to the scene. All other first-alarm companies should stay a block or two away until they are given orders to do otherwise. Level II staging is used on multiple-alarm responses; all companies coming on the greater alarms should respond to a designated area and stand by until they are assigned a final destination and duty.

10-91

13. Fire apparatus usually travels slower than the normal flow of traffic and the use of warning lights and sirens may create traffic conditions that actually slow the fire unit's response.
14. Fire apparatus should be placed between the flow of traffic and the firefighters working on the incident to act as a shield.

CHAPTER 5

1. False. Liquids have a specific volume, while gases do not.
2. False. They require an outside means of priming, usually by a small positive displacement pump.
3. True
4. True
5. A
6. C
7. C
   B
   A
   E
   D
8. A. Naturally
   B. Chemically
   C. Mechanically
9. The following are correct: they are self-priming; they are susceptible to wear; a relief valve must be installed, tested, and maintained; the output of a piston pump contains some surges and requires some type of pressure chamber if an even pressure is desired; all positive displacement pumps have a positive action, which requires that the pump be turning for any water to pass through it; when the rotary pump is in operation, the water being pumped provides a seal between the rotary element and the case, thus compensating for irregularities caused by wear.
10. A. The amount of water being discharged
    B. The speed at which the impeller is turning
    C. The intake pressure
11. The following are correct: auxiliary engine, cross-mounted engine, power take-off, midship transfer drive.
12. A. Positive displacement pumps
    B. Exhaust primers
    C. Vacuum primers
13. A. Compound gauge
    B. Master pressure gauge

10-91

14. A. Differential
    B. Paddlewheel
    C. Spring Probe
    D. Turbine
15. The following are correct: speedometer, tachometer, oil pressure, ammeter, voltmeter, air pressure, water temperature, and fuel gauge.
16. impeller, casing
17. stage
18. volume
19. pressure
20. It expands.
21. Water hammer is the force created by the sudden acceleration or deceleration of water.
22. A positive action takes place, forcing all the water and air out of the pump body each time one cycle of operation is completed.

23. Centrifugal force creates the water velocity needed to attain the required pump discharge pressure.
24. Anytime more than half of the required large outlets are in use, the pump should be in the parallel position.
25. Two identical impellers are mounted on the same shaft. When in the PARALLEL position, both impellers are supplying water directly to the discharges for maximum volume. When in the SERIES position, one impeller discharges water to the second impeller in order to produce maximum pressure.
26. The following are correct: they do not prime themselves, they may be damaged by contaminated water or cavitation, discharges may be shut down for short periods while the pump is turning with no damage, they may take advantage of intake pressures, change in the amount of water being discharged will change the discharge pressure of the pump.
27. Its sensitivity to pressure change and its ability to relieve excessive pressure within the pump discharge.
28. When over-pressurization occurs, the engine throttle adjusts to the necessary level to efficiently supply the required flow.
29. Drain valves provide a means of relieving pressure from the hoselines after the discharge valve has been closed.
30. The bleeder valve facilitates removal of air from hoselines coming into the gated intake while the pumper is supplying attack lines from its booster tank.
31. It measures the pump discharge pressure.
32. The primary function of an auxiliary cooler is to control the temperature of cooling water in the apparatus engine during pumping operations.
33. Paddlewheel

10-91

CHAPTER 6

1. False. The pumper with the largest capacity pump should be positioned at the water supply source.
2. True
3. D
4. C
5. A
6. B
7. C
8. C
9. A. The pump's own water tank
   B. Pressurized sources
   C. Static sources
10. The following are correct: hose streams and gauges will fluctuate; the pump makes a sputtering or popping noise, or in severe cases it sounds like gravel is passing through it; lack of reaction on the pressure gauge when the setting of the throttle is changed.
11. A. Bring the apparatus to a full stop and set the brake. Allow the engine to idle.
    B. Shift the drive transmission into neutral (both standard and automatic transmissions).
    C. Operate the pump shift lever to transfer power from the drive axle to the pump drive.
    D. Shift the truck transmission into the proper gear for pumping and lock the lever in place.
    E. If the apparatus is equipped with a standard transmission, engage the clutch slowly.
    F. In apparatus equipped with automatic transmission, depress the accelerator to ensure the shift has been completed and the apparatus will not "drive away."
12. The following are correct: open a booster line, open a discharge drain valve, partially open the tank fill valve, use a bypass or circulator valve if so equipped.
13. The following are correct: the stability of the ground, convenience of connecting hoselines, safety of the operator.
14. The following are correct: open discharge, pump is in SERIES (PRESSURE) position, open intake bleeder valve, open discharge drain.
15. A. The amount of water needed
    B. The distance from the source
16. PRESSURE (SERIES)
17. 24 inches (600 mm)
18. PARALLEL (VOLUME)
19. 150 psi (1 050 kPa)

20. Cavitation occurs when water is being discharged from the pump faster than it is coming in; this is sometimes expressed as "the pump running away from the water."

21. Partially open the pump-to-tank line so water will circulate and keep the pump cool.

22. Open the bleeder valve on the intake line.

23. A. Position the apparatus in a safe position and immobilize it by setting the parking brake and blocking the wheels.

B. Engage the pump and select the proper gear
10-91 in the road transmission. Lock it in place.

C. Open the tank-to-pump valve.

D. Set the transfer valve to the SERIES position (if necessary).

E. Increase the throttle setting to obtain the desired pressure, priming if necessary.

F. Set the relief valve or pressure governor.

G. Open the circulator valve or partially open the tank fill valve.

H. When an external supply becomes available, open the intake valve while closing the tank-to-pump valve. As soon as an adequate supply of water is available, divert enough of it through the tank fill line to replenish the supply in the tank.

24. Make sure the hydrant is opened fully.

25. A blockage is developing in the intake line.

26. A relay operation is required anytime the source of water is located more than a few hundred feet from the fire.

27. From Table 6.1, we see that each pumper can pump 500 gpm (2 000 L/min) through 3,000 feet (870 m) of hose. So the number of pumpers
10-91 needed can be determined as follows:

$$\frac{6{,}000}{3{,}000} = 2 + 1 = 3 \text{ pumpers needed}$$

or

$$\frac{1{,}800}{870} = 2.07 + 1 = 3.07 \text{ or } 4 \text{ pumpers needed}$$

CHAPTER 7

1. False. Only qualified service personnel should perform these tasks.

2. True

3. False. This is of no benefit to the engine and may clog fuel injectors.

4. The following are correct: check crankcase oil for proper level; check radiator water level; check all batteries; check all visible and audible warning signals; check fuel level (maintain a full tank at all times); visually check water tank level; check each tire for cuts, breaks, and proper inflation; pressure test all brakes by operating foot pedal; clean apparatus windows; wash and dry entire vehicle if needed; operate changeover valves on two-stage pumps.

5. The following are correct: check transmission oil level; check differential oil level; check power steering fluid level; check brake system when equipped with hydraulic brakes; check master cylinder brake fluid level and watch for any wheel cylinder or hose leaks; check air system for leaks and bleed moisture from air tanks (on apparatus equipped with air brakes); check fan belt and generator or alternator belts; check battery terminals and cables; operate valves in cooling system; check drains and all hose connections for security; check drive shaft and universal joints; start motor and observe oil pressure and idling speed after engine is warm; clean underneath chassis; clean the engine and electrical motors; check for loose nuts, studs, and pin.

6. The following are correct: check all possible extinguishers by lifting and checking gauges, check hose loads for correct finish, inventory all nozzles and appliances and operate all valves, check air pressure in protective breathing equipment, examine regulator, check all handlights and flashlights.

7. The following are correct: remove all hose from fire apparatus, check condition, and reload, changing the bends; service test all fire hose (annually); remove all ladders from their racks, and clean them; lubricate all motor-, gear-, or hand-operated appliances according to manufacturer's specifications; flush the pump and water tank (monthly).

8. The following are correct: rotating lights, hose rewind reel, windshield wipers, apparatus controls, heater and defroster fan.
9. The following are correct: open all pump drains and flush out sediment (weekly); check and clean intake strainers (weekly or after each use); check pump gear box for proper oil level and traces of water; operate pump primer with all pump valves closed; operate changeover valve while pumping from booster tank (weekly); check packing glands for excessive leaks; operate all valves, including relief valve (weekly); check all gauges for proper operation; recalibrate flowmeter as per manufacturer's instructions.
10. manufacturer
11. Maintenance means to keep something in a state of usefulness or readiness. Repair means to restore something that has become inoperable.
12. Engine, batteries, and undercarriage
13. Effective lubrication depends upon the use of a proper grade of recommended lubricant, the frequency of lubrication, the amount used, and the method of lubrication.
14. Its viscosity

CHAPTER 8

1. False. These tests are usually performed by the manufacturer and/or independent testing personnel.
2. True
3. B
4. C
5. D 5/94
6. A
7. B
8. The following are correct: vacuum test, three-hour pump test, automatic pump pressure control tests, tank-to-pump flow rate test.
9. The following are correct:

   The apparatus must accelerate to 35 mph (56 km/h) from a standing start within 25 seconds. Apparatus equipped with water tanks greater than 800 gallons (3 200 L) capacity are allowed 30 seconds.

   The apparatus must accelerate from 15 mph (24 km/h) to 35 mph (56 km/h) within 30 seconds without shifting.

   The apparatus must achieve a minimum top speed of 50 mph (80 km/h). This top speed may be lower for apparatus designed for slower speeds.

   The apparatus must come to a full stop from 20 mph (32 km/h) within 30 feet (9.1 m).

   The apparatus parking brake must control the rear wheels or all wheels, be mechanically actuated, and it must be securely held in position when applied.
10. A. Underwriter's Laboratories, *Guide for Certification of Fire Department Pumpers*

    B. NFPA's, Fire Apparatus Maintenance by Robert Ely
11. The following are correct: transmission in wrong gear, high gear lockup not functioning (automatic transmission), clutch slipping, engine overheating, muffler clogged, tachometer inaccurate, engine governor malfunctioning, intake hose too small, intake strainer submerged incorrectly (for example, too close to surface, too close to bottom), intake screens clogged, lift too high, intake hose clogged or inner lining collapsed, excessive air leaks at intake side of pump impellers clogged, pump or intake hose not fully primed, relief valve or pressure governor malfunctioning, transfer valve in wrong position, inaccurate gauges, pitot tube partially clogged, nozzle too large, seized turbocharger.
12. A. Dry vacuum test

    B. Priming test

    C. Capacity Test

    D. Tachometer and engine rpm check

    E. Ignition check

    F. Pressure governor or relief valve test

    G. Overload test

    H. 200 psi (1 350 kPa) test

    I. 250 psi (1 700 kPa) test

    J. Tank-to-pump flow test

13. preservice and service
14. Preservice testing assures the purchaser that the pump and pump components will operate properly under normal use.
15. The purchaser can specify preservice tests in the apparatus specifications.
16. At least once a year or whenever the apparatus has had extensive repair.
17. Pres. Corr. (U.S.) $= \frac{5 + 10}{2.3} = 6.52$ psi

    Pres. Corr. (Metric) $= \frac{1.5 + 3.3}{0.1} = 48$ kPa

CHAPTER 9

1. False. Local laws govern advertisements.
2. False. The apparatus committee should make any changes necessary and review the specifications line by line before sending them out.
3. True
4. False. The apparatus committee should evaluate the whole process they just completed. They should also monitor the performance of the new equipment to provide information for future apparatus purchases.
5. A. Present equipment
   B. Required equipment
   C. Future equipment needed
6. The following are correct: pump, aerial device, gradability, equipment to be carried, climate, tank capacity.
7. The following are correct: learn of new developments, check on technology limits, get accurate cost estimates, consider trade-offs among various performance options and/or design features, avoid misunderstandings and inconsistencies in the specifications, obtain several expert (if conflicting) opinions on the merits of options under consideration, minimize the number of "exceptions" or deviations from the specifications.
8. apparatus committee, apparatus manufacturer
9. NFPA 1901, *Standard on Automotive Fire Apparatus*
10. Federal Emergency Management Organization (FEMA)

Index

IFSTA MANUALS AND FPP PRODUCTS

For a current catalog describing these and other products, call or write your local IFSTA distributor or Fire Protection Publications, IFSTA Headquarters, Oklahoma State University, Stillwater, OK 74078-0118. Phone: 1-800-654-4055

AWARENESS LEVEL TRAINING FOR HAZARDOUS MATERIALS

prepares fire, police, EMS, and public utilities to recognize and identify the presence of hazardous materials at an emergency scene. Addresses the requirements in NFPA 472, Chapter 2: Competencies for First Responders at the Awareness Level. includes responsibilities of the first responder, identification systems, types of containers, and personal protective equipment. 1st Edition (1995), 152 pages.

STUDY GUIDE AWARENESS LEVEL TRAINING FOR HAZARDOUS MATERIALS

The companion study guide in question and answer format. (1995), 184 pages.

FIRE DEPARTMENT AERIAL APPARATUS

includes information on the driver/operator's qualifications; vehicle operation; types of aerial apparatus; positioning, stabilizing, and operating aerial devices; tactics for aerial devices; and maintaining, testing, and purchasing aerial apparatus. Detailed appendices describe specific manufacturers' aerial devices. 1st Edition (1991), 386 pages, addresses NFPA 1002.

STUDY GUIDE FOR AERIAL APPARATUS

The companion study guide in question and answer format. 1991, 140 pages.

AIRCRAFT RESCUE AND FIRE FIGHTING

comprehensively covers commercial, military, and general aviation. Subjects covered include personal protective equipment, apparatus and equipment, extinguishing agents, engines and systems, fire fighting procedures, hazardous materials, and fire prevention. It also contains a glossary and review questions with answers. 3rd Edition (1992), 247 pages, addresses NFPA 1003.

BUILDING CONSTRUCTION RELATED TO THE FIRE SERVICE

helps firefighters become aware of the many construction designs and features of buildings found in a typical first alarm district and how these designs serve or hinder the suppression effort. Subjects include construction principles, assemblies and their resistance to fire, building services, door and window assemblies, and special types of structures. 1st Edition (1986), 166 pages, addresses NFPA 1001 and NFPA 1031, levels I & II.

CHIEF OFFICER

explains the skills necessary to plan and maintain an efficient and cost-effective fire department. The combination of an ever-increasing fire problem, spiraling personnel and equipment costs, and the development of new technologies for decision making require far more than expertise in fire suppression. Today's chief officer must possess the ability to plan and administrate as well as have political expertise. 1st Edition (1985), 211 pages, addresses NFPA 1021, level VI.

SELF-INSTRUCTION FOR CHIEF OFFICER

The companion study guide in question and answer format. 1986, 142 pages.

FIRE DEPARTMENT COMPANY OFFICER

focuses on the basic principles of fire department organization, working relationships, and personnel management. For the firefighter aspiring to become a company officer, or a company officer wishing to improve management skills. 2nd Edition (1990), 278 pages, addresses NFPA 1021, levels I, II, & III.

COMPANY OFFICER STUDY GUIDE

The companion study guide in question and answer format. Includes problem applications and case studies. 1991, 243 pages.

ESSENTIALS OF FIRE FIGHTING

is the "bible" on basic firefighter skills and is used throughout the world. The easy-to-read format is enhanced by 1,600 photographs and illustrations. Topics covered include: personal protective equipment, building construction, firefighter safety, fire behavior, portable extinguishers, SCBA, ropes and knots, rescue, forcible entry, ventilation, communications, water supplies, fire streams, hose, fire cause determination, public fire education and prevention, fire suppression techniques, ladders, salvage and overhaul, and automatic sprinkler systems. 3rd Edition (1992), 590 pages, addresses NFPA 1001.

STUDY GUIDE FOR 3rd EDITION OF ESSENTIALS OF FIRE FIGHTING

The companion learning tool for the new 3rd edition of the manual. It contains questions and answers to help you learn the important information in the book. 1992, 322 pages.

PRINCIPLES OF FOAM FIRE FIGHTING

Covers both Class A and Class B fires and foams. Includes information on portable foam extinguishers, foam concentrates, portable foam proportioning equipment, foam and ARFF apparatus, and fixed foam extinguishing systems. Also discusses the latest compressed air foam systems. Provides detailed information on using foam to control structural, wildland, industrial, petrochemical facility, and aircraft fires. 1st Edition (1996).

PRINCIPLES OF EXTRICATION

leads you step-by-step through the procedures for disentangling victims from cars, buses, trains, farm equipment, and industrial situations. Fully illustrated with color diagrams and more than 500 photographs. It includes rescue company organization, protective clothing, and evaluating resources. Review questions with answers at the end of each chapter. 1st Edition (1990), 365 pages.

FIRE AND LIFE SAFETY EDUCATOR

describes the role of today's educator and provides direction for planning fire and life safety education programs. This manual addresses topics such as working cooperatively with other community agencies, working within the legislative process, and finding and using resources (including requesting program funds through grants and in-kind contributions), and it also provides the technical background information that the fire and life safety educator needs. Also addressed are data sources and using data for public fire and life safety education planning.

The manual outlines the curriculum development process and gives general guidelines for planning successful presentations. It familiarizes you with information on what motivates people to learn and how people learn differently. Other topics include educational materials selection and program evaluation.

This manual introduces the fire and life safety educator to today's media resources and covers techniques for working with the media. Four special topics — evacuation plans, burn prevention, station tours and equipment demonstrations, and approaches to juvenile firesetting — are also addressed. This manual is intended to be the first document on the fire and life safety educator's bookshelf.

FIRE CAUSE DETERMINATION
gives you the information necessary to make on-scene fire cause determinations. You will know when to call for a trained investigator, and you will be able to help the investigator. It includes a profile of firesetters, finding origin and cause, documenting evidence, interviewing witnesses, and courtroom demeanor. 1st Edition (1982), 159 pages, addresses NFPA 1021, Fire Officer I, and NFPA 1031, levels I & II.

FORCIBLE ENTRY
reflects the growing concern for the reduction of property damage as well as firefighter safety. Contains technical information about forcible entry tactics, tools, and methods, as well as door, window, and wall construction. Tactics discuss the degree of danger to the structure and leaving the building secure after entry. Includes a section on locks and through-the-lock entry. 7th Edition (1987), 270 pages, helpful for NFPA 1001.

GROUND COVER FIRE FIGHTING PRACTICES
explains the dramatic difference between structural fire fighting and wildland fire fighting. It discusses the apparatus, equipment, and extinguishing agents used to combat wildland fires. Outdoor fire behavior and how fuels, weather, and topography affect fire spread are explained. Also covers personnel safety, management, and suppression methods. It contains a glossary, sample fire operation plan, fire control organization system, fire origin and cause determination, and water expansion pump systems. 2nd Edition (1982), 152 pages.

FIRE SERVICE GROUND LADDER PRACTICES
addresses NFPA 1001, *Standard for Fire Fighter Professional Qualifications*, (ladders). Ladders provides a broad scope of understanding of ground ladders and their use. Ladders familiarizes you with design construction, maintenance, and service testing. It shows procedures for handling, arising, and climbing ladders and includes methods for special uses. Ladders contains a glossary, review questions, and a sample testing repair form. 9th Edition (1996), 203 pages.

HAZARDOUS MATERIALS FOR FIRST RESPONDERS
covers the objectives for First Responder at the Awareness and Operational levels contained in NFPA 472. Includes information on properties of hazardous materials, recognizing and identifying hazardous materials, personal protective equipment, emergency scene command and control, incident control tactics and strategies, and decontamination. 2nd Edition (1994), 241 pages.

STUDY GUIDE FOR IFSTA HAZARDOUS MATERIALS FOR FIRST RESPONDERS
The companion study guide in question and answer format. 2nd Edition (1994), 253 pages.

HAZARDOUS MATERIALS: MANAGING THE INCIDENT
addresses OSHA 1910.120 and NFPA 472, *Standard for Professional Competence of Responders to Hazardous Materials Incidents*. Provides the reader with a logical, systematic process for responding to and managing hazardous materials emergencies. It is directed toward the haz mat technician, incident commander, the off-site specialty employee, and haz mat response team members. Includes numerous charts, diagrams, scan sheets, checklists, and reference information. Topics include haz mat management system, health and safety, ICS, politics of haz mat incident management, hazard and risk evaluation, decontamination, and more! 2nd Edition (1994)

STUDENT WORKBOOK FOR HAZARDOUS MATERIALS: MANAGING THE INCIDENT
The companion study guide in question and answer format. 2nd Edition (1994)

INSTRUCTOR'S GUIDE FOR HAZARDOUS MATERIALS: MANAGING THE INCIDENT
Provides lessons based on each chapter. 2nd Edition (1994)

HAZ MAT RESPONSE TEAM LEAK AND SPILL GUIDE
contains articles by Michael Hildebrand reprinted from *Speaking of Fire*'s popular Hazardous Materials Nuts and Bolts series. Two additional articles from *Speaking of Fire* and the hazardous material incident SOP from the Chicago Fire Department are also included. 1st Edition (1984), 57 pages.

EMERGENCY OPERATIONS IN HIGH-RACK STORAGE
is a concise summary of emergency operations in the high-rack storage area of a warehouse. It explains how to develop a pre-emergency plan, the equipment needed to implement the plan, type and amount of training personnel need to handle an emergency, and interfacing with various agencies. Includes consideration questions, and trial scenarios. 1st Edition (1981), 97 pages.

HOSE PRACTICES
is the most comprehensive single source about hose and its use. The manual details basic methods of handling hose, including large diameter hose. It is fully illustrated with photographs showing loads, evolutions, and techniques. This complete and practical book explains the national standards for hose and couplings. 7th Edition (1988), 245 pages, addresses NFPA 1001.

FIRE PROTECTION HYDRAULICS AND WATER SUPPLY ANALYSIS
covers the quantity and pressure of water needed to provide adequate fire protection, the ability of existing water supply systems to provide fire protection, the adequacy of a water supply for a sprinkler system, and alternatives for deficient water supply systems. 1st Edition (1990), 340 pages.

INCIDENT COMMAND SYSTEM (ICS)
was developed by a multiagency task force. Using this system, fire, police, and other government groups can operate together effectively under a single command. The system is modular and can be used to meet the requirements of both day-to-day and large-incident operations. It is the approved basic command system taught at the National Fire Academy. 1st Edition (1983), 220 pages, helpful for NFPA 1021.

INDUSTRIAL FIRE BRIGADE TRAINING: INCIPIENT LEVEL

assists management in complying with applicable laws and regulations, primarily NFPA 600 and 29 CFR 1910, and to assist them in training those who provide incipient level fire protection for industrial occupancies. It is also intended to serve as a reference and training resource for individual emergency responders. 1st Edition (1995), 184 pages.

FIRE INSPECTION AND CODE ENFORCEMENT

is a comprehensive guide to the principles and techniques of inspection for both uniformed and civilian inspectors. Text includes information on how fire travels, electrical hazards, and fire resistance requirements. It covers storage, handling, and use of hazardous materials; fire protection systems; and building construction for fire and life safety. 5th Edition (1987), 316 pages, addresses NFPA 1001 and NFPA 1031, levels I & II.

STUDY GUIDE FOR FIRE INSPECTION AND CODE ENFORCEMENT

The companion study guide in question and answer format with case studies. 1989, 272 pages.

FIRE SERVICE INSTRUCTOR

explains the characteristics of a good instructor, shows how to determine training requirements, and teach to the level of your class. It discusses principles and procedures of teaching and learning, and covers the use of training aids and devices. Also covers the principles of testing as well as test construction. Included are chapters on safety, legal considerations, and computers. 5th Edition (1990), 326 pages, addresses NFPA 1041, levels I & II.

LEADERSHIP IN THE FIRE SERVICE

was created from the series of lectures given by Robert F. Hamm to assist in leadership development. It provides the foundation for getting along with others, explains how to gain the confidence of your personnel, and covers what is expected of an officer. Includes information on supervision, evaluations, delegating, and teaching. Topics include: the successful leader today, a look into the past may reveal the future, and self-analysis for officers. 1st Edition (1967), 132 pages.

FIRE SERVICE ORIENTATION AND TERMINOLOGY

Fire Service Orientation and Indoctrination has been revised. It has a new name and a new look. Keeping the best of the old — traditions, history, and organization — this new manual provides a complete dictionary of fire service terms. To be used in conjunction with **Essentials of Fire Fighting** and the other IFSTA manuals. 3rd Edition (1993), addresses NFPA 1001.

PRIVATE FIRE PROTECTION AND DETECTION

Discusses ways in which fires may be prevented or attacked in their incipient phase until the fire brigade or public fire protection arrives. Covers information on automatic sprinkler systems, hose standpipe systems, fixed fire pump installations, portable fire extinguishers, fixed special agent extinguishing systems, and fire alarm and detection systems. Information on the design, operation, maintenance, and inspection of these systems and equipment is provided. 2nd Edition (1994).

FIRE DEPARTMENT PUMPING APPARATUS

is the Driver/Operator's encyclopedia on operating fire pumps and pumping apparatus. It covers pumpers, tankers (tenders), brush apparatus, and aerials with pumps. Explains safe driving techniques, getting maximum efficiency from the pump, and basic water supply. It includes specification writing, apparatus testing, and extensive appendices of pump manufacturers. 7th Edition (1989), 374 pages, addresses NFPA 1002.

STUDY GUIDE FOR PUMPING APPARATUS

The companion study guide in question and answer format. 1990, 100 pages.

FIRE SERVICE RESCUE

provides fire and rescue personnel with information and procedures related to victim removal from hazardous or life-threatening situations. Covers management and incident command, rope rescue, fireground search and rescue, rescue situations involving structural collapse, elevators, confined space, water/ice, trench, caves and mines, and grain storage vessels. 6th Edition (1996).

RESIDENTIAL SPRINKLERS A PRIMER

outlines U.S. residential fire experience, system components, engineering requirements, and issues concerning automatic and fixed residential sprinkler systems. Written by Gary Courtney and Scott Kerwood, reprinted from *Speaking of Fire*. Supplements **Private Fire Protection.** 1st Edition (1986), 16 pages.

FIRE DEPARTMENT OCCUPATIONAL SAFETY

addresses the basic responsibilities for a safety officer and the requirements and procedures for a safety and health program. Includes an overview of establishing and implementing a safety program, physical fitness and health considerations, safety in training, fire station safety, tool and equipment safety and maintenance, personal protective equipment, en-route hazards and response, emergency scene safety, and special hazards. 2nd Edition (1991), 366 pages, addresses NFPA 1500, 1501.

SALVAGE AND OVERHAUL

covers salvage operations, equipment selection and care, as well as techniques for using salvage equipment to minimize fire damage caused by water, smoke, heat, and debris. The overhaul section includes methods for finding hidden fire, protection of fire cause evidence, safety during overhaul operations, and restoration of property and fire protection systems after a fire. 7th Edition (1985), 225 pages, addresses NFPA 1001.

SELF-CONTAINED BREATHING APPARATUS

contains all the basics of SCBA use, care, testing, and operation. Special attention is given to safety and training. The chapter on Emergency Conditions Breathing has been completely revised to incorporate safer emergency methods that can be used with newer models of SCBA. Also included are appendices describing regulatory agencies and donning and doffing procedures for nine types of SCBA. Covers NFPA, OSHA, ANSI, and NIOSH regulations and standards as they pertain to SCBA. 2nd Edition (1991), 360 pages, addresses NFPA 1001.

THE SOURCEBOOK FOR FIRE COMPANY TRAINING EVOLUTIONS

provides volunteer and career training officers and company officers with ideas for presenting more than 50 weekly or monthly training sessions. Each session contains information on the standards covered, equipment needed, outlines for the presentations and practical exercises, and a listing of pertinent resources and training materials. The sessions cover basic fire fighting, apparatus operation, company evolutions, indoor sessions for rainy days, and competitive exercises with a practical training value. 1st Edition (1994) 238 pages.

STUDY GUIDE FOR SELF-CONTAINED BREATHING APPARATUS

The companion study guide in question and answer format. 1991, 131 pages.

FIRE STREAM PRACTICES

This carefully written text covers the physics of fire and water; the characteristics, requirements, and principles of good streams; and fire fighting foams. **Streams** includes formulas for the application of fire fighting hydraulics, as well as actions and reactions created by applying streams under a variety of circumstances. The friction loss equations and answers are included, and review questions are located at the end of each chapter. 7th Edition (1989), 464 pages, addresses NFPA 1001 and NFPA 1002.

GASOLINE TANK TRUCK EMERGENCIES

provides emergency response personnel with background information, general procedures, and guidelines to be followed when responding to and operating at incidents involving MC-306/DOT 406 cargo tank trucks. Specific topics include: incident management procedures, site safety considerations, methods of product transfer, and vehicle uprighting considerations. 1st Edition (1992), 51 pages, addresses NFPA 472.

FIRE SERVICE VENTILATION

describes and illustrates the safe operations related to ventilation, products of combustion, elements and situations that influence the ventilation process, ventilation methods and procedures, and tools and mechanized equipment used in ventilation. The manual includes chapter reviews, a glossary, and applicable safety considerations. 7th Edition (1994), addresses NFPA 1001, 311 pages.

WATER SUPPLIES FOR FIRE PROTECTION

acquaints you with the principles, requirements, and standards used to provide water for fire fighting. Rural water supplies as well as fixed systems are discussed. It includes requirements for size and carrying capacity of mains, hydrant specifications, maintenance procedures conducted by the fire department, and relevant maps and record-keeping procedures. Review questions at the end of each chapter. 4th Edition (1988), 268 pages, addresses NFPA 1001, NFPA 1002, and NFPA 1031, levels I & II.

CURRICULUM PACKAGES

FIRE OFFICER I AND FIRE OFFICER II CURRICULUM PACKAGES

Developed by the Maryland Fire and Rescue Institute, the Fire Officer I and Fire Officer II curriculum was designed to provide aspiring officers and officer candidates with an educationally sound, objective-based course of instruction.

ESSENTIALS CURRICULUM PACKAGE

A competency-based teaching package with 19 chapters and 22 lessons as well as classroom and practical activities to teach the student the information and skills needed to qualify for the position of Fire Fighter I or II. Corresponds to **Essentials of Fire Fighting**, 3rd Edition.

The Package includes the Essentials Instructor's Guide (the how, what, and when to teach); the Student Guide (a workbook for group instruction); and 445 full-color overhead transparencies.

LEADERSHIP

A complete teaching package that assists the instructor in teaching leadership and motivational skills at the Company Officer level. Each lesson gives an outline of the subject matter to be covered, approximate time required to teach the material, specific learning objectives, and references for the instructor's preparation. Sources for suggested films and videotapes are included.

INCIPIENT INDUSTRIAL CURRICULUM FOR INDUSTRIAL FIRE BRIGADE TRAINING

designed to aid those in industry and those who teach industrial fire brigade training, this comprehensive curriculum can be used for instructor-led group instruction or for instructor-monitored self-instruction. Includes Instructor's Guide, 106 full-color transparencies, and Student Applications Workbook.

TRANSLATIONS

PRACTICAS Y TEORIA PARA BOMBEROS

is a direct translation of **Fire Service Practices for Volunteer and Small Community Fire Departments**, 6th edition. Please contact your distributor or FPP for shipping charges to addresses outside U.S. and Canada. 347 pages.

OTHER ITEMS

TRAINING AIDS

Fire Protection Publications carries a complete line of videos, overhead transparencies, and slides. Call for a current catalog.

NEWSLETTER

The nationally acclaimed and award-winning newsletter, *Speaking of Fire*, is published quarterly and available to you free. Call today for your free subscription.

COMMENT SHEET

DATE ____________________ NAME ____

ADDRESS ____

ORGANIZATION REPRESENTED ____

CHAPTER TITLE ________________________________ NUMBER __________

SECTION/PARAGRAPH/FIGURE ________________________ PAGE __________

1. Proposal (include proposed wording or identification of wording to be deleted), OR PROPOSED FIGURE:

2. Statement of Problem and Substantiation for Proposal:

RETURN TO: IFSTA Editor
Fire Protection Publications
Oklahoma State University
930 N. Willis
Stillwater, OK 74078-8045

SIGNATURE ______________________________

Use this sheet to make any suggestions, recommendations, or comments. We need your input to make the manuals as up to date as possible. Your help is appreciated. Use additional pages if necessary.